Chromatin Regulation and Dynamics

Chromatin Regulation and Dynamics

Edited by

Anita Göndör

Department of Microbiology, Tumor and Cell Biology,
Karolinska Institutet, Stockholm, Sweden

AMSTERDAM • BOSTON • HEIDELBERG • LONDON
NEW YORK • OXFORD • PARIS • SAN DIEGO
SAN FRANCISCO • SINGAPORE • SYDNEY • TOKYO
Academic Press is an imprint of Elsevier

ELSEVIER

Library of Congress Cataloging-in-Publication Data
A catalog record for this book is available from the Library of Congress

British Library Cataloguing-in-Publication Data
A catalogue record for this book is available from the British Library

ISBN: 978-0-12-803395-1

For information on all Academic Press publications
visit our website at https://www.elsevier.com/

Working together
to grow libraries in
developing countries

www.elsevier.com • www.bookaid.org

Publisher: Mica Haley
Acquisition Editor: Peter Linsley
Editorial Project Manager: Lisa Eppich
Production Project Manager: Karen East and Kirsty Halterman
Designer: Mark Rogers

Typeset by Thomson Digital

Table of Contents

List of Contributors

P. Agarwal Department of Molecular Biosciences, Institute for Molecular and Cellular Biology, University of Texas at Austin, Austin, TX, United States

D. Bade Hubrecht Institute, Uppsalalaan, Utrecht, The Netherlands

A.J. Bannister The Gurdon Institute, University of Cambridge, Cambridge, United Kingdom

W.J. Belden Department of Animal Sciences, School of Environmental and Biological Sciences, Rutgers, The State University of New Jersey, New Brunswick, NJ, United States

M. Berdasco Cancer Epigenetics Group, Cancer Epigenetics and Biology Program (PEBC), Bellvitge Biomedical Research Institute (IDIBELL), Barcelona, Catalonia, Spain

C. Brossas Institut Jacques–Monod, CNRS, Paris Diderot University, Paris, France

S. Cacchione Department of Biology and Biotechnology, Istituto Pasteur Italia - Fondazione Cenci Bolognetti, Sapienza University of Rome, Rome, Italy

G. Castelo-Branco Laboratory of Molecular Neurobiology, Department of Medical Biochemistry and Biophysics, Karolinska Institutet, Stockholm, Sweden

A. Cicconi Department of Biology and Biotechnology, Istituto Pasteur Italia - Fondazione Cenci Bolognetti, Sapienza University of Rome, Rome, Italy

D. Doenecke Institute for Molecular Biology, University of Göttingen, Göttingen, Lower Saxony, Germany

M.E. Donohoe Burke Medical Research Institute, White Plains; Department of Neuroscience, Department of Cell and Developmental Biology, Brain Mind Research Institute, Weill Cornell Medical College, New York, NY, United States

B. Duriez Institut Jacques–Monod, CNRS, Paris Diderot University, Paris, France

S. Erhardt ZMBH, DKFZ-ZMBH-Alliance; Cell Networks Excellence Cluster, University of Heidelberg, Im Neuenheimer Feld, Heidelberg, Germany

M. Esteller Cancer Epigenetics Group, Cancer Epigenetics and Biology Program (PEBC), Bellvitge Biomedical Research Institute (IDIBELL); Department of Physiological Sciences II, School of Medicine, University of Barcelona; Catalan Institution for Research and Advanced Studies (ICREA), Barcelona, Catalonia, Spain

A.M. Falcão Laboratory of Molecular Neurobiology, Department of Medical Biochemistry and Biophysics, Karolinska Institutet, Stockholm, Sweden

A. Fiszbein Institute of Physiology, Molecular Biology and Neurosciences (IFIBYNE-CONICET) and Department of Physiology, Molecular and Cell, Faculty of Natural Sciences, University of Buenos Aires, Ciudad Universitaria, Buenos Aires, Argentina

A. Galati Department of Biology and Biotechnology, Istituto Pasteur Italia - Fondazione Cenci Bolognetti, Sapienza University of Rome, Rome, Italy

M.A. Godoy Herz Institute of Physiology, Molecular Biology and Neurosciences (IFIBYNE-CONICET) and Department of Physiology, Molecular and Cell, Faculty of Natural Sciences, University of Buenos Aires, Ciudad Universitaria, Buenos Aires, Argentina

L.I. Gomez Acuña Institute of Physiology, Molecular Biology and Neurosciences (IFIBYNE-CONICET) and Department of Physiology, Molecular and Cell, Faculty of Natural Sciences, University of Buenos Aires, Ciudad Universitaria, Buenos Aires, Argentina

A. Göndör Department of Microbiology, Tumor and Cell Biology, Karolinska Institutet, Stockholm, Sweden

A.R. Kornblihtt Institute of Physiology, Molecular Biology and Neurosciences (IFIBYNE-CONICET) and Department of Physiology, Molecular and Cell, Faculty of Natural Sciences, University of Buenos Aires, Ciudad Universitaria, Buenos Aires, Argentina

A. Lennartsson Department of Biosciences and Nutrition, Karolinska Institutet, Huddinge, Sweden

M. Lezzerini Integrated Cardio Metabolic Centre, Department of Medicine, Karolinska Institutet, Huddinge, Sweden

S.J. Linder Program in Biological and Biomedical Sciences, Harvard Medical School; Massachusetts General Hospital Cancer Center, Boston, MA, United States

R. Margueron Curie Institute; INSERM; CNRS, Paris, France

M. Martino Department of Microbiology, Tumor and Cell Biology, Karolinska Institutet, Stockholm, Sweden

E. Micheli Department of Biology and Biotechnology, Istituto Pasteur Italia - Fondazione Cenci Bolognetti, Sapienza University of Rome, Rome, Italy

L. Millán-Ariño Department of Medicine, Karolinska University Hospital, Stockholm, Sweden

K.M. Miller Department of Molecular Biosciences, Institute for Molecular and Cellular Biology, University of Texas at Austin, Austin, TX, United States

R. Mostoslavsky   Program in Biological and Biomedical Sciences, Harvard Medical School; Massachusetts General Hospital Cancer Center, Boston, MA, United States

A-.K. Östlund Farrants   Department of Molecular Biosciences, The Wenner–Gren Institute, Stockholm University, Stockholm, Sweden

M-.N. Prioleau   Institut Jacques–Monod, CNRS, Paris Diderot University, Paris, France

C.G. Riedel   Integrated Cardio Metabolic Centre, Department of Medicine, Karolinska Institutet, Huddinge, Sweden

B.A. Scholz   Department of Microbiology, Tumor and Cell Biology, Karolinska Institutet, Stockholm, Sweden

I. Tzelepis   Department of Microbiology, Tumor and Cell Biology, Karolinska Institutet, Stockholm, Sweden

A-.L. Valton   Institut Jacques–Monod, CNRS, Paris Diderot University, Paris, France; Program in Systems Biology, Department of Biochemistry and Molecular Pharmacology, University of Massachusetts Medical School, University of Massachusetts, Worcester, MA, United States

M. Wassef   Curie Institute; INSERM; CNRS, Paris, France

Preface

Chromatin research is a quickly developing field, which is in the center of cell differentiation, development, stem cell biology and aging. Moreover, almost all complex diseases display chromatin changes, many of which have proved to be causally linked to the disease, or to have diagnostic and/or prognostic values. This book provides a comprehensive overview on the most recent scientific achievements of chromatin-based processes, and is divided into 17 chapters that build on each other, but can also be consulted independently.

The first 11 chapters cover the basic principles of chromatin regulation and introduce the reader to the language of chromatin marks, their dynamics throughout the cell cycle and their context-dependent functions in various nuclear processes including transcriptional regulation, DNA replication, splicing, DNA repair, and ribosomal RNA transcription. These chapters discuss novel principles with a cross-disciplinary perspective, explain chromatin-based processes in the context of development, and present links to numerous human diseases.

Building on the information introduced in the first 11 chapters of the book, Chapters 12–14 discuss the mechanism by which heritable chromatin states contribute to the function of specialized regions of the genome, such as telomeres and centromeres, as well as to the formation of repressed states on the inactive X chromosome; processes that are all central to development and often deregulated in diseases.

As the activity of chromatin-modifying enzymes is influenced by the levels of intermediary metabolites that act as cofactors or substrates for the enzymatic reactions, Chapters 15 and 16 integrate various aspects of chromatin biology with cellular metabolic states. Chromatin states as well as many of the cellular metabolic pathways that influence chromatin modifiers are under the regulation of circadian clocks. Circadian chromatin transitions are, in turn, central in regulating the oscillating expression of gene products that control metabolism, establishing a two-way relationship between daily oscillations in metabolic processes and chromatin states. In line with the role of circadian regulation in adaptation to changes in the environment, deregulation of circadian rhythm predisposes to a wide range of complex diseases, such as metabolic and psychiatric disorders, as well as cancer.

The recent years have witnessed an explosion of research suggesting that deregulated chromatin states are central to tumor development. Unstable chromatin states in tumor cells have thus been suggested to maintain phenotypic heterogeneity and plasticity within the tumor tissue, enabling the selection of the fittest clones under continuously changing selective pressure, and thereby promoting tumor evolution. Chapter 17 thus explores how chromatin states regulate cellular phenotypic plasticity in health, aging and diseases, such as cancer. Interestingly, the stability of chromatin states and their role in the maintenance of cellular phenotypes in health and disease are influenced by the three-dimensional organization of the genome within the nuclear space. Introduced already in Chapter 1, several chapters of this book discuss how spatial compartmentalization of nuclear functions and dynamic physical interactions between distant regulatory elements affect chromatin states, transcription, replication, and DNA repair. Chapter 17 provides, moreover, an overview of these features and presents novel hypotheses on how deregulated three-dimensional nuclear architecture and genome organization might contribute to the emergence of major tumor hallmarks, such as increased phenotypic plasticity.

Taken together, the chapters of the book are written by prominent experts in the field with the ambition to integrate a broad range of topics on chromatin research to promote crosstalk between basic sciences and their applications in medicine. The book is thus targeted toward biological and medical scientists, as well as undergraduate and PhD students in biology or medicine with an interest in chromatin regulation and in how chromatin-mediated processes contribute to development, aging and complex diseases.

Finally, I would like to express my gratitude to all the contributors and coauthors whose work has made it possible to bring this book into existence. I would like to thank professor Trygve Tollefsbol for organizing the *Translational Epigenetics* series and inviting me to participate in this ambition. Many thanks are given to Catherine Van Der Laan, Lisa Eppich, and the production team at Elsevier for their dedication, encouragement, and assistance in printing this book. Appreciation and thanks are also given to the anonymous reviewers of the chapters for their valuable comments. I would like to recognize the contribution of my group members and colleagues, especially Drs Barbara A. Scholz and Lluís Millan-Ariño, as well as Mirco Martino and Ilias Tzelepis. Finally, I thank my family for their support and patience during the preparation of this book.

Anita Göndör

CHAPTER 1

A Brief Introduction to Chromatin Regulation and Dynamics

I. Tzelepis[a], M. Martino[a], A. Göndör

Department of Microbiology, Tumor and Cell Biology,
Karolinska Institutet, Stockholm, Sweden

1.1 INTRODUCTION TO BASIC CONCEPTS OF CHROMATIN REGULATION

The distinct cell types of a multicellular organism have stable, characteristic phenotypes and perform specialized functions; despite that they contain, with few exceptions, the very same DNA sequence. The existence of a system that regulates the cell type–specific use of the genetic material and provides cellular memories of gene expression patterns over time has been long recognized [1,2]. At the same time, such a mechanism has to display a considerable level of flexibility and responsiveness to environmental cues, enabling the cells to change their phenotypes during adaptive responses [1,2]. The complexity of the molecular mechanism that regulates when and how genes should be expressed has only recently been elucidated, long after the first description of the biological processes and phenotypes they regulate. In the nucleus DNA is thus organized in chromatin structure that includes an array of histone and nonhistone proteins, their posttranslational modifications (PTMs), RNA components, as well as DNA modifications [1,2]. Chromatin regulates not only the efficient packaging of the genome into the nuclear space, but also influences the accessibility of the underlying DNA to *trans*-acting factors, and provides a platform for the regulated recruitment of enzymatic functions and proteins that orchestrate various genomic functions [1,2]. Chromatin thus plays essential roles in all nuclear processes templated by the genetic material, including transcription (Chapters 1–6, 8–10), RNA splicing (Chapter 8), DNA replication (Chapters 5–6), and DNA repair (Chapter 11), with far-reaching consequences on human health [1,2]. It is not surprising therefore that chromatin research has had an increasing influence on a wide range of research

CONTENTS

---

[a]These authors contributed equally to the manuscript.

fields, such as developmental biology, aging, and various monogenic, as well as complex diseases.

Chromatin marks can be divided into open, transcriptionally permissive, euchromatin modifications and compact, repressive, heterochromatin marks [1,2]. The identity and function of the various histone PTMs and DNA modifications is discussed in details in Chapters 2–3. An important feature of histone modifications is their reversible nature, which reflects that these PTMs are the result of opposing enzymatic activities [1,2]. Even modifications of the DNA, such as cytosine (C) methylation, are reversible, although active demethylation requires collaboration between multiple enzymes and DNA repair factors [1,2] (see also Chapter 3). A key question in chromatin biology is therefore how chromatin-modifying activities are targeted to specific sites of the genome to maintain gene expression patterns or bring about a change in the expression of specific genes. Although not completely understood, this process is regulated by interactions between chromatin modifiers and sequence-specific, DNA-binding proteins/transcription factors or other existing chromatin components [3]. Environmental cues that regulate the expression and/or function of transcription factors or directly modify chromatin and chromatin modifiers play thus important roles in the regulation of chromatin states and the expressivity of the genome [3,4].

A subset of the dynamic chromatin modifications have been shown to be heritable during mitosis and sometimes even during meiosis [1,2]. These heritable chromatin marks are called epigenetic marks, which can propagate the effects of transient environmental signals, developmental cues and cellular metabolic states on gene expression long after the exposure to the initial stimulus [1,2]. Several mechanisms have evolved to enable epigenetic modifications to be copied and maintained during cell divisions, thereby providing cellular memories that maintain specific states of differentiation [1,2]. Finally, even heritable epigenetic states can be altered at a genome-wide level during certain stages of development, upon specific signals and in diseases, as discussed further. One of the enigmas of chromatin regulation is how two opposing features, namely the stability and plasticity of chromatin states, is fine-tuned in response to internal and external signals. Regulation of chromatin dynamics is thus central to our understanding of the mechanisms that balance on one hand, the robustness of cellular phenotypes against perturbations and on the other hand, adaptive phenotypic plasticity [4,5] (discussed in Chapter 17).

In this chapter we will start by presenting the early experiments that highlighted the existence of cellular memories of gene expression patterns over time, and inspired investigations to uncover the molecular mechanism of epigenetic inheritance. As mentioned earlier, it has long been recognized that even these heritable cellular states can be reprogrammed during certain developmental windows and in diseases. We will thus briefly present the major reprogramming events that take place during development and in cancer. We will then provide an overview of the history of chromatin research that eventually uncovered the link between cellular memories and chromatin states. As all chromatin-templated functions take place in the three-dimensional (3D) space of the nucleus, we will introduce the basic concepts of 3D genome organization and its consequences on genomic functions, such as transcription and replication. We will end by a modern definition of chromatin regulation that views chromatin states as platforms for the integration of internal and external cues over time during development and during adaptation to environmental signals, as well as in diseases.

1.2 EPIGENETIC PHENOMENA: HERITABILITY OF CHROMATIN STATES DURING CELL DIVISION

The term "epigenetic" has been introduction by Conrad Waddington in 1942 [2,6,7] (Fig. 1.1), and has undergone many different interpretations since then. Waddington used it to describe all the regulated processes that lead to the development of the adult organism from the zygote, and suggested that this process required interactions between the genotype, epigenotype, and the environment [7]. In his famous metaphor (Fig. 1.2A), he thus described cell differentiation as a ball rolling down on the "epigenetic landscape" toward well-defined valleys representing mature cell states. In this representation, "canalization" of the rolling ball by the valleys toward specific directions, or in other words "buffering," refers to the maintenance of stable developmental outcomes despite environmental perturbations [7]. On the contrary, "developmental plasticity" refers to the generation of multiple cellular phenotypes from the same genotype. Hence, the concept of a regulatory layer that interacts with both the genotype and the environment has been proposed before the discovery of the chromatin-based mechanisms of gene regulation [2]. A more modern use of epigenetics builds on the knowledge about the existence of dynamic and heritable chromatin modifications, and was proposed by Riggs and Porter in 1996 to include the "mitotically and/or meiotically stable changes in gene function that cannot be explained by changes in the DNA sequence" [8]. Epigenetic phenomena thus refer to the cellular memories of chromatin states, which can be surprisingly stable during the lifetime of the organism, sometimes even in between generations. These epigenetic features are thus essential for the stable maintenance of cell type–specific gene expression patterns and normal development [2].

1.2.1 Inheritance of Stable, Cell Type–Specific Gene Expression Patterns

Early experiments providing evidence for the existence of stable gene expression patterns through cell divisions were performed already in the 1960s. Hadorn

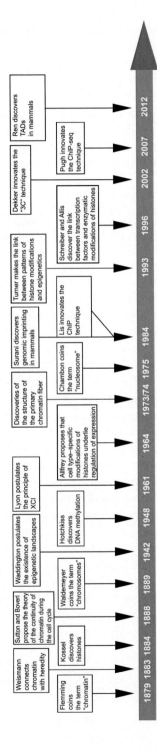

FIGURE 1.1 Milestones of chromatin research.

The image represents the timeline of the key discoveries in chromatin research from 1879 to the present. The list is not complete as several key discoveries were achieved over long periods of time, involving many laboratories. *ChIP*, Chromatin immunoprecipitation; *TADs*, topological-associated domain; *XCI*, X-chromosome inactivation.

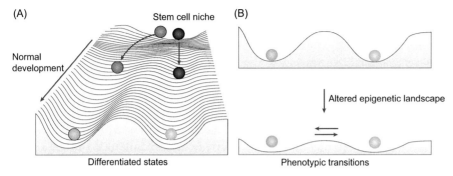

FIGURE 1.2 Waddington landscape of epigenetic regulation during differentiation and reprogramming.

(A) The Waddington landscape of development is a metaphor that compares cell differentiation to balls rolling down on the "epigenetic landscape" toward well-defined valleys representing mature cellular states. In this representation, "canalization" of the rolling ball by the valleys toward specific directions, or in other words "buffering," refers to the maintenance of stable developmental outcomes despite external or internal perturbations. Conversely, the generation of multiple cellular phenotypes from the same genotype reflects "phenotypic plasticity." Differentiation is viewed as a hierarchical process, where the developmental potential of stem cells *(gray balls on the top of the hill)* is continuously restricted *(colored balls rolling down the hill)*. Moreover, Waddington proposed that changes in the landscape can be induced by mutations in the genes.
(B) Differentiation and the process of continuous restriction in developmental potential can be reversed by external or internal signals that can reconfigure the epigenetic landscape and reduce barriers against *trans*-differentiation *(as illustrated by the depth of the valleys)* or barriers against dedifferentiation (alter the the slope of the hill, not shown). Cell fates can thus be interconvertible in vitro, and cells can, upon specific cues, *trans*-differentiate into other related cell types originating from the same or different germ layers. The images depict a transition state between two cell fates *(colored balls)*, before the establishment of positive feedback or feed-forward loops to sustain a particular new cell fate. Moreover, epigenetic marks are globally erased and reprogrammed during two stages of early development, namely after fertilization and during the development of primordial germ cells (PGCs). This process is likely governed by signaling cues and pioneer transcription factors that can reprogram epigenetic states in collaboration with chromatin modifiers. *Source Adapted from Feinberg AP, Koldobskiy MA, Gondor A. Epigenetic modulators, modifiers and mediators incancer aetiology and progression. Nat Rev Genet 2016;17:284–99. [4]*

has thus shown that specific epithelial cells of the imaginal discs of *Drosophila* larvae maintained their differentiation stage even after transplanting them into adult females, where they could proliferate without differentiation [9]. Imaginal discs contain epithelial cell clusters destined to develop into specific external structures and appendages of the fly after metamorphosis. Following long-term culture in adult females, these epithelial cells originating from imaginal discs could thus give rise to the expected structures when transplanted back to larvae [9]. This property has later been linked to chromatin regulation by the evolutionary conserved Polycomb (PcG) and Trithorax (TrxG) proteins, which serve to lock in the transcriptionally repressed and active states, respectively, providing memories of gene expression states [10] (discussed in Chapter 7).

1.2.2   Genomic Imprinting

Another example of cellular memories is represented by the phenomenon of genomic imprinting (Fig. 1.1), which refers to the stable parent of origin–specific, monoallelic expression of the so-called imprinted genes [11]. Stable inheritance of parent of origin–specific features were first discovered by Helen Crouse in the 1960s in the insect, mealybug [12]. In male embryos the paternal set of haploid chromosomes thus becomes silenced and packaged into compact chromatin structure, suggestive of a paternal-specific epigenetic memory established in the male germline, which is then maintained throughout the mitotic cell divisions of the organism. The existence of genomic imprinting has been demonstrated also in mice [13]. Using inbred mice, Surani et al. have devised experiments that uncovered that both the paternal and the maternal pronuclei are necessary for normal embryonic and fetal development, because the parental chromosomes display functional differences that are not encoded in the DNA sequence [13]. Hence, nuclear transplantation experiments of either two paternal pronuclei or two maternal pronuclei, alternatively a paternal and a maternal pronucleus into an enucleated, activated oocyte documented that although the paternal and maternal pronuclei contained the same DNA sequence, only those zygotes developed to term that contained both a paternal and a maternal genome [13]. It has been suggested already in the early 1980s that this functional difference between the genetically identical parental chromosomes reflects the reversible and heritable marking or "imprinting" of the mouse genome during male and female gametogenesis [13]. This imprint is then maintained during development even after the activation of transcription programs in the developing embryo [14]. Only later has it been established that the molecular mechanism of imprinting is linked to specific chromatin marks [15]. Finally, the first imprinted gene was discovered by Denis Barlow and coworkers in the 1990s [16]. We now know of around hundred potentially imprinted genes in the human and mouse genomes [11]. Several of these genes have been extensively investigated and found to be regulated by imprinting control regions, which are differentially marked during male and female germline development [11]. The resulting parental-specific marks are generally stably maintained during the lifetime of the offspring, and manifest as parent of origin–specific monoallelic expression [11]. The importance of imprinting is highlighted by the existence of the so-called imprinting disorders [11], which emerge upon genetic or epigenetic disruption of imprinted expression patterns, and include developmental defects and predisposition to cancer and psychiatric disorders [17]. Furthermore, nonequivalent expression of imprinted genes from the paternal and maternal chromosomes acts as barrier to parthenogenesis [18]. Interindividual differences in the stringency of imprinted gene expression in twinpairs and the alterations observed in imprinted expression within an individual during aging indicate, however, that even stable imprinting marks can be affected by environmental factors, resulting in allelic variation in gene expression [19,20].

1.2.3 Random Monoallelic Gene Expression

The second category of heritable monoallelic expression is established in somatic cells as opposed to germline, where the choice of which allele is expressed is random. For example, as a consequence of X-chromosome inactivation (XCI) (Fig. 1.1), almost all genes on one of the two X chromosomes of the somatic cells in female mammals become packaged into heterochromatin and undergo stable repression to enable dosage compensation [21]. Mary F. Lyon was the first to suggest a unifying theory for the experiments performed in the 1950s, and proposed the hypothesis that random XCI (discussed in Chapter 14) occurs early during development [22]. This hypothesis facilitated the progress of the research field of epigenetic inheritance and explained the phenotype of X-linked diseases from new perspectives. An intriguing question is how the randomness of XCI is regulated and shielded from environmental signals that might skew this process. Indeed, longitudinal twin studies of skewed XCI highlighted that the randomness of this process shows only moderate stability, likely increasing the phenotypic discordance between female monozygotic twins [23].

Other forms of random monoallelic expression that can be stably maintained during cell division include the phenomenon of allelic exclusion, which restricts the expression of certain cell surface receptors, such as the subunits of B- and T-cell receptors and the olfactory receptors (ORs) in olfactory neurons, to one allele per cell [24,25]. Frank M. Burnet has recognized the importance of this process already in the 1950s, when he postulated the "clonal selection theory of acquired immunity" [26]. Allelic exclusion thus ensures that only one allele of the antigen receptor subunits is expressed on each B and T cells, the choice of which is then maintained during cell divisions. This process provides unique specificity of antigen recognition for each B and T cell, and forms the basis for clonal selection, that is, expansion of the B and T cells that recognize a specific antigen. Similarly, in each olfactory neuron, only one allele of the OR genes is expressed to ensure specific odorant sensing in each olfactory neuron [24]. Although the discovery of the phenomenon of allelic exclusion [27,28] goes back to the middle of the last century, it has only been recognized recently that this process also involves the differential regulation of chromatin states and accessibility of the underlying DNA, which is then heritable during cell division [25,29,30].

Recent estimates suggest that at any given time up to 1000 genes of the human genome might be expressed in a monoallelic fashion, which might be heritable through several cell divisions via epigenetic mechanisms [31,32].

The result is the diversification of expression patterns by increasing the potential combinations of epialleles with different sequence polymorphisms in a cell population. Such diversity likely contributes to phenotypic differences among cells, providing selectable features during development in response to environmental cues [31,32]. Indeed, one of the central questions of developmental biology is how genetically identical cells respond differentially to environmental cues, which has already been linked to cellular differences in transcriptionally permissive and repressive heritable chromatin modifications (reviewed in [33]).

1.2.4   Other Epigenetic Phenomena

Other epigenetic phenomena in mammalian cells include the position effect variegation [34], which refers to the variable silencing of genes located near heterochromatin regions. This process was discovered in *Drosophila* translocation mutants already in the 1950s [34], and was linked to the variable spreading of heterochromatin over the translocated regions. Such variegated spreading of heterochromatin can induce phenotypic variation among genetically identical cells, and has recently been shown to be sensitive to environmental signals, such as stress and MAPK signaling in yeast [35]. Finally, the position, function, and inheritance of the centromeres are also determined largely by epigenetic factors in mammalian cells [36], as reviewed in Chapter 12. Neocentromere formation, that is, formation of functional centromeres at ectopic sites, is a potentially hazardous process that can result in chromosome breakage and cell cycle arrest [37], which has to be therefore tightly regulated.

The identity and function of the various epigenetic marks that form stable cellular memories and their inheritance is still under extensive investigation, and are discussed in detail in Chapters 2, 3, 5–7 of this book. The precise mechanisms by which developmental cues signal to chromatin and initiate the formation epigenetic states are also just being elucidated. Moreover, even heritable epigenetic states can be reprogrammed during certain developmental windows, upon specific signals or in vitro to enable a change in developmental potential.

1.3   REPROGRAMMING OF EPIGENETIC STATES DURING DEVELOPMENT AND IN DISEASES

Recent advances have uncovered that the hierarchical process of differentiation and continuous restriction of developmental potential presented in Waddington's epigenetic landscape can be reversed (reviewed in [38]) (Fig. 1.2B). Experiments in the 1950s documented that the transfer of advanced blastula cell nuclei into enucleated oocytes [39] enabled the development of an embryo. Furthermore, in 1996 a sheep was cloned via nuclear transfer from cells of an established cell line [40]. These experiments suggest that epigenetic marks that define different cellular phenotypes can be reprogrammed—in the presence of certain signals—to totipotency, that is, to a cell state that supports the development of an organism. Furthermore, cell fusion between pluripotent cells and differentiated cells [41] leads to the reprogramming of the differentiated cell nucleus to pluripotency, a self-renewing stem cell state that can be induced to differentiate into all three germ layers upon exposure to the appropriate signals. Similarly, in 2006 Yamanaka and Takahashi generated the first induced pluripotent stem cells from differentiated cells by transcription factor–based reprogramming [42], the efficiency of which is modulated by factors that affect chromatin [43].

During the last 3 decades several experiments have, moreover, shown that cell fates can be interconvertible, that is, differentiated cells can *trans*-differentiate into other related cell types originating from the same germ layer, but also into cell types of different germ layers (reviewed in [38]). Direct intra- and intergerm layer cell fate conversions and indirect cell fate conversions that involve transient acquisition of pluripotency during this process appear to be governed by signaling cues and pioneer transcription factors that can bind DNA even when it is packaged into heterochromatin, and collaborate with chromatin modifiers to reprogram barriers against dedifferentiation and establish positive feedback or feed-forward loops to sustain a particular cell fate [38].

1.3.1   Reprogramming of Epigenetic Marks During Early Development

Cellular memories are also reprogrammed on a genome-wide scale during key stages of development with a common denominator of an increase in developmental potential, defined as the range of cell types a cell can give rise to [1,4]. Hence, it has been long recognized that the transcriptomes of the sperm and egg have to be reprogrammed in the zygote to support the emergence of a totipotent state and to establish differentiation potential toward both embryonic and extraembryonic lineages [44]. Experiments in the 1980s started to address the molecular mechanisms of the reprogramming events during early embryonic development, and established that these events are paralleled by temporal, genome-scale changes in DNA methylation levels [45]. Shortly after it became clear that such reprogramming events do not interfere, however, with the maintenance of parental imprinting [46]. The transcriptional changes during early lineage specification [47] and the molecular mechanism of epigenetic reprogramming and differentiation are under extensive investigation and still poorly understood (discussed in Chapter 3 and reviewed in [48,49]).

A second reprogramming event takes place during primordial germ cell (PGC) specification, which also involves major epigenetic changes including DNA demethyaltion [45,50] (Chapter 3). Early experiments that uncovered the functional differences between parental chromosomes have already hinted at a reprogramming event during germline development, that has to reset the inherited paternal and maternal imprints in the offspring's germline to represent the gender of the offspring [13]. Cell fusion experiments between PGC-derived embryonic germ cells and somatic cells have demonstrated the reprogramming potential of embryonic germ cells already in the 1990s. The somatic cell nuclei in the hybrid cells thus underwent extensive epigenetic reprogramming that also involved the removal of parental imprints [51]. It is now well established that PGCs derive from postimplantation epiblast cells under the control of bone morphogenetic protein (BMP) and WNT signaling pathways. It is, however, still unclear how a subset of postimplantation epiblast cells develops readiness to respond to such signals, that is, acquires developmental competence to become PGCs [52]. In the cells destined to become PGCs, somatic differentiation programs thus become repressed, and multiple layers of chromatin modifications acquired during previous developmental stages are erased [50]. Moreover, even parental imprinting marks are erased in these cells to enable the deposition of correct imprinting marks in the developing germ cells according to the gender of the organism [50]. It is now well established, that reprogramming events during germline development leads to the transient reacquisition of some features of pluripotency, while maintaining commitment to gametogenesis. Male and female germline development then proceeds toward the generation of mature egg and sperm, including the establishment of parental-specific imprinting marks [50].

1.3.2   Epigenetic Reprogramming in Tumor Development

Finally, extensive epigenetic reprogramming has been suggested to underlie the emergence of cancer stem cells states [53–55], which display increased developmental potential, fueling the emergence of more or less differentiated tumor cells and the growth and resilience of tumor tissue [4]. Cancers stem cells often reexpress pluripotency factors [56–58], such as OCT4 and NANOG—expressed normally only in embryonic stem cells—to destabilize or prevent the emergence of differentiated cellular phenotypes [4]. Such factors, which were recently termed epigenetic mediators [4], collaborate with chromatin modifiers to increase the phenotypic plasticity of tumor tissue and drive cancer evolution under selective pressure [4] (Chapter 17). As their expression in various tumors is linked with worse prognosis and metastasis formation, the mechanism of their reactivation is the subject of extensive research [4] (Chapter 17).

In the following paragraphs we describe the milestones of chromatin research (Fig. 1.1), which uncovered the molecular mechanism of epigenetic memory formation and reprogramming. These include the discovery of the basic chromatin unit, the nucleosome, the identification of the first histone and DNA modifications, and the discovery of their role in transcriptional regulation. We will then end by presenting a modern view of epigenetic regulation that places chromatin-based processes in the 3D space of the nucleus and at the interface between the environment and the genotype.

1.4 EARLY PERIOD OF CHROMATIN RESEARCH

The initial phase of chromatin research (Fig. 1.1) has been facilitated by the innovation of the microscope that opened up a new world for the lucky ones who had the first view of the nucleus. Although Leeuwenhook invented the first microscope in 1659 enabling the discovery of crude histological features, a microscope capable of observing nuclear substructures emerged only in 1826 [59]. Approximately 50 years later Walter Fleming coined the word "chromatin" to describe the threads stained by aniline dyes that he could see in mitotic cells under the microscope [59]. While Walther Flemming was the first to make the connection between chromatin and nucleic acid and discovered mitosis, it was Wilhelm Waldemeyer who in 1889 created the term "chromosomes" to describe "the dynamically turning over threads" first visualized by Flemming during mitosis. A missing link was provided by August Weismann who proposed a connection between chromatin and later on chromosomes and the genetic material in 1883 and 1891 [59]. The difference between interphase chromatin and chromosomes was hotly debated during those times until Walter Sutton and Theodor Boveri proposed the theory of continuity of chromosomes during the entire cell cycle despite the disappearance of mitotic chromosome structures in the interphase [59]. The implications of this realization were first advocated by Edmund B. Wilson who hypothesized that chromatin/chromosomes cannot consist of a homogenous substance, and suggested that they were instead complex systems that underwent cell cycle–specific changes in morphology [59]. Thus, without knowing almost anything about the chromatin and mitotic chromosome structure the pioneers managed to predict their relationships. Just a few years earlier, in 1884, Albrecht Kossel had discovered the histones [59], although the importance of this observation became clear much later.

1.5 DISCOVERY OF THE NUCLEOSOME AND NUCLEOSOME POSITIONING

During the next 7 decades many key discoveries were made, such as the demonstration that a biochemical entity, the deoxyribonucleic acid, constitutes the heritable genetic material [59,60]. However, during this time very little research has been performed to increase our understanding of chromatin, until a paradigm change unraveled the basic principles of chromatin organization about 4 decades ago. Again a technical innovation, the electron microscope, laid down the fundament for the discovery that chromatin is organized by single-repeating units, termed the nucleosome [61]. Thus by spreading out nuclear material, the Olins couple and C. Woodcock uncovered that histones did not cover the DNA uniformly, but instead formed the nucleosomes, which were organized as beads on a string with globular structures separated by what appeared to be as spacer regions [61]. Following a race, biochemists, such as Roger Conrberg and Jean O. Thomas, in the early 1970s managed to purify mononucleosomes and determine that these consist of DNA sequences wrapped around a core octamer consisting of two copies of the four major histones, H2A, H2B, H3, and H4 [61]. Further chromatin compaction was then found to be regulated by the binding of the linker histone H1 to linker regions between nucleosomes [62].

It is now well established, that the localization of nucleosomes along the length of the DNA can be influenced by a combination of factors, such as sequence features, that repel or attract nucleosomes, transcription factors, the RNA pol II machinery, and chromatin-remodeling enzymes that can evict, slide, or re-model nucleosomes (Chapter 4) (reviewed in [63]). While certain regions tend to display precise and reproducible nucleosome positioning in a cell population, others have varying levels of nucleosome positioning up to no positioning at all [63]. Nucleosome positioning in turn dictates which portion of the genome is generally more available for transcription factors, which often occupy binding sites localized within the linker regions [63]. Certain transcription factors, called pioneer transcription factors, can bind DNA even when it is tightly wrapped around the nucleosome [63]. Such factors are thus able to remodel, even compact, repressed chromatin states in collaboration with chromatin-remodeling enzymes, and play important roles during lineage commitment, differentiation, and epigenetic reprogramming. A term related to, but distinct from nucleosome positioning is nucleosome occupancy [63]. While most of the genome is occupied by nucleosomes, certain active regulatory regions, such as enhancers, promoters, and terminators, show low-nucleosome occupancy or are nucleosome-free to enable the efficient binding of the transcription machinery [63]. Nucleosome positioning, spacing, and occupancy thus emerge as important regulators of cell type–specific gene expression patterns, which can undergo dynamic changes during differentiation [64]. Taken together, the discovery of the nucleosome has revolutionized chromatin research by opening up fundamentally novel questions, such as how nucleosome positioning (Chapter 4), DNA and histone modifications (Chapters 2–3, 5–6), and the 3D folding of the primary chromatin fiber (Chapters 7–14, 17) collaborate to affect genomic functions.

1.6   HISTONE MODIFICATIONS: DISCOVERY AND FUNCTION

Given that the structure of chromosomes undergoes cell cycle–specific morphological changes, Edmund B. Wilson predicted that the packaging of the genome in chromatin might be dynamically regulated [59]. Preceding the discovery of nucleosomes, there was biochemical evidence present for the existence of cell type–specific difference in chromatin composition and the presence of PTMs on histones [65,66]. For example, E. Stedman proposed in 1950 that the histones displayed cell type–specific differences in their arginine content, which were prophetically suggested to be linked with cell type–specific gene expression [65]. Moreover, in 1964 Vincent Allfrey and Alfred E. Mirsky documented that histones can be acetylated and methylated after the completion of translation, and proposed that these PTMs could switch on or off different genes in cell type–specific manners [66] (Fig. 1.1). Many others, such as C. Crane–Robinson and B. Turner have followed suit, to strengthen the initial contention by Allfrey and coworkers. However, the efforts to realize this important idea into hard evidence took another few decades, when two different groups could conclusively determine that histone modifications provided information that could determine transcriptional activation and repression. Thus, using an affinity column Stuart Schreiber's lab purified a transcriptional repressor, Rpd3, that displayed histone deacetylase (HDAC) activity [67]. Conversely, work from David Allis' lab, uncovered that a positive regulator of transcription, Gcn5p, had histone acetyl transferase activity [68] (Fig. 1.1).

1.6.1 The Language of Histone Modifications

The invention of a set of relatively recent techniques uncovered numerous PTMs present on histones and other cellular proteins, as well as their genomic distributions (Chapter 2). Milestones include the invention of the chromatin immunoprecipitation (ChIP) method to capture DNA sequences occupied by a protein of interest by John T. Lis and David Gilmour in 1984 [69], as well as the adaptation of mass spectrometric analysis to the detection of histone PTMs in 1995 (Fig. 1.1). The perception that different histone modifications might act in combinatorial manner to establish transcriptionally permissive and repressive domains in the genome was first formalized by B. Turner [70] (Fig. 1.1) and Perez–Ortin [71] in 1993. This concept was then developed further by C.D. Allis [72] and T. Jenuwein [73], who predicted that the repertoire of PTMs on the same and/or adjacent histones formed the so-called "histone code" to regulate cell type–specific gene expression patterns [74]. The advent of high-throughput sequencing techniques and their adaptation to chromatin research, such as the invention of ChIP sequencing first described by the Pugh group [75], provided genome-wide views on how different chromatin proteins and histone PTMs regulate the expressivity of the genome. An important observation of these experiments was that the effect of chromatin marks is context-specific and depends on the genomic location, the density of the mark, and the surrounding chromatin marks, leading to the modification of the initial "histone code" hypothesis that assigned different chromatin modifications a universal meaning [76]. We now discriminate between an array of "chromatin writers" that catalyze the PTMs, a similarly diverse group of "chromatin readers" that interpret the PTMs, and "chromatin erasers" that remove the PTMs. Their collaboration, often in response to environmental cues, establish active or inactive chromatin states in cell type–specific manners. Furthermore, systematic mapping of chromatin proteins and PTMs genome wide in several species and cell types have documented that chromosomes have domain organization, displaying qualitative and quantitative differences in chromatin composition and transcriptional regulation within the different domains [77,78]. This organization of chromosomes likely reflects the existence of extensive "cross talk" between histone modifications to reinforce and spread or weaken the transcriptionally permissive or repressive states [79], as discussed further and in several other chapters of this book. Moreover, histone modifications often collaborate with DNA modifications in the regulation of genomic functions. These chromatin modifications form multiple layers of self-reinforcing chromatin states that define cell type–specific gene expression patterns and need to be removed during reprogramming events to increase developmental potential [79]. In the following paragraphs we will discuss the discovery of DNA methylation and its various cross talk with histone modifications with consequences on nuclear functions.

1.7   DISCOVERY OF DNA METHYLATION: FUNCTIONS AND CROSS TALK WITH HISTONE MODIFICATIONS

1.7.1   Discovery of DNA Methylation

In 1948 it was discovered that the cytosine residue (C) of the DNA itself could be methylated at the 5th carbon position (5-methylcytosine; mC) when followed by a guanine residue (G) [80] (Fig. 1.1). The machinery that methylates the mammalian genome at cytosine includes three different enzymes (Chapter 3). The first enzyme discovered to be involved in DNA methylation in vivo is the de novo maintenance DNA methyltransferase (DNMT) 1, DNMT1 [81], which functions in the maintenance of DNA methylation states during cell divisions [82–85]. The major role of DNMT1 is thus to copy the methylation state of the parental DNA strand to the newly synthesized DNA strand during S phase. Two additional enzymes, DNMT3A and B, were found to be responsible for de novo methylation of previously unmethylated sequences [82,86]. An important consequence of this system is that CpG methylation can be heritable during cell division. In line with its heritability, DNA methylation plays important roles in many epigenetic phenomena, such as genomic imprinting, X inactivation, and other forms of monoallelic expression [82,83,87] (discussed in Chapter 3).

Although the correlation between increased DNA methylation and gene repression in vivo and its effect in vitro had been known already in the late 1980s [88], the earliest observations providing proof that DNA methylation can alter gene expression in vivo was documented by Siegfried in 1999 [85]. Hence, reverse epigenetic experiments confirmed that insertion of a mutant CpG island in the vicinity of a gene promoter leads not only to de novo DNA methylation but also to gene repression in vivo. Recently, it has became evident that the effect of CpG methylation on transcription is dependent on the local CpG density [89]. Hence, while DNA methylation of promoters with moderate to high CpG-content is linked to gene repression, methylation of promoters with low CpG-content does not affect transcription. As discussed in Chapter 3, DNA methylation might modulate the accessibility of chromatin directly, by affecting the binding of sequence-specific transcription factors, or indirectly, via the recruitment of methyl-binding proteins and histone-modifying enzymes (reviewed in [1,90]). Much later it was discovered that in ESCs and in the brain, cytosine methylation could take place also in other sequence contexts [91], where C is not followed by a G (mCH). Similarly to mCpG, non-CpG methylation has been shown to affect the binding of transcription factors and has the potential to regulate gene expression [91].

Interestingly, most of the genome-wide de novo DNA methylation takes place right after implantation, paralleling the decrease in developmental potential and stabilizing cell type–specific patterns of gene repression (reviewed in [1,87]). After this stage, somatic cells in general maintain their global DNA methylation patterns in a rather stable manner, and reflect the events taking place in early development. Although changes in DNA methylation after implantation do take place, they are mainly local events that involve sequence- and cell type–specific changes, such as the local methylation or demethylation of developmentally regulated genes paralleling their repression and activation, respectively [1]. In line with the role of DNA methylation in the stable maintenance of cell type–specific gene expression patterns, global erasure of epigenetic marks after fertilization and during germline development involve extensive genome-wide DNA demethylation events, as discussed in Chapter 3.

1.7.2 DNA Methylation in Tumor Development

Altered DNA methylation patterns have also been observed in the context of tumor development. Moreover, epigenetic instability has been hypothesized to underlie the emergence of "cellular amnesia," which refers to the dedifferentiated phenotypes and cellular states of varying stages of differentiation within the tumor [79]. The earliest observations linking hypomethylation of oncogenes and cell surface receptors genes to increased gene expression in cancer were documented already in the 1980s [92,93]. Not until recently was it, however, understood that the nature of epigenetic disruption in cancer goes beyond the aberrant activation or repression of a particular set of genes, such as oncogenes and tumor suppressor genes, respectively [4,5,94,95] (Chapter 17). Understanding the contribution of deregulated DNA methylation to tumor development was greatly enhanced by high-throughput, genome-wide techniques that enabled the detection of CpG methylation in the genome of cancer cells and their normal counterparts in an unbiased manner [5]. These experiments uncovered the context-specific function of DNA methylation and pinpointed the genomic regions that are particularly sensitive to environmental perturbations with consequences on cellular phenotypes and tumor development [5]. The novel findings of these observations formed the basis of the hypothesis that epigenetic instability and increased variation in chromatin states, in particular DNA methylation, at vulnerable chromatin domains play central roles in tumor development [4,5,79]. Epigenetic instability is thus suggested to maintain transcriptional and phenotypic heterogeneity among the tumor cells, which heterogeneity enables the selection of the fittest clones and drives tumor evolution under changing selective pressure [4,5,79] (Chapter 17).

1.7.3   Cross Talk Between DNA Methylation and Histone Modifications

Despite of the clear functional link between gene expression and DNA methylation in development and disease, experimental evidence indicates that this process might be more complex than initially considered (Chapter 3). DNA methylation thus appears to be rather the consequence than the cause of repression, and might serve as an additional layer of stable repression once transcription has been inhibited by histone modifications [96]. The basis of its cross talk with histone marks includes the recruitment of the de novo DNA methyltransferase, DNMT3A, to regions enriched in certain repressive histone modifications [97] (Chapter 3). In turn, methylated DNA–binding proteins can attract repressive histone-modifying enzymes, such as HDACs to methylated CpGs [98], which induce histone deacetylation to enable the subsequent deposition of repressive histone methylation marks [98]. As methylated DNA–binding proteins can also interact with histone methyltransferases [98] to establish heterochromatin, the cross talk between DNA methylation and repressive histone modifications can form multiple layers of self-reinforcing epigenetic state, that is, heritable during cell division. Such chromatin states can be found at repetitive elements, on regions of the inactive X chromosome, as well as at stably repressed pluripotency genes and other developmentally repressed genes in differentiated somatic cells [79].

Histone- and DNA methylation–based repression is, however, not the only mechanism by which cells can propagate cell type–specific patterns of gene repression. Inactivation of lineage-specific genes in pluripotent cells, for example, requires a more plastic form of heritable gene regulation, which is mediated by repressor proteins forming the PcG complexes (Chapter 7 and reviewed in [99]). In the 1940s, PcG silencing was initially discovered in *Drosophila*, which has little or no DNA methylation [100]. Mutations of PcG proteins resulted in homeotic transformation: morphological alterations where a body part is replaced by another body part normally found elsewhere [9,100]. These changes were caused by the deregulated expression of developmental regulators, such as the *HOX* genes, that specify the anterior–posterior axis [99]. Recently it has become clear that PcG proteins are evolutionary conserved repressor proteins providing a memory of gene repression in mammalian cells also [99].

PcG complexes act antagonistically to activating TrxG proteins [99] and maintain the silenced state of repressed target genes by both modifying chromatin states and affecting the recruitment of the RNA Pol II machinery [99]. At the same time, PcGs actively prevent the de novo DNA methylation of their target genes to ensure the plasticity of the repressed states [101]. Moreover, PGCs can also form bivalent promoter states together with TrxG proteins mainly, but not exclusively, in embryonic stem cells [99]. Such bivalent states are characterized by the presence of both activating, as well as repressive histone modifications, and represent a poised state for future gene activation [99]. These observations highlight that multiple layers of epigenetic modifications formed by DNA and histone modifications contribute to the maintenance of cell type–specific gene expression patterns of differentiated cells, which have to be reprogrammed via multiple parallel mechanisms during early development and can be deregulated in diseases (Chapter 17).

1.8 REPLICATION TIMING: POTENTIAL VEHICLE FOR EPIGENETIC INHERITANCE AND REPROGRAMMING

For chromatin modifications to serve as epigenetic memory, they have to be faithfully copied to the newly replicated DNA [1] (Chapters 5–6). The S phase can thus not only ensure the maintenance of chromatin marks, but has long been recognized also as a window of opportunity for reprogramming [1]. For example, while the association of DNMT1 to the replication fork ensures the stable inheritance of DNA methylation patterns, downregulation of DNMT1 or the inhibition of its association to the newly synthesized DNA facilitates passive DNA demethylation genome wide [1].

Apart from DNA methylation, repressive and activating histone modifications can also possess epigenetic inheritance that can be a target for regulation during reprogramming events [1]. For example, both TrxG and PcG proteins have been shown to remain associated with their targets during S phase to reestablish cell type–specific chromatin states in the daughter cells [102,103]. Furthermore, inheritance of other histone modifications has been suggested to rely on chromatin-modifying enzymes recruited to parental histone marks via chromatin reader modules to mirror these chromatin modifications on the newly replicated DNA [103]. Coupling between replication and epigenetic inheritance is thus a target of extensive investigation and is facilitated by the recruitment of chromatin-modifying enzymes and chromatin-remodeling enzymes to the replication fork [1]. Moreover, while certain histone modifications are reestablished on newly incorporated histones right after replication, others are restored by continuous modifications over several cell generations [103].

Inheritance of chromatin modifications is linked also to the timing of replication during S phase [104,105]. The term replication timing (discussed in Chapter 7) refers to that certain parts of the genome are consistently replicated early in the S phase, while others are replicated during the mid- or late S phases. Although not a complete division, it was observed already in 1999 that early replicating regions depict generally active domains, whereas late replicating regions contain primarily inactive regions [104,105]. It is not fully understood what determines when a region will replicate, that is, either early or late, even though active and inactive chromatin modifications have been clearly documented to play a role [104,106] (reviewed in [1,107]). Intriguingly, it was discovered in the 1960s that mammalian chromosomes are replicated in a regional manner forming replication time zones, where multiple replication origins fire simultaneously, in a coordinated manner [108,109]. It is thus plausible that change in replication timing at a certain region requires modification of chromatin states over large domains, and is regulated by the coordination of replication timing change at multiple neighboring replication origins prior to overt replication.

An interesting observation suggests that the timing of replication can in turn regulate chromatin states on newly formed DNA. Hence, nuclear microinjection of reporter plasmid during early S phase or late S phase resulted in different chromatin conformation and transcriptional activity, where early replication correlated with transcription and late replication with repression [106,106a]. Moreover, replication time zones seem to be regulated in a cell type– and differentiation stage–specific manner, which are, in many cases, determined before the establishment of cell type–specific transcriptional states [107,110–117], suggesting that replication timing might regulate transcription and has an epigenetic memory. Although direct evidence for the role of replication timing in the regulation of gene expression in vivo, at endogenous loci is still missing, it is an appealing hypothesis. It thus raises the possibility of the existence of a two-way relationship between chromatin states and replication timing in the propagation of epigenetic information [1,107], which is a likely target of reprogramming events to bring about global changes in chromatin structure [106].

1.9 CHROMATIN FOLDING IN 3D: BASIC PRINCIPLES OF GENOME ORGANIZATION

Chromatin modifications and nuclear functions act in the larger context of chromosomes that are folded in the 3D space of the nucleus. It has thus been documented already in the 1950s by E.B. Lewis in *Drosophila*, that homologous alleles of a locus can influence each other, a process that in most of the cases depended on the pairing, that is, physical proximity between the two alleles (reviewed in [118]). This process often involved enhancer action in *trans*, and was termed transvection. Although homologous pairing and transvection are characteristic features of the *Drosophila* genome, similar processes have been observed rarely also in other organisms, including mammals [119].

1.9.1 Cross Talk Between Regulatory Elements

Another example of context-dependent gene regulation is represented by the widely accepted notion that enhancers and silencers, as well as insulator elements that separate enhancers from their target genes can act over large distances (Chapter 17). Prior to the discovery of enhancers in 1981, gene regulation in mammals was considered to be similar to the bacterial system [120]. By a sheer coincidence, enhancers were discovered by Walter Schaffner when he placed a portion of the SV40 genome in front of a globin reporter gene, which led to a more than 100-fold enhancement of gene expression. Moreover, the same effect could be demonstrated when the SV40 enhancer was placed downstream of the reporter gene, illustrating that enhancers can act over distances and in an orientation-independent manner [121]. A few years later, the first enhancer from the mammalian genome—the immunoglobulin enhancer—was identified by the Schaffner [114] and Tonegawa [115] groups. Not long after, Schaffner and coworkers have also proven that enhancers can activate their target genes in a *trans* configuration, that is, even if the enhancer is not linked to the promoter of its target gene by a DNA backbone [122]. In this pioneering experiment two DNA fragments, containing either the promoter or the enhancer, were labeled with avidin or biotin, respectively. The high affinity between avidin and biotin was then expected to enable complex formation between the enhancer and the promoter in the same cell after transient transfection [122]. Indeed this experiment showed that tethering a reporter gene to an enhancer via a protein bridge can strongly activate transcription, and forms the fundament for our understanding that enhancers can activate their target genes over large distances by forming chromatin loops. As enhancers regulate cell type–specific gene expression (reviewed in [123]), one of the major questions of chromatin biology deals with the factors that regulate cell type– and differentiation stage–specific formation of enhancer–promoter interactions. Apart from the molecular ties of enhancer–promoter contacts, this process is likely influenced by and influences the folding of the genome in the 3D space of the nucleus [124].

1.9.2 Technical Innovations to Detect 3D Genome Organization

Valuable insight into these processes has been provided by the so-called "C" techniques, which were initiated by the chromosome conformation capture

(3C) technique [125] (Figs. 1.1 and 1.3). This approach is thus designed to analyze interactions between regions that are predicted to be proximal to each other. It is based on formaldehyde cross-linking of chromatin to stabilize the dynamic contacts between distant loci that formed interactions in vivo. This step is followed by restriction enzyme digestion and intramolecular ligation of the cohesive ends within chromatin fiber complexes under very dilute conditions. Finally, the resulting chimeric chromatin fragments representing regions that were in close physical proximity in the living cell are amplified with PCR (Fig. 1.3). While the 3C assay has uncovered several basic principles of chromatin folding, it is based on educated guesses that particular sequences might

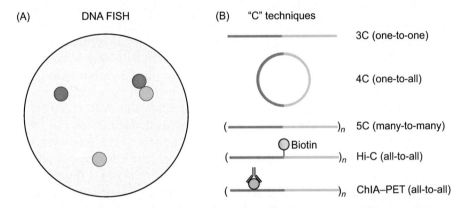

FIGURE 1.3 Schematic overview of the techniques designed to detect 3D genome organization.
(A) The DNA fluorescent in situ hybridization *(FISH)* assay measures the frequency of proximity between specific genomic regions, as well as between genomic regions and nuclear hallmarks, such as the lamina. (B) The chromosome conformation capture *(3C)* technique is designed to detect interactions between two regions that are predicted to be proximal to each other in the 3D space of the nucleus *(one-to-one approach)*. It is based on formaldehyde cross-linking of chromatin to stabilize the dynamic contacts between distant loci present in the living cells. This step is followed by restriction enzyme digestion and intramolecular ligation of the cohesive ends within chromatin fiber complexes, which results in the formation of chimeric DNA fragments containing regions that were in close physical proximity in the living cell. The circular 3C *(4C)* technique enables the identification of genome-wide interaction frequencies between the genome and a particular bait region without any preconceived assumption beyond the bait itself *(one-to-all approach)*. This is achieved by PCR amplification of unknown sequences with inverse primers positioned within the known bait. The 3C carbon copy *(5C)* technique is an extension of the 3C technique, but with a more systematic analysis of all possible interactions within subdomains of chromosomes *(many-to-many approach)*. The *Hi-C* technique is designed to detect the relative frequency of all-to-all chromatin fiber contacts, averaged over millions of cells. The assay involves the labeling of ligation junctions with biotin, followed by the selective enrichment of biotin-labeled chimeric sequences and high-throughput sequencing. The chromatin interaction analysis by paired-end tag sequencing *(ChIA–PET)* assay is designed to detect all possible combinations of interactions within a cell population, which take place in the presence of a particular protein epitope.

or might not interact with each other. This limitation was partially neutralized by the innovation of the circular 3C (4C) technique that enabled the identification of genome-wide interaction frequencies between the genome and a particular bait(s) (Fig. 1.3). This is achieved by employing extensive PCR amplifications of unknown sequences with inverse primers strategically positioned within the known bait [126,127]. The 5C technique that followed in the evolution of "C" techniques is an extension of the 3C technique, but with a more systematic analysis of all possible interactions within smaller subdomains of chromosomes [128] (Fig. 1.3). All of these techniques provide, however, only limited information and detect only a fraction of all possible interactions in the genome within a cell population. The advent of the so-called Hi-C technique opened up a more rich landscape of chromatin fiber interactions [129], as it is designed to detect the relative frequency of all-to-all chromatin fiber contacts, averaged over millions of cells (Fig. 1.3). These experiments eventually led to the discovery of that the genome is organized into topological-associated domains (TADs) [130] (Figs. 1.1 and 1.3). Chromatin fiber interactions thus preferentially occur within TADs but not between TADs to provide an organizational chromatin unit much beyond the primary chromatin fiber, and to establish constrains of enhancer–promoter interactions [130]. Although the mechanism underlying the organization of TADs is still an open question, chromatin architectural proteins have been implicated in the regulation of TAD boundary strength from yeast to mammals [124,131] (Chapter 17). As these Hi-C experiments were performed on large cell populations, TADs might represent the sum of numerous snapshots, which might not exist simultaneously within individual cells. As a consequence, we remain ignorant if these chromatin loops are stable or continuously formed and re-formed, and to what extent they contribute to chromatin folding at different parts of the cell cycle. For example, mitotic chromosomes have already been shown to display a very different chromosome organization with a homogenous, compact folding state [132]. Even with these significant advances of our knowledge on chromatin folding, these experiments were not informative about the protein mediators of 3D interactions. This has been partially solved by the invention of chromatin interaction analysis by paired-end tag sequencing (ChIA–PET) (Fig. 1.3), designed to capture chromatin fiber interactions that take place in the presence of a particular protein, such as RNA Pol II [133]. Pol II ChIA–PET analyses revealed, for example, the complexity of enhancer–promoter interactions, showing that multiple enhancers units located at distant parts of the genome can impinge on a shared set of target genes in large cell populations, highlighting the complexity of transcriptional regulation in 3D (see Chapter 17).

The diversity of chromatin fiber contacts in a cell population has raised the question, whether chromatin fiber contacts and genome conformation display large cell-to-cell variation. Single-cell Hi-C experiments [134] have confirmed these initial proposals, making the interpretation of average genome conformations obtained from large cell population complex and challenging [135]. One of the major questions in 3D chromatin organization is how are chromatin fiber interactions regulated by marks of the primary chromatin fiber, and how do they influence genomic functions and cell-to-cell variations in gene expression. Answering these questions will require the quantitative, high-throughput detection of chromatin contact frequencies, chromatin marks, and gene expression levels in small cell populations or single cells. Challenges include the low resolution and sensitivity of the single-cell Hi-C assay, which currently enables the identification of only a small fraction of the existing chromatin fiber interactions [134]. Statistical analyses of cell-to-cell variations in genome folding would thus require measurements from thousands of cells. Other assays that can detect 3D chromatin folding include the fluorescent in situ hybridization (FISH) assay (Fig. 1.3) to measure proximities between defined genomic regions [136]. As the resolution of DNA FISH is much lower than that of the "C" techniques, it is only suitable to score for proximities as estimates for the probability of interactions (reviewed in [136]). A recent microscopy-based assay termed chromatin in situ proximity (ChRISP) [137,138] was invented to overcome the poor resolution of DNA FISH. It is thus built on in situ proximity ligation principles to translate tight proximity between specific genomic regions and/or a specific region and chromatin marks into fluorescent signals.

1.9.3 Compartmentalization of Nuclear Functions

Nevertheless, both Hi-C and single-cell Hi-C experiments have provided valuable information about global genome organization, and documented that chromatin states of the primary chromatin fiber are linked with the probability of chromatin fiber contacts in 3D [139]. Hence, active regions tend to interact with other active regions, and inactive chromatin domains tend to similarly contact other inactive domains. The spatial separation between transcriptionally permissive and repressive chromatin modifications in the nucleus has been proposed to facilitate the maintenance of cell type–specific gene expression patterns by reducing variations in chromatin states and therefore transcriptional noise [5,106] (Chapter 17). The mechanism of compartmentalization of different transcriptional states is under extensive investigation. For example, the nuclear periphery emerges as a key coordinator of transcriptional silencing at cell type–specific and constitutively repressed regions by attracting both the chromatin modifiers that establish compact, silent chromatin states and the repressed chromatin domains [140] (Chapter 17).

The 3D organization of nuclear functions is also reflected in the spatiotemporal regulation of replication timing [141–143], where early replicating chromatin domains tend to replicate in small replication factories distributed throughout the nuclear interior, followed by the replication of the peripheral heterochromatin in mid- to late S phase and finally the replication of centromeric and other heterochromatin states in very large transcription factories in the nuclear interior. Such a distribution of replication foci also correlates with the 3D contact maps within and between chromosomes generated by Hi-C [144]. TAD boundaries have thus been shown to correspond to replication time zones, and interchromsomal contacts were significantly more frequent between regions that replicate at similar times during S phase [144]. Further studies will be required to explore the potential cause-and-effect relationship between 3D genome folding and replication timing, as well as the exact mechanism how reprogramming of replication time zones are integrated with the reprogramming of 3D chromatin cross talk upon environmental signals [145]. The important role of the 3D nuclear architecture in the regulation of nuclear functions implies that factors known to reprogram chromatin states and gene expression repertoires will likely also affect 3D genome organization as part of their action (discussed further in Chapter 17).

1.10   SIGNAL INTEGRATION AT THE LEVEL OF CHROMATIN

As predicted by Waddington, chromatin-based mechanisms provide an interface between the genotype and the environment to regulate development and adaptation (Fig. 1.2). Chromatin has recently been proposed to act as a platform for the integration of diverse signaling and metabolic pathways, and propagate their effects on nuclear functions over time, long after the disappearance of the initial stimuli [3]. Numerous environmental signals are thus known to induce transient or long-lived changes in chromatin states genome wide or locally, at regulatory elements [123], which can subsequently be inherited during the cell cycle. These signals include, for example, cues that activate or repress signaling pathways [3,146] and metabolic pathways [3,147], alter the pH of the environment [148], and initiate mechanical [149] or bioelectric signaling [150]. However, it remains to be elucidated why and how certain signals can exert these effects on chromatin states, while other signals are buffered against.

Signaling pathways can modulate chromatin states by directly altering the PTMs of histones, chromatin modifiers, reader and remodeling proteins, as well as by affecting the level of DNA methylation at various stages of differentiation [3]. For example, kinase cascades can induce short-lived phosphorylation of histone H3 at Serine 10 (H3S10ph), which in turn can elicit multiple indirect effects in different cell types [3]. By preventing the recruitment of HDACs, H3S10ph can facilitate the accumulation of acetylation on histones and the consequent recruitment of bromodomain-containing factor 4

(BRD4), a chromatin reader that recognizes acetylated histones enriched at active enhancers and promoters, and facilitates productive transcriptional elongation [151]. c-Jun NH(2)-terminal kinase (JNK)–mediated phosphorylation of H3S10 during neuronal development has also been correlated with gene activation, while inhibiting JNK resulted in the repression of its target genes [152]. Importantly, BRD4 remains bound to chromatin throughout the cell cycle, including mitotic chromosomes, and thus provides an epigenetic memory of chromatin states [153]. Stress signals can redistribute BRD4 from its target sites toward stress-inducible genes by activating phosphatases, such as PP1 [151]. The PP1-mediated removal of H3S10ph thus enables HDAC activity to deacetylate histones, leading to the release of BRD4 from chromatin, and enabling its subsequent targeting to stress-inducible genes [151].

Signaling pathways can also directly phosphorylate and modulate the activity of chromatin modifiers, and thereby influence the heritability of epigenetic marks [3]. Well-documented examples include the phosphatidylinositide-3-kinase (PI3K)–AKT pathway–mediated phosphorylation of DNMT1 (reviewed in [3]), which interferes with the binding of DNMT1 to the replication fork, and counteracts maintenance DNA methylation [154]. Similarly, the phosphorylation of the PcG proteins, EZH2 and BMI1, inhibits their functions [155]. Prosurvival signals, for example hormones and growth factors, could thus interfere with the maintenance of epigenetic states by targeting chromatin modifiers [3].

Finally, not only signaling but also metabolic pathways can affect chromatin states, enabling the cell to rewire its transcriptome and adapt to cellular metabolic states [147] (Chapter 15). Several intermediary metabolites have thus been shown to act as substrates or cofactors for chromatin-modifying enzymes, connecting metabolism to gene regulation. A seminal example of the interplay between chromatin states, environmental signals, and cellular metabolic states is represented by the circadian regulation (discussed in Chapters 16–17). Self-running circadian oscillations in behavior and physiology were likely selected during the course of evolution because it enabled the organism to prepare for changes in its environment [156]. Cell autonomous cellular clocks regulate circadian chromatin transitions, which contribute to daily oscillations in the transcriptome, proteome, and thus phenotype. In line with the oscillations in cellular protein composition and modifications, circadian rhythm has already been shown to affect cellular responsiveness to environmental signals [157], likely influencing the enigmatic mechanisms of "buffering" against perturbations. At the same time, cellular clocks can be entrained, that is, their phase can be reset, by certain environmental cues, such as light [158], food intake [158], high-fat diet [159], insulin [160], TGFβ signaling [161], and horse serum shock [162]. Entrainment by brief exposure to horse serum shock has recently been shown to involve large-scale changes in chromatin structure and 3D genome organization [163]. Misalignment between the geophysical time and entrainment cues, and the disruption of daily rhythm has been linked to aging and the development of complex diseases, such as diabetes and cancer [164].

These examples illustrate that the partial reversibility and heritability of chromatin marks enable the propagation of beneficial or harmful environmental signals during adaptive and maladaptive processes, respectively [165]. These chromatin features have thus been proposed to contribute to the development of multifactorial diseases that are caused by the complex interplay between the genotype, epigenotype, and harmful environmental signals over long periods of time [165]. Similarly, the same features are also expected to modify the phenotype of monogenic diseases [165]. Moreover, aging-related epigenetic drift, that is, alterations in chromatin states over time, likely contributes to the age-related functional decline of the organism [166–168]. An intriguing question is how are chromatin states buffered against perturbations to enable the stable outcome of developmental processes and the maintenance of cellular phenotypes. Answer to this question will likely require an understanding of the various levels of cross talk among various chromatin modifications, and the mechanisms of self-reinforcing feedback and feed-forward loops that might be in action to preserve and maintain chromatin states. Finally, the compartmentalization of nuclear functions and the conformation of the genome within the 3D architecture of the nucleus are emerging as important regulators of the stability of epigenetic states and cellular phenotypes (Chapter 17 and reviewed in [4,123]).

1.11 OUTLOOK

Despite the achievements described earlier and throughout this book, the fundamental questions that were raised in the previous century on development and adaptation to environmental change are still not completely understood. Addressing the mechanism of signal integration at the level of chromatin will likely require technical innovations. Both the signaling pathways and the changes they confer to the chromatin might thus be very dynamic, and the various signals might affect multiple chromatin marks in a combinatorial fashion. As current high-throughput assays provide merely snapshots of chromatin marks in large cell populations, novel approaches will be required to detect the simultaneous presence of multiple proteins or dynamic chromatin modifications in small cell populations or in single cells, over time.

Similarly, it is still not completely elucidated how the interphase chromosomes are folded in 3D, and how 3D genome organization regulates cell type–specific expression repertoires. Moreover, identifying cause-and-effect relationships between chromatin structures and nuclear functions in response to environmental cues is challenged by the variable and dynamic character of these processes due to, for example, random cell-to-cell variations in chromatin accessibility and the dynamic binding of transcription factors to DNA [169]. This endeavor will require new techniques that can simultaneously detect 3D chromatin structures and nuclear functions in single cells. As both the ChIP-seq and RNA-seq techniques are designed to analyze large cell populations, conclusions about correlations between gene activity and chromatin modifications will likely need to be reassessed by novel techniques at the single-cell level. A recent innovation is the ChRISP [137,138] assay, which can combine the detection of both the structural features of chromatin and ongoing nuclear processes in single cells.

Furthermore, very little is known about how signaling and metabolic pathways influence 3D genome organization to affect phenotypes during development and in diseases. As described in Chapter 17, reprogramming of 3D nuclear architecture has been linked to transitions in cellular phenotypes and cancer development [4]. It will be, for example, important to determine how chromatin mobility between transcriptionally repressive and permissive compartments is regulated, and how compartmentalization itself can be altered by external cues. Several nuclear envelope proteins have thus been documented to bind to or sequester members of signaling pathways with potential consequences on the outcome of signal transduction, as well as chromatin–lamina interactions [146].

Finally, an exciting new era has opened now for the investigation of the utility of epigenetic marks in the diagnosis, prognosis, and treatment of a wide range of diseases. For example, epigenetic marks that are stable over time and associate with certain phenotypes have already been shown to predict disease risk in epigenome-wide association studies [170]. Moreover, several avenues of epigenetic therapy for the treatment of cancer, as well as other diseases are in the experimental phase, and will likely bring about a paradigm shift in clinical practice [4,165].

Abbreviations

AKT	Protein kinase B (PKB)
BMP	Bone morphogenetic protein
BRD4	Bromodomain-containing 4
3C	Chromosome conformation capture
4C	Circular chromosome conformation capture
5C	Chromosome conformation capture carbon copy
ChIA–PET	Chromatin interaction analysis by paired-end tag sequencing
DNMT1	De novo maintenance DNA methyltransferase 1
DNMT3A	De novo DNA methyltransferase 3A
DNMT3B	De novo DNA methyltransferase 3B
H3S10ph	H3 Serine 10 phosphorylation
JNK	c-Jun NH(2)-terminal kinase
OR	Olfactory receptors

PcG	Polycomb group
PGC	Primordial germ cell
PI3K	Phosphatidylinositide-3-kinase
PTM	Posttranslational modification
TAD	Topological-associated domain
TrxG	Trithorax group
XCI	X-chromosome inactivation

Acknowledgments

This work has been supported by the grant from Karolinska Institutet to AG.

Conflict of Interest

The authors declare no conflict of interest.

References

[1] Almouzni G, Cedar H. Maintenance of epigenetic information. Cold Spring Harb Perspect Biol 2016;8(5):a019372.

[2] Felsenfeld G. A brief history of epigenetics. Cold Spring Harb Perspect Biol 2014;6(1):a018200.

[3] Badeaux AI, Shi Y. Emerging roles for chromatin as a signal integration and storage platform. Nat Rev Mol Cell Biol 2013;14:211–24.

[4] Feinberg AP, Koldobskiy MA, Gondor A. Epigenetic modulators, modifiers and mediators in cancer aetiology and progression. Nat Rev Genet 2016;17:284–99.

[5] Pujadas E, Feinberg AP. Regulated noise in the epigenetic landscape of development and disease. Cell 2012;148:1123–31.

[6] Waddington CH. Epigenetics and evolution. Sym Soc Exp Biol 1953;7:186–99.

[7] Noble D. Conrad Waddington and the origin of epigenetics. J Exp Biol 2015;218:816–8.

[8] Riggs AD, Porter TN. Overview of epigenetic mechanisms. In: Russo VEA, editor. Epigenetic mechanisms of gene regulation. New York: Cold Spring Harbor Laboratory Press, Cold Spring Harbor; 1996. p. 29–45.

[9] Hadorn, E. Transdetermination in cells. Sci Am 1968;219:110–114.

[10] Lee N, Maurange C, Ringrose L, Paro R. Suppression of Polycomb group proteins by JNK signalling induces transdetermination in *Drosophila* imaginal discs. Nature 2005;438: 234–7.

[11] Peters J. The role of genomic imprinting in biology and disease: an expanding view. Nat Rev Genet 2014;15:517–30.

[12] Gerbi SA. Helen Crouse (1914–2006): imprinting and chromosome behavior. Genetics 2007;175:1–6.

[13] Surani MA, Barton SC, Norris ML. Development of reconstituted mouse eggs suggests imprinting of the genome during gametogenesis. Nature 1984;308:548–50.

[14] Surani MA, Barton SC, Norris ML. Nuclear transplantation in the mouse: heritable differences between parental genomes after activation of the embryonic genome. Cell 1986;45:127–36.

[15] Reik W, Collick A, Norris ML, Barton SC, Surani MA. Genomic imprinting determines methylation of parental alleles in transgenic mice. Nature 1987;328:248–51.

[16] Stoger R, et al. Maternal-specific methylation of the imprinted mouse Igf2r locus identifies the expressed locus as carrying the imprinting signal. Cell 1993;73:61–71.

[17] Soellner L, et al. Recent advances in imprinting disorders. Clin Genet 2016. Epub ahead of print.

[18] Kono T. Genomic imprinting is a barrier to parthenogenesis in mammals. Cytogenet Genome Res 2006;113:31–5.

[19] Schneider E, et al. Spatial, temporal and interindividual epigenetic variation of functionally important DNA methylation patterns. Nucleic Acids Res 2010;38:3880–90.

[20] Ollikainen M, Craig JM. Epigenetic discordance at imprinting control regions in twins. Epigenomics 2011;3:295–306.

[21] Payer B. Developmental regulation of X-chromosome inactivation. Semin Cell Dev Biol 2016. (Epub ahead of print).

[22] Lyon MF. Gene action in the X-chromosome of the mouse (*Mus musculus* L.). Nature 1961;190:372–3.

[23] Wong CC, et al. A longitudinal twin study of skewed X chromosome-inactivation. PLoS One 2011;6:e17873.

[24] Rodriguez I. Singular expression of olfactory receptor genes. Cell 2013;155:274–7.

[25] Levin-Klein R, Bergman Y. Epigenetic regulation of monoallelic rearrangement (allelic exclusion) of antigen receptor genes. Front Immunol 2014;5:625.

[26] Burnet FM. The clonal selection theory of acquired immunity. Cambridge: The University Press; 1959.

[27] Nossal GJ, Lederberg J. Antibody production by single cells. Nature 1958;181:1419–20.

[28] Pernis B, Chiappino G, Kelus AS, Gell PG. Cellular localization of immunoglobulins with different allotypic specificities in rabbit lymphoid tissues. J Exp Med 1965;122:853–76.

[29] Cedar H, Bergman Y. Choreography of Ig allelic exclusion. Curr Opin Immunol 2008;20:308–17.

[30] Spicuglia S, Pekowska A, Zacarias-Cabeza J, Ferrier P. Epigenetic control of Tcrb gene rearrangement. Semin Immunol 2010;22:330–6.

[31] Ohlsson R, Genetics. Widespread monoallelic expression. Science 2007;318:1077–8.

[32] Gimelbrant A, Hutchinson JN, Thompson BR, Chess A. Widespread monoallelic expression on human autosomes. Science 2007;318:1136–40.

[33] Tee WW, Reinberg D. Chromatin features and the epigenetic regulation of pluripotency states in ESCs. Development 2014;141:2376–90.

[34] Hannah A. Localization and function of heterochromatin in *Drosophila melanogaster*. Adv Genet 1951;4:87–125.

[35] Mazor Y, Kupiec M. Developmentally regulated MAPK pathways modulate heterochromatin in Saccharomyces cerevisiae. Nucleic Acids Res 2009;37:4839–49.

[36] Westhorpe FG, Straight AF. The centromere: epigenetic control of chromosome segregation during mitosis. Cold Spring Harb Perspect Biol 2015;7:a015818.

[37] Sato H, Saitoh S. Switching the centromeres on and off: epigenetic chromatin alterations provide plasticity in centromere activity stabilizing aberrant dicentric chromosomes. Biochem Soc Trans 2013;41:1648–53.

[38] Ladewig J, Koch P, Brustle O. Leveling Waddington: the emergence of direct programming and the loss of cell fate hierarchies. Nat Rev Mol Cell Biol 2013;14:225–36.

[39] Briggs R, King TJ. Transplantation of living nuclei from blastula cells into enucleated frogs' eggs. Proc Natl Acad Sci USA 1952;38:455–63.

[40] Campbell KH, McWhir J, Ritchie WA, Wilmut I. Sheep cloned by nuclear transfer from a cultured cell line. Nature 1996;380:64–6.

[41] Cowan CA, Atienza J, Melton DA, Eggan K. Nuclear reprogramming of somatic cells after fusion with human embryonic stem cells. Science 2005;309:1369–73.

[42] Takahashi K, Yamanaka S. Induction of pluripotent stem cells from mouse embryonic and adult fibroblast cultures by defined factors. Cell 2006;126:663–76.

[43] Lee J, et al. Activation of innate immunity is required for efficient nuclear reprogramming. Cell 2012;151:547–58.

[44] Rossi M, Augusti-Tocco G, Monroy A. Differential gene activity and segregation of cell lines: an attempt at a molecular interpretation of the primary events of embryonic development. Q Rev Biophys 1975;8:43–119.

[45] Monk M, Boubelik M, Lehnert S. Temporal and regional changes in DNA methylation in the embryonic, extraembryonic and germ cell lineages during mouse embryo development. Development 1987;99:371–82.

[46] Brandeis M, et al. The ontogeny of allele-specific methylation associated with imprinted genes in the mouse. EMBO J 1993;12:3669–77.

[47] Petropoulos S, et al. Single-cell RNA-seq reveals lineage and X chromosome dynamics in human preimplantation embryos. Cell 2016;165:1012–26.

[48] Burton A, Torres-Padilla ME. Chromatin dynamics in the regulation of cell fate allocation during early embryogenesis. Nat Rev Mol Biol Rev 2014;15:723–34.

[49] Borsos M, Torres-Padilla ME. Building up the nucleus: nuclear organization in the establishment of totipotency and pluripotency during mammalian development. Genes Dev 2016;30:611–21.

[50] Hackett JA, Zylicz JJ, Surani MA. Parallel mechanisms of epigenetic reprogramming in the germline. Trends Genet 2012;28:164–74.

[51] Tada M, Tada T, Lefebvre L, Barton SC, Surani MA. Embryonic germ cells induce epigenetic reprogramming of somatic nucleus in hybrid cells. EMBO J 1997;16:6510–20.

[52] Gunesdogan U, Magnusdottir E, Surani MA. Primordial germ cell specification: a context-dependent cellular differentiation event [corrected]. Philos Trans R Soc Lond B 2014;369:20130543.

[53] Visvader JE. Cells of origin in cancer. Nature 2011;469:314–22.

[54] Friedmann-Morvinski D, Verma IM. Dedifferentiation and reprogramming: origins of cancer stem cells. EMBO Rep 2014;15:244–53.

[55] Medema JP. Cancer stem cells: the challenges ahead. Nat Cell Biol 2013;15:338–44.

[56] Nagata T, et al. KLF4 and NANOG are prognostic biomarkers for triple-negative breast cancer. Breast Cancer 2016. (Epub ahead of print).

[57] Yong X, et al. *Helicobacter pylori* upregulates Nanog and Oct4 via Wnt/beta-catenin signaling pathway to promote cancer stem cell-like properties in human gastric cancer. Cancer Lett 2016;374:292–303.

[58] Schwitalla S, et al. Intestinal tumorigenesis initiated by dedifferentiation and acquisition of stem-cell-like properties. Cell 2013;152:25–38.

[59] Deichmann U. Chromatin: its history, current research, and the seminal researchers and their philosophy. Perspect Biol Med 2015;58:143–64.

[60] Avery OT, Macleod CM, McCarty M. Studies on the chemical nature of the substance inducing transformation of pneumococcal types: induction of transformation by a desoxyribonucleic acid fraction isolated from *Pneumococcus* type III. J Exp Med 1944;79:137–58.

[61] Olins DE, Olins AL. Chromatin history: our view from the bridge. Nat Rev Mol Cell Biol 2003;4:809–14.

[62] Thoma F, Koller T, Klug A. Involvement of histone H1 in the organization of the nucleosome and of the salt-dependent superstructures of chromatin. J Cell Biol 1979;83:403–27.

[63] Struhl K, Segal E. Determinants of nucleosome positioning. Nat Struct Mol Biol 2013;20:267–73.

[64] Jiang C, Pugh BF. Nucleosome positioning and gene regulation: advances through genomics. Nat Rev Genet 2009;10:161–72.

[65] Stedman E. Cell specificity of histones. Nature 1950;166:780–1.

[66] Allfrey VG, Faulkner R, Mirsky AE. Acetylation and methylation of histones and their possible role in the regulation of RNA synthesis. Proc Natl Acad Sci USA 1964;51:786–94.

[67] Taunton J, Hassig CA, Schreiber SL. A mammalian histone deacetylase related to the yeast transcriptional regulator Rpd3p. Science 1996;272:408–11.

[68] Brownell JE, et al. Tetrahymena histone acetyltransferase A: a homolog to yeast Gcn5p linking histone acetylation to gene activation. Cell 1996;84:843–51.

[69] Gilmour DS, Lis JT. Detecting protein–DNA interactions in vivo: distribution of RNA polymerase on specific bacterial genes. Proc Natl Acad Sci USA 1984;81:4275–9.

[70] Turner BM. Decoding the nucleosome. Cell 1993;75:5–8.

[71] Tordera V, Sendra R, Perez-Ortin JE. The role of histones and their modifications in the informative content of chromatin. Experientia 1993;49:780–8.

[72] Strahl BD, Allis CD. The language of covalent histone modifications. Nature 2000;403:41–5.

[73] Jenuwein T, Allis CD. Translating the histone code. Science 2001;293:1074–80.

[74] Turner BM. Histone acetylation and an epigenetic code. Bioessays 2000;22:836–45.

[75] Albert I, et al. Translational and rotational settings of H2A.Z. nucleosomes across the *Saccharomyces cerevisiae* genome. Nature 2007;446:572–6.

[76] Gardner KE, Allis CD, Strahl BD. Operating on chromatin, a colorful language where context matters. J Mol Biol 2011;409:36–46.

[77] de Graaf CA, van Steensel B. Chromatin organization: form to function. Curr Opin Genet Dev 2013;23:185–90.

[78] Bickmore WA, van Steensel B. Genome architecture: domain organization of interphase chromosomes. Cell 2013;152:1270–84.

[79] Ohlsson R, et al. Epigenetic variability and the evolution of human cancer. Adv Cancer Res 2003;88:145–68.

[80] Hotchkiss RD. The quantitative separation of purines, pyrimidines, and nucleosides by paper chromatography. J Biol Chem 1948;175:315–32.

[81] Bestor TH. Cloning of a mammalian DNA methyltransferase. Gene 1988;74:9–12.

[82] Iyer LM, Abhiman S, Aravind L. Natural history of eukaryotic DNA methylation systems. Prog Mol Biol Transl Sci 2011;101:25–104.

[83] Li E, Beard C, Jaenisch R. Role for DNA methylation in genomic imprinting. Nature 1993;366:362–5.

[84] Li E, Bestor TH, Jaenisch R. Targeted mutation of the DNA methyltransferase gene results in embryonic lethality. Cell 1992;69:915–26.

[85] Siegfried Z, et al. DNA methylation represses transcription in vivo. Nat Genet 1999;22:203–6.

[86] Okano M, Xie S, Li E. Cloning and characterization of a family of novel mammalian DNA (cytosine-5) methyltransferases. Nat Genet 1998;19:219–20.

[87] Cantone I, Fisher AG. Epigenetic programming and reprogramming during development. Nat Struct Mol Biol 2013;20:282–9.

[88] Yisraeli J, et al. Muscle-specific activation of a methylated chimeric actin gene. Cell 1986;46:409–16.

[89] Weber M, et al. Distribution, silencing potential and evolutionary impact of promoter DNA methylation in the human genome. Nat Genet 2007;39:457–66.

[90] Jin J, et al. The effects of cytosine methylation on general transcription factors. Sci Rep 2016;6:29119.

[91] Kinde B, Gabel HW, Gilbert CS, Griffith EC, Greenberg ME. Reading the unique DNA methylation landscape of the brain: non-CpG methylation, hydroxymethylation, and MeCP2. Proc Natl Acad Sci USA 2015;112:6800–6.

[92] Feinberg AP, Vogelstein B. Hypomethylation distinguishes genes of some human cancers from their normal counterparts. Nature 1983;301:89–92.

[93] Kaneko Y, et al. Hypomethylation of c-myc and epidermal growth factor receptor genes in human hepatocellular carcinoma and fetal liver. Jpn J Cancer Res 1985;76:1136–40.

[94] Feinberg AP. Phenotypic plasticity and the epigenetics of human disease. Nature 2007;447:433–40.

[95] Feinberg AP. The epigenetic basis of common human disease. Trans Am Clin Climatol Assoc 2013;124:84–93.

[96] Feldman N, et al. G9a-mediated irreversible epigenetic inactivation of Oct-3/4 during early embryogenesis. Nat Cell Biol 2006;8:188–94.

[97] Lehnertz B, et al. Suv39h-mediated histone H3 lysine 9 methylation directs DNA methylation to major satellite repeats at pericentric heterochromatin. Curr Biol 2003;13:1192–200.

[98] Fuks F, et al. The methyl-CpG-binding protein MeCP2 links DNA methylation to histone methylation. J Biol Chem 2003;278:4035–40.

[99] Grossniklaus U, Paro R. Transcriptional silencing by polycomb-group proteins. Cold Spring Harb Perspect Biol 2014;6:a019331.

[100] Marx J. Developmental biology. Combing over the Polycomb group proteins. Science 2005;308:624–6.

[101] de la Cruz CC, et al. The polycomb group protein SUZ12 regulates histone H3 lysine 9 methylation and HP1 alpha distribution. Chromosome Res 2007;15:299–314.

[102] Petruk S, et al. TrxG and PcG proteins but not methylated histones remain associated with DNA through replication. Cell 2012;150:922–33.

[103] Alabert C, et al. Two distinct modes for propagation of histone PTMs across the cell cycle. Genes Dev 2015;29:585–90.

[104] McNairn AJ, Gilbert DM. Epigenomic replication: linking epigenetics to DNA replication. Bioessays 2003;25:647–56.

[105] Dimitrova DS, Gilbert DM. The spatial position and replication timing of chromosomal domains are both established in early G1 phase. Mol Cell 1999;4:983–93.

[106] Gondor A, Ohlsson R. Replication timing and epigenetic reprogramming of gene expression: a two-way relationship? Nature Rev Genet 2009;10:269–76.

[106a] Zhang J, Xu F, Hashimshony T, Keshet I, Cedar H. Establishment of transcriptional competence in early and late S phase. Nature 2002;420(6912):198–202.

[107] Rivera-Mulia JC, Gilbert DM. Replication timing and transcriptional control: beyond cause and effect-part III. Curr Opin Cell Biol 2016;40:168–78.

[108] Huberman JA, Riggs AD. Autoradiography of chromosomal DNA fibers from Chinese hamster cells. Proc Natl Acad Sci USA 1966;55:599–606.

[109] Huberman JA, Riggs AD. On the mechanism of DNA replication in mammalian chromosomes. J Mol Biol 1968;32:327–41.

[110] Ensminger AW, Chess A. Coordinated replication timing of monoallelically expressed genes along human autosomes. Human Mol Gen 2004;13:651–8.

[111] Mostoslavsky R, et al. Asynchronous replication and allelic exclusion in the immune system. Nature 2001;414:221–5.

[112] Julienne H, Audit B, Arneodo A. Embryonic stem cell specific "master" replication origins at the heart of the loss of pluripotency. PLoS Comput Biol 2015;11:e1003969.

[113] Hiratani I, et al. Genome-wide dynamics of replication timing revealed by in vitro models of mouse embryogenesis. Genome Res 2010;20:155–69.

[114] Banerji J, Olson L, Schaffner W. A lymphocytespecific cellular enhancer is located downstream of the joining region in immunoglobulin heavy chain genes. Cell 1983;33:729–40.

[115] Gillies SD, Morrison SL, Oi VT, Tonegawa S. A tissue-specific transcription enhancer element is located in the major intron of a rearranged immunoglobulin heavy chain gene. Cell 1983;33:717–28.

[116] Santos J, et al. Differences in the epigenetic and reprogramming properties of pluripotent and extra-embryonic stem cells implicate chromatin remodelling as an important early event in the developing mouse embryo. Epigenet Chromatin 2010;3:1.

[117] Farago M, et al. Clonal allelic predetermination of immunoglobulin-kappa rearrangement. Nature 2012;490:561–5.

[118] Duncan IW. Transvection effects in *Drosophila*. Annu Rev Genet 2002;36:521–56.

[119] Sandhu KS, et al. Nonallelic transvection of multiple imprinted loci is organized by the H19 imprinting control region during germline development. Genes Dev 2009;23:2598–603.

[120] Schaffner W. Enhancers—from their discovery to today's universe of transcription enhancers. Biol Chem 2015;396:311–27.

[121] Banerji J, Rusconi S, Schaffner W. Expression of a beta-globin gene is enhanced by remote SV40 DNA sequences. Cell 1981;27:299–308.

[122] Mueller-Storm HP, Sogo JM, Schaffner W. An enhancer stimulates transcription in trans when attached to the promoter via a protein bridge. Cell 1989;58:767–77.

[123] Jin F, Li Y, Ren B, Natarajan R. Enhancers: multi-dimensional signal integrators. Transcription 2011;2:226–30.

[124] Cubenas-Potts C, Corces VG. Architectural proteins, transcription, and the three-dimensional organization of the genome. FEBS Lett 2015;589:2923–30.

[125] Dekker J, Rippe K, Dekker M, Kleckner N. Capturing chromosome conformation. Science 2002;295:1306–11.

[126] Gondor A, Rougier C, Ohlsson R. High-resolution circular chromosome conformation capture assay. Nat Protoc 2008;3:303–13.

[127] Zhao Z, et al. Circular chromosome conformation capture (4C) uncovers extensive networks of epigenetically regulated intra- and interchromosomal interactions. Nat Genet 2006;38:1341–7.

[128] Dostie J, et al. Chromosome conformation capture carbon copy (5C): a massively parallel solution for mapping interactions between genomic elements. Genome Res 2006;16:1299–309.

[129] Belton JM, et al. Hi-C: A comprehensive technique to capture the conformation of genomes. Methods 2012;58:268–76.

[130] Dixon JR, et al. Topological domains in mammalian genomes identified by analysis of chromatin interactions. Nature 2012;485:376–80.

[131] Rao SS, et al. A 3D map of the human genome at kilobase resolution reveals principles of chromatin looping. Cell 2014;159:1665–80.

[132] Naumova N, et al. Organization of the mitotic chromosome. Science 2013;342:948–53.

[133] Li G, et al. Extensive promoter-centered chromatin interactions provide a topological basis for transcription regulation. Cell 2012;148:84–98.

[134] Nagano T, et al. Single-cell Hi-C for genome-wide detection of chromatin interactions that occur simultaneously in a single cell. Nat Protoc 2015;10:1986–2003.

[135] Tjong H, et al. Population-based 3D genome structure analysis reveals driving forces in spatial genome organization. Proc Natl Acad Sci USA 2016;113:E1663–72.

[136] Fraser J, Williamson I, Bickmore WA, Dostie J. An overview of genome organization and how we got there: from FISH to Hi-C. Microbiol Mol Biol Rev 2015;79:347–72.

[137] Chen X, et al. The visualization of large organized chromatin domains enriched in the H3K9me2 mark within a single chromosome in a single cell. Epigenetics 2014;9:1439–45.

[138] Chen X, et al. Chromatin in situ proximity (ChrISP): single-cell analysis of chromatin proximities at a high resolution. Biotechniques 2014;56:117–8.

[139] Lieberman-Aiden E, et al. Comprehensive mapping of long-range interactions reveals folding principles of the human genome. Science 2009;326:289–93.

[140] Mattout A, Cabianca DS, Gasser SM. Chromatin states and nuclear organization in development—a view from the nuclear lamina. Genome Biol 2015;16:174.

[141] Leonhardt H, et al. Dynamics of DNA replication factories in living cells. J Cell Biol 2000;149:271–80.

[142] O'Keefe RT, Henderson SC, Spector DL. Dynamic organization of DNA replication in mammalian cell nuclei: spatially and temporally defined replication of chromosome-specific alpha-satellite DNA sequences. J Cell Biol 1992;116:1095–110.

[143] Lemaitre C, Bickmore WA. Chromatin at the nuclear periphery and the regulation of genome functions. Histochem Cell Biol 2015;144:111–22.

[144] Dileep V, et al. Topologically associating domains and their long-range contacts are established during early G1 coincident with the establishment of the replication-timing program. Genome Res 2015;25:1104–13.

[145] Foti R, et al. Nuclear architecture organized by Rif1 underpins the replication-timing program. Mol Cell 2016;61:260–73.

[146] Barton LJ, Soshnev AA, Geyer PK. Networking in the nucleus: a spotlight on LEM-domain proteins. Curr Opin Cell Biol 2015;34:1–8.

[147] Etchegaray JP, Mostoslavsky R. Interplay between metabolism and epigenetics: a nuclear adaptation to environmental changes. Mol Cell 2016;62:695–711.

[148] McBrian MA, et al. Histone acetylation regulates intracellular pH. Mol Cell 2013;49:310–21.

[149] Rabineau M, et al. Cell guidance into quiescent state through chromatin remodeling induced by elastic modulus of substrate. Biomaterials 2015;37:144–55.

[150] Tseng AS, Levin M. Transducing bioelectric signals into epigenetic pathways during tadpole tail regeneration. Anat Rec 2012;295:1541–51.

[151] Hu X, et al. Histone cross-talk connects protein phosphatase 1alpha (PP1alpha) and histone deacetylase (HDAC) pathways to regulate the functional transition of bromodomain-containing 4 (BRD4) for inducible gene expression. J Biol Chem 2014;289:23154–67.

[152] Tiwari VK, et al. A chromatin-modifying function of JNK during stem cell differentiation. Nat Genet 2012;44:94–100.

[153] Dey A, Nishiyama A, Karpova T, McNally J, Ozato K. Brd4 marks select genes on mitotic chromatin and directs postmitotic transcription. Mol Biol Cell 2009;20:4899–909.

[154] Hervouet E, et al. Disruption of Dnmt1/PCNA/UHRF1 interactions promotes tumorigenesis from human and mice glial cells. PLoS One 2010;5:e11333.

[155] Voncken JW, et al. MAPKAP kinase 3pK phosphorylates and regulates chromatin association of the polycomb group protein Bmi1. J Biol Chem 2005;280:5178–87.

[156] Gerhart-Hines Z, Lazar MA. Circadian metabolism in the light of evolution. Endocr Rev 2015;36:289–304.

[157] Janich P, et al. The circadian molecular clock creates epidermal stem cell heterogeneity. Nature 2011;480:209–14.

[158] Masri S, Orozco-Solis R, Aguilar-Arnal L, Cervantes M, Sassone-Corsi P. Coupling circadian rhythms of metabolism and chromatin remodelling. Diabetes Obes Metab 2015;17(Suppl. 1): 17–22.

[159] Gallardo CM, Gunapala KM, King OD, Steele AD. Daily scheduled high fat meals moderately entrain behavioral anticipatory activity, body temperature, and hypothalamic c-Fos activation. PLoS One 2012;7:e41161.

[160] Yamajuku D, et al. Real-time monitoring in three-dimensional hepatocytes reveals that insulin acts as a synchronizer for liver clock. Sci Rep 2012;2:439.

[161] Kon N, et al. Activation of TGF-beta/activin signalling resets the circadian clock through rapid induction of Dec1 transcripts. Nat Cell Biol 2008;10:1463–9.

[162] Balsalobre A, Damiola F, Schibler U. A serum shock induces circadian gene expression in mammalian tissue culture cells. Cell 1998;93:929–37.

[163] Zhao H, et al. PARP1- and CTCF-mediated interactions between active and repressed chromatin at the lamina promote oscillating transcription. Mol Cell 2015;59:984–97.

[164] Dominguez LJ, Barbagallo M. The biology of the metabolic syndrome and aging. Curr Opin Clin Nutr Metab Care 2016;19:5–11.

[165] Feinberg AP, Fallin MD. Epigenetics at the crossroads of genes and the environment. JAMA 2015;314:1129–30.

[166] Vandiver AR, et al. Age and sun exposure-related widespread genomic blocks of hypomethylation in nonmalignant skin. Genome Biol 2015;16:80.

[167] Marioni RE, et al. DNA methylation age of blood predicts all-cause mortality in later life. Genome Biol 2015;16:25.

[168] Martino D, et al. Longitudinal, genome-scale analysis of DNA methylation in twins from birth to 18 months of age reveals rapid epigenetic change in early life and pair-specific effects of discordance. Genome Biol 2013;14:R42.

[169] Voss TC, Hager GL. Dynamic regulation of transcriptional states by chromatin and transcription factors. Nat Rev Genet 2014;15:69–81.

[170] Flanagan JM. Epigenome-wide association studies (EWAS): past, present, and future. Methods Mol Biol 2015;1238:51–63.

CHAPTER 2

Histone Modifications and Histone Variants in Pluripotency and Differentiation

A.J. Bannister*, A.M. Falcão, G. Castelo-Branco****

*The Gurdon Institute, University of Cambridge, Cambridge, United Kingdom
**Laboratory of Molecular Neurobiology, Department of Medical Biochemistry
and Biophysics, Karolinska Institutet, Stockholm, Sweden

2.1 HISTONES AND THE REGULATION OF TRANSCRIPTION

2.1.1 Histones

Albrecht Kossel first described histones in 1884 following his studies on avian red blood cells [1]. Since then, we have learned that histones help package DNA into the nuclei of eukaryotic cells. Although superficially this may seem a rather trivial function, it is actually a complex undertaking. The histones allow approximately 2 meters of DNA to be compacted within the confines of a spherical nucleus of approximately 10 μm in diameter, resulting in a macromolecular DNA–protein complex, which is referred to as chromatin. Chromatin not only packages the DNA but also allows regulated access to it. This is important because many factors, such as transcription, repair and replication factors, require tightly controlled access to the genetic material.

2.1.1.1 Core and Linker Histones

There are four core histones, H2A, H2B, H3, and H4. Two of each core histone form a central octameric complex around which 146 bp of DNA is wrapped. This relatively compact structure is referred to as a nucleosome and it represents the fundamental unit of chromatin. Nucleosomes are connected together by "linker" DNA and a fifth histone, the linker histone H1, which induces further compaction by altering the entry/exit angle of the DNA into each nucleosome. This serves to compact the nucleosomal "beads on a string" chromatin into 30 nm fibers, at least in in vitro studies. Ultimately, supercoiling and twisting maximally compacts the chromatin into complex structures, such as mitotic chromosomes [2]. Broadly speaking, chromatin can be subdivided into two architectural forms: (1) heterochromatin that is highly condensed, contains mostly inactive genes and gene-poor regions and replicates relatively late; and

CONTENTS

Chromatin Regulation and Dynamics. http://dx.doi.org/10.1016/B978-0-12-803395-1.00002-2

(2) euchromatin that adopts a relatively open arrangement, contains most of the active genes and replicates relatively early.

Early studies concluded that chromatin was very stable with little histone turnover exceptfor the natural dilution that occurs as a consequence of semi-conservative DNA replication [3,4]. However, it is now known that histone turnover does happen, albeit in a highly regulated fashion. The prototypical (or canonical) core- and linker histones are synthesized at the beginning of the S phase of the cell cycle and are deposited into chromatin in a DNA replication-dependent manner. Their genes do not contain introns and the respective mRNAs lack polyadenylation, having instead a specialized stem-loop structure at their 3′ end. In contrast, some histones are synthesized throughout the cell cycle and are deposited into chromatin in a replication-independent manner [5] (Table 2.1). These histones are referred to as replacement histones or "variants" and they are deposited into chromatin by specific histone chaperones. Their genes contain introns allowing, at least in some cases, for alternative splicing. They are highly conserved between species, indicating that they have specialized functions that canonical histones cannot perform [6].

2.1.1.2 Histone Variants

In higher eukaryotes, the canonical core histone H3 is more specifically named H3.1. An abundant variant, named H3.3, only differs from H3.1 by five amino acid substitutions and preferentially replaces H3.1 at active genes throughout the cell cycle. In contrast, H3.1 is deposited throughout the genome only during DNA replication [5]. It is not yet fully understood how the functions of the core and variant histones differ, but the presence of one or more variants within a particular nucleosome generally affects the overall stability of that

Table 2.1 A List of the Main Mammalian Histone Variants and Their Functions

Histone Variants		Comments
Histone H2A	H2A.Z	Nucleosome positioning and context-dependent gene regulation
	H2A.X	Protection from double strand DNA damage
	MacroH2A	X-chromosome inactivation and poising genes for activation
	H2A.Bdb	Draws nucleosomes together, gene activation, and spermatogenesis
Histone H2B	H2B1A	Testis-specific
	H2BWT	Mitotic preservation of telomeres
Histone H3	H3.3	Replication-independent deposition, and transcription activation
	CENP-A	Centromeric H3 and kinetochore assembly

nucleosome [7]. Moreover, specific variants can be generally associated with a particular chromatin state. For example, H3.3 is linked to transcriptionally active or permissive chromatin, whereas another H3 variant, CENP-A, is linked to repressive chromatin at centromeric regions. However, there are no strict rules governing the relationship between the presence of a particular variant and transcriptional activity. H2A.Z is, for example, linked to both transcriptional activity and repression. The differential activity of variant histones can be explained, at least in part, by their various posttranslational modifications (PTMs). For example, at active genes acetylated H2A.Z is enriched within and immediately around promoter sequences, whereas in repressed genomic loci, such as in heterochromatin, H2A.Z is found ubiquitylated (reviewed in [6]). As we discuss later, PTMs of both canonical and variant histones represent a major mechanism regulating DNA-based processes, such as transcription.

The critical importance of histones in regulating DNA-templated processes is underscored by the finding that specific residues are mutated in cancers. Recurrent somatic heterozygotic mutations in genes encoding canonical H3.1 and the variant H3.3 have been identified in paediatric glioblastomas [8,9]. They result in amino acid substitutions at two closely spaced N-terminal residues in H3, lysine 27 to methionine and glycine 34 to arginine or valine. Since there are many H3 genes in the mammalian genomes, the mutated H3 proteins can only comprise a small fraction of the total H3 in a cell, especially as the mutations are heterozygotic. This indicates that the mutated H3 proteins probably gain a dominant negative function. In the case of the lysine to methionine mutation at position 27, it likely locks one of the modification enzymes, EZH2, into a nonproductive complex with the effect of diminishing the enzyme's global H3K27 methyltransferase activity [10]. This will have disastrous consequences for cellular processes, such as differentiation and cell cycle control.

In addition to disease as discussed earlier, histones and their variants are very tightly linked to differentiation and development, which will be explored in greater detail later in this chapter. All histones are capable of fine-tuning transcriptional programs via a series of very carefully orchestrated PTMs that combinatorially regulate their function. In addition to PTMs, changes in the turnover and exchange of nucleosomal histones and their variants regulate development. For instance, an increase in the level of macroH2A1 at specific genes associated with liver metabolism leads to a decrease in their expression in the liver [11]. Furthermore, the dynamic turnover of H3.3-containing nucleosomes regulates neuronal- and glial-specific genes apparently independently of H3.3 PTMs [12].

2.1.2 Histone Posttranslational Modifications

As already eluded to, histones are subject to many distinct PTMs (Table 2.2) that mainly occur on amino acid side chains. A basic understanding of how

Table 2.2 Histone Modifications and their Nomenclature

Modification	Amino Acid Modified	Single Letter Amino Acid	Degree of Modification	Abbreviation	Example of Modification
Acetylation	Lysine	K	Mono	ac	H3K14ac
Methylation	Lysine	K	Mono	me1	H3K9me1
			Di	me2	H3K9me2
			Tri	me3	H3K9me3
	Arginine	R	Mono	me1	H3R2me1
			Disymmetric	me2s	H3R2me2s
			Diasymmetric	me2as	H3R2me2as
Phosphorylation	Serine, threonine, or tyrosine	S/T/Y	Mono	ph	H3S10ph
Ubiquitylation	Lysine	K	Mono	ub1	H2BK20ub1
Sumoylation	Lysine	K	Mono	su	H2BK6su
ADP ribosylation	Glutamate	E	Mono	ar1	H2BE2ar1
			Poly	arn	H2BE2arn
Propionylation	Lysine	K	Mono	pr	H4K5pr
Butyrylation	Lysine	K	Mono	bu	H4K5bu
Crotonylation	Lysine	K	Mono	cr	H3K4cr
Malonylation	Lysine	K	Mono	ma	H2BK116ma
Succinylation	Lysine	K	Mono	su	H3K14su
Formylation	Lysine	K	Mono	fo	H3K23fo
Citrullination	Arginine	R	Mono	cit	H4R3cit
Hydroxylation	Tyrosine	Y	Mono	oh	H4Y51oh
O-GlcNAcylation	Serine or threonine	S/T	Mono	og	H3T32og

Modified and updated from Xhemalce B, Dawson MA, Bannister, AJ. Histone modifications. In: Meyers RA, editor. Encyclopedia of molecular cell biology and molecular medicine: epigenetic regulation and epigenomics. Weinheim, Germany: Wiley-VCH Verlag GmbH & Co. KGaA; 2011 [14].

these modifications affect chromatin function was attained in 1997, when the first high-resolution X-ray crystal structure of the nucleosome was solved [13]. The structure indicated that histone N-terminal tails extend beyond their own nucleosome and directly contact adjacent nucleosomes. Consequently, it was suggested that these contacts were involved in the regulation of higher-order chromatin structure and chromatin function.

There is now a list of at least 15 distinct modifications that occur on histones (Table 2.2) (reviewed in [14,15]). The nomenclature for describing specific residues within histones and relevant modifications is to firstly reference the histone (e.g., H3) followed by the modified amino acid in single letter amino acid code (e.g., K9 for lysine 9) and then, if appropriate, the modification itself (e.g., me; for methylation) [16]. For modifications that can occur multiple times on a single amino acid side chain (e.g., lysine methylation) the number of modifications is indicated at the end of the description (e.g., H3K9me3 for H3K9 trimethylation) (Table 2.2).

Like histones themselves, histone modifications are highly conserved between species. This emphasizes the fundamental importance of not only the linear amino acid sequence of each histone protein but also of the PTMs that are added as a consequence of various signal transduction pathways. The very large number of possible modifications, which may occur in isolation or more typically in various combinations with other modifications, generates a huge combinatorial complexity. Moreover, histone marks can regulate chromatin structure not only by merely providing steric hindrance, but also by recruiting various protein complexes, such as ATP-dependent chromatin remodeling complexes or histone-modifying enzymes, which can further modify the local chromatin state. In this review, we will focus on histone acetylation and methylation, as these modifications are the most studied and best understood, and there are numerous examples highlighting the critical roles that these modifications play in developmental processes. However, for a more comprehensive discussion of these and other histone modifications, we refer the reader to more specialized reviews [14,15].

2.1.2.1 Histone Acetylation

First discovered by Vincent Allfrey in 1964 [17], histone acetylation mainly occurs on lysine side chains, where it neutralizes the positive charge of the amino acid. Due to this, acetylated lysines can no longer take part in electrostatic interactions with DNA, which promotes the decompaction of chromatin. Consequently, highly acetylated regions of the genome tend to be relatively less compact.

Histone acetylation is a highly dynamic modification regulated by the opposing actions of two families of enzymes, namely histone acetyltransferases (HATs) and histone deacetylases (HDACs) [14]. All HATs utilize acetyl CoA as cofactor in the enzymatic reaction, and they can be divided into two classes defined by their cellular localization. Cytoplasmic HATs, or type B enzymes, acetylate free histones, particularly H4K5 and H3K12, in an organismal-specific pattern [18] to facilitate the deposition of histones into chromatin, after which these modifications are removed. Nuclear HATs, or type A enzymes, acetylate chromatinized histones predominantly on their exposed N-terminal tails (Fig. 2.1) [14]. The nuclear enzymes can be grouped based on amino acid sequence and structural similarities into at least three families: (1) GNAT, (2) MYST, and (3) p300/CREB-binding protein (CBP) families. Generally HATs exist in large protein complexes that confer substrate specificity and allow accurate genomic targeting, often via interactions with a transcription factors (TFs) that bind to specific DNA sequences.

Histone acetylation is removed via the action of HDACs. Based on their amino sequences, HDACs can be divided into four classes. Members of class I HDACs show similarity to yeast Rpd3, while those in class II show similarity to yeast Hda1. There is only a single member of class IV, namely HDAC11. The deacetylases within these three classes share a common catalytic mechanism [20]. However, class III HDACs, which are similar to yeast Sir2 and are referred to as sirtuins, require NAD+ as a cofactor for their activity [21]. Similar to HATs, HDACs reside within large complexes, often containing more than one HDAC enzyme. This feature, coupled with the fact that they possess relatively low-substrate specificity, has made it difficult to dissect the specific functions of the different HDACs in vivo. However, they do undertake specific roles, as embryonic stem (ES) cell differentiation is, for example, controlled by HDAC1 but not HDAC2 [22,22a]. Furthermore, neuronal differentiation from neural stem cells (NSCs) is repressed specifically by HDAC3, while oligodendrocyte differentiation is repressed by HDAC 2 [23]. The specific roles of HDACs is in part regulated by the overall composition of the specific HDAC complex they reside in, which is important for their genomic localization [14].

Consistent with the role of acetylation in neutralizing the positive charge of lysine side chains, histone acetylation is very tightly linked with transcriptional competence, and it is enriched at the transcriptional start site (TSS) of active genes [24]. Its mode of action certainly involves direct structural perturbation, either by preventing electrostatic interactions to occur between the histone N-termini and the adjacent nucleosomes or, when present in the core of the nucleosome (e.g., at H3K122), by directly destabilizing the nucleosomal structure [25]. In addition, histone acetylation also provides a binding site for proteins with specialized acetyl-lysine -binding domains called bromodomains (Fig. 2.1) [26]. Interestingly, many of the HATs, such as p300 and GCN5, contain bromodomains. Presumably, this allows the acetyltransferases to be locally enriched at acetylated genomic sites, thereby promoting further acetylation in a positive-feedback manner. Histone acetylation also forms a docking site for other classes of epigenetic effector complexes. The catalytic subunit of the SWI/SNF chromatin remodeling complex that employs energy derived from ATP hydrolysis to reposition nucleosomes uses its bromodomain to bind DNA damage–induced acetylated H3 to facilitate efficient DNA double strand–break repair [27]. Indeed, this mechanism of recruiting chromatin remodeling activities to specific genomic loci is widely used throughout the genome to regulate many processes including gene regulation.

Notwithstanding that mentioned previously, our understanding of the fine details of how acetylation exerts its effects on transcription is far from complete. Although recent studies targeting HATs to specific genomic loci indicate that acetylation is indeed casual to transcriptional activation [28,29], it seems that the presence of acetylated lysines per se is insufficient for active transcription, and the mark needs to be actively turned over for efficient transcription, at least at some genes [30]. Furthermore, global inhibition of HDACs leads to significant transcriptional change only at 2.4% of the genes, 33% of which were actually downregulated [31]. The molecular mechanism(s) underlying these findings remain to be determined.

2.1.2.2 Histone Methylation

The main sites of methylation within histones are the side chains of arginines and lysines (Fig. 2.1). In contrast to acetylation, methylation of these basic side chains does not alter their charge. However, the complexity of this modification is higher, as the lysine residue may be mono-, di-, or trimethylated (me1, me2, or me3), whereas arginine may be mono- or dimethylated either symmetrically (me2s) or asymmetrically (me2as) [14].

All histone lysine methyltransferases (PKMTs) catalyze the transfer of a methyl group from S-adenosylmethionine (SAM) to lysine ε-amino groups, and most of them contain a conserved enzymatic domain referred to as the SET domain. Overall, this class of enzymes tends to be specific with respect to substrate target as well as the degree of methylation laid down (i.e., mono-, di-, or tri-). For example, SUV39H1, the first PKMT identified, di- and trimethylates H3K9, establishing modifications that are highly enriched in repressive heterochromatin [32]. The yeast Set1 enzyme specifically targets H3K4 and is capable of catalyzing all three methylation states, whereas in mammalian cells, Set7/9 similarly targets H3K4 but only catalyzes the reaction to the monomethyl state [33].

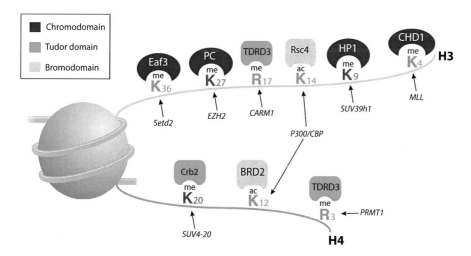

FIGURE 2.1 Diagram illustrating some of the major methylated *(me)* and acetylated *(ac)* sites within histone N-terminal tails, examples of the modifying enzymes (italicized) and examples of proteins with domains that specifically bind to the modified histone.

Modified sites generally correlated with active transcription are shown in *green*, while those generally correlated with a repressed chromatin state are shown in *red*. *Source Adapted from Bannister AJ, Kouzarides T. Regulation of chromatin by histone modifications. Cell Res 2011;21(3):381–95 [19].*

As both enzymes are transcriptional coactivators, these observations highlight that the site of histone methylation and the degree of methylation are both important factors for biological output.

Histone protein arginine methyltransferases (PRMTs) catalyze the transfer of a methyl group from SAM to an arginine's ω-guanidino group. The family consists of at least 11 enzymes that can be divided into 2 groups: (1) type 1 enzymes catalyze the formation of Rme1 and Rme2as and (2) type 2 enzymes catalyze formation of Rme1 and Rme2s. The most relevant PRMTs with respect to histone methylation are PRMTs 1, 4, 5, and 6 (reviewed in [34]). As with the PKMTs, the biological role of histone–arginine methylation depends on many factors including which arginine is methylated, the degree of methylation and the surrounding chromatin architecture.

Histone methylation was, for a long time, considered to be a highly stable modification. Indeed, it seemed to possess the ideal characteristics required of a mark capable of carrying epigenetic information. However, evidence was present in the literature early on, albeit sparsely, suggesting that active turnover of histone methylation does occur. This led to the proposition of two potential chemical reactions for enzymatic demethylation in 2002 [35]. Both reactions were eventually shown to occur in vivo.

Lysine-specific demethylase 1 (LSD1) was the first true histone demethylase identified [36]. The enzyme utilizes FAD as a cofactor and functions via an amine-oxidation pathway that requires a protonated nitrogen. Hence, LSD1 only demethylates mono- and dimethylated lysines. Furthermore, its substrate specificity depends on associated cofactors. When complexed with the androgen receptor, it demethylates H3K9me2/1, a mark of repressed chromatin. In this case, its activity is consistent with that of a transcriptional activator. However, when associated with the co-REST complex, it demethylates H3K4me2/1, which is an activity consistent with a transcriptional repressor (reviewed in [37]). The different LSD1 complexes have important roles in differentiation and development. For instance, the LSD1/co-REST complex times progenitor differentiation by controlling the stability of H3K4 methylation [38].

Lysine demethylation may also occur via an oxidative mechanism and radical attack involving Fe (II) and α-ketoglutarate. Enzymes employing this mechanism can potentially demethylate all three states of methylated lysine, although many appear to have specific preferences in this regard (reviewed in [39]). All enzymes in this class contain a catalytic jumonji domain and they are critical in many developmental processes, such as neural development (reviewed in [40]).

Demethylation via radical attack is compatible, at least in theory, with demethylation of arginine side chains, although this has yet to be convincingly demonstrated. However, arginine is subject to deimination generating citrulline in place of the original arginine [41]. One of the enzymes responsible, Peptidyl

Arginine Deiminase 4 (PADI4), catalyzes the deimination of arginine and monomethylated arginine giving rise to citrulline in each case. Although not a "true" demethylase in that it does not generate unmethylated arginine, PADI4 nevertheless removes monomethylated arginine from histones. Interestingly, PADI4 regulates pluripotency via a mechanism involving H1 citrullination and chromatin decompaction [42].

2.1.2.3 *Context-Dependent Role of Histone Modifications*

Generally speaking, histone methylations can be grouped depending on their biological effects. Lysine methylations at H3K4, H3K36, and H3K79 are tightly associated with active genes, whereas methylations at H3K9, H3K27, and H4K20 are tightly associated with transcriptional repression (Fig. 2.1). Importantly though, within these "groups" the spatial distributions of the methylated sites can be dramatically different. For instance, H3K4me3 generally peaks sharply around the TSS of active genes, whereas H3K36me3 is enriched throughout the transcribed region of active genes [43]. Likewise, not all repressed heterochromatin is the same. For example, pericentric heterochromatin is highly enriched with H3K9me3, whereas facultative heterochromatin, such as that found on the mammalian inactive X-chromosome, is enriched with H3K27me3. In addition, there is an intriguing gray area found in ES cells, which contains so-called "bivalent domains" in the promoters of developmentally regulated genes [44]. Here both H3K4me3 (active mark) and H3K27me3 (repressive mark) appear to coexist, which arrangement helps to maintain genes in a transcriptionally competent, so-called poised state. When a cell enters a lineage requiring expression of a particular gene with a bivalent chromatin state, H3K27me3 is lost and H3K4me3 is retained in its promoter, and vice versa if the gene is to be maintained inactive. In contrast to lysines, relatively little is known concerning the genomic location of histone–arginine methylations. We do know, however, that H3R17me and H3R26me are typically found at active genes and H3R8me at inactive genes.

In addition to the genes themselves, regulatory elements, such as enhancers, are also decorated with specific histone modifications, which depend on the overall activity of the element. For instance, a poised enhancer typically displays H3K4me1, H3K27me3 and low or absent H3K27ac. An active enhancer on the other hand contains H3K4me1, high levels of H3K27ac, H3K122ac and absent or low levels of H3K27me3 [25,45]. Exactly how the modifications exert their effects is not certain, but the mechanism certainly involves recruitment of further chromatin modifiers, such as chromatin remodeling activities. Additionally, the modification may directly destabilize nucleosomes, as exemplified by H3K122ac [25].

The addition of one or more methyl groups to arginine or lysine side chains does not affect the overall charge of the relevant side chain nor does it perturb the overall protein structure. Their addition does, however, affect hydrophobicity, and the methylated sites generate binding platforms that can be recognized by specialized protein domains. For methylated lysines, such domains include chromodomains, PWWP domains, Tudor domains (also bind methylated arginines), MBT domains and certain PHD fingers (reviewed in [14]). These domains are found within a wide range of proteins with multifaceted cellular roles including development, which not only recognize specific methylated lysines but also discriminate between different methylated states. For instance, the Polycomb group (PcG) protein, L3MBTL1, integrates positional information to repress developmentally regulated genes. This factor harbors three MBT domains that simultaneously bind to H4K20me1/2 and H1bK26me1/2 to compact the local chromatin structure [46]. Furthermore, L3MBTL1 binds to the HP1 protein, which itself binds to H3K9me2/3 via its chromodomain, thereby generating a highly repressive multivalent chromatin architecture. L3MBTL1-mediated repression is important in the maintenance of stem cell potency, because it represses, for example, the ability of stem cells to drive hematopoietic-specific transcriptional programs [47].

2.1.2.4 *Small Molecule Inhibitors of Chromatin Regulators*

Given the importance of histone modifications in differentiation, development and disease, much effort is now being placed on identifying small molecule inhibitors of the enzymes and chromatin readers that mediate the relevant biological outcomes.

These compounds are proving highly valuable assets in endeavors to reprogram cells to a more progenitor-like phenotype and in the fight against diseases, such as cancer. For example, inhibition of the H3K9 methyltransferase G9a by the small molecule BIX01294 reprograms bone marrow cells to a cardiac competent progenitor phenotype [48]. Furthermore, the small molecule I-BET prevents specific bromodomain-containing proteins (BRDs), such as BRD4, to activate certain oncogenes, and shows good efficacy against MLL-driven leukemias [49]. Indeed, I-BET is currently being assessed in clinical trials. Intriguingly though, the efficacy of this class of inhibitors is not limited to an anticancer role, because a very similar inhibitor that targets the same BRDs as I-BET has been shown to effectively suppress murine cardiomyocyte hypertrophy in vitro and pathological cardiac remodelling in vivo [50]. In addition, the inhibitor perturbs murine spermatogenesis in mice leading to sterility [51]. The finding that cessation of drug administration reverses the process, enabling the mice to become fertile again, has obvious implications for using I-BET-like drugs as contraceptives. Moreover, the wide spectrum of physiological responses achieved with a single inhibitor highlights how useful this approach is likely going to be in the study of developmental biology and more generally in human medicine. These findings also emphasize the importance of chromatin modifications in critical cellular processes. Indeed, the ability of at least some of the modifications to be passed from one cell generation to the next has underpinned the suggestion that they may be capable of carrying non-DNA encoded heritable (epigenetic) information. There is thus obvious potential for the information to be "edited" via the administration and consequent action of small molecular drugs, such as I-BET.

2.2 HISTONE PTMS AND VARIANTS DURING DEVELOPMENT

2.2.1 Epigenetic Landscapes During Development

Given that all cells in the adult organism derive from one single cell, the zygote, and therefore share an (nearly) identical genome, the remarkable diversity of cells typessuggests that regulatory mechanisms must exist that extend beyond genetics, invoking an"epi"genetic level. The term "epigenetics" was first coined by Conrad Hal Waddington in the 1940s (Fig. 2.2) [52]. Waddington's description of an epigenetic landscape compares a cell during development to a marble rolling down a hill, changing its character along the way and following "irreversible" paths (canalization) within valleys (determined by gene activity), ultimately leading to terminal specialization and differentiation. The niche or environment in which the cell resides provides signalling cues to determine which path the cell takes and which phenotypes it acquires. These cues will ultimately lead to differential accessibility of the RNA polymerase

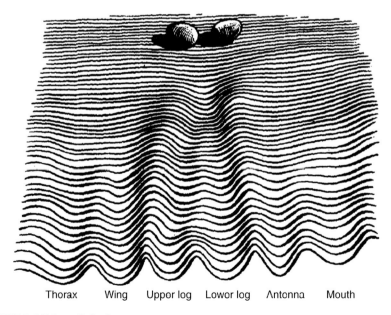

Thorax   Wing   Uppor log   Lowor log   Antonna   Mouth

FIGURE 2.2 Epigenetic landscape.
Source adapted from Gilbert SF. Epigenetic landscaping—Waddington use of cell fate bifurcation diagrams. Biol Philos 1991;6(2):135–54 [52].

machinery to specific locations in the genome and consequently to the establishment of diverse transcriptional programs. Berger, Kouzarides, Shiekhattar, and Shilatifard [53] recently suggested a novel nomenclature for the classification of epigenetic mechanisms. Highlighting the position of epigenetic states at the interface between the genotype and the environment, extracellular or environmental stimuli capable of affecting chromatin were termed epigenators, which would trigger a cascade of intracellular signals (termed epigenetic initiators including TFs and noncoding RNAs) ultimately leading to the localization of epigenetic effector machineries at specific sites in the genome, thereby modulating chromatin and transcription locally. When considered genome wide, the coordinate actions of epigenators and epigenetic initiators together with the epigenetic effector machinery operating in a cell at a certain time and space would determine the transcriptome of the cell and its phenotype. Furthermore, the epigenetic state of a cell could be preserved even in the absence of the original epigenators. This would be achieved by the action of epigenetic maintainers, of which histone (and DNA) modifiers, certain histone PTMs, and histone variants are essential components [53].

Most cells within a certain lineage are constantly exposed to changing environments during development, as they progress from stem cell states through progenitor cell states, until they finally reach terminally differentiated states. For instance, cells within the inner cell mass (ICM) of the blastocyst are pluripotent, and will be specified during development to cells of the different germ layers, ectoderm, mesoderm, and endoderm. Morphogenesis and patterning will define differential identities of regions along the anterior-posterior and dorsal-ventral axes of the embryo. The position of the cells at these axes will ultimately determine their cell fate within a certain embryonic lineage. A subset of stem/progenitor cells will maintain their phenotype and self-renewal capacity even in adulthood. In order to maintain their unique epigenetic state, these cells are present in specific stem cell niches, under the surveillance of epigenators and epigenetic initiators, responsible for the localization of the epigenetic effector machinery. When the niche is altered transiently, some of these progenitor/stem cells are able to maintain their phenotype despite the presence of other sets of epigenators that would normally lead to their differentiation and even transition to an unrelated cell state. Epigenetic initiators and maintainers are likely to be essential for the inheritance of an epigenetic state during such events in an "adverse" niche [53].

2.2.2   Histone PTMs and Variants in Pluripotency and Reprogramming

Pluripotent cells, such as ES cells isolated from the ICM of blastocysts and induced pluripotent (iPS) cells generated by reprogramming from lineage-committed cells, have the unique potential to differentiate into all cell types present in the adult body and can also selfrenew in vitro. The epigenetic state of pluripotent cells is maintained by an intricate pluripotency transcriptional machinery, which is ultimately controlled by TFs, such as, Nanog, Oct4, and Sox2 [54]. Pluripotent cells are able to rapidly acquire different fates depending on the stimuli (epigenators) they are exposed to. In order to preserve this high degree of plasticity, pluripotent cells have a unique epigenetic landscape.

2.2.2.1 Unique Properties of the Chromatin of Pluripotent Cells

Pluripotent cells display a unique three-dimensional genome organisation where pluripotency factors and chromatin architectural proteins (such as cohesin and the mediator complex) establish specific enhancer–promoter loops and long-range interactions to maintain the pluripotent transcriptional state [55]. In addition, the chromatin of pluripotent cells displays a more open configuration when compared to differentiated cells, enabling the RNA polymerase machinery to access genes involved in developmental processes and cell specification/differentiation [56]. Indeed, RNA polymerase is already bound to the promoters of these genes in mouse ES cells [57]. Moreover, these genes are bivalently associated with both active (H3K4me3)- and repressive (H3K27me3) histone marks, and reside in a repressed but poised transcriptional state that can be rapidly activated [44]. During the differentiation of pluripotent cells into the neural lineage, active neural genes lose H3K27me3 and retain the activatory H3K4me3 mark, while nonneural genes lose H3K4me3 while retaining H3K27me3 and remain inactive [58]. Dissolution of the pluripotency network and resolution of bivalency at developmentally regulated genes during differentiation occurs at specific stages of the cell cycle (G2/S and G1, respectively), through the action of cell cycle regulators, such as cyclin B1 and CDK2, and chromatin modifiers, such as the H3K4me3 methyltransferase MLL2/KMT2B [59,60].

In contrast to differentiated cells, pluripotent cells have hyperdynamic chromatin, where structural chromatin proteins, such as histone H1, display hyperdynamic binding to DNA [61]. This feature is regulated in part by citrullination of H1 at Arg-54, a modification that is catalyzed by the chromatin writer PADI4 and leads to global chromatin decondensation [42]. Inhibiting the hyperdynamic binding of histone H1 to chromatin impairs the self-renewal capacity of ES cells [61], highlighting its functional importance in the maintenance of pluripotency. In line with this observation, histone H1 promotes stable repression of pluripotency genes during differentiation, and is therefore required for the proper execution of differentiation programs [62].

2.2.2.2 How is a Pluripotent Epigenetic State Established?

Pioneering work from John Gurdon, Helen Blau and Shinya Yamanaka showed that a terminally differentiated somatic cell can be reprogrammed to a pluripotent state by nuclear transfer, cell fusion or TF transduction [63,63a]. Strikingly, the work of Yamanaka et al. showed that the ectopic presence of four epigenetic initiators (the TFs OCT4, SOX2, KLF4 and C-MYC) in a differentiated nucleus was sufficient to reset the epigenetic landscape to pluripotency. Many other TF combinations have since been shown to lead to epigenetic reprogramming (reviewed in [64]), which process involves several transitions between intermediate epigenetic states. The first wave of reprogramming events involves the upregulation of genes regulating cell proliferation, metabolism and cytoskeleton, and the downregulation of cell type-specific and developmentally regulated genes. The second wave of reprogramming leads then to the upregulation of pluripotency genes [65]. These changes in transcriptome profiles coincide with genomewide changes in the epigenetic landscape, which inculde the initial widespread erasure of H3K27me3 marks to establsih an open chromatin state, the gradual acquisition of H3K4me3/H3K27me3 bivalency at developmentally regulated genes, and DNA demethylation in late stages of reprogramming [66,67].

TFs contain domains that target them to specific loci within the genome. TF binding motifs are particularly enriched at regulatory regions, such as gene promoters, classical enhancers and superenhancers (dense clusters of transcriptionally active enhancers) [68], which are defined by the high density of DNA sequence–specific binding motifs. As such, TFs act as nucleators for genome "reading" and are ultimately prototypical epigenetic initiators. TFs recruit a myriad of interactor proteins including histone-modifying enzymes, chromatin remodelers and adaptor proteins to specific genomic locations to modulate transcriptional activity in the vicinity of their binding sites. Certain TFs can also act as pioneer factors, which can bind to nucleosomal DNA as single factors and either prime heterochromatin toward a euchromatin state and cause subsequent transcriptional activation [69,70] or initiate the formation of heterochromatin at sites of active transcription [71]. While some TFs coordinate their action, often having binding sites in proximity to each other in regulatory regions of key genes, other TFs have mutually exclusive domains of expression and actively repress each other`s function. In the developing spinal cord, such exclusive domains of TF expression determine the cell type (i.e., subtype of neurons or glia) to be specified from neural progenitor cells along the dorsal-ventral axis [72]. Similarly, TFs of the homeobox family (including the Hox family [73]) play important roles not only in regionalization, but also in cell specification along the anterior-posterior axis..

Proteomic studies have indicated that pluripotency TFs are present in complexes with several different histone modifiers/chromatin remodelers [74–78], many of which have been shown to play important roles in the establishment of the epigenetic landscape in pluripotent cells (reviewed in [64] and [79]). For instance, transduction of the corepressor RCOR2 or of the H3K36me2/3 histone demethylase KDM2B/JHDM1A/1B, or knockdown of the H3K79 methyltransferase DOT1L can replace specific pluripotency TFs in reprogramming paradigms [64]. Further emphasising the importance of chromatin in defining specific cellular states, several other chromatin modulators and chromatin modifiers have been found to be required for efficient reprogramming of differentiated cells to the pluipotent state. Examples include WDR5 that regulates the deposition of H3K4me3, the H3K27me3 demethylase UTX and members of the Polycomb complex [64,79]. In addition, treatment with chemical modulators of histone modifiers including valproic acid (HDAC inhibitor), BIX-01294 (inhibitor of G9a H3K9me2 methyltransferase), and vitamin C (activator of H3K36me2/3 histone demethylase KDM2B/JHDM1A/1B) lead to a more efficient reprogramming (reviewed in [80]).

2.2.2.3　Histone Variants in Pluripotency

Histone variants also play important and complex roles in the regulation of pluripotency (further reviewed in [81] and [82]). For example, histone Macro H2A isoforms bind to H3K27me3-marked regulatory regions of pluripotency genes in fibroblasts and prevent reprogramming by contributing to the maintenance of repression [83–85]. Other histone variants can have differential functions in pluripotency, which could be dependent on the specific PTMs they present. Histone H3.3 is required for nuclear reprogramming [86] and silencing of endogenous retrotransposons in pluripotent cells [87], but it is also associated with the maintenance of an epigenetic memory of the cell of origin during reprogramming [88]. H2A.Z is required for ES cell self-renewal but also for differentiation [89]. H2A.X negatively regulates ES cell proliferation [90], while being required for self-renewal [82].

2.2.2.4　Histone Modifications as Barriers Against Reprogramming

Only a subset of cells expressing the pluripotency TFs accomplish full reprogramming, which led to the hypothesis that either there is a predetermined subset of (elite) cells capable of undergoing reprogramming or that reprogramming is mainly a stochastic process [91]. Single cell RNA-Seq experiments suggest that both models might be valid, with a first stochastic phase of gene activation ultimately leading to the activation of endogenous SOX2, a master regulator of the pluripotent state, that will determine a transcriptional program compatible with pluripotency in a subset of cells [92]. In both the elite and the stochastic models, epigenetic barriers would prevent reprogramming of the cellular identity. In line with this hypothesis, global alterations in histone modifications associated with the activation of innate immunity triggered by foreign DNA and viral delivery of the Yamanaka TFs greatly enhance the efficiency of reprogramming [93]. Moreover, it has recently been shown that the NuRD complex that mediates gene repression by recruiting HDAC and chromatin remodeling activities might be one of these epigenetic barriers, as knocking out one of its components, MBD3, in fibroblasts led to a reprogramming efficiency of nearly 100% [94]. Nevertheless, removal of MBD3 from NSCs has the opposite effect on reprogramming [95], suggesting the existence of different epigenetic barriers depending on the cell of origin.

A myriad of histone PTMs and variants have been implicated in the establishment of pluripotency by TFs, modulating the epigenetic landscape both on a global level and at a specific cohort of genes. Interestingly, although elevated expression of exogenous pluripotency TFs appears to be required to kick start reprogramming, their continued expression beyond the initial stages might be actually detrimental [96,97] or lead to alternative cell states [98]. Considering that histone modifications themselves can be docking sites for the enzymes that deposit them, they might not require TFs to maintain epigenetic information. As such, after the initial nucleation event by a TF, chromatin modifiers would be able to self-maintain the local epigenetic landscape, which has to be counteracted during reprogramming. While the maintenance of H3K27me3 during DNA replication has been documented [99], it is still unclear what mechanisms would be operational for other histone modifications and whether this would be a commonly used process to achieve epigenetic inheritance. Tetramer splitting of homotypic nucleosomes, with semiconservative inheritance of the tetramers that would then serve as templates for modification deposition on newly incorporated histones, was also recently hypothesized as an alternative mechanism for epigenetic inheritance [79]. Finally, pioneer TFs, such as ASCL1, might be able to recognize a trivalent chromatin signature (composed of H3K9me3, H3K4me1, and H3K27ac) at enhancers [100]. As the localisation of trivalent chromatin states in the genome is cell type-specific, this putative alternative mechanism of recruitment of TFs to poised chromatin could be involved in epigenetic inheritance, and could explain why pioneer TFs, such as ASCL1, more efficiently reprogram certain cell types than others.

2.2.3   Histone PTMs/Variants in Cell Specification and Differentiation in the Developing Central Nervous System

The development of the mammalian nervous system progresses through a series of steps, including first the induction of the ectoderm lineage and folding of the neural tube [101]. The ventricles of the neural tube are subsequently lined up with neuroepithelial progenitors (NSCs) that will, depending on their position along the axes of the neural tube undergo proliferation/self-renewal, migration, specification, and differentiation into the three major cell types of the central nervous system (CNS), that is, neurons and the two glial cell types, astrocytes and oligodendrocytes, followed by their functional maturation in the neural circuits. These developmental processes are determined by region-specific exposure to epigenators, such as sonic hedgehog (SHH), fibroblast growth factors (Fs), bone morphogenetic proteins (BMPs), retinoic acid, among many others [101]. NSC specification and differentiation is accompanied by remarkable changes in gene expression patterns that ultimately define the cellular morphology and function. This process involves intricate relationships between programs that silence genes associated with multipotency and self-renewal and that activate cell type–specific genes. Unique epigenetic programs involving DNA and histone modifications together with epigenetic initiators and epigenators coordinate these rearrangements in the chromatin.

2.2.3.1 *Histone (De)Methylation in Neural Development*

As previously mentioned, histone methylation has been associated with both gene activation and -silencing. Chromatin modifiers from the same family of enzymes can display opposite functions and be required for both NSC maintenance and neural commitment and differentiation depending on the modifications they catalyze. The PKMT EZH2 (enhancer of zeste homolog 2) trimethylates H3K27 and prevents premature differentiation of NSCs [102]. Cortical progenitors lacking EZH2 present elevated expression of genes involved in neurogenesis and neuronal differentiation and produce more neurons at the expense of self-renewal [102]. Surprisingly, EZH2 is still highly expressed in cells of the glial lineage, such as oligodendrocytes, and when over-expressed in NSCs, EZH2 promotes differentiation toward the oligodendrocyte lineage [103,104]. Other histone methylation marks have also been linked to neuronal gene repression. The PKMT G9a is recruited by the RE1-silencing TF (REST)/neuron-restrictive silencer factor (NRSF) in nonneuronal cells to silence neuronal genes through H3K9 methylation [105]. In a similar manner, H3K9 histone methyltransferases are fundamental to prevent the expression of neuronal lineage genes during oligodendrocyte differentiation [106]. Genome-wide analysis revealed that H3K9me3 repressive marks are indeed enriched in neuronal lineage–related genes during oligodendrocyte differentiation [106].

Histone methylation marks are dynamically regulated by the opposing actions of PKMTs and histone demethylases (HDMs). The HDM family member jumonji domain-containing 3 (JMJD3), a histone H3K27-specific demethylase, is necessary for neural lineage commitment from mouse EC cells [107] and promotes neuronal differentiation from NSCs [108]. MLL1 (mixed lineage leukemia 1), a PKMT that methylates H3K4, is also required for neuronal differentiation from postnatal NSCs possibly through modulating a histone H3K27-specific demethylase [109]. In MLL1-deficient NSCs the promoter of *Dlx2*, a gene encoding an important neurogenic TF, thus does not lose H3K4me3 but rather acquires K3K27me3, suggesting a possible crosstalk between MLL1 and a H3K27me3 demethylase [109]. In the absence of MLL1, *Dlx2* becomes bivalently marked by H3K4me3 and H3K27me3 leading to failure of transcriptional activation [109]. Conversely, LSD1, which demethylates mono- or dimethylated H3K4 or H3K9, is required for neural stem cell proliferation. Inhibition of LSD1 activity and gene expression knockdown thus leads to a reduction of NSC proliferation [110]. Interestingly, a recent report shows that a specific isoform of LSD1, LSD1+8a that has an additional exon 8a, is essential for neuronal maturation. LSD1+8a functions as a gene activator by demethylating H3K9me2 at promoters of genes essential for neuronal differentiation [111]. Demethylation of H3K4me3 is also involved in neuronal maturation, as SMCX/JARID1C is required for development of dendrites in rat granule neurons and for neuronal survival during zebrafish development [112].

2.2.3.2 Histone (De)Acetylation in Neural Development

Histone acetylation, mediated by HATs, plays key roles in the NSC self-renewal and neuronal as well as glial specification. QKF, a transcriptional activator from the MYST family of HATs, is required for adult NSC self-renewal and adult neurogenesis. Qkf-mutant mice exhibited a reduced number of NSCs and failed to generate a normal number of neurons in the adult olfactory bulb [113]. Mutations in the HATs CBP and p300 have been identified as the cause for the Rubinstein–Taybi syndrome [114], which is characterized by cognitive dysfunction. In mice, knockdown of CBP resulted in inhibition of neurogenesis and gliogenesis. CBP binds to neuronal and glial promoters and regulates cortical precursor differentiation [115]. Other studies showed that CBP is recruited to the complex NGN2/RAR to activate motor neuron gene expression in the spinal cord. Deletion of CBP and p300 specifically in the developing spinal cord reduces histone H3 acetylation and severely impairs motor neuron specification [116]. P300 was also reported to drive astrocytic differentiation from neural progenitors through the formation of a complex with SMAD1 and STAT3 that activates GFAP gene promoter [117].

Chromatin relaxation promoted by HATs can be reversed by HDACs, resulting in gene silencing. Different HDACs have been shown to have a crucial role in both NSC self-renewal and maintenance and in neural cell fate decisions. It was previously reported that interaction of HDACs 3 and 5 with TLX TF is crucial to maintain NSC self-renewal and proliferation in the adult brain [118]. TLX thus recruits HDACs 3 and 5 to the promoter regions of *p21* and *Pten* involved in the regulation of NSC proliferation. Inhibition of HDAC activity induces the expression of these genes and reduces NSC proliferation [118]. Furthermore, inhibition of HDACs in adult NSCs results in an HDAC-dependent upregulation of neuron-specific genes, suggesting that neural lineage progression requires maintenance of acetylation in NSCs [119]. HDACs 1 and 2 were reported to be essential for the progression of neuronal progenitors to mature neurons, as ablation of these HDACs in the mouse CNS led to cerebellar and hippocampal abnormalities [120].

HDACs reside in large regulatory repressive complexes, such as REST, NCoR1, SMRT (also known as NCoR2) [121], NuRD, and SIN3 [122]. In fact, the first studies placing HDACs to promoters of neuronal genes were performed in nonneuronal cells, where the transcriptional repressor REST/NRSF is essential for the suppression of these genes [123,124]. Subsequent reports have show that REST and its corepressors dissociate from neuronal genes as progenitor cells differentiate into neurons [125]. Additionally, SMRT complexes were reported to prevent the progression of NSCs to neurons by repressing the expression of the histone H3K27-specific demethylase JMJ3, an activator of the neurogenic program [108]. Transcriptional repression mediated by NCoR1 is necessary to inhibit glial differentiation from NSCs, as NCoR1 gene disruption impairs NSC self-renewal and triggers spontaneous differentiation into astrocytes [126].

HDAC activity has also been described to play key roles in glial cell lineages, such as in oligodendrocyte specification and differentiation [23,127–131]. In the hindbrain of HDAC1-mutant zebrafish, the oligodendrocyte TFs, OLIG2 and SOX10, are absent and specification fails to occur [127]. Similarly, progenitor and mature oligodendrocytes are lacking in the HDAC1/HDAC2 double-mutant mice, as HDAC1/HDAC2 suppress WNT/β–catenin signaling that inhibits oligodendrocyte differentiation [128]. Interestingly, these HDACs were also found at the promoters of oligodendrocyte differentiation repressors in the cuprizone-induced model of remyelination in young mice, but not in old mice [131]. In this model, mice are fed with the copper chelator, cuprizone, leading to oligodendrocyte depletion and subsequent replacement/remyelination through the differentiation of adult oligodendrocyte progenitors. The inefficient recruitment of HDACs to the promoters of oligodedrocyte inhibitors and neural stem cell markers in aged mice prevents the induction of myelin gene expression, providing an explanation for the observed decline in the efficiency of remyelination with age [131]. In contrast, another study highlighted that HDAC2 can also act as a repressor of oligodendrocyte differentiation of NSCs in a thyroid hormone (T3)-dependent manner, by reducing the expression of SOX10, a critical TF for oligodendrocyte differentiation and myelination [23].

Histone methylation and acetylation have been associated with several aspects of neural development. Interestingly, the same histone PTM can in some instances regulate the same process in opposite ways, suggesting that determining the specific temporal and special developmental context (the range of epigenators the cells are exposed to) will be essential to allow prediction of which genomic loci will be targeted by the epigenetic modifiers and whether it will induce or repress a certain developmental event. It also remains unclear whether histone modifiers will present unique expression profiles in a similar manner as TFs, or whether they will present broader and overlapping patterns of expression. In the latter case, the specificity of HDAC action might be determined by TFs (or other epigenetic initiators, such as noncoding RNAs) that would be able to recruit specific enzymatic activities to their target sites, influenced by the cell type-specific epigenetic landscape. Indeed, epigenetic footprinting has recently been used to identify TFs that are key for neural lineage progression from ES cells to neural progenitors [132].

2.2.4 Terminally Differentiated Neural Cell States: Histones in the Regulation of Functional States

A terminally differentiated cell in the adult CNS is specialized to execute certain functions. This functional specialization would be incompatible with the cell fate plasticity occurring earlier during differentiation. As such, the epigenetic landscape of terminally differentiated CNS cells prevents the expression of genes involved in multipotency and genes characteristic of other cell lineages, while preserving the expression of cell type–specific genes, allowing the maintenance of their identity. Nonetheless, differentiated brain cells are constantly exposed to extracellular cues that change their epigenetic signatures and dictate their functional states, rather than potential.

2.2.4.1 *Histone Modifiers in Synaptic Development*

One of the main functions of the cells in the CNS is to transmit chemical and electrical information through an intricate network of connections. This information is in many instances canalized through the synapse, the most well-studied specialized connection between adjacent neural cells and neighboring postsynaptic neurons. Epigenators, epigenetic initiators and chromatin modifying activities play key roles in the establishment of neuronal connectivity during brain development. The NuRD chromatin remodeling complex is required for synaptic differentiation by repressing a set of genes that inhibit presynaptic differentiation. In vivo NuRD depletion impairs the synapses between the parallel fibres of the granule neuronsand the Purkinje cells in the cerebellar cortex [133]. In contrast, another study has documented that inhibition of HDAC1 and HDAC2 in immature hippocampal neurons promotes excitatory synapse maturation and an increase in synapse number in vitro [134]. Deletion of HDAC1 and HDAC2 in mouse neurons during early synaptic development also results in enhanced excitatory synapse maturation. Interestingly, depletion of HDAC2 alone, but not HDAC1, decreases synaptic activity in mature hippocampal neurons, suggesting that the roles of HDAC1 and 2 in synapse maturation and function are dependent on the maturation state of the neuron [134]. Furthermore, neural activity was also shown to be required for the translocation of HDAC9 from the nucleus to the cytoplasm in cortical neurons. This translocation, or HDAC9 knockdown, induces the expression of gene produts that lead to an increase in the dendritic growth of cortical neurons [135].

2.2.4.2 *Histone Modifiers in Neural Plasticity, Learning, and Memory*

The connections between neurons and neural cells are changing over time in response to general environmental stimuli. Neural plasticity involved in learning and long-term memory formation requires modifications in the neural circuits that can occur at the level of the synapse (synaptic plasticity), and also elicits the formation of new synaptic connections by changing the excitatory properties of individual neurons. Most of the long-term structural changes of synaptic connectivity are dependent on the stable regulation of gene expression. Histone H3 acetylation in the CA1 region of the hippocampus (brain structure involved in memory) is increased during formation of long-term memory in vivo. Inhibition of HDACs enhanced long-term potentiation (long-lasting strengthening in the synaptic efficacy following a stimulation) of synapses in the CA1 area [136]. Additional studies report HDAC2 as a negative modulator of learning and memory, as overexpression of HDAC2 in neurons decreases synaptic plasticity and impairs memory formation. Furthermore, genomewide analysis demonstrated HDAC2 binding to the promoters of genes implicated in synaptic plasticity and memory formation [137]. As such, HDAC2 not only controls synaptic development [134], but also synaptic plasticity.

HATs have also been implicated in synaptic plasticity and memory. Transgenic mice expressing CBP lacking acetyltransferase activity thus exhibit memory deficits. This phenotype is, moreover, reverted by the HDAC inhibitor TSA [138]. Rubinstein–Taybi syndrome is an inheritable disorder caused by mutations in the gene encoding the CBP, and is characterized by mental retardation and skeletal abnormalities. Importantly, CBP heterozygous mice that serve as a model system for Rubinstein-Taybi syndrome display deficient long-term potentiation that is reverted by the inhibition of histone deacetylase activity [139]. Likewise, it was reported that CREB is required for the enhancement of memory and synaptic plasticity by HDAC inhibition, suggesting novel avenues of therapeutic interventions [140].

Little is known about histone methylation in the regulation of synaptic plasticity and memory formation. However, a recent study reports that neuronal ablation of the H3K4-specific methyltransferase KMT2A/MLL1 in the mouse postnatal brain results in major cognitive dysfunctions and severely impaired synaptic plasticity of pyramidal neurons, suggesting that these prefrontal cortex neurons are highly dependent on MLL1-mediated H3K4 methylation to regulate key genes for cognition and emotion [141].

2.2.4.3 Histone Modifiers in Brain Function Deregulation

A myriad of studies have demonstrated the role of stress in brain malfunction. Long-lasting and irreversible changes in biological processes caused by stress involve alterations in gene expression and a significant modification of the chromatin landscape [142]. In a stress-vulnerable mouse strain, for example, stress triggers epigenetic modifications at the glial cell-derived neurotrophic factor (*Gdnf*) promoter impairing its expression and consequently promoting depression-related behaviors [143]. Chromatin remodeling also occurs in diseases, such as depression, where the brain displays altered levels of histone acetylation and methylation (reviewed in [144]). Epigenetic down-regulation of RAS-related C3 botulinum toxin substrate 1 (RAC1), a small Rho GTPase that regulates the synaptic structure, was shown to modulate synaptic spines in a mouse model of depression and in postmortem brains from subjects with depression [145].

The role of physical exercise in chromatin remodeling in the brain has also been demonstrated. In a recent study physical exercise counteracted the effects of stress by epigenetically modulating the expression levels of brain-derived neurotrophic factor (BDNF). The increase in *Bdnf* gene expression was accounted for by an upregulation of H3 acetylation at its promoter [146]. Also chronic drug exposure has been shown to induce long-lasting effects in brain neural networks through chromatin remodeling (reviewed in [147]). In fact, *Bdnf* expression plays a key role in the behavioral plasticity induced by opiates. Mice chronically exposed to opiates thus display a downregulation of *Bdnf* expression in the ventral tegmental area, which is due to suppressed phospho-CREB bindingand increased H3K27me3 levels [148].

It is widely described that permanent disruption of normal transcriptional activity in several neurodegenerative disorders is mediated by alterations in the epigenome (reviewed in [149]). Tau-mediated neurodegeneration (that occurs in tauopathies, such as Alzheimer's disease) is promoted by global chromatin relaxation. The loss of heterochromatin was observed in tau transgenic mice and *Drosophila* but also in human brains of Alzheimer patients where H3K9me3 was depleted from diseased neurons [150]. Furthermore, in Ataxia telangiecta-sia (A-T), a neurodegenerative disorder caused by a mutation in the *ATM* gene, the repressive marker H3K27me3 accumulates in the brain due to the increased stability of the EZH2 histone methyltransferase. EZH2 is thus a novel target of ATM-mediated phosphorylation that reduces its stability. In transgenic mice lacking ATM the levels of the EZH2 methyltransferase and H3K27me3 are thus high, whereasknockdown of EZH2 prevents neurodegeneration, demonstrating the key role of EZH2 in the disease mechanism of A-T [151].

2.3 CONCLUSIONS

PTMs of canonical nucleosomal histones and histone variants regulate several key biological processes, and contribute to cell fate decisions during development and to functional outcomes in terminally differentiated cells. Surprisingly, a recent study indicated that H3.3 is the dominant histone H3 isoform in adult neurons, and its turnover is critical for appropriate transcriptional regulation and neuronal plasticity [12]. As described previously, the importance of histone variants during development is also emerging, adding another layer of complexity to the establishment of epigenetic landscapes. The interplay and contribution of long-lived canonical histones and more dynamic histone variants in development and in functional outcomes in terminally differentiated cells is far from being understood and will constitute a fascinating area of research in the coming years. Small molecules that alter histone PTMs or modulate binding of effector proteins to histone PTMs have already been used in epigenetic-based therapies in animal models of disease [49,152]. Thus, the development of new small molecules targeting histone PTMs/variants has the potential to lead to novel "epigenetic" therapeutic strategies for a wide range of diseases, including neurological disorders.

ABBREVIATIONS

BRDs	Bromodomain-containing proteins
ES cells	Embryonic stem cells
HATs	Histone acetyltransferases
HDACs	Histone deacetylases
HDMs	Histone demethylases
ICM	Inner cell mass
IPs cells	Induced pluripotent cells
LSD1	Lysine-specific demethylase 1
NSCs	Neural stem cells
PcG	Polycomb group
PKMTs	Histone lysine methyltransferases
PRMTs	Protein arginine methyltransferases
PTMs	Posttranslational modifications
SAM	S-adenosylmethionine
TFs	Transcription factors
TSS	Transcriptional start site

Acknowledgments

A.J.B acknowledges support from Cancer Research UK (CRUK) Grant no. C7/A10827. A.M.F. was supported by a European Committee for Treatment and Research of Multiple Sclerosis (ECTRIMS) postdoctoral fellowship. G.C.-B. was supported by Swedish Research Council, European Union (FP7/Marie Curie Integration Grant, EPIOPC), Swedish Brain Foundation, Swedish Society of Medicine, Åke Wiberg foundation, Clas Groschinsky foundation, Petrus och Augusta Hedlunds foundation, and Karolinska Institutet.

References

[1] Kossel A. Uber einen peptoartigen Bestandteil des Zellkern. Z Physiol Chem 1884;8:511–5.

[2] Kornberg RD. Chromatin structure: a repeating unit of histones and DNA. Science 1974;184(4139):868–71.

[3] Byvoet P, Shepherd GR, Hardin JM, Noland BJ. The distribution and turnover of labeled methyl groups in histone fractions of cultured mammalian cells. Arch Biochem Biophys 1972;148(2):558–67.

[4] Duerre JA, Lee CT. In vivo methylation and turnover of rat brain histones. J Neurochem 1974;23(3):541–7.

[5] Talbert PB, Henikoff S. Histone variants—ancient wrap artists of the epigenome. Nat Rev Mol Cell Biol 2010;11(4):264–75.

[6] Monteiro FL, et al. Expression and functionality of histone H2A variants in cancer. Oncotarget 2014;5(11):3428–43.

[7] Biterge B, Schneider R. Histone variants: key players of chromatin. Cell Tissue Res 2014;356(3):457–66.

[8] Schwartzentruber J, et al. Driver mutations in histone H3.3 and chromatin remodelling genes in paediatric glioblastoma. Nature 2012;482(7384):226–31.

[9] Wu G, et al. Somatic histone H3 alterations in pediatric diffuse intrinsic pontine gliomas and non-brainstem glioblastomas. Nat Genet 2012;44(3):251–3.

[10] Lewis PW, et al. Inhibition of PRC2 activity by a gain-of-function H3 mutation found in pediatric glioblastoma. Science 2013;340(6134):857–61.

[11] Changolkar LN, et al. Developmental changes in histone macroH2A1-mediated gene regulation. Mol Cell Biol 2007;27(7):2758–64.

[12] Maze I, et al. Critical role of histone turnover in neuronal transcription and plasticity. Neuron 2015;87(1):77–94.

[13] Luger K, Mader AW, Richmond RK, Sargent DF, Richmond TJ. Crystal structure of the nucleosome core particle at 2.8 A resolution. Nature 1997;389(6648):251–60.

[14] Xhemalce B, Dawson MA, Bannister, AJ. Histone modifications. In: Meyers RA, editor. Encyclopedia of molecular cell biology and molecular medicine: epigenetic regulation and epigenomics. Weinheim, Germany: Wiley-VCH Verlag GmbH & Co. KGaA; 2011.

[15] Kouzarides T. Chromatin modifications and their function. Cell 2007;128(4):693–705.

[16] Turner BM. Reading signals on the nucleosome with a new nomenclature for modified histones. Nat Struct Mol Biol 2005;12(2):110–2.

[17] Allfrey VG, Faulkner R, Mirsky AE. Acetylation and methylation of histones and their possible role in the regulation of RNA synthesis. Proc Natl Acad Sci USA 1964;51:786–94.

[18] Parthun MR. Hat1: the emerging cellular roles of a type B histone acetyltransferase. Oncogene 2007;26(37):5319–28.

[19] Bannister AJ, Kouzarides T. Regulation of chromatin by histone modifications. Cell Res 2011;21(3):381–95.

[20] de Ruijter AJ, van Gennip AH, Caron HN, Kemp S, van Kuilenburg AB. Histone deacetylases (HDACs): characterization of the classical HDAC family. Biochem J 2003;370(Pt. 3):737–49.

[21] Saunders LR, Verdin E. Sirtuins: critical regulators at the crossroads between cancer and aging. Oncogene 2007;26(37):5489–504.

[22] Dovey OM, Foster CT, Cowley SM. Histone deacetylase 1 (HDAC1), but not HDAC2, controls embryonic stem cell differentiation. Proc Natl Acad Sci USA 2010;107(18):8242–7.

[22a] Dovey OM, Foster CT, Cowley SM. Histone deacetylase 1 (HDAC1), but not HDAC2, controls embryonic stem cell differentiation. Proc Natl Acad Sci USA 2010;107(18):8242–7.

[23] Castelo-Branco G, et al. Neural stem cell differentiation is dictated by distinct actions of nuclear receptor corepressors and histone deacetylases. Stem Cell Reports 2014;3(3):502–15.

[24] Wang Z, et al. Combinatorial patterns of histone acetylations and methylations in the human genome. Nat Genet 2008;40(7):897–903.

[25] Tropberger P, et al. Regulation of transcription through acetylation of H3K122 on the lateral surface of the histone octamer. Cell 2013;152(4):859–72.

[26] Mujtaba S, Zeng L, Zhou MM. Structure and acetyl-lysine recognition of the bromodomain. Oncogene 2007;26(37):5521–7.

[27] Lee HS, Park JH, Kim SJ, Kwon SJ, Kwon J. A cooperative activation loop among SWI/SNF, gamma-H2AX and H3 acetylation for DNA double-strand break repair. EMBO J 2010;29(8):1434–45.

[28] Hilton IB, et al. Epigenome editing by a CRISPR-Cas9-based acetyltransferase activates genes from promoters and enhancers. Nat Biotechnol 2015;33(5):510–7.

[29] Heller EA, et al. Locus-specific epigenetic remodeling controls addiction- and depression-related behaviors. Nat Neurosci 2014;17(12):1720–7.

[30] Hazzalin CA, Mahadevan LC. Dynamic acetylation of all lysine 4-methylated histone H3 in the mouse nucleus: analysis at c-fos and c-jun. PLoS Biol 2005;3(12):e393.

[31] Boudadi E, et al. The histone deacetylase inhibitor sodium valproate causes limited transcriptional change in mouse embryonic stem cells but selectively overrides polycomb-mediated Hoxb silencing. Epigenetics Chromatin 2013;6(1):11.

[32] Rea S, et al. Regulation of chromatin structure by site-specific histone H3 methyltransferases. Nature 2000;406(6796):593–9.

[33] Xiao B, et al. Structure and catalytic mechanism of the human histone methyltransferase SET7/9. Nature 2003;421(6923):652–6.

[34] Wolf SS. The protein arginine methyltransferase family: an update about function, new perspectives and the physiological role in humans. Cell Mol Life Sci 2009;66(13):2109–21.

[35] Bannister AJ, Schneider R, Kouzarides T. Histone methylation: dynamic or static? Cell 2002;109(7):801–6.

[36] Shi Y, et al. Histone demethylation mediated by the nuclear amine oxidase homolog LSD1. Cell 2004;119(7):941–53.

[37] Klose RJ, Zhang Y. Regulation of histone methylation by demethylimination and demethylation. Nat Rev Mol Cell Biol 2007;8(4):307–18.

[38] Lee MC, Spradling AC. The progenitor state is maintained by lysine-specific demethylase 1-mediated epigenetic plasticity during *Drosophila* follicle cell development. Genes Dev 2014;28(24):2739–49.

[39] Mosammaparast N, Shi Y. Reversal of histone methylation: biochemical and molecular mechanisms of histone demethylases. Annu Rev Biochem 2010;79:155–79.

[40] Fueyo R, Garcia MA, Martinez-Balbas MA. Jumonji family histone demethylases in neural development. Cell Tissue Res 2015;359(1):87–98.

[41] Cuthbert GL, et al. Histone deimination antagonizes arginine methylation. Cell 2004;118(5):545–53.

[42] Christophorou MA, et al. Citrullination regulates pluripotency and histone H1 binding to chromatin. Nature 2014;507(7490):104–8.

[43] Bannister AJ, Kouzarides T. Reversing histone methylation. Nature 2005;436(7054):1103–6.

[44] Bernstein BE, et al. A bivalent chromatin structure marks key developmental genes in embryonic stem cells. Cell 2006;125(2):315–26.

[45] Heinz S, Romanoski CE, Benner C, Glass CK. The selection and function of cell type-specific enhancers. Nat Rev Mol Cell Biol 2015;16(3):144–54.

[46] Trojer P, et al. L3MBTL1, a histone-methylation-dependent chromatin lock. Cell 2007;129(5):915–28.

[47] Perna F, et al. The polycomb group protein L3MBTL1 represses a SMAD5-mediated hematopoietic transcriptional program in human pluripotent stem cells. Stem Cell Reports 2015;4(4):658–69.

[48] Mezentseva NV, et al. The histone methyltransferase inhibitor BIX01294 enhances the cardiac potential of bone marrow cells. Stem Cells Dev 2013;22(4):654–67.

[49] Dawson MA, et al. Inhibition of BET recruitment to chromatin as an effective treatment for MLL-fusion leukaemia. Nature 2011;478(7370):529–33.

[50] Anand P, et al. BET bromodomains mediate transcriptional pause release in heart failure. Cell 2013;154(3):569–82.

[51] Matzuk MM, et al. Small-molecule inhibition of BRDT for male contraception. Cell 2012;150(4):673–84.

[52] Gilbert SF. Epigenetic landscaping—Waddington use of cell fate bifurcation diagrams. Biol Philos 1991;6(2):135–54.

[53] Berger SL, Kouzarides T, Shiekhattar R, Shilatifard A. An operational definition of epigenetics. Genes Dev 2009;23(7):781–3.

[54] Marson A, et al. Connecting microRNA genes to the core transcriptional regulatory circuitry of embryonic stem cells. Cell 2008;134(3):521–33.

[55] Apostolou E, et al. Genome-wide chromatin interactions of the Nanog locus in pluripotency, differentiation, and reprogramming. Cell Stem Cell 2013;12(6):699–712.

[56] Gaspar-Maia A, Alajem A, Meshorer E, Ramalho-Santos M. Open chromatin in pluripotency and reprogramming. Nat Rev Mol Cell Biol 2011;12(1):36–47.

[57] Min IM, et al. Regulating RNA polymerase pausing and transcription elongation in embryonic stem cells. Genes Dev 2011;25(7):742–54.

[58] Mohn F, et al. Lineage-specific polycomb targets and de novo DNA methylation define restriction and potential of neuronal progenitors. Mol Cell 2008;30(6):755–66.

[59] Gonzales KA, et al. Deterministic restriction on pluripotent state dissolution by cell-cycle pathways. Cell 2015;162(3):564–79.

[60] Singh AM, et al. Cell-cycle control of bivalent epigenetic domains regulates the exit from pluripotency. Stem Cell Reports 2015;5(3):323–36.

[61] Meshorer E, et al. Hyperdynamic plasticity of chromatin proteins in pluripotent embryonic stem cells. Dev Cell 2006;10(1):105–16.

[62] Zhang YZ, et al. Histone H1 depletion impairs embryonic stem cell differentiation. Plos Genet 2012;8(5):e1002691.

[63] Yamanaka S, Blau HM. Nuclear reprogramming to a pluripotent state by three approaches. Nature 2010;465(7299):704–12.

[63a] Firas J, Liu X, Polo JM. Epigenetic memory in somatic cell nuclear transfer and induced pluripotency: evidence and implications. Differentiation 2014;88(1):29–32.

[64] Theunissen TW, Jaenisch R. Molecular control of induced pluripotency. Cell Stem Cell 2014;14(6):720–34.

[65] Papp B, Plath K. Epigenetics of reprogramming to induced pluripotency. Cell 2013;152(6):1324–43.

[66] Polo JM, et al. A molecular roadmap of reprogramming somatic cells into iPS cells. Cell 2012;151(7):1617–32.

[67] Hussein SM, et al. Genome-wide characterization of the routes to pluripotency. Nature 2014;516(7530):198–206.

[68] Whyte WA, et al. Master transcription factors and mediator establish super-enhancers at key cell identity genes. Cell 2013;153(2):307–19.

[69] Soufi A, Donahue G, Zaret KS. Facilitators and impediments of the pluripotency reprogramming factors' initial engagement with the genome. Cell 2012;151(5):994–1004.

[70] Soufi A, et al. Pioneer transcription factors target partial DNA motifs on nucleosomes to initiate reprogramming. Cell 2015;161(3):555–68.

[71] Bulut-Karslioglu A, et al. A transcription factor-based mechanism for mouse heterochromatin formation. Nat Struct Mol Biol 2012;19(10):1023–30.

[72] Briscoe J, Novitch BG. Regulatory pathways linking progenitor patterning, cell fates and neurogenesis in the ventral neural tube. Philos Trans R Soc Lond B Biol Sci 2008;363(1489):57–70.

[73] Narita Y, Rijli FM. Hox genes in neural patterning and circuit formation in the mouse hind-brain. Curr Top Dev Biol 2009;88:139–67.

[74] Wang J, et al. A protein interaction network for pluripotency of embryonic stem cells. Nature 2006;444(7117):364–8.

[75] Pardo M, et al. An expanded Oct4 interaction network: implications for stem cell biology, development, and disease. Cell Stem Cell 2010;6(4):382–95.

[76] van den Berg DL, et al. An Oct4-centered protein interaction network in embryonic stem cells. Cell Stem Cell 2010;6(4):369–81.

[77] Costa Y, et al. NANOG-dependent function of TET1 and TET2 in establishment of pluripotency. Nature 2013;495(7441):370–4.

[78] Ding J, et al. Tex10 coordinates epigenetic control of super-enhancer activity in pluripotency and reprogramming. Cell Stem Cell 2015;16(6):653–68.

[79] Tee WW, Reinberg D. Chromatin features and the epigenetic regulation of pluripotency states in ESCs. Development 2014;141(12):2376–90.

[80] Pasque V, Jullien J, Miyamoto K, Halley-Stott RP, Gurdon JB. Epigenetic factors influencing resistance to nuclear reprogramming. Trends Genet 2011;27(12):516–25.

[81] Skene PJ, Henikoff S. Histone variants in pluripotency and disease. Development 2013;140(12):2513–24.

[82] Turinetto V, Giachino C. Histone variants as emerging regulators of embryonic stem cell identity. Epigenetics 2015;10(7):563–73.

[83] Gaspar-Maia A, et al. MacroH2A histone variants act as a barrier upon reprogramming towards pluripotency. Nat Commun 2013;4:1565.

[84] Pasque V, et al. Histone variant macroH2A marks embryonic differentiation in vivo and acts as an epigenetic barrier to induced pluripotency. J Cell Sci 2012;125(Pt. 24):6094–104.

[85] Buschbeck M, et al. The histone variant macroH2A is an epigenetic regulator of key developmental genes. Nat Struct Mol Biol 2009;16(10):1074–9.

[86] Jullien J, et al. HIRA dependent H3.3 deposition is required for transcriptional reprogramming following nuclear transfer to *Xenopus* oocytes. Epigenetics Chromatin 2012;5(1):17.

[87] Elsasser SJ, Noh KM, Diaz N, Allis CD, Banaszynski LA. Histone H3.3 is required for endogenous retroviral element silencing in embryonic stem cells. Nature 2015;522(7555):240–4.

[88] Ng RK, Gurdon JB. Epigenetic memory of an active gene state depends on histone H3.3 incorporation into chromatin in the absence of transcription. Nat Cell Biol 2008;10(1):102–9.

[89] Hu G, et al. H2A.Z. facilitates access of active and repressive complexes to chromatin in embryonic stem cell self-renewal and differentiation. Cell Stem Cell 2013;12(2):180–92.

[90] Andang M, et al. Histone H2AX-dependent GABA(A) receptor regulation of stem cell proliferation. Nature 2008;451(7177):460–4.

[91] Yamanaka S. Elite and stochastic models for induced pluripotent stem cell generation. Nature 2009;460(7251):49–52.

[92] Buganim Y, et al. Single-cell expression analyses during cellular reprogramming reveal an early stochastic and a late hierarchic phase. Cell 2012;150(6):1209–22.

[93] Lee J, et al. Activation of innate immunity is required for efficient nuclear reprogramming. Cell 2012;151(3):547–58.

[94] Rais Y, et al. Deterministic direct reprogramming of somatic cells to pluripotency. Nature 2013;502(7469):65–70.

[95] dos Santos RL, et al. MBD3/NuRD facilitates induction of pluripotency in a context-dependent manner. Cell Stem Cell 2014;15(1):102–10.

[96] Silva J, et al. Promotion of reprogramming to ground state pluripotency by signal inhibition. PLoS Biol 2008;6(10):e253.

[97] Niwa H, Miyazaki J, Smith AG. Quantitative expression of Oct-3/4 defines differentiation, dedifferentiation or self-renewal of ES cells. Nat Genet 2000;24(4):372–6.

[98] Tonge PD, et al. Divergent reprogramming routes lead to alternative stem-cell states. Nature 2014;516(7530):192–7.

[99] Hansen KH, et al. A model for transmission of the H3K27me3 epigenetic mark. Nat Cell Biol 2008;10(11):1291–300.

[100] Wapinski OL, et al. Hierarchical mechanisms for direct reprogramming of fibroblasts to neurons. Cell 2013;155(3):621–35.

[101] Kandel ER. Principles of neural science. 5th ed. New York: McGraw-Hill; 2013. pp. l, 1709 p.

[102] Pereira JD, et al. Ezh2, the histone methyltransferase of PRC2, regulates the balance between self-renewal and differentiation in the cerebral cortex. Proc Natl Acad Sci USA 2010;107(36):15957–62.

[103] Sher F, et al. Differentiation of neural stem cells into oligodendrocytes: involvement of the polycomb group protein Ezh2. Stem Cells 2008;26(11):2875–83.

[104] Sher F, Boddeke E, Olah M, Copray S. Dynamic changes in Ezh2 gene occupancy underlie its involvement in neural stem cell self-renewal and differentiation towards oligodendrocytes. PloS One 2012;7(7):e40399.

[105] Roopra A, Qazi R, Schoenike B, Daley TJ, Morrison JF. Localized domains of G9a-mediated histone methylation are required for silencing of neuronal genes. Mol Cell 2004;14(6): 727–38.

[106] Liu J, et al. Chromatin landscape defined by repressive histone methylation during oligodendrocyte differentiation. J Neurosci 2015;35(1):352–65.

[107] Burgold T, et al. The histone H3 lysine 27-specific demethylase Jmjd3 is required for neural commitment. PloS One 2008;3(8):e3034.

[108] Jepsen K, et al. SMRT-mediated repression of an H3K27 demethylase in progression from neural stem cell to neuron. Nature 2007;450(7168):415–9.

[109] Lim DA, et al. Chromatin remodelling factor Mll1 is essential for neurogenesis from postnatal neural stem cells. Nature 2009;458(7237):529–33.

[110] Sun G, et al. Histone demethylase LSD1 regulates neural stem cell proliferation. Mol Cell Biol 2010;30(8):1997–2005.

[111] Laurent B, et al. A specific LSD1/KDM1A isoform regulates neuronal differentiation through H3K9 demethylation. Mol Cell 2015;57(6):957–70.

[112] Iwase S, et al. The X-linked mental retardation gene SMCX/JARID1C defines a family of histone H3 lysine 4 demethylases. Cell 2007;128(6):1077–88.

[113] Merson TD, et al. The transcriptional coactivator Querkopf controls adult neurogenesis. J Neurosci 2006;26(44):11359–70.

[114] Roelfsema JH, Peters DJ. Rubinstein–Taybi syndrome: clinical and molecular overview. Expert Rev Mol Med 2007;9(23):1–16.

[115] Wang J, et al. CBP histone acetyltransferase activity regulates embryonic neural differentiation in the normal and Rubinstein–Taybi syndrome brain. Dev Cell 2010;18(1): 114–25.

[116] Lee S, Lee B, Lee JW, Lee SK. Retinoid signaling and neurogenin2 function are coupled for the specification of spinal motor neurons through a chromatin modifier CBP. Neuron 2009;62(5):641–54.

[117] Nakashima K, et al. Synergistic signaling in fetal brain by STAT3-Smad1 complex bridged by p300. Science 1999;284(5413):479–82.

[118] Sun G, Yu RT, Evans RM, Shi Y. Orphan nuclear receptor TLX recruits histone deacetylases to repress transcription and regulate neural stem cell proliferation. Proc Natl Acad Sci USA 2007;104(39):15282–7.

[119] Hsieh J, Nakashima K, Kuwabara T, Mejia E, Gage FH. Histone deacetylase inhibition-mediated neuronal differentiation of multipotent adult neural progenitor cells. Proc Natl Acad Sci USA 2004;101(47):16659–64.

[120] Montgomery RL, Hsieh J, Barbosa AC, Richardson JA, Olson EN. Histone deacetylases 1 and 2 control the progression of neural precursors to neurons during brain development. Proc Natl Acad Sci USA 2009;106(19):7876–81.

[121] Guenther MG, Barak O, Lazar MA. The SMRT and N-CoR corepressors are activating cofactors for histone deacetylase 3. Mol Cell Biol 2001;21(18):6091–101.

[122] Ahringer J. NuRD and SIN3 histone deacetylase complexes in development. Trends Genet 2000;16(8):351–6.

[123] Chong JA, et al. REST: a mammalian silencer protein that restricts sodium channel gene expression to neurons. Cell 1995;80(6):949–57.

[124] Schoenherr CJ, Anderson DJ. The neuron-restrictive silencer factor (NRSF): a coordinate repressor of multiple neuron-specific genes. Science 1995;267(5202):1360–3.

[125] Ballas N, Grunseich C, Lu DD, Speh JC, Mandel G. REST and its corepressors mediate plasticity of neuronal gene chromatin throughout neurogenesis. Cell 2005;121(4):645–57.

[126] Hermanson O, Jepsen K, Rosenfeld MG. N-CoR controls differentiation of neural stem cells into astrocytes. Nature 2002;419(6910):934–9.

[127] Cunliffe VT, Casaccia-Bonnefil P. Histone deacetylase 1 is essential for oligodendrocyte specification in the zebrafish CNS. Mech Dev 2006;123(1):24–30.

[128] Ye F, et al. HDAC1 and HDAC2 regulate oligodendrocyte differentiation by disrupting the beta-catenin–TCF interaction. Nat Neurosci 2009;12(7):829–38.

[129] Marin-Husstege M, Muggironi M, Liu A, Casaccia-Bonnefil P. Histone deacetylase activity is necessary for oligodendrocyte lineage progression. J Neurosci 2002;22(23):10333–45.

[130] Shen S, Li J, Casaccia-Bonnefil P. Histone modifications affect timing of oligodendrocyte progenitor differentiation in the developing rat brain. J Cell Biol 2005;169(4):577–89.

[131] Shen S, et al. Age-dependent epigenetic control of differentiation inhibitors is critical for remyelination efficiency. Nature Neurosci 2008;11(9):1024–34.

[132] Ziller MJ, et al. Dissecting neural differentiation regulatory networks through epigenetic footprinting. Nature 2015;518(7539):355–9.

[133] Yamada T, et al. Promoter decommissioning by the NuRD chromatin remodeling complex triggers synaptic connectivity in the mammalian brain. Neuron 2014;83(1):122–34.

[134] Akhtar MW, et al. Histone deacetylases 1 and 2 form a developmental switch that controls excitatory synapse maturation and function. JJ Neurosci 2009;29(25):8288–97.

[135] Sugo N, et al. Nucleocytoplasmic translocation of HDAC9 regulates gene expression and dendritic growth in developing cortical neurons. Eur J Neurosci 2010;31(9):1521–32.

[136] Levenson JM, et al. Regulation of histone acetylation during memory formation in the hippocampus. J Biol Chem 2004;279(39):40545–59.

[137] Guan JS, et al. HDAC2 negatively regulates memory formation and synaptic plasticity. Nature 2009;459(7243):55–60.

[138] Korzus E, Rosenfeld MG, Mayford M. CBP histone acetyltransferase activity is a critical component of memory consolidation. Neuron 2004;42(6):961–72.

[139] Alarcon JM, et al. Chromatin acetylation, memory, and LTP are impaired in CBP+/− mice: a model for the cognitive deficit in Rubinstein–Taybi syndrome and its amelioration. Neuron 2004;42(6):947–59.

[140] Vecsey CG, et al. Histone deacetylase inhibitors enhance memory and synaptic plasticity via CREB: CBP-dependent transcriptional activation. J Neurosci 2007;27(23):6128–40.

[141] Jakovcevski M, et al. Neuronal Kmt2a/Mll1 histone methyltransferase is essential for prefrontal synaptic plasticity and working memory. J Neurosci 2015;35(13):5097–108.

[142] Stankiewicz AM, Swiergiel AH, Lisowski P. Epigenetics of stress adaptations in the brain. Brain Res Bull 2013;98:76–92.

[143] Uchida S, et al. Epigenetic Status Of Gdnf In The Ventral Striatum Determines Susceptibility And Adaptation To Daily Stressful Events. Neuron 2011;69(2):359–72.

[144] Sun HS, Kennedy PJ, Nestler EJ. Epigenetics of the depressed brain: role of histone acetylation and methylation. Neuropsychopharmacol 2013;38(1):124–37.

[145] Golden SA, et al. Epigenetic regulation of RAC1 induces synaptic remodeling in stress disorders and depression. Nat Med 2013;19(3):337–44.

[146] Ieraci A, Mallei A, Musazzi L, Popoli M. Physical exercise and acute restraint stress differentially modulate hippocampal brain-derived neurotrophic factor transcripts and epigenetic mechanisms in mice. Hippocampus 2015;25(11):1380–92.

[147] Schmidt HD, McGinty JF, West AE, Sadri-Vakili G. Epigenetics and psychostimulant addiction. Cold Spring Harb Perspect Med 2013;3(3). a012047.

[148] Koo JW, et al. Epigenetic basis of opiate suppression of Bdnf gene expression in the ventral tegmental area. Nature Neurosci 2015;18(3):415–22.

[149] Landgrave-Gomez J, Mercado-Gomez O, Guevara-Guzman R. Epigenetic mechanisms in neurological and neurodegenerative diseases. Front Cell Neurosci 2015;9.

[150] Frost B, Hemberg M, Lewis J, Feany MB. Tau promotes neurodegeneration through global chromatin relaxation. Nature Neurosci 2014;17(3):357–66.

[151] Li JL, et al. EZH2-mediated H3K27 trimethylation mediates neurodegeneration in ataxia-telangiectasia. Nature Neurosci 2013;16(12):1745–53.

[152] Castelo-Branco G, et al. Acute treatment with valproic acid and l-thyroxine ameliorates clinical signs of experimental autoimmune encephalomyelitis and prevents brain pathology in DA rats. Neurobiol Dis 2014;71:220–33.

CHAPTER 3

Dynamics and Function of DNA Methylation During Development

A. Lennartsson

Department of Biosciences and Nutrition, Karolinska Institutet, Huddinge, Sweden

3.1 ESTABLISHMENT OF THE DNA METHYLATION LANDSCAPE

Holliday, Pugh, and Riggs suggested in 1975 that the cytosine moiety of CG dinucleotides could be methylated and function as an epigenetic mark through cell division [1,2]. A CG dinucleotide, that is, a cytosine followed by a guanine, is commonly referred to as CpG, which means that the cytosine and guanine are only separated by a linking phosphate group "—C—phosphate—G—." Regions with high CpG density are called CpG islands (CGIs). Although many definitions have been suggested, the most common definition for a CGI is a minimum of 500 bp region with more than 50% C + G content and a CpG observed/expected ratio >0.6. Methylated cytosines (mCs) spontaneously deaminate to thymines that cause substitution mutations, resulting in the underrepresentation of CpGs in the genome (21% of that expected in the human genome) [3]. Similarly, CGIs may be preserved in the genome since they are generally unmethylated in early development.

DNA methylation is established and maintained by three DNA methyltransferases (DNMTs); DNMT1, DNMT3A, and DNMT3B [4,5] (Fig. 3.1A). DNMT1 is mainly active in maintaining DNA methylation during S phase by methylating the newly synthesized DNA strand at hemimethylated sequences arising after the semiconservative DNA replication. DNMT1 is recruited to DNA by UHRF that recognizes and binds to hemimethylated DNA via its SET- and RING-associated (SRA) domain [6,7]. In contrast to DNMT1, DNMT3A, and DNMT3B are de novo methyltransferases, acting on previously unmethylated DNA [5](Fig. 3.1B). DNMT3A and DNMT3B form complexes with the catalytically inactive DNMT3L, which stabilizes their interaction with DNA and promotes methylation. The functional division between the DNMTs is not absolute. The different DNMTs also have some overlapping function. It was recently shown that both DNMT3A and DNMT3B, in addition to DNMT1, are also important to maintain DNA methylation patterns during cell division [8].

CONTENTS

(Continued)

Chromatin Regulation and Dynamics. http://dx.doi.org/10.1016/B978-0-12-803395-1.00003-4

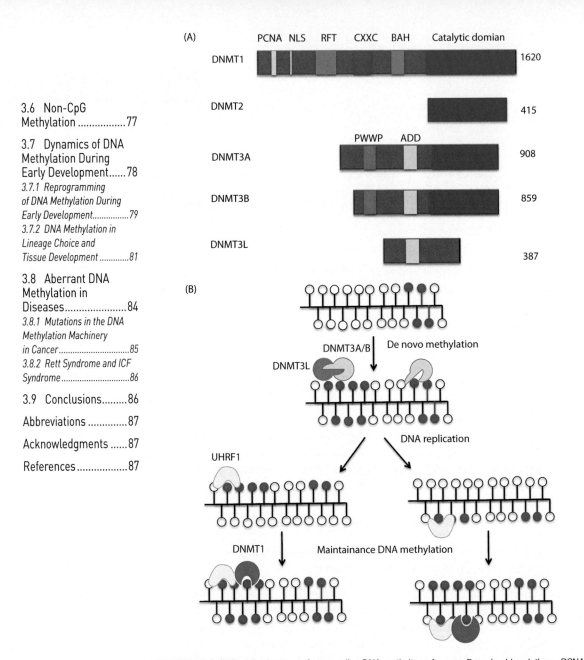

FIGURE 3.1 (A) Protein structure of mammalian DNA methyltransferases. Domain abbreviations: *PCNA*, PCNA-interacting domain; *NLS*, nuclear localization signal; *RFT*, replication foci-targeting domain; *CXXC*, a cysteine-rich domain implicated in binding DNA sequences containing CpG dinucleotides; *BAH*, bromo-adjacent homology domain implicated in protein–protein interactions; *PWWP*, a domain containing a highly conserved "proline–tryptophan–tryptophan–proline" motif involved in heterochromatin association; *ADD*, an ATRX-related cysteine-rich region containing a C2–C2 zinc finger and an atypical PHD domain implicated in protein–protein interaction. (B) DNA methylation pathways. Unmethylated DNA is de novo methylated by DNMT3A and DNMT3B either in complex with DNMT3L or alone. After DNA replication, hemimethylated DNA is recognized by UHRF1, which recruits DNMT1 that methylates the unmethylated strand. Thereby, the parental DNA methylation pattern is maintained after cell division in the daughter cells.

For example, when murine *Dnmt3a/b* are deleted, the levels of hemimethylated DNA in repeat sequences increases, suggesting that they are involved in methylating hemimethylated DNA [9].

DNA methylation patterns are established via a complex interactions between the DNMTs and DNA sequences, which are determined by CpG density, chromatin structure and histone modifications, other epigenetic regulators, and transcription factors that can either protect the DNA from methylation or recruit DNMTs and target DNA methylation to specific sites. Those interactions together contribute to that some genomic regions display more variable DNA methylation levels than other regions, both between different tissues and between healthy and cancer tissues [10].

3.2   DNA METHYLATION IN CpG RICH VERSUS CpG POOR REGIONS

Approximately 60% of the genes in the human genome are associated with a promoter proximal CGI. Approximately 40% of those genes are expressed in a cell type–specific pattern [11,12]. In general, however, promoter proximal CGIs are located in ubiquitously expressed genes, which are in an open, permissive chromatin state (Fig. 3.2). It has been shown that genes with CGI-containing promoters do not require the SWI/SNF complex for chromatin remodeling to be activated, while non-CGI promoters do [13]. The SWI/SNF

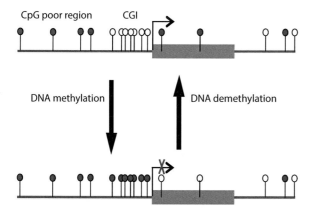

FIGURE 3.2 CpG poor and CpG rich regions.
CpG rich regions, so called CpG islands (CGIs) are frequently located in promoter regions in an unmethylated state, but recurrently becomes methylated, for example, in cancer. CpG poor regions are more commonly methylated than CGIs. The effect of DNA methylation on transcription depends, however, on the CpG density. Unmethylated CGIs are associated with transcriptionally permissive chromatin, while CGI methylation correlate with repressed transcription. The correlation with transcriptional repression is less in CpG poor regions than in CpG intermediate or rich regions.

chromatin remodeling complexes utilize ATP to disassemble or slide nucleosomes to create a more accessible chromatin state, which is not required in the CGI promoters, as they already display a permissive chromatin state. In vitro experiments have shown that nucleosome assembly is less efficient in CGIs than in non-CGI sequences [13]. The lower nucleosome assembly efficiency may thus contribute to the more permissive chromatin state at CGIs. The permissive chromatin state results in a broad (50–100 bp) region of transcriptional initiation with multiple transcription start sites [14]. In addition, CGI promoters often lack TATA-box architecture [14].

CGIs, especially the promoter-proximal CGIs, are mostly unmethylated in somatic cells, whereas CpGs outside of CGIs are generally methylated. Silenced CGI promoters are often not repressed by DNA methylation, but instead by trimethylation of histone 3 lysine 27 (H3K27me3), which is established by the Poly-Comb repressive complex (Chapter 7) [15–17]. Unmethylated CGI promoters are marked by H3K4me3, independently of whether the promoter is active or not [18]. The H3K4me3-labeling of CGI-promoters is a way to protect the CGI from DNA methylation. H3K4me3 thus inhibits the binding of DNMT3A/B/L, and thereby prevents DNA methylation [19]. Consequently, H3K4me3 modifications at CGIs participate in maintaining the unmethylated state. Methylation of CGIs in healthy cells is restricted to promoters that should be stably silenced for long term. Long-term repressed genes are, for example, the silent allele of imprinted genes, genes located on the inactivated X-chromosome or genes that are exclusively expressed in germ cells and silenced in somatic cells.

Approximately 50% of the CGIs are located in promoters, and the remaining CGIs are either intragenic, that is, located in gene bodies, or intergenic, and are more frequently methylated. For example, 34% of all intragenic CGIs are methylated in the human brain [20]. Although the function of these CGIs is not completely clear, they may represent regulatory regions, such as enhancers or insulators. A subset of them has been suggested to constitute previously undetected promoters whose transcriptional activity and DNA methylation levels are dynamically regulated during development [20,21]. Those transcription start sites may either belong to mRNAs or long noncoding RNAs (lnRNA).

In general, the function of DNA methylation in gene bodies is less clear, than DNA methylation in promoter regions. It has even been suggested to positively correlate with active transcription, in contrast to promoter methylation that represses transcription. Furthermore, DNA methylation in gene bodies has been hypothesized to stabilize the chromatin and inhibit nucleosome displacement resulting from the elongating RNA polymerase. This model suggests that DNA methylation has a function similar to that of H3K36 methylation in gene bodies. H3K36me3 thus recruits histone deacetylases (HDACs) that deacetylate histones and consequently cause a more compact chromatin structure. The compact chromatin structure then inhibits transcription from cryptic initiation sites within the gene body. In *Saccharomyces cerevisiae*, for example, disruption of the H3K36me3 pathway by deletion of the lysine H3K36 causes enhanced chromatin accessibility after transcription and increased cryptic transcription [22]. DNA methylation in the gene body may thus repress cryptic transcription in a similar manner as H3K36me3 [23].

DNA methylation in gene bodies has also been suggested to have a role in splicing (Chapter 8), based on the observation that exons have higher methylation levels than introns [24–26]. Nucleosome-bound DNA has higher methylation levels than flanking DNA, which might be explained by the observations that DNMTs have a preference for nucleosome-bound DNA [24]. Besides higher levels of DNA methylation, exons are thus enriched in nucleosomes, in particular at exon–intron and intron–exon boundaries [27]. In addition, exons also have a different histone modification pattern than introns, and are enriched in H3K36me3 modifications to potentially regulate chromatin structure. Furthermore, binding of MeCP2/MBDs to the methylated exons causes recruitment of HDACs that maintain local hypoacetylation [26]. Although further evidence and mechanistic insight is still needed, such chromatin-mediated cross talk with RNA polymerase II at exon–intron borders may regulate splicing.

3.3 DNA DEMETHYLATION AND HYDROXY- METHYLATION

Initially, DNA demethylation was thought to occur exclusively as a passive process, where lack of or inefficient maintenance during replication leads to dilution of methyl groups during several cell cycles. However, recently two different main mechanisms for active DNA demethylation have been described: (1) deamination via the activation-induced cytidine deaminase (AID), or the apolipoprotein B mRNA-editing enzyme (APOBEC) and (2) oxidation of mCs via ten eleven translocation (TET) dioxygenases.

APOBEC was first discovered to deaminate cytosine (C) to uracil (U) on mRNA molecules. Later several APOBEC proteins have been shown to be able to deaminate cytosines on DNA as well. AID is another member of the deaminase enzyme family that targets cytosines on DNA. The resulting U:G/T:G base mismatch is then cleaved by DNA glycosylases, such as thymine–DNA glycosylase (TDG), between the mC and the deoxyribose ring. The TDG cleavage creates an abasic site (AP) and initiates a base excision repair (BER) reaction that finally converts mC to C [28–30].

The second DNA demethylation pathway is mediated by the TET enzyme family that has three members (TET1-3). All TET family members can convert mC to hydroxymethyl cytosine (hmC) via an oxidation reaction [31,32]. Whether

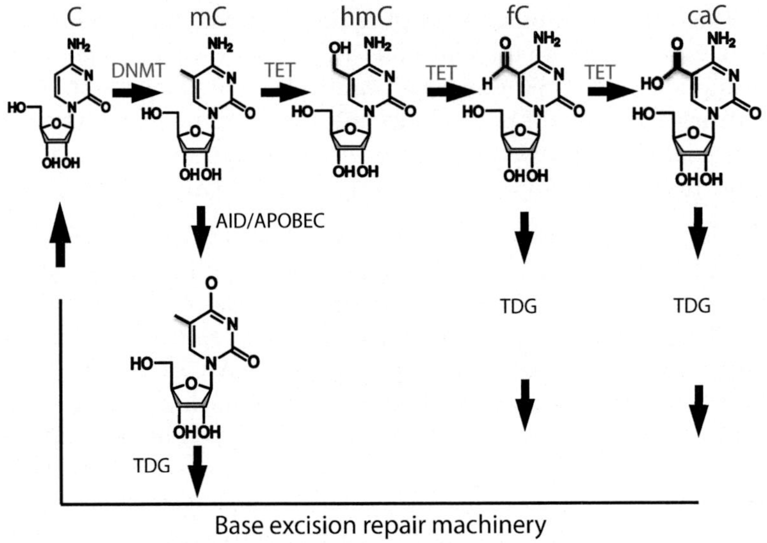

FIGURE 3.3 Mechanisms for active DNA demethylation.
DNA is demethylated by stepwise oxidation (TET) or deamination (mediated by AID/APOBEC) or other pathways.

hmC has a function of its own or if it is a mere step toward DNA demethylation is still not clear. An increasing amount of evidence suggests, however, that hmC indeed has a function as a separate DNA modification. For example, some of the proteins that bind mC have been shown to also bind hmC. It has been found to be present in higher levels in some tissues than in others, which suggests an independent function- and tissue-specific regulation [31]. For example, hmC levels are especially high in the brain.

The TET-mediated oxidation proceeds in a stepwise fashion from hmC to generate formylcytosine (fC) and carboxylcytosine (caC) [33,34] (Fig. 3.3). Finally, fC or caC is recognized and excised by TDG. This leads to DNA demethylation, since an unmethylated cytosine is inserted into the DNA in exchange for the fC or caC via the BER machinery [33]. Even though fC and caC are transition states, fC is sufficiently abundant in embryonic stem cell (ESC) to be detected by mass spectrometry, albeit at very low levels [35]. Interestingly, the level of fC decreases with differentiation and cannot be detected in neurons, although the amount of hmC in neurons is relatively high compared to that in ESC [35]. However, it should be noted that the levels of hmC and fC are very low compared to mC, that is, less than 1 and 0.1%, respectively [35].

The TET enzymes appear to have both overlapping and unique functions. Triple deletion studies using CRISPR/Cas9 technology have shown that deletion of *Tet1-3* leads to increased methylation levels, in agreement with their role in DNA demethylation [36]. Loss of *Tet1* or *Tet2* leads to genome-wide loss of hmC. In mouse ESC, *Tet2* deletion causes larger reduction of hmC than *Tet1* deletion [37]. Although the entire TET family can oxidize mC to hmC, they may have some nonoverlapping specificity when it comes to genomic targets. TET2 has been suggested to have a higher DNA demethylation effect on enhancer regions than TET1, at least in mouse ESC [37]. As DNA hypermethylation at enhancer regions reduces the enhancer activity [37,38], TET2 may play an important role in regulating cell type–specific gene expression change during differentiation. Indeed, loss of *Tet2* leads to the development of acute myeloid leukemia (AML) in a mouse model, accompanied by enhancer hypermethylation, while hypermethylation was not observed at CGIs and promoters in this model [39]. Aberrant enhancer methylation resulted in repression of multiple tumor suppressor genes, suggesting that TET2 is essential to protect the cells from leukemic transformation by keeping enhancers hypomethylated [39]. Importantly, a subset of enhancers become demethylated during human myelopoieis and may be important for normal differentiation [38]. The observed *TET2* mutations in AML patients may therefore disturb the myeloid differentiation program by causing aberrant enhancer activity.

In addition, the deamination (AID/APOBEC) and oxidation (TET) DNA demethylation pathways have been suggested to interact with each other at different stages. TET-produced hmC can, for example, be processed by AID that consequently promotes demethylation of hmC [40] (Fig. 3.3).

3.4 DNA METHYLATION AND TRANSCRIPTIONAL SILENCING

Although the majority of the human genome is noncoding, it has transcriptional potential that needs to be repressed. Pericentromeric repeats and retrotransposons, for example, belong to the genomic regions that are stably silenced by DNA methylation. Gene transcription can also be regulated by DNA methylation. By reinforcing transcriptional programs during development, DNA methylation thus constitutes a mechanism for the cell to remember its identity during cell division. Whether DNA methylation represses transcription via active transcriptional silencing or via interference with transcriptional activation is still debated, but it probably depends on the precise genomic location and context of the methylated CpGs. As the majority of CGIs that gain methylation during differentiation are already transcriptionally silent in ESCs [16], silencing likely precedes DNA methylation that stabilizes the silenced state. Importantly, the effect of DNA methylation on gene transcription has been suggested to depend on CpG density. Hence, the cytosine methylation of low CpG-content promoters correlates less with decreased mRNA expression levels than than that of intermediate or high CpG density promoters [18].

DNA methylation has been suggested to represses transcription by two major mechanisms: by either effecting transcription factor binding or by recruiting proteins with methyl-binding domains (MBDs) that in turn mediate repression.

3.4.1   Interaction With Transcription Factors

Mammalian transcription factor–binding sites are enriched in CpG sites [41]. It has thus been proposed that mC can act as a "fifth base" and thereby broaden the binding site specificity of a transcription factor [42]. Although some transcription factors are sensitive for methylation in their binding sites [43], the effect of DNA methylation on transcription factor–binding varies between different transcription factors. For example, it has been demonstrated that DNA methylation can decrease the affinity of certain transcription factors, such as YY1 and E2F, to their binding sites [44]. Also binding of CTCF, which is important for 3D chromosomal interactions that mediate enhancer and insulator function (Chapter 17), is inhibited by DNA methylation at certain genomic loci [45,46]. However, the correlation between transcription factor–binding and DNA methylation is not always negative. DNA methylation has also been shown to increase the affinity of certain factors, such as that of C/EBP, to DNA. Finally, KLF4 has been proposed to bind to both methylated and unmethylated motifs in different sequence context in ESC [42]. These observations demonstrate that DNA methylation adds complexity to transcription factor networks.

Binding of transcription factors may also actively contribute to downstream changes in DNA methylation. The transcription factor PU.1, which is important for monocyte and osteoclast differentiation, has been suggested to contribute to both hyper- and hypomethylation. PU.1 binding leads to the recruitment of either DNMT3B or TET2, which causes hyper- or hypomethylation, respectively [47]. However, what dictates whether PU.1 interacts with DNMT3B or TET2 remains to be shown. One could speculate that either posttranslational modification of PU.1 or combinatorial binding of neighboring transcription factors may affect the interaction partners of PU.1. Binding of a transcription factor to a specific site can also protect the site from DNA methylation, as demonstrated for CTCF [48]. In addition, it has been shown that several transcription factors can bind to methylated CG poor regions and cause active DNA demethylation, either via TET-interaction, as suggested for PU.1, or via interaction with members of the DNA repair pathway [49,50]. Finally, deletion of the transcription factor REST in mouse ESCs leads to increased DNA methylation at its binding sites, whereas reintroduction of it into REST negative ESCs causes demethylation. Whether the protection from DNA methylation is an active or simply a passive process by steric hindrance remains to be shown.

The tight interaction between transcription factors and DNA methylation suggests that DNA methylation protects cells from aberrant transcription factor activity, and at the same time transcription factor binding can regulate DNA methylation. Through this cross talk, DNA methylation and transcription factors together create a more stable and controlled transcriptional regulation than they would do separately.

3.4.2 Interaction With Methyl-CpG Binding Domain Containing Proteins

Methylated CpGs are recognized and bound by proteins that contain a MBD, which in turn act to repress transcription via chromatin compaction [51–53]. There are several known proteins that contain a MBD domain, such as MECP2, MBD1-4. These proteins recruit transcriptional silencing machinery, such as HDACs that repress transcription by compacting the chromatin. Methylated promoters have a nucleosome located at the transcription start site that blocks the initiation of transcription [54]. Apart from HDACs, MBD proteins that bind to methylated DNA can also recruit H3K9 methyltransferasesand stabilize nucleosome positioning [55]. By examining the mechanism of repression of the *NANOG* promoter and *OCT4* enhancer during differentiation, it was shown that higher nucleosome density appears prior to DNA methylation [56]. This would suggest that transcriptional repression occurs before DNA methylation and that DNA methylation stabilizes an already repressed state.

3.5 INTERACTIONS BETWEEN DNA METHYLATION AND HISTONE MODIFICATIONS

DNA is wrapped around nucleosomes inside the nucleus. The different enzymes of the DNA methylation machinery interact with chromatin components, which interactions constitute an additional level of gene regulation. For example, DNMT3A/B are compartmentalized by interactions with different nuclear components, such as nucleosomes, and are not available as free soluble enzymes [57]. Several of those interactions and cross-talk between DNA methylation and histones will be discussed below, see Fig. 3.4.

3.5.1 Protein Binding to Unmethylated CpGs

Unmethylated CpGs are recognized by proteins that contain a CXXC domain [56]. The CXXC domain is found in several different proteins, such as the CXXC finger protein 1 (CFP1) [58], histone methylases MLL1 and MLL2 and the histone demethylases KDM2A and KDM2B [59], see Table 3.1. The first CXXC protein that was discovered was CFP1, a protein essential for early mouse development [60]. Most CGI promoters are enriched in CFP1 binding independent of the promoter activity [58]. CFP1 is part of the H3K4 methyltransferase complex SETD1, together with SETD1A/B, ASH2L, RbBP5, WDR and DPY-30 [61]. CFP1 has been proposed to target H3K4methylation to CpG promoters

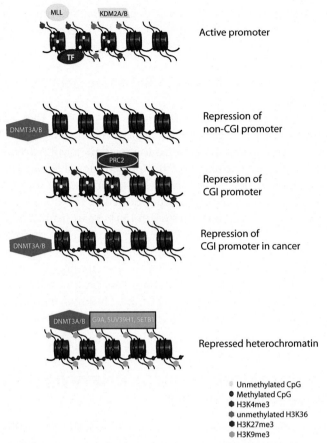

FIGURE 3.4 Interaction between DNA methylation and histone modifications.
Actively transcribed genes: MLL methylates H3K4 that inhibits DNMT3A/B binding, and KDM2A demethylates H3K36 to maintain a transcriptionally permissive chromatin state. Repression of gene transcription at non-CGI promoters: DNMT3A/B bind to unmethylated H3K4, which leads to methylated CpGs and repression of transcription. Repression of transcription at CGI: PolyComb regulated CGI promoters are silenced by PRC2 and H3K27 methylation during normal development. However, these CGI promoters are frequently DNA methylated in cancer, which is paralleled by the loss of PRC2 binding and lack of H3K27me3 modifications. Repressed heterochromatin: constitutive heterochromatin is characterized by H3K9 methylation. The H3K9 methyltransferases recruit DNMT3A/B that causes DNA methylation and further represses gene expression.

via recruitment of H3K4 methyltransferases. Indeed, deletion of *Cfp1* in mouse ESC causes loss of H3K4me3 at transcription start sites and mistargeting of H3K4me3 to intergenic regions [62].

Since the discovery of CFP1, several CXXC domain-containing proteins have been identified, see Table 3.1. The MLL H3K4 methyltransferase family consists of five members (MLL1–5). The CXXC domains in MLL1 and MLL2 have already

Table 3.1 CXXC Domain Containing Proteins [59]

Gene Name	Aliases
KDM2A	CXXC8, FBL11, FBL7, JHDM1A, LILNA
KDM2B	CXXC2, FBXL10, JHDM1B, PCCX2
FBXL19	CXXC11, FBL19, JHDM1C
CXXC1	CFP1, CGBP, PCCX1, PHF18, SPP1, ZCGPC1
DNMT1	ADCADN, AIM, CXXC9
KMT2A	MLL, MLL1, ALL-1, CXXC7, HRX, TRX1, WDSTS
KMT2D	MLL2, MLL4, AAD10, ALR, CAGL114, KABUK1, KMS, TNRC21
MBD1	CXXC3, PCM1, RFT
TET1	CXXC6, LCX
TET3	
CXXC4	IDAX
CXXC5	CF5, HSPS195, RINF, WID

been suggested to have a role in MLL1/2 targeting to unmethylated sites [63]. However, in contrast to KDM2A and CFP1, MLL1 is only targeted to transcriptionally active CGI promoters. The more restricted binding pattern of MLL1 might be regulated by other chromatin-binding domains in MLL, such as N-terminal AT hooks or the PHD domain. In the case of several CXXC proteins, including the previously mentioned MBD1, TET1, and TET3, the interaction with nonmethylated DNA is however, less clear. None less, the fact that TET2 does not contain a CXXC-domain while TET1 does may provide an explanation why TET2 preferably interact with enhancers, which are often CpG poor.

The H3K36 demethylases KDM2A and KDM2B also contain CXXC domains. Binding of KDM2A to nonmethylated CpGs leads to H3K36 demethylation. As previously discussed in this chapter, H3K36 methylation is enriched in gene bodies of expressed genes where it inhibits transcriptional initiation. However, KDM2A also binds to CGIs in promoters, regardless of their transcriptional state. KDM2A binding at transcription start sites removes H3K36 methylation and thereby creates a permissive chromatin structure for transcriptional initiation. KDM2A binding through its CXXC domain is blocked by DNA methylation, ensuring that the chromatin of promoters with methylated CGIs remains compacted [64]. KDM2B has, at least partly, a nonoverlapping function to that of KDM2A. KDM2B binds to a subset of developmentally-regulated CGI promoters that are silenced by the PolyComb complex. These KDM2B target genes are devoid of KDM2A binding [65]. Binding of KDM2B to unmethylated CpGs has been proposed to contribute to the recruitment of the PolyComb repressive complex 1 (PRC1), which ubiquitinates H2A119, leading to transcriptional repression [65].

Different proteins have different functional use of their CXXC domain. For example, DNMT1 contains several different protein domains including a CXXC domain. It may seem contradictory that DNMT1 has a CXXC domain, since most CGI remain unmethylated. However, the structure of DNMT1–DNA complexes revealed an auto inhibitory function of the CXXC domain. When DNMT1 binds to nonmethylated CpGs, the CXXC domain blocks the interaction between the catalytic site of DNMT1 and the CpG. Consequently, the CpG remains unmethylated, despite the binding of DNMT1 [66]. The function of the CXXC domain has thus been suggested to provide an auto inhibitory mechanism that ensures that only hemimethylated CGs become methylated and not other unmethylated CGs [66].

3.5.2   Interactions Between DNMTs and Histones via the ATRX-DNMT3-DNMT3L Domain

Several mechanisms have been suggested to explain how the CGIs remain unmethylated. Importantly, all three DNMT3s contain the ATRX-DNMT3-DNMT3L (ADD) domain that binds to H3, but only when H3K4 is unmethylated [19,67]. Unmethylated H3K4 is associated with transcriptionally inactive genes. H3K4me3 at nucleosomes that are flanking transcription start sites consequently inhibits the binding of the DNMT3L complex and thereby blocks de novo methylation. Lack of H3K4 methylation, on the other hand, acts as a signal to recruit DNAMT3A/B and DNMT3L. DNMT3L forms complexes with DNMT3A/B, which dock and stabilize the interaction between DNMT3A/B with their target sites [19]. Indeed mutation of DNMT3L in ESC leads to abnormal DNA methylation pattern and hypomethylation at specific promoters.

Interestingly, genes that harbor CGIs in their promoters have H3K4me3 at the TSS flanking nucleosomes regardless of the transcriptional activity. This could, for example, be a measure to maintain the CGIs unmethylated during development. The importance of the cross talk between DNA methylation and H3K4 methylation for the establishment of correct DNA methylation patterns during development is strengthened by the observation that the H3K4 demethylases LSD1 and LSD2 are essential to obtain correct cell type–specific DNA methylation patterns [68,69]. Finally, actively transcribed genes also often contain the histone variant H2A.Z at the TSS, which is strongly anticorrelated with DNA methylation [70,71].

In contrast to H3K4 methylation, H3K9 methylation, and H3K9 methyltransferases recruit DNMTs, and H3K9me3 and DNA methylation are highly correlated genome wide [72]. H3K9me3 is associated with transcriptional silencing and heterochromatin formation, for example, at retrotransposons. Silencing and inactivation of retrotransposons maintains genome integrity, which is one of the main functions of H3K9me2/3 and DNA methylation. Mammals have several H3K9 methyltransferases that generate specific H3K9 methylation levels. PRDM3 and PRDM16 are H3K9me1-specific methylases, [73]. G9A and GLP form stoichiometric heteromeric complexes that methylate H3K9me1 to H3K9me2, while

SUV39H1, SUV39H2, and SETDB1 are H3K9 trimethyltransferases [74–76]. There seems to be a complex interplay between H3K9 methyltransferases and DNA methylation. On one hand, several H3K9 methyltransferases interact with and recruit DNMTs. For example, G9A/GLP recruits DNMT3A and DNMT3B to early-embryonic genes, causing de novo DNA methylation [77–79]. Thus, the G9A/GLP-dependent deposition of H3K9me2 appears to predate DNA methylation at these genomic sites, with the consequent DNA methylation stabilizing the repressed state. Separate knockout studies of *Suv39H1/2*, *G9a*, and *Setdb1* in mouse ESCs all resulted in DNA hypomethylation at different regions [75,77,80]. However, deletions of *Dnmt1* and *Dnmt3A/B* had no effect on H3K-9me3 levels [81,82], suggesting the existence of DNA methylation-independent pathways for H3K9 methylation in ESCs. On the contrary, the H3K9 methylation pattern changes when the DNA methylation is perturbed in human cancer cells [83,84]. Whether the difference in interaction hierarchy in these studies reflects difference between normal and cancer cells, or between ESC and more mature cells remains to be shown. Potential pathways for DNA methylation–induced histone modifications might involve MBD proteins, such as MBD1, which can recruit SUV39H1 to methylated sequences [85].

Also other silencing epigenetic factors play important roles in the regulation of DNA methylation. For example, heterochromatin protein 1 (HP1) that recognizes H3K9me3 modifications and binds HDACs has been suggested to recruit DNMTs to chromatin [86,87]. Furthermore, the chromatin remodeling enzyme, HELLS/LSH [88,89] promotes de novo DNA methylation at repetitive sequences. PolyComb complexes have also been suggested to have a role in the regulation of DNA methylation. The PolyComb Repressive complex 2 contains the H3K27 methyltransferase EZH2. Interestingly, the PolyComb complex tends to be targeted to CGI-containing regions of the genome [90], whereas non-CGI promoters are not often associated with PolyComb or H3K27me3 [16]. CGIs located in promoters of repressed genes that are not silenced by DNA methylation often carry the H3K27me3 mark that may also actively protect these CGIs from DNA methylation, and vice versa [91]. However, promoters with H3K27me3 in ESC are more than 4 times more likely to acquire DNA methylation during development, supporting the idea that there is a relationship of some sort between PolyComb and DNA methylation [16].

3.6 NON-CpG METHYLATION

CpG methylation is by far the predominant form of DNA methylation, but recent studies have shown that methylation of non-CpGs, where C is not followed by a G, but instead A, C, or T (CpH, H = A, C, or T) also occurs [92]. CpH methylation is predominantly found in ESC, induced pluripotent stem cells (iPSC), oocytes prospermatogonia, neurons, and glia cells [93]. However, CpH methylation levels have not yet been comprehensively mapped in different cell types. A study in the mouse brain recently showed that CpH methylation is enriched in CpG poor regions and like mC, anticorrelate with transcription [93]. The transcriptional repressive effect may be mediated via binding of methyl-CpG-binding proteins, such as MeCP2 that has been shown to bind also to methylated CpH [93]. In contrast to CpG methylation that is established and maintained by DNMT1, DNMT3a, and DNMT3B, CpH has been suggested to become methylated primarily by DNMT3A and DNMT3B [92]. Non-CpG methylation is a new research area and a lot remains to be discovered concerning its function and distribution.

3.7   DYNAMICS OF DNA METHYLATION DURING EARLY DEVELOPMENT

The role of DNA methylation in development was discovered long before DNA methylation as a phenomenon was known. Linnaeus discovered in year 1742 a peculiar specimen of the plant *Linaria vulgaris* [Linnaeus, C. De Peloria (Diss. Ac. Amoenitates Academicae III, Uppsala, 1749)]. The new specimen had five spurs instead of one that *L. vulgaris* normally has. The finding was against the theory that species had arisen from an act of original creation and remained unchanged since then. Linnaeus called the plant "Peloria," which is Greek for "monster" [94]. More than 250 years later, it was shown that Peloria was an epimutation of *L. vulgaris* [95]. Due to this epimutation, Peloria has thus lower expression of the *Lcyc* gene than *L. vulgaris*, which regulates floral asymmetry. No mutations in the DNA sequence were found, but *Lcyc* was heavily methylated in Peloria, which silenced its expression and consequently caused the Peloria phenotype [95]. However, the epimutation is not completely stable, and can revert during somatic development through DNA demethylation. Thus the role of DNA methylation in development has puzzled scientist for hundreds of years and still do, even if we have much more knowledge now than Linnaeus had.

Historically the focus has been on DNA methylation of promoters, since CGIs in promoters often are aberrantly methylated in cancer. However, new methodologies, such as methylated DNA immunoprecipitation (MeDIP-seq) and Whole Genome Bisulfite Sequencing (WGBS) have made it possible to study the methylome in a more unbiased manner [90]. WGBS is based on the treatment of DNA with bisulfite, which converts unmethylated cytosine to uracil, while DNA methylation protects the cytosine from the conversion. In this way mCs can be distinguished from unmethylated cytosines after sequencing. An important disadvantage of bisulfite treatment is that hydroxymethylation also protects the cytosine from uracil conversion. Consequently, DNA methylation and hydroxymethylation cannot be distinguished. However, recent technical development with a preceding oxidation step has made it possible to differentiate between methylation and hydroxymethylation with WGBS. The aforementioned new methodologies have lead to the discovery that changes in CpG methylation during development and in diseases often occur also in CpG poor regions, such as gene bodies, insulators, enhancers, intergenic and gene-poor regions.

3.7.1 Reprogramming of DNA Methylation During Early Development

Proper DNA methylation regulation is crucial for normal cell differentiation and development. The importance of DNA methylation is clearly illustrated by the observations that deletion of any of the DNMTs *Dnmt1*, *Dnmt3A*, or *Dnmt3B* in mouse embryos is lethal [5,96]. Although ESCs can maintain their stem cell properties in the absence of *Dnmt1*, *Dnmt3A*, and *Dnmt3B*, they are incapable to differentiate [97]. The DNMT-deficient ESCs thus fail to silence pluripotency factors, which blocks their differentiation [97].

Studies in mice and humans have shown that there are two waves of global DNA demethylation during development, which parallel major changes in developmental potential (Fig. 3.5). Hence, there is a major epigenetic reprogramming

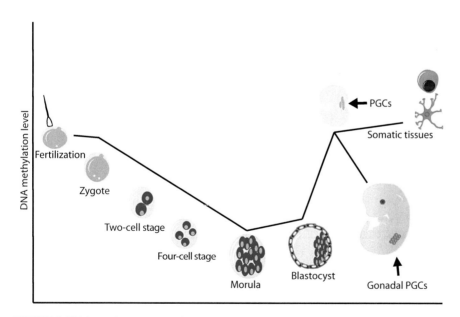

FIGURE 3.5 Epigenetic reprogramming during embryogenesis.
There are two waves of global DNA demethylation during embryogenesis. The first wave resets the epigenome for pluripotency at the blastocyst stage, with a steady decrease in DNA methylation from the zygote stage to the morula/blastocyst stage. The second wave of DNA demethylation occurs in the primordial germ cells *(PGCs)*. The DNA methylation is genome wide erased to restore germ line potency.

event in the preimplanted embryo and a second global demethylation event in primordial germ cells (PGCs) that are the precursors of sperms and oocytes undergoing their sex specification in the gonads [98].

After fertilization, the first wave of epigenetic reprogramming resets the gametic epigenome for totipotency to enable subsequent differentiation both in mice and humans [99]. It has been shown that the most dramatic DNA demethylation occurs at the zygotic stage, with a milder, continuous demethylation until the blastocyst stage [99,100]. The demethylation of the paternal genome occurs earlier in development than that of the maternal genome. Importantly, the genome-wide reprogramming events that take place after fertilization leave imprinted autosomal genes unaffected [45]. Imprinting is defined as the deposition of parent-of-origin specific chromatin marks at imprinting control regions, that takes place during male and female germ line development, and can be manifested as parent-of-origin specific monoallelic gene expression pattern of this specific set of genes in the offspring.

Establishment of the imprinted chromatin marks in the PGCs starts during the second wave of epigenetic reprogramming, when first all the parental-specific chromatin marks as well as global DNA methylation are erased, and then the correct maternal- or paternal-specific imprints and germ line–specific chromatin states are established during female and male germ line development, respectively [101–103]. Human PGCs are formed at the onset of gastrulation, which takes place at developmental week 2 in the posterior epiblast. During weeks 3–5 the definitive PGCs migrate from the yolk sac wall to the developing gonads. Recently, the first studies of human PGC development [101–103] have shown that at week 4 the genome-wide DNA methylation levels of PGCs are very low compared to somatic cells, indicating that the first DNA demethylation wave has already occurred at that time. The DNA methylation levels remain low until week 19, suggesting that the remethylation must occur somewhat later in development. Despite some differences, DNA methylation patterns in humans between weeks 4–19 resemble the patterns previously observed in mice between embryonic days 10.5–13.5 [99,103]. Establishment and maintenance of parental-specific DNA methylation patterns is important for the stable maintenance of imprinted expression [45]. Different factors, such as PGC7/Stella and ZFP57 have been proposed to have a role in maintaining the DNA methylation at imprinted genes during DNA demethylation waves after fertilization [45]. Many more factors are likely to be involved and how this protective process occurs in detail is still unknown.

In general, these observations suggest that reprogramming of the DNA methylation patterns is conserved during evolution and is a fundamental process among mammals. Although the mechanism of DNA demethylation during early development is not fully known, it seems to involve a combination of several processes. For example, the DNA methylation maintaining factors, such as DNMT1, DNMT3A, PRDM14, and UHRF1, are expressed at very low levels in PGCs, leading to passive demethylation due to cell proliferation. In addition, the expression of enzymes that are involved in DNA demethylation pathways, such as TET1 and TET2, are upregulated [101–103]. TET3 has a slightly different expression pattern and is highly expressed at the zygote stage [104]. Hydroxymethylation has indeed been suggested to be involved in epigenetic reprogramming in the zygote [105]. Also AID is involved in the DNA demethylation waves. Mouse PGCs deficient in AID thus show up to threefold increase in DNA methylation, suggesting that AID plays a crucial role in the PGC DNA demethylation program [106]. Thus several parallel DNA demethylation mechanisms contribute to epigenetic reprogramming during embryogenesis.

The global DNA demethylation wave during germ line development and in the zygote is not completely unspecific. Some regions maintain slightly higher remaining methylation levels than others. For example, gene bodies have slightly higher remaining methylation levels than neighboring intergenic regions. Moreover, loci that can escape the DNA demethylation waves during germ line development are enriched for genes associated with neurological and metabolic disorders, such as obesity-related traits, schizophrenia, and multiple sclerosis [100,101]. Although the methylation levels at these regions do not show correlation with gene expression, these findings suggest that they might be involved in transgenerational epigenetic inheritance [98,99,101].

Transposable elements are normally DNA methylated to ensure stable repression. In PGCs, certain evolutionarily young, still active retrotransposons, retain DNA methylation in a higher degree than the evolutionarily older transposons of the same family [100–102]. Interestingly, a punctuated pattern of the silencing heterochromatin mark H3K9me3 exists in all stages of PGC development, suggesting that H3K9me3 represses the constitutive heterochromatic regions in PGC in the absence of DNA methylation. Moreover, regions that are found to be resistant for complete DNA demethylation are also enriched for H3K9me3 [103]. This suggests that DNA demethylation is not completely unspecific and that certain loci are more resistant to demethylation than others.

3.7.2 DNA Methylation in Lineage Choice and Tissue Development

A cell modifies its identity and transcriptome during cell differentiation and maturation. The pattern of DNA methylation changes concordantly and locks the transcriptome in a cell type and differentiation stage–specific state. This occurs at all stages of development from embryogenesis to adult tissue–specific stem cell differentiation. For example, DNA methylome analysis of 19 different blood and skin cells representing various differentiation stages and lineages has shown that each cell type can be distinguished based on its DNA methylation pattern [107]. The different cell types could still be separated when additional tissues (brain, liver, and ESCs) were added to the analysis. Importantly, these experiments document that different cell types and differentiation stages have individual DNA methylation landscapes.

Both DNA methylation and demethylation has been described during differentiation in several different cell differentiation systems. For example, DNA methylation has been shown to increase during mouse ESC differentiation to neuronal progenitors [16]. In contrast, a major demethylation was demonstrated in CG poor regions during human ESC differentiation into endodermal lineage when analyzed with MeDIP-sequencing [108]. Furthermore, regions displaying loss of DNA methylation during neutrophil differentiation were enriched in enhancer elements important for the regulation of neutrophil-specific gene expression patterns [38]. It is, however, important to notice that it is only a few percent of the CpGs that change during a specific differentiation path. DNA methylation levels at most of the CpGs remain unaltered during differentiation [25,38]. In general, the methylation of CpG poor regions is less stable and varies more during cell differentiation. The fact that CG density and genomic location are important factors for how DNA methylation patterns are changed within a genomic region during cell differentiation makes it difficult to compare different studies that have used different methods with different sensitivity and coverage. Therefore, different methylation patterns described in different cell types and differentiation models may not necessarily depend on the cell type, but on the method used. General interpretations comparing different studies thus must be made with caution.

DNA methylation is an essential modification for the maintenance of stemness, and the establishment of proper lineage choice and differentiation [90]. One of the most interesting questions is, however, whether DNA methylation is important for the establishment or the stabilization of developmental decisions, or for both. The hematopoiesis is one of the best-characterized differentiation systems, where hematopoietic stem cells (HSCs) give rise to the lymphoid and myeloid lineages. It has been demonstrated that the level of DNA methylation is important for the lineage choice between myeloid and lymphoid lineages. Reduced DNMT1 activity in HSCs causes, for example, self-renewal defects and a differentiation bias toward myelopoiesis [109]. A study in early mouse hematopoiesis shows that the lymphoid lineage choice depends on the acquisition of DNA methylation, while myelopoiesis is associated with loss of DNA methylation [110]. Hence, there are twice as many regions with higher methylation levels in early lymphoid progenitors compared to early myeloid progenitors, than vice versa [107]. Lymphoid progenitors display an increased methylation at the loci and binding sites of myeloid transcription factors that may inhibit the binding of the myeloid factors to block myeloid differentiation in lymphoid cells [104]. This could explain the bias toward the myeloid lineage when the DNA methylation machinery is deficient and, as a consequence, the myeloid lineage-specific transcription factor–binding sites become demethylated during early hematopoietic development. DNA methylation may therefore protect the evolutionary younger lymphoid lineage to enter the "default" myeloid lineage [107].

Also within myelopoiesis, DNA demethylation may play a role in differentiation pathway choice [38]. Changes in DNA methylation is thus predominantly observed in the branch points of lineage choice, such as granulocyte/monocyte contra erythrocytes and granulocytes contra monocytes. The DNA methylation is, however, stable after the promyelocyte stage and onward [38]. Moreover, no further changes in DNA methylation are observed when all lineage choices are made. This suggests that DNA methylation is either important when a cell makes a lineage choice toward myelopoieis or that DNA methylation plays a more prominent role in locking in the cell into a chosen differentiation path.

Active DNA demethylation mediated by TET2 is also important for normal hematopoiesis. HSCs deficient in TET2 have decreased hmC levels, but increased self-renewal capacity, which consequently leads to an enlarged HSC/progenitor pool and enhanced hematopoiesis in both lymphoid and myeloid lineages. However, TET2 deficiency also skews the hematopoiesis toward monocytes and macrophages on the expense of lymphocytes and the myeloid erythrocytes and granulocytes [111, 112]. Moreover, TET2 has also been suggested to be important for late monocyte/macrophage differentiation. *TET2* knockdown thus causes reduced upregulation of macrophage markers in C/EBPα-induced transdifferentiaion assays from pre-B cells into macrophages [113].

DNA methylation is essential to maintain tissue homeostasis not only in hematopoiesis. Similar mechanisms have been discovered in other tissues, for example, in neurogenesis. For example, DNMT1 deletion in neuronal progenitors activates glial differentiation–related genes, but not neuronal-specific genes [114]. The pervious examples demonstrate that timed DNA demethylation of enhancers and promoters of lineage-specific genes are required for appropriate differentiation and lineage choice in many different systems. Conversely genes promoting stemness become methylated during the differentiation process [16, 115]. For example, reduction of DNA methylation in intestinal epithelium by deletion of *Dnmt1* causes an increase in the size of the stem cell compartment, and interferes with the balance of differentiation and self-renewal by perturbing the timely repression of genes regulating stemness [115]. Correct regulation of DNA methylation is thus required for both stemness and appropriate differentiation. Consequently, when the fine-tuned DNA methylation balance is disturbed via different mechanisms, it may lead to the development of different diseases.

3.8 ABERRANT DNA METHYLATION IN DISEASES

DNA methylation is essential for normal development and cellular homeostasis. Several genes that are involved in the DNA methylation machinery are mutated in cancer and in developmental diseases (Table 3.2) [116]. Cancer is probably the most studied disease where aberrant DNA methylation pattern has been well documented [117]. In the cancer methylome CGIs tend to be hypermethylated, while other regions in the genome are hypomethylated. The hypomethylation may contribute to chromosomal instability [118,119] and consequently to the development of cancer. The field of cancer epigenetics has historically been focused on CGIs in promoter regions, where increased methylation has been described at many tumor suppressor genes. Hypermethylation of CGIs in promoter regions of tumor suppressors contribute to their repression and, as a consequence, to the uncontrolled proliferation that is seen in tumors [120]. In addition, promoters of developmentally regulated genes that normally are silenced by the PolyComb repressor complex 2 (PRC2) in ESCs are more likely to be repressed by DNA methylation in cancer [121].

During the last years the focus has shifted toward the importance of CpG poor regions of the cancer genome. Studies, for example, in colon cancer have shown that most changes in DNA methylation occur distant from CGIs, either in the so called CGI shores that surround CGIs, or in "oceans" distant from both CGIs and shores [10]. The differently methylated regions that distinguish cancer tissues from their normal counterparts overlap with regions where the DNA methylation levels also differ between different normal tissues [10]. Hence, there may be specific regions in the genome where the DNA methylation is more variable and sensitive for perturbation. The regions where the DNA methylation variation is high are located in the vicinity of genes that are involved in development and morphogenesis [122]. Moreover, the DNA

Table 3.2 Mutated Genes Involved in the DNA Methylation Machinery [116]

Gene	Disease
DNMT1	Autosomal dominant cerebellar ataxia, deafness and narcolepsy (ADCA-DN), hereditary sensory and autonomic neuropathy with dementia and hearing loss type IE (HSAN IE)
DNMT3A	AML, ALL, myelodysplastic syndrome (MDS), myeloproliferative neoplasms (MPN), angioimmunoblastic, overgrowth syndrome
DNMT3B	ICF syndrome type 1
ZBTB24	ICF syndrome type 2
TET1	AML and ALL
TET2	AML, MDS, MPN, chronic myelomonocytic leukemia
MECP2	Rett syndrome

methylation in those regions correlates with the expression of the neighboring genes. This has been suggested to lead to variable cellular phenotypes to enable adaptation to the changing microenvironment during development [122]. Similarly, increased methylation variation in cancer cells might contribute to cancer cell phenotypic heterogeneity and tumor evolution under changing selective pressure [122].

3.8.1 Mutations in the DNA Methylation Machinery in Cancer

Somatic mutations in genes that are involved in the DNA methylation machinery are documented in several tumors. In AML, mutations in DNA methylation regulators, such as *DNMT3A, TET2, IDH1*, and *IDH2*, are frequent [123]. Those mutations cause an aberrant DNA methylation pattern that contributes to leukomogenesis. Loss of function of *TET2* and *DNMT3A* are early events in leukomogenesis [124]. Early mutations in *TET2* and *DNMT3A* and other genes lead to clonal expansion in approximately 5–6% of the healthy elderly population (older than 65 years of age) [125]. In the younger population (younger than 50 years of age) the incidence is only 0.7%. Interestingly, the three most commonly mutated genes in healthy elderly population are the epigenetic regulators *DNMT3A, TET2*, and *ASXL1* that is part of the PolyComb complex. All of them are frequently mutated in leukemia, suggesting that clones with those mutations are likely preleukemic. Indeed, people with clonal hematopoiesis were substantially at higher risk to be diagnosed with leukemia 6 months or more after the DNA sample was analyzed [125].

It has been proposed that epigenetic alterations are very early events during tumor development, which preceed somatic mutations in stem and progenitor cells [126]. Although this model has been difficult to prove, there is evidence that it might correctly capture the sequence of events at least in certain tumors. Ependymomas (childhood brain tumors), represent one example of early epigenetic perturbation in combination with extremely low mutation levels and none somatic single nucleotide variants [127]. There are two types of hindbrain ependymomas located within the posterior fossa of the skull. One subtype mainly affects infants, while the other subtype occurs in older children. The subtype that affects infants has a higher level of CGI methylation than the other subtype, and exhibits a CpG island methylator phenotype (CIMP) with increased methylation at PRC2 target genes without any detected mutations in the DNA methylation machinery or elsewhere in the genome [127]. Although it has been described in many other cancer types that genes that are normally regulated by PRC2 are more commonly silenced by DNA methylation in cancer, the mechanism of this methylation gain is not known [16,128]. The characteristics of ependymomas thus support the model that epigenetic aberrations can occur without any preceding genetic mutations.

3.8.2 Rett Syndrome and ICF Syndrome

Proper DNA methylation is required for normal development as described previously. Mutations in various genes involved in DNA methylation pathways lead to disturbed development. This is exemplified by the Rett syndrome (RTT) and immunodeficiency, centromeric instability and facial anomalies (ICF) syndrome, which both involve mutations in genes encoding for DNA methylation regulatory factors, *MECP2* and *DNMT3B*, respectively.

Mutations in the X-linked *MECP2* that binds to methylated CpGs, cause the neurological disorder called RTT [129]. The symptoms for RTT are for example impaired motor skills, loss of speech. RTT has a gender bias and affects almost exclusively women. Mutations in the only X-linked copy of *MECP2* in men lead to early lethality. However, mutations that do not give complete loss of function in MECP2 can be found in boys, although very rarely.

Correct response to DNA methylation is a very delicate balance. Also enhanced MECP2 activity caused by gene duplication leads to mental retardation and other complications [130]. The molecular consequence of deregulated MECP2 is still not completely elucidated. MECP2 has been shown to interact with other repressive complexes, such as NCOR-SMRT [131]. Despite that MECP2 binds to methylated DNA, it is involved in both transcriptional activation and repression, and may also have posttranscriptional functions [129]. The transcriptional activating function has been suggested to involve recruitment of cyclic AMP-responsive element-binding protein 1 (CREB1) to promoters [132]. Furthermore, MECP2 may function as a transcriptional activator when it is bound to hmC, which has been shown to correlate with active transcription [133]. The different functions of MECP2 make the downstream effects of MECP2 mutations complex.

Also ICF syndrome is characterized by perturbed DNA methylation machinery. More than half of the ICF patients carry mutations in *DNMT3B* [5,134]. ICF syndrome is a rare autosomal recessive disorder that is characterized by Immunodeficiency, Centromeric instability and Facial anomalies. ICF leads to several developmental defects in the blood and brain. ICF patients often die due to infections at a young age, because of defects in B-cell development [135].

3.9 CONCLUSIONS

The reversible nature of epigenetic marks including DNA methylation makes the epigenome a very promising therapeutic target. Knowledge about the function of DNA methylation and how it is connected to different phenotypes will dramatically increase in the coming years. The reduced cost for WGBS and MeDIP-seq analyses makes it possible to study DNA methylation at a much higher resolution than before. This will increase our understanding of how DNA methylation contributes to the development of diseases. Some drugs are already approved and used in the clinic both in Europe and in the United States. For example, Azacytidine is the most commonly used DNA methylation inhibitory drug, which is the first choice of treatment for high-risk myeloid dysplastic syndrome. However, detailed information on how DNA methylation regulates cellular homeostasis and development will be crucial to be able to use DNA methylation as a safe drug target.

Abbreviations

caC	Carboxylcytosine
ADD	ATRX-DNMT3-DNMT3L domain
AID	Activation-induced cytidine deaminase
AML	Acute myeloid leukemia
APOBEC	Apolipoprotein B mRNA-editing enzyme
BER	Base excision repair
CFP1	CXXC finger protein 1
CGI	CpG islands
CIMP	CpG island methylator phenotype
DNMT	DNA methyltransferases
ESC	Embryonic stem cell
fC	Formylcytosine
HDAC	Histone deacetylase
hmC	Hydroxymethyl cytosine
HSC	Hematopoietic stem cells
ICF	Immunodeficiency, centromeric instability and facial anomalies
MBD	Methyl-binding domain
mC	Methylated cytosine
PGC	Primordial germ cells
PRC1	PolyComb repressive complex 1
PRC2	PolyComb repressive complex 2
RTT	Rett syndrome
TDG	Thymine-DNA glycosylase
TET	Ten eleven translocation

Acknowledgments

I would like to thank Esteban Ballestar and Michelle Rönnerblad for critical reading of the text.

References

[1] Holliday R, Pugh JE. DNA modification mechanisms and gene activity during development. Science 1975;187(4173):226–32.

[2] Riggs AD. X inactivation, differentiation, and DNA methylation. Cytogenet Cell Genet 1975;14(1):9–25.

[3] Lander ES, Linton LM, Birren B, Nusbaum C, Zody MC, Baldwin J, et al. Initial sequencing and analysis of the human genome. Nature 2001;409(6822):860–921.

[4] Bestor T, Laudano A, Mattaliano R, Ingram V. Cloning and sequencing of a cDNA encoding DNA methyltransferase of mouse cells. The carboxyl-terminal domain of the mammalian enzymes is related to bacterial restriction methyltransferases. J Mol Biol 1988;203(4):971–83.

[5] Okano M, Bell DW, Haber DA, Li E. DNA methyltransferases Dnmt3a and Dnmt3b are essential for de novo methylation and mammalian development. Cell 1999;99(3):247–57.

[6] Avvakumov GV, Walker JR, Xue S, Li Y, Duan S, Bronner C, et al. Structural basis for recognition of hemi-methylated DNA by the SRA domain of human UHRF1. Nature 2008;455(7214):822–5.

[7] Sharif J, Muto M, Takebayashi S, Suetake I, Iwamatsu A, Endo TA, et al. The SRA protein Np95 mediates epigenetic inheritance by recruiting Dnmt1 to methylated DNA. Nature 2007;450(7171):908–12.

[8] Jones PA, Liang G. Rethinking how DNA methylation patterns are maintained. Nat Rev Genet 2009;10(11):805–11.

[9] Liang G, Chan MF, Tomigahara Y, Tsai YC, Gonzales FA, Li E, et al. Cooperativity between DNA methyltransferases in the maintenance methylation of repetitive elements. Mol Cell Biol 2002;22(2):480–91.

[10] Irizarry RA, Ladd-Acosta C, Wen B, Wu Z, Montano C, Onyango P, et al. The human colon cancer methylome shows similar hypo- and hypermethylation at conserved tissue-specific CpG island shores. Nat Genet 2009;41(2):178–86.

[11] Larsen F, Gundersen G, Lopez R, Prydz H. CpG islands as gene markers in the human genome. Genomics 1992;13(4):1095–107.

[12] Zhu J, He F, Hu S, Yu J. On the nature of human housekeeping genes. Trends Genet 2008;24(10):481–4.

[13] Ramirez-Carrozzi VR, Braas D, Bhatt DM, Cheng CS, Hong C, Doty KR, et al. A unifying model for the selective regulation of inducible transcription by CpG islands and nucleosome remodeling. Cell 2009;138(1):114–28.

[14] Carninci P, Sandelin A, Lenhard B, Katayama S, Shimokawa K, Ponjavic J, et al. Genome-wide analysis of mammalian promoter architecture and evolution. Nat Genet 2006;38(6):626–35.

[15] Lynch MD, Smith AJ, De Gobbi M, Flenley M, Hughes JR, Vernimmen D, et al. An interspecies analysis reveals a key role for unmethylated CpG dinucleotides in vertebrate polycomb complex recruitment. EMBO J 2012;31(2):317–29.

[16] Mohn F, Weber M, Rebhan M, Roloff TC, Richter J, Stadler MB, et al. Lineage-specific polycomb targets and de novo DNA methylation define restriction and potential of neuronal progenitors. Mol Cell 2008;30(6):755–66.

[17] Tanay A, O'Donnell AH, Damelin M, Bestor TH. Hyperconserved CpG domains underlie polycomb-binding sites. Proc Natl Acad Sci USA 2007;104(13):5521–6.

[18] Weber M, Hellmann I, Stadler MB, Ramos L, Paabo S, Rebhan M, et al. Distribution, silencing potential and evolutionary impact of promoter DNA methylation in the human genome. Nat Genet 2007;39(4):457–66.

[19] Ooi SK, Qiu C, Bernstein E, Li K, Jia D, Yang Z, et al. DNMT3L connects unmethylated lysine 4 of histone H3 to de novo methylation of DNA. Nature 2007;448(7154):714–7.

[20] Maunakea AK, Nagarajan RP, Bilenky M, Ballinger TJ, D'Souza C, Fouse SD, et al. Conserved role of intragenic DNA methylation in regulating alternative promoters. Nature 2010;466(7303):253–7.

[21] Illingworth RS, Gruenewald-Schneider U, Webb S, Kerr AR, James KD, Turner DJ, et al. Orphan CpG islands identify numerous conserved promoters in the mammalian genome. PLoS Genet 2010;6(9):e1001134.

[22] Carrozza MJ, Li B, Florens L, Suganuma T, Swanson SK, Lee KK, et al. Histone H3 methylation by Set2 directs deacetylation of coding regions by Rpd3S to suppress spurious intragenic transcription. Cell 2005;123(4):581–92.

[23] Suzuki MM, Kerr AR, De Sousa D, Bird A. CpG methylation is targeted to transcription units in an invertebrate genome. Genome Res 2007;17(5):625–31.

[24] Chodavarapu RK, Feng S, Bernatavichute YV, Chen PY, Stroud H, Yu Y, et al. Relationship between nucleosome positioning and DNA methylation. Nature 2010;466(7304):388–92.

[25] Laurent L, Wong E, Li G, Huynh T, Tsirigos A, Ong CT, et al. Dynamic changes in the human methylome during differentiation. Genome Res 2010;20(3):320–31.

[26] Maunakea AK, Chepelev I, Cui K, Zhao K. Intragenic DNA methylation modulates alternative splicing by recruiting MeCP2 to promote exon recognition. Cell Res 2013;23(11):1256–69.

[27] Schwartz S, Meshorer E, Ast G. Chromatin organization marks exon-intron structure. Nat Struct Mol Biol 2009;16(9):990–5.

[28] Bhutani N, Brady JJ, Damian M, Sacco A, Corbel SY, Blau HM, Reprogramming towards pluripotency requires AID-dependent DNA demethylation. . Nature 2010;463(7284): 1042–7.

[29] Rai K, Huggins IJ, James SR, Karpf AR, Jones DA, Cairns BR. DNA demethylation in zebrafish involves the coupling of a deaminase, a glycosylase, and gadd45. Cell 2008;135(7):1201–12.

[30] Zhu JK. Active DNA demethylation mediated by DNA glycosylases. Annu Rev Genet 2009;43:143–66.

[31] Kriaucionis S, Heintz N. The nuclear DNA base 5-hydroxymethylcytosine is present in Purkinje neurons and the brain. Science 2009;324(5929):929–30.

[32] Tahiliani M, Koh KP, Shen Y, Pastor WA, Bandukwala H, Brudno Y, et al. Conversion of 5-methylcytosine to 5-hydroxymethylcytosine in mammalian DNA by MLL partner TET1. Science 2009;324(5929):930–5.

[33] He YF, Li BZ, Li Z, Liu P, Wang Y, Tang Q, et al. Tet-mediated formation of 5-carboxylcytosine and its excision by TDG in mammalian DNA. Science 2011;333(6047):1303–7.

[34] Ito S, Shen L, Dai Q, Wu SC, Collins LB, Swenberg JA, et al. Tet proteins can convert 5-methylcytosine to 5-formylcytosine and 5-carboxylcytosine. Science 2011;333(6047):1300–3.

[35] Pfaffeneder T, Hackner B, Truss M, Munzel M, Muller M, Deiml CA, et al. The discovery of 5-formylcytosine in embryonic stem cell DNA. Angew Chem Int Ed 2011;50(31):7008–12.

[36] Lu F, Liu Y, Jiang L, Yamaguchi S, Zhang Y. Role of Tet proteins in enhancer activity and telomere elongation. Genes Dev 2014;28(19):2103–19.

[37] Hon GC, Song CX, Du T, Jin F, Selvaraj S, Lee AY, et al. 5mC oxidation by Tet2 modulates enhancer activity and timing of transcriptome reprogramming during differentiation. Mol Cell 2014;56(2):286–97.

[38] Ronnerblad M, Andersson R, Olofsson T, Douagi I, Karimi M, Lehmann S, et al. Analysis of the DNA methylome and transcriptome in granulopoiesis reveals timed changes and dynamic enhancer methylation. Blood 2014;123(17):e79–89.

[39] Rasmussen KD, Jia G, Johansen JV, Pedersen MT, Rapin N, Bagger FO, et al. Loss of TET2 in hematopoietic cells leads to DNA hypermethylation of active enhancers and induction of leukemogenesis. Genes Dev 2015;29(9):910–22.

[40] Guo JU, Su Y, Zhong C, Ming GL, Song H. Hydroxylation of 5-methylcytosine by TET1 promotes active DNA demethylation in the adult brain. Cell 2011;145(3):423–34.

[41] Deaton AM, Bird A. CpG islands and the regulation of transcription. Genes Dev 2011;25(10):1010–22.

[42] Hu S, Wan J, Su Y, Song Q, Zeng Y, Nguyen HN, et al. DNA methylation presents distinct binding sites for human transcription factors. Elife 2013;2:e00726.

[43] Liu Y, Zhang X, Blumenthal RM, Cheng X. A common mode of recognition for methylated CpG. Trends Biochem Sci 2013;38(4):177–83.

[44] Campanero MR, Armstrong MI, Flemington EK. CpG methylation as a mechanism for the regulation of E2F activity. Proc Natl Acad Sci USA 2000;97(12):6481–6.

[45] Barlow DP, Bartolomei MS. Genomic imprinting in mammals. Cold Spring Harb Perspect Biol 2014;6(2):1–20.

[46] Watt F, Molloy PL. Cytosine methylation prevents binding to DNA of a HeLa cell transcription factor required for optimal expression of the adenovirus major late promoter. Genes Dev 1988;2(9):1136–43.

[47] de la Rica L, Rodriguez-Ubreva J, Garcia M, Islam AB, Urquiza JM, Hernando H, et al. PU.1 target genes undergo Tet2-coupled demethylation and DNMT3b-mediated methylation in monocyte-to-osteoclast differentiation. Genome Biol 2013;14(9):R99.

[48] Feldmann A, Ivanek R, Murr R, Gaidatzis D, Burger L, Schubeler D. Transcription factor occupancy can mediate active turnover of DNA methylation at regulatory regions. PLoS Genet 2013;9(12):e1003994.

[49] Kress C, Thomassin H, Grange T. Active cytosine demethylation triggered by a nuclear receptor involves DNA strand breaks. Proc Natl Acad Sci USA 2006;103(30):11112–7.

[50] Stadler MB, Murr R, Burger L, Ivanek R, Lienert F, Scholer A, et al. DNA-binding factors shape the mouse methylome at distal regulatory regions. Nature 2011;480(7378):490–5.

[51] Baubec T, Ivanek R, Lienert F, Schubeler D. Methylation-dependent and -independent genomic targeting principles of the MBD protein family. Cell 2013;153(2):480–92.

[52] Hendrich BD, Plenge RM, Willard HF. Identification and characterization of the human XIST gene promoter: implications for models of X chromosome inactivation. Nucleic Acids Res 1997;25(13):2661–71.

[53] Meehan RR, Lewis JD, McKay S, Kleiner EL, Bird AP. Identification of a mammalian protein that binds specifically to DNA containing methylated CpGs. Cell 1989;58(3):499–507.

[54] Lin JC, Jeong S, Liang G, Takai D, Fatemi M, Tsai YC, et al. Role of nucleosomal occupancy in the epigenetic silencing of the MLH1 CpG island. Cancer Cell 2007;12(5):432–44.

[55] Wade PA, Wolffe AP. ReCoGnizing methylated DNA. Nat Struct Biol 2001;8(7):575–7.

[56] You JS, Kelly TK, De Carvalho DD, Taberlay PC, Liang G, Jones PA. OCT4 establishes and maintains nucleosome-depleted regions that provide additional layers of epigenetic regulation of its target genes. Proc Natl Acad Sci USA 2011;108(35):14497–502.

[57] Jeong S, Liang G, Sharma S, Lin JC, Choi SH, Han H, et al. Selective anchoring of DNA methyltransferases 3A and 3B to nucleosomes containing methylated DNA. Mol Cell Biol 2009;29(19):5366–76.

[58] Thomson JP, Skene PJ, Selfridge J, Clouaire T, Guy J, Webb S, et al. CpG islands influence chromatin structure via the CpG-binding protein Cfp1. Nature 2010;464(7291):1082–6.

[59] Long HK, Blackledge NP, Klose RJ. ZF-CxxC domain-containing proteins, CpG islands and the chromatin connection. Biochem Soc Trans 2013;41(3):727–40.

[60] Carlone DL, Skalnik DG. CpG binding protein is crucial for early embryonic development. Mol Cell Biol 2001;21(22):7601–6.

[61] Lee JH, Skalnik DG. CpG-binding protein (CXXC finger protein 1) is a component of the mammalian Set1 histone H3-Lys4 methyltransferase complex, the analogue of the yeast Set1/COMPASS complex. J Biol Chem 2005;280(50):41725–31.

[62] Clouaire T, Webb S, Skene P, Illingworth R, Kerr A, Andrews R, et al. Cfp1 integrates both CpG content and gene activity for accurate H3K4me3 deposition in embryonic stem cells. Genes Dev 2012;26(15):1714–28.

[63] Birke M, Schreiner S, Garcia-Cuellar MP, Mahr K, Titgemeyer F, Slany RK. The MT domain of the proto-oncoprotein MLL binds to CpG-containing DNA and discriminates against methylation. Nucleic Acids Res 2002;30(4):958–65.

[64] Blackledge NP, Zhou JC, Tolstorukov MY, Farcas AM, Park PJ, Klose RJ. CpG islands recruit a histone H3 lysine 36 demethylase. Mol Cell 2010;38(2):179–90.

[65] Farcas AM, Blackledge NP, Sudbery I, Long HK, McGouran JF, Rose NR, et al. KDM2B links the Polycomb Repressive Complex 1 (PRC1) to recognition of CpG islands. Elife 2012;1:e00205.

[66] Song J, Rechkoblit O, Bestor TH, Patel DJ. Structure of DNMT1-DNA complex reveals a role for autoinhibition in maintenance DNA methylation. Science 2011;331(6020):1036–40.

[67] Zhang Y, Jurkowska R, Soeroes S, Rajavelu A, Dhayalan A, Bock I, et al. Chromatin methylation activity of Dnmt3a and Dnmt3a/3L is guided by interaction of the ADD domain with the histone H3 tail. Nucleic Acids Res 2010;38(13):4246–53.

[68] Ciccone DN, Su H, Hevi S, Gay F, Lei H, Bajko J, et al. KDM1B is a histone H3K4 demethylase required to establish maternal genomic imprints. Nature 2009;461(7262):415–8.

[69] Wang J, Hevi S, Kurash JK, Lei H, Gay F, Bajko J, et al. The lysine demethylase LSD1 (KDM1) is required for maintenance of global DNA methylation. Nat Genet 2009;41(1):125–9.

[70] Conerly ML, Teves SS, Diolaiti D, Ulrich M, Eisenman RN, Henikoff S. Changes in H2A.Z occupancy and DNA methylation during B-cell lymphomagenesis. Genome Res 2010;20(10):1383–90.

[71] Zilberman D, Coleman-Derr D, Ballinger T, Henikoff S. Histone H2A.Z and DNA methylation are mutually antagonistic chromatin marks. Nature 2008;456(7218):125–9.

[72] Meissner A, Mikkelsen TS, Gu H, Wernig M, Hanna J, Sivachenko A, et al. Genome-scale DNA methylation maps of pluripotent and differentiated cells. Nature 2008;454(7205):766–70.

[73] Pinheiro I, Margueron R, Shukeir N, Eisold M, Fritzsch C, Richter FM, et al. Prdm3 and Prdm16 are H3K9me1 methyltransferases required for mammalian heterochromatin integrity. Cell 2012;150(5):948–60.

[74] Rea S, Eisenhaber F, O'Carroll D, Strahl BD, Sun ZW, Schmid M, et al. Regulation of chromatin structure by site-specific histone H3 methyltransferases. Nature 2000;406(6796):593–9.

[75] Schultz DC, Ayyanathan K, Negorev D, Maul GG, Rauscher F III. SETDB1: a novel KAP-1-associated histone H3, lysine 9-specific methyltransferase that contributes to HP1-mediated silencing of euchromatic genes by KRAB zinc-finger proteins. Genes Dev 2002;16(8):919–32.

[76] Tachibana M, Sugimoto K, Fukushima T, Shinkai Y. Set domain-containing protein, G9a, is a novel lysine-preferring mammalian histone methyltransferase with hyperactivity and specific selectivity to lysines 9 and 27 of histone H3. J Biol Chem 2001;276(27):25309–17.

[77] Dong KB, Maksakova IA, Mohn F, Leung D, Appanah R, Lee S, et al. DNA methylation in ES cells requires the lysine methyltransferase G9a but not its catalytic activity. EMBO J 2008;27(20):2691–701.

[78] Epsztejn-Litman S, Feldman N, Abu-Remaileh M, Shufaro Y, Gerson A, Ueda J, et al. De novo DNA methylation promoted by G9a prevents reprogramming of embryonically silenced genes. Nat Struct Mol Biol 2008;15(11):1176–83.

[79] Esteve PO, Chin HG, Smallwood A, Feehery GR, Gangisetty O, Karpf AR, et al. Direct interaction between DNMT1 and G9a coordinates DNA and histone methylation during replication. Genes Dev 2006;20(22):3089–103.

[80] Lehnertz B, Ueda Y, Derijck AA, Braunschweig U, Perez-Burgos L, Kubicek S, et al. Suv39h-mediated histone H3 lysine 9 methylation directs DNA methylation to major satellite repeats at pericentric heterochromatin. Curr Biol 2003;13(14):1192–200.

[81] Matsui T, Leung D, Miyashita H, Maksakova IA, Miyachi H, Kimura H, et al. Proviral silencing in embryonic stem cells requires the histone methyltransferase ESET. Nature 2010;464(7290):927–31.

[82] Tsumura A, Hayakawa T, Kumaki Y, Takebayashi S, Sakaue M, Matsuoka C, et al. Maintenance of self-renewal ability of mouse embryonic stem cells in the absence of DNA methyltransferases Dnmt1, Dnmt3a and Dnmt3b. Genes Cells 2006;11(7):805–14.

[83] Espada J, Ballestar E, Fraga MF, Villar-Garea A, Juarranz A, Stockert JC, et al. Human DNA methyltransferase 1 is required for maintenance of the histone H3 modification pattern. J Biol Chem 2004;279(35):37175–84.

[84] Nguyen CT, Weisenberger DJ, Velicescu M, Gonzales FA, Lin JC, Liang G, et al. Histone H3-lysine 9 methylation is associated with aberrant gene silencing in cancer cells and is rapidly reversed by 5-aza-2′-deoxycytidine. Cancer Res 2002;62(22):6456–61.

[85] Fujita N, Watanabe S, Ichimura T, Tsuruzoe S, Shinkai Y, Tachibana M, et al. Methyl-CpG binding domain 1 (MBD1) interacts with the Suv39h1-HP1 heterochromatic complex for DNA methylation-based transcriptional repression. J Biol Chem 2003;278(26):24132–8.

[86] Honda S, Selker EU. Direct interaction between DNA methyltransferase DIM-2 and HP1 is required for DNA methylation in *Neurospora crassa*. Mol Cell Biol 2008;28(19):6044–55.

[87] Robertson KD, Ait-Si-Ali S, Yokochi T, Wade PA, Jones PL, Wolffe AP. DNMT1 forms a complex with Rb, E2F1 and HDAC1 and represses transcription from E2F-responsive promoters. Nat Genet 2000;25(3):338–42.

[88] Ren J, Briones V, Barbour S, Yu W, Han Y, Terashima M, et al. The ATP binding site of the chromatin remodeling homolog Lsh is required for nucleosome density and de novo DNA methylation at repeat sequences. Nucleic Acids Res 2015;43(3):1444–55.

[89] Yu W, McIntosh C, Lister R, Zhu I, Han Y, Ren J, et al. Genome-wide DNA methylation patterns in LSH mutant reveals de-repression of repeat elements and redundant epigenetic silencing pathways. Genome Res 2014;24(10):1613–23.

[90] Ku M, Koche RP, Rheinbay E, Mendenhall EM, Endoh M, Mikkelsen TS, et al. Genomewide analysis of PRC1 and PRC2 occupancy identifies two classes of bivalent domains. PLoS Genet 2008;4(10):e1000242.

[91] Brinkman AB, Gu H, Bartels SJ, Zhang Y, Matarese F, Simmer F, et al. Sequential ChIP-bisulfite sequencing enables direct genome-scale investigation of chromatin and DNA methylation cross-talk. Genome Res 2012;22(6):1128–38.

[92] He Y, Ecker JR. Non-CG methylation in the human genome. Annu Rev Genomics Hum Genet 2015;16:55–77.

[93] Guo JU, Su Y, Shin JH, Shin J, Li H, Xie B, et al. Distribution, recognition and regulation of non-CpG methylation in the adult mammalian brain. Nat Neurosci 2014;17(2):215–22.

[94] Gustafsson A. Linnaeus' Peloria: The history of a monster. Theor Appl Genet 1979;54(6):241–8.

[95] Cubas P, Vincent C, Coen E. An epigenetic mutation responsible for natural variation in floral symmetry. Nature 1999;401(6749):157–61.

[96] Li E, Bestor TH, Jaenisch R. Targeted mutation of the DNA methyltransferase gene results in embryonic lethality. Cell 1992;69(6):915–26.

[97] Jackson M, Krassowska A, Gilbert N, Chevassut T, Forrester L, Ansell J, et al. Severe global DNA hypomethylation blocks differentiation and induces histone hyperacetylation in embryonic stem cells. Mol Cell Biol 2004;24(20):8862–71.

[98] Gkountela S, Li Z, Vincent JJ, Zhang KX, Chen A, Pellegrini M, et al. The ontogeny of cKIT+ human primordial germ cells proves to be a resource for human germ line reprogramming, imprint erasure and in vitro differentiation. Nat Cell Biol 2013;15(1):113–22.

[99] Smith ZD, Chan MM, Humm KC, Karnik R, Mekhoubad S, Regev A, et al. DNA methylation dynamics of the human preimplantation embryo. Nature 2014;511(7511):611–5.

[100] Guo H, Zhu P, Yan L, Li R, Hu B, Lian Y, et al. The DNA methylation landscape of human early embryos. Nature 2014;511(7511):606–10.

[101] Gkountela S, Zhang KX, Shafiq TA, Liao WW, Hargan-Calvopina J, Chen PY, et al. DNA demethylation dynamics in the human prenatal germline. Cell 2015;161(6):1425–36.

[102] Guo F, Yan L, Guo H, Li L, Hu B, Zhao Y, et al. The transcriptome and DNA methylome landscapes of human primordial germ cells. Cell 2015;161(6):1437–52.

[103] Tang WW, Dietmann S, Irie N, Leitch HG, Floros VI, Bradshaw CR, et al. A unique gene regulatory network resets the human germline epigenome for development. Cell 2015;161(6):1453–67.

[104] Gu TP, Guo F, Yang H, Wu HP, Xu GF, Liu W, et al. The role of Tet3 DNA dioxygenase in epigenetic reprogramming by oocytes. Nature 2011;477(7366):606–10.

[105] Wossidlo M, Nakamura T, Lepikhov K, Marques CJ, Zakhartchenko V, Boiani M, et al. 5-Hydroxymethylcytosine in the mammalian zygote is linked with epigenetic reprogramming. Nat Commun 2011;2:241.

[106] Popp C, Dean W, Feng S, Cokus SJ, Andrews S, Pellegrini M, et al. Genome-wide erasure of DNA methylation in mouse primordial germ cells is affected by AID deficiency. Nature 2010;463(7284):1101–5.

[107] Bock C, Beerman I, Lien WH, Smith ZD, Gu H, Boyle P, et al. DNA methylation dynamics during in vivo differentiation of blood and skin stem cells. Mol Cell 2012;47(4):633–47.

[108] Chavez L, Jozefczuk J, Grimm C, Dietrich J, Timmermann B, Lehrach H, et al. Computational analysis of genome-wide DNA methylation during the differentiation of human embryonic stem cells along the endodermal lineage. Genome Res 2010;20(10):1441–50.

[109] Broske AM, Vockentanz L, Kharazi S, Huska MR, Mancini E, Scheller M, et al. DNA methylation protects hematopoietic stem cell multipotency from myeloerythroid restriction. Nat Genet 2009;41(11):1207–15.

[110] Ji H, Ehrlich LI, Seita J, Murakami P, Doi A, Lindau P, et al. Comprehensive methylome map of lineage commitment from haematopoietic progenitors. Nature 2010;467(7313):338–42.

[111] Ko M, Bandukwala HS, An J, Lamperti ED, Thompson EC, Hastie R, et al. Ten-Eleven-Translocation 2 (TET2) negatively regulates homeostasis and differentiation of hematopoietic stem cells in mice. Proc Natl Acad Sci USA 2011;108(35):14566–71.

[112] Moran-Crusio K, Reavie L, Shih A, Abdel-Wahab O, Ndiaye-Lobry D, Lobry C, et al. Tet2 loss leads to increased hematopoietic stem cell self-renewal and myeloid transformation. Cancer Cell 2011;20(1):11–24.

[113] Kallin EM, Rodriguez-Ubreva J, Christensen J, Cimmino L, Aifantis I, Helin K, et al. Tet2 facilitates the derepression of myeloid target genes during CEBPalpha-induced transdifferentiation of pre-B cells. Mol Cell 2012;48(2):266–76.

[114] Fan G, Martinowich K, Chin MH, He F, Fouse SD, Hutnick L, et al. DNA methylation controls the timing of astrogliogenesis through regulation of JAK-STAT signaling. Development 2005;132(15):3345–56.

[115] Sheaffer KL, Kim R, Aoki R, Elliott EN, Schug J, Burger L, et al. DNA methylation is required for the control of stem cell differentiation in the small intestine. Genes Dev 2014;28(6):652–64.

[116] Hamidi T, Singh AK, Chen T. Genetic alterations of DNA methylation machinery in human diseases. Epigenomics 2015;7(2):247–65.

[117] Baylin SB, Jones PA. A decade of exploring the cancer epigenome—biological and translational implications. Nat Rev Cancer 2011;11(10):726–34.

[118] Dodge JE, Okano M, Dick F, Tsujimoto N, Chen T, Wang S, et al. Inactivation of Dnmt3b in mouse embryonic fibroblasts results in DNA hypomethylation, chromosomal instability, and spontaneous immortalization. J Biol Chem 2005;280(18):17986–91.

[119] Gaudet F, Hodgson JG, Eden A, Jackson-Grusby L, Dausman J, Gray JW, et al. Induction of tumors in mice by genomic hypomethylation. Science 2003;300(5618):489–92.

[120] Jones PA, Baylin SB. The epigenomics of cancer. Cell 2007;128(4):683–92.

[121] Easwaran H, Johnstone SE, Van Neste L, Ohm J, Mosbruger T, Wang Q, et al. A DNA hypermethylation module for the stem/progenitor cell signature of cancer. Genome Res 2012;22(5):837–49.

[122] Feinberg AP, Irizarry RA. Stochastic epigenetic variation as a driving force of development, evolutionary adaptation, and disease. Proc Natl Acad Sci USA 2010;107:1757–64.

[123] Eriksson A, Lennartsson A, Lehmann S. Epigenetic aberrations in acute myeloid leukemia: early key events during leukemogenesis. Exp Hematol 2015;43(8):609–24.

[124] Corces-Zimmerman MR, Hong WJ, Weissman IL, Medeiros BC, Majeti R. Preleukemic mutations in human acute myeloid leukemia affect epigenetic regulators and persist in remission. Proc Natl Acad Sci USA 2014;111(7):2548–53.

[125] Genovese G, Kahler AK, Handsaker RE, Lindberg J, Rose SA, Bakhoum SF, et al. Clonal hematopoiesis and blood-cancer risk inferred from blood DNA sequence. N Engl J Med 2014;371(26):2477–87.

[126] Feinberg AP, Ohlsson R, Henikoff S. The epigenetic progenitor origin of human cancer. Nat Rev Genet 2006;7(1):21–33.

[127] Mack SC, Witt H, Piro RM, Gu L, Zuyderduyn S, Stutz AM, et al. Epigenomic alterations define lethal CIMP-positive ependymomas of infancy. Nature 2014;506(7489):445–50.

[128] Ohm JE, McGarvey KM, Yu X, Cheng L, Schuebel KE, Cope L, et al. A stem cell-like chromatin pattern may predispose tumor suppressor genes to DNA hypermethylation and heritable silencing. Nat Genet 2007;39(2):237–42.

[129] Lyst MJ, Bird A. Rett syndrome: a complex disorder with simple roots. Nat Rev Genet 2015;16(5):261–75.

[130] Van Esch H. MECP2 duplication syndrome. Mol Syndromol 2012;2(3–5):128–36.

[131] Lyst MJ, Ekiert R, Ebert DH, Merusi C, Nowak J, Selfridge J, et al. Rett syndrome mutations abolish the interaction of MeCP2 with the NCoR/SMRT co-repressor. Nat Neurosci 2013;16(7):898–902.

[132] Chahrour M, Jung SY, Shaw C, Zhou X, Wong ST, Qin J, et al. MeCP2, a key contributor to neurological disease, activates and represses transcription. Science 2008;320(5880):1224–9.

[133] Mellen M, Ayata P, Dewell S, Kriaucionis S, Heintz N. MeCP2 binds to 5hmC enriched within active genes and accessible chromatin in the nervous system. Cell 2012;151(7):1417–30.

[134] Xu GL, Bestor TH, Bourc'his D, Hsieh CL, Tommerup N, Bugge M, et al. Chromosome instability and immunodeficiency syndrome caused by mutations in a DNA methyltransferase gene. Nature 1999;402(6758):187–91.

[135] Ehrlich M, Sanchez C, Shao C, Nishiyama R, Kehrl J, Kuick R, et al. ICF, an immunodeficiency syndrome: DNA methyltransferase 3B involvement, chromosome anomalies, and gene dysregulation. Autoimmunity 2008;41(4):253–71.

CHAPTER 4

ATP-Dependent Chromatin Remodeling: From Development to Disease

M. Lezzerini, C.G. Riedel

Integrated Cardio Metabolic Centre, Department of Medicine,
Karolinska Institutet, Huddinge, Sweden

4.1 INTRODUCTION

The nuclear DNA of eukaryotic cells is packaged into higher order nucleoprotein structures referred to as chromatin, which not only compacts the genome, but also maintains its integrity and regulates the accessibility of the underlying DNA to nuclear processes. The simplest unit of chromatin is the nucleosome, where approximately 150 bp of DNA are wrapped around an octamer of core histone proteins (two copies each of H2A, H2B, H3, and H4) [1]. A fifth histone protein, the linker histone H1, eventually helps to fold arrays of these nucleosomes into higher order chromatin structures, as described in previous chapters of this book.

Packaging the DNA into chromatin and the formation of higher order chromatin structures can restrict the accessibility of DNA sequences, to many factors [2]. Hence processes, such as, transcription, DNA replication, homologous recombination, or DNA repair require enzymes that can change DNA accessibility and expose (or sometimes occlude) the sequences required for binding by transcription factors, repair enzymes, etc. To this purpose, cells have evolved several mechanisms that can alter DNA sequence accessibility. On the one hand, they can add methylation marks to their DNA (Chapter 3) or add a variety of post-translational modifications to their histones (Chapter 2). These modifications are recognized by specific effector proteins, which, once recruited, enable chromatin rearrangements [3,4]. In addition to these two mechanisms, noncoding RNA molecules and the RNA interference machinery have been shown to contribute to the formation of inaccessible chromatin states [5,6]. However, in this book chapter we will focus on yet another mechanism, namely ATP-dependent chromatin remodeling, conferred by a variety of enzymes, which have the ability to change DNA sequence accessibility by changing the position or composition of histone octamers on the DNA [7].

CONTENTS

(Continued)
95

Chromatin Regulation and Dynamics. http://dx.doi.org/10.1016/B978-0-12-803395-1.00004-6

Table 4.1 Subunit Composition of the Different Human Chromatin Remodeling Complexes

SWI/SNF Family

ATPase	BRM or BRG1	BRG1
Complex	BAF	PBAF
Unique subunits	BAF250, SS18	BAF180, BAF200, BRD7
Common subunits	BAF45a–d, BAF47, BAF53a/b, BAF57, BAF60a–c, BAF155, BAF170, β-Actin	

INO80 Family

ATPase	INO80	SRCAP	p400
Complex	INO80	SRCAP	TIP60–p400
Unique subunits	ARP5/8, IES2/6, MCRS1, Amida, CCDC95, NFRKB, FLJ20309, UCH37, YY1	ARP6, ZnF–HIT1, XPG	TIP60, TRRAP, BRD8, MRG15/X, FLJ11730, MRGBP, ING3, EPC1/EPC-like; YL1, DMAP1, GAS41
Common subunits	BAF53a–b, RUVBL1/2, β-Actin		

ISWI Family

ATPase	SNF2H						SNF2H/L	SNF2L
Complex	ACF	CHRAC	NoRC	ToRC	RSF	WICH	CERF	NURF
Unique subunits	ACF1	ACF1, CHRAC15/17	TIP5	TIP5, CtBP	RSF1	WSTF	CERC2	BPTF, RbAp46/48

CHD Family

ATPase	CHD1/2	CHD3–5 (Mi-2α/β)	CHD6–9
Complex		NuRD	
Unique subunits		MBD2/3, MTA1–3, RbAp46/48, p66a/b, HDAC1/2, CDK2AP1	

This table is predominantly based on the following sources: Clapier CR, Cairns BR. The biology of chromatin remodeling complexes. Annu Rev Biochem 2009;78:273–304 [7] and Morrison AJ, Shen X. Chromatin remodelling beyond transcription: the INO80 and SWR1 complexes. Nat Rev Mol Cell Biol 2009;10(6):373–84 [12].

4.2 ATP-DEPENDENT CHROMATIN REMODELERS

ATP-dependent chromatin remodelers are proteins that use the energy provided by ATP hydrolysis to either reposition nucleosomes, eject them, or influence their assembly and composition. These remodelers consist of a catalytic subunit and, in most cases, also a set of accessory proteins. Accessory proteins can influence the activity and (by mediating protein–protein or protein–DNA interactions) also the localization and substrate specificity of the complex [7].

The catalytic subunits of ATP-dependent chromatin remodelers all fall into the Snf2 family of helicase-like proteins. They carry an ATPase domain, which is comprised of two lobes, a DExx-containing- and a HELICase superfamily c-terminal (HELICc) domain. Flanking these domains we find a number of domains that are unique to specific remodeler families, many of which have the ability to recognize posttranslational histone modifications, which helps the targeting of the chromatin remodelers to specific regions in the genome [8,9] (see also Section 4.3.2). Based on the unique domains in their catalytic subunits most chromatin remodelers can be grouped into four major families: SWI/SNF, ISWI, CHD, and INO80 [9–11]. We provide an overview of these ATP-dependent chromatin remodeler families and their human catalytic and accessory subunits in Table 4.1.

4.2.1 SWI/SNF Family

SWItch/Sucrose Non-Fermenting (SWI/SNF) family chromatin remodelers are large complexes of about 2 MDa in size. Their catalytic subunits are defined by two unique protein domains that flank their ATPase domain: a bromodomain and an HSA domain [HSA stands for "Helicase/Switch-defective protein 3 (Swi3)–Adaptor 2 (Ada2)–Nuclear receptor CoRepressor (N-CoR)–Transcription Factor (TF)IIIB (SANT)-Associated", of which the first binds acetylated histones while the second binds proteins related to actin [13,14]. Two main classes of SWI/SNF complexes exist: BRG1-Associated Factor (BAF) and Polybromo-containing BAF (PBAF) [7,15–18]. In humans, BAF contains one of two paralogous ATPase subunits, either BRG1 or BRM, while PBAF only contains BRG1. While they share the majority of their accessory subunits, they are further distinguished by specific signature subunits: most importantly BAF250 for the BAF complex and the bromodomain-containing proteins BRD7, BAF180, and BAF200 for the PBAF complex (Table 4.1). In conjunction with a few cell type–specific subunits, all these different subunit combinations lead to a variety of SWI/SNF remodeling complexes with distinct expression patterns, specificities, and thus physiological roles [7,19].

In general, SWI/SNF complexes reposition or eject nucleosomes and cooperate with transcription factors in promoting the transcriptional activation and thus expression of target genes [7,17,20,21].

4.2.2 ISWI Family

The catalytic subunits of the Imitation SWItch (ISWI) family of chromatin remodelers are characterized by the presence of two C-terminal domains, a SANT, and a SANT-Like ISWI (SLIDE) domain. Together they mediate binding of ISWI to DNA and to unmodified histone H4 tails [7,22]. ISWI can then associate with a wide variety of alternative accessory subunits, making it part of a diverse set of chromatin remodeling complexes with different specificities. Much of this specificity is achieved by accessory subunits carrying different kinds of chromatin-binding domains, leading to distinct targeting of the complexes. Some of the most prominent ISWI complexes are the ATP-utilizing Chromatin assembly and remodeling Factor (ACF), the NUcleosome Remodeling Factor (NURF), and the CHRomatin Accessibility Complex (CHRAC). However, a number of yet additional complexes have later been identified [i.e., the Nucleolar Remodeling Complex (NoRC), the Toutatis-containing chromatin Remodeling Complex (ToRC), the WSTF–ISWI CHromatin remodeling complex (WICH), the Remodeling and Spacing Factor (RSF), and the CERC2-containing Remodeling Factor (CERF)] [7,9,23]. Looking at the more prominent complexes, CHRAC and ACF are very similar to each other. They include a large Acf1 subunit, which contains a bromodomain, two Plant HomeoDomain (PHD) fingers, and a WSTF/Acf1/Cbp146 (WAC) domain necessary for efficient binding to DNA and for chromatin assembly [24–26]. In contrast, NURF complexes are characterized by the BPTF subunit, which mediates binding to and thus synergizes with a variety of sequence-specific transcription factors [27,27a]. ISWI family complexes have generally been implicated in the control of nucleosome spacing and thereby chromatin assembly, which predominantly promotes gene silencing but also DNA replication and chromosome segregation [28]. For example, CHRAC participates in *Polycomb*-mediated gene repression in *Drosophila melanogaster* [29], and the mammalian NoRC complex mediates heterochromatin formation and thereby silences ribosomal, as well as, centromeric DNA [30,31]. Somewhat distinct is the NURF complex, which is thought to disrupt regular nucleosome spacing. This in turn can lead to nucleosome displacement from regulatory DNA sequences and thus gene activation [32].

4.2.3 CHD Family

Defining feature of the Chromodomain Helicase DNA-binding (CHD) family of chromatin remodelers are two chromodomains positioned in tandem in the N-terminal part of the catalytic subunit. These domains enable binding to DNA or to methylated lysines on histone tails. Presence of additional motifs further divides the CHD family into three subfamilies [9,11,33]: The Chd1 subfamily comprises CHD1 and 2, and is defined by a C-terminal SANT/SLIDE-like DNA-binding domain with particular affinity for AT-rich regions [34,35]; the Mi-2 subfamily comprises CHD3 and 4 (also known as Mi-2α and β) and CHD5, and is defined by the presence of two N-terminal PHD domains [33,36]; and the KISMET/CHD7 subfamily comprises CHD6 to 9, and is defined by the presence of SANT, as well as, BRahma and Kismet (BRK) domains [33].

In contrast to the SWI/SNF and ISWI families of chromatin remodelers, complex composition and also functions of the CHD chromatin remodelers in vivo are rather diverse. While several of the CHD family members function as monomers, others assemble into large complexes in which accessory subunits can provide additional chromatin-binding domains or even enzymatic activities. While some CHD chromatin remodelers displace nucleosomes and can activate transcription, others fulfill transcriptional repressive roles [9,33]. For example, CHD1 is thought to function as a monomer and to control nucleosome spacing, similar to ISWI [37]. It often cooperates with transcription factors and the transcription machinery to activate target gene expression [38,39]. Quite distinct from CHD1 is the well-characterized Nucleosome Remodeling and Deacetylase (NuRD) complex of the Mi-2 subfamily. As ATPase subunit, it carries either CHD3, 4, or 5, which then associate with a number of accessory subunits—most notably Methyl CpG–Binding Domain (MBD) proteins, MeTastasis-Associated (MTA) proteins, p66, and Retinoblastoma-Associated Protein (RbAP), as well as, the histone deacetylases HDAC1 or 2 [40–45]. Its ATP-dependent chromatin remodeling activity in combination with the histone-deacetylating activity from its HDAC subunit makes the NuRD complex predominantly a transcriptional repressor.

4.2.4 INO80 Family

The fourth family of ATP-dependent chromatin remodelers is the INOsitol-requiring 80 family (INO80), which in humans includes the ATPase subunits INO80, Snf2-Related CBP Activator Protein (SRCAP), and p400. Each of them function as part of larger multisubunit complexes, namely the INO80, SRCAP, and TIP60–p400 complexes, respectively [12].

Similar to the other three remodeler families, chromatin remodeling complexes of the INO80 family also can reposition and eject nucleosomes [46]. However, several features make this family distinct. Structurally, the ATPase domains of their catalytic subunits have long sequence insertions that are required for correct assembly of these complexes. Functionally, these complexes exhibit DNA-helicase activity (due to presence of the helicase-like accessory subunits RUVBL1/2), and they show affinity for DNA structures found at replication forks and at sites undergoing homologous recombination. Moreover, they have the ability to bind to and exchange dimers of H2A/H2B histones and histone variants from already assembled nucleosomes [12,46,47]. Especially this last function is important, because it controls the deposition of the histone variant H2AZ. Interestingly, the INO80 complex can remove the histone variant H2AZ from nucleosomes, while SRCAP and TIP60–p400 complexes deposit H2AZ [12,47,48]. Presence of H2AZ in chromatin eventually influences transcription, DNA replication, and has the ability to impair homologous recombination [47–49].

As a consequence of all these properties, INO80 family chromatin remodelers fulfill a multitude of functions in vivo: They all regulate transcription, mostly acting as transcriptional activators; they participate in DNA damage response and telomere maintenance (will be discussed in more detail in Section 4.5); and specifically the INO80 complex can also alleviate DNA replication stress and thus help assure DNA replication under genotoxic conditions [12,46].

4.3   MECHANISM OF ACTION AND TARGETING OF ATP-DEPENDENT CHROMATIN REMODELING COMPLEXES

4.3.1   Mechanism of ATP-Dependent Chromatin Remodeling

Chromatin or nucleosome remodeling factors can reposition (or "slide") nucleosomes, facilitate the binding of transcription factors to nucleosomal DNA, exchange histones and histone variants, evict histone octamers, contribute to nucleosome assembly, or regularly space arrays of nucleosomes [7,50,51]. Many studies have investigated the mechanisms by which chromatin remodelers catalyze these reactions. Although there is no model that would fully describe the series of events leading to the remodeled products, a lot of valuable insights have been gained from these studies. For instance regarding the minimal substrates, it is well established that SWI/SNF family enzymes can remodel in vitro–assembled core nucleosomes [52], while ISWI and CHD family remodelers need DNA to protrude outside of the nucleosome for stable binding and remodeling [53–58]. These differences are consistent with data suggesting that, when the SWI/SNF complex binds to the nucleosome, it encircles it in its central cavity [59,60], while ISWI and CHD proteins use their SANT–SLIDE domains to bind to extranucleosomal DNA [35,54,61]. For a remodeling reaction by SWI/SNF or ISWI to occur, the ATPase domain of the remodeler first binds about 20 bp away from the pseudodyad of the nucleosome. This event leads to conformational changes in the DNA and/or to the detachment ("looping" or "bulging") of DNA from the histones, as suggested by the observation that the nucleosomal DNA now becomes more accessible to nuclease digestion [62–64]. Displacement of DNA away from H2A at the entry [65] and at the exit [66] sides of the nucleosome has also been observed upon ACF binding to the nucleosome, prior to ATP hydrolysis. Similar observations have been made for the RSC chromatin remodeling complex in yeast [59]. Finally, ATP-dependent DNA translocation occurs [64,67–69], forces in the piconewton range are generated [70], and histone–DNA contacts are rearranged, placing the histone octamer in its new location on the DNA double strand. In the cases of SWI/SNF, such remodeling cycle shifts the nucleosome by about 50 bp and in the case of ISWI by about 10 bp [62].

Despite such knowledge, much remains to be learned about these nucleosome remodeling reactions. An eagerly pursued goal in the field is to uncover the high resolution structures of remodeling factors bound to the nucleosome, which may answer some of the remaining questions.

4.3.2 Targeting of ATP-Dependent Chromatin Remodeling Complexes

Besides their catalytic activity, another important aspect of ATP-dependent chromatin remodeling complexes is their targeting to the intended substrate regions in the genome. Unlike transcription factors, which usually bind very specific DNA motifs, chromatin remodeling factors only show moderate DNA sequence preferences [71]. So, how are remodelers targeted to specific loci? It is thought to be the combined action of several different targeting mechanisms that determines which genomic regions will be remodeled. Maybe the most specific means of targeting chromatin remodelers is their recruitment via transcription factors: Here the transcription factor provides the DNA sequence specificity and recruits the remodeler to elicit chromatin changes that influence transcription at the site. Examples for such recruitment include the targeting of the SWI/SNF complex by steroid receptors [21,72], the recruitment of the NURF complex by factors that regulate the boundaries between active and inactive chromatin domains [27,27a], or the binding of the NuRD complex to factors that regulate erythroid development [73]. Targeting by other DNA-bound proteins can also occur. For example, the preinitiation complex is known to recruit CHD1 to sites of transcription [38], or PCNA recruits SWI/SNF [74] and CAF1 recruits ACF [75] to sites of DNA replication.

A second important targeting mechanism comes from the fact that the remodeling ATPases and their accessory subunits frequently contain domains with affinity for histone marks, in particular bromodomains that bind acetylated histones, or chromodomains and PHD fingers that each have affinity for methylated histones [76]. Some remodelers can also recognize methylated DNA, in particular the NuRD complex via its MBD2 subunit [77]. Interestingly, long nucleosome-free stretches of DNA attract binding by the yeast INO80 family remodeler SWR1 [78].

While all of the above mechanisms are mediated by protein–protein and protein–DNA interactions, recent work has shown that even chromatin-associated non-coding RNAs bind and thus may help to target chromatin remodelers [79]. This illustrates the plethora of mechanisms that can be involved in guiding chromatin remodelers to specific sites in the genome, and it is usually their combination and synergy that determines the genomic regions where remodeling reactions occur.

4.4 THE ROLE OF CHROMATIN REMODELING COMPLEXES DURING DEVELOPMENT

At this point, we have discussed the structures and mechanisms of ATP-dependent chromatin remodeling complexes. But what are their physiological roles? Many studies in different model organisms, ranging from simple model organisms, such as, worms or flies all the way to mammals have tried to address this question and found a variety of crucial roles, most commonly during metazoan development [19].

4.4.1   Functions of SWI/SNF Complexes in Development

Of all the chromatin remodeling complexes, SWI/SNF complexes are the best studied and have proven essential for proper development. Historically, the first metazoan SWI/SNF genes were discovered in *D. melanogaster*, which acted as suppressors of patterning defects caused by *Polycomb* gene mutations [80]. These genes were named *brahma (brm)*, *moira (mor)*, and *osa*—resembling human BRG1/BRM, BAF155/170, and BAF250, respectively (Table 4.1). In *D. melanogaster*, maternal as well as zygotic *brm* are necessary for proper embryogenesis. Embryos completely void of *brm* thus die in late embryogenesis, whereas heterozygous mutants display a variety of developmental defects [81].

Also in mammals, the SWI/SNF family of remodelers plays crucial developmental roles. Their functions are diverse, being involved at many stages from genome activation in the zygote all the way to the terminal differentiation of tissues [19]. In order to assure these diverse functions, SWI/SNF complexes assemble in a combinatorial fashion with many of their subunits being specifically expressed only in certain cell types or physiological contexts. As already mentioned in Section 4.2.1, mammals express two alternative ATPase subunits, BRG1 and BRM, and there are two classes of SWI/SNF complexes that differ by signature accessory subunits, namely the BAF and PBAF complexes. Several studies demonstrated that BRG1 and BRM, as well as, the BAF and PBAF complexes have different expression patterns and functions. For example, in mice, maternal *Brg1* is essential for embryogenesis. Depletion of *Brg1* results in zygotic genome activation failure and arrest of embryonic development within already the very first cell divisions [82,83]. On the other hand, mouse *Brm* is largely dispensable during development, with its loss only leading to mild cell overproliferation in specific tissues [84].

Such early developmental requirement of *Brg1* is further supported by observations made in Embryonic Stem Cells (ESCs). ESCs specifically express an esBAF complex, which contains BRG1 and makes exclusive use of the accessory subunit BAF155, while being void of the closely related BAF170 [85,86]. esBAF is required for the self-renewal and pluripotency of ESCs [87,88].

Interestingly, differentiation of ESCs into dedicated cell types likewise can depend on BAF complexes and their composition. A prominent example is the differentiation into neurons, which requires the exchange of multiple BAF subunits: Lineage entry begins with exclusion of BAF60b from the BAF complex, to be replaced by BAF60a and c, and also occasional incorporation of BRM, to yield neuronal progenitors. In the later course of terminal differentiation, BAF45a and BAF53a need to be replaced by BAF45b/45c and BAF53b, respectively [89].

A second prominent role of SWI/SNF complexes lies in the development of the heart; here BAF complexes containing BAF60c instead of BAF60a cooperate with transcription factors of the cardiac lineage to promote differentiation into cardiomyocytes [90]. At the same time, PBAF (as addressed by mutation of BAF180) is required in the epicardium for coronary vessel and heart chamber development [91,92]. Beyond neuronal and cardiac development, SWI/SNF is involved in the development of skeletal muscle and T cells [19], with presumably even more roles yet to be discovered.

Importantly, two human syndromes caused by mutations in SWI/SNF subunits have been described: Coffin–Siris and Nicolaides–Baraitser syndromes [93]. Both are characterized by developmental defects, in particular morphological abnormalities and intellectual disability.

4.4.2 Functions of ISWI Complexes in Development

Similarly to SWI/SNF, metazoan ISWI also was first identified in *D. melanogaster*. The loss of ISWI thus leads to lethality during late larval or early pupal development [94]. It is required for chromosome structure and condensation, likely due to a role in facilitating the incorporation of histone H1 into chromatin [95].

Also for ISWI, mammals have diversified the portfolio of subunits that can be incorporated into ISWI complexes: There are thus two catalytic ATPase subunits, SNF2L and SNF2H, each with distinct expression patterns and functions. SNF2L exclusively incorporates into NURF or CERF complexes, while SNF2H is found in ACF, CHRAC, NoRC, ToRC, RSF, CERF, or WICH complexes [19,28] (Table 4.1). Knowledge on the physiological roles of all these complexes is still incomplete. Nevertheless, existing data show some essential roles during development: For example, loss of NURF activity in mice by mutation of its BPTF subunit leads to defects in germ-layer formation and eventual lethality around day 8 of embryonic development [96]. Consistently, in ESCs the NURF complex is required for differentiation into endoderm and mesoderm lineages. Mutation of the second SNF2L-containing complex, CERF, also leads to embryonic defects, in particular during neurulation [23].

Similarly to the loss of SNF2L-containing complexes, deletion of SNF2H leads to early embryonic arrest in mice [97]. In humans, mutations in the WICH subunit WSTF cause Williams–Beuren syndrome, which leads to cognitive, cardiac, and other morphological defects [98].

4.4.3 Functions of CHD Complexes in Development

The CHD family contains nine different ATPase subunits, most of which have been characterized for their roles in development. The Chd1 subfamily has been well studied in *D. melanogaster*, where Chd1 is required for the incorporation of the histone variant H3.3 into chromatin. *Chd1* loss leads to defects in oogenesis, spermatogenesis, as well as, early development [99,100]. In mice, CHD1 is required to maintain pluripotency of ESCs, while its loss promotes differentiation into neural lineages and prevents the formation of the endoderm [101].

In the Mi-2 subfamily, NuRD complexes have been studied extensively. Similarly to SWI/SNF or ISWI, they assemble in a variety of subunit combinations, depending on tissue and physiological context. This gives them diverse, mostly transcriptionally repressive, context-dependent functions [41,77]. For example in mice, the NuRD accessory subunit MBD3 is required for silencing of pluripotency genes, and its loss leads to lethality shortly after blastocyst implantation [102,103].

In the case of KISMET/CHD7 subfamily, research was initiated with the discovery of *kismet* in *D. melanogaster*. Similarly to the SWI/SNF ATPase *brahma*, *kismet* was identified as a suppressor of *Polycomb* gene mutations during early development [80]. Research in mammals has focused on the closest Kismet ortholog, CHD7, that appears to be required for the transcriptional activation of tissue-specific genes during differentiation [104–106]. In humans, mutations in this chromatin remodeler cause CHARGE syndrome, characterized by a multitude of developmental phenotypes, in particular growth and morphological retardation, as well as, heart and eye defects [107]. More recently, CHD7 has been shown to bind and cooperate with the transcription factor SOX2, jointly regulating expression of a variety of target genes causal to developmental disorders, indicating that CHD7 may even influence a yet broader panel of syndromes [108]. Finally, CHD8 prevents p53-mediated apoptosis during early embryogenesis in mice by promoting the assembly of p53-inhibiting CHD8–p53–histone H1 complexes on DNA [109].

4.4.4 Functions of INO80 Complexes in Development

This brings us to the INO80 family of ATP-dependent chromatin remodelers, which likewise appear to be important for mammalian development. In mouse ESCs, the TIP60–p400 complex has been shown to repress genes that otherwise lead to ESC differentiation [110]. Similarly, INO80 has been shown to promote pluripotency by activating the transcription of many pluripotency genes, including *Nanog*, *Oct4*, and *Sox2* [111].

4.5 THE ROLE OF CHROMATIN REMODELING COMPLEXES IN CANCER

In light of the many roles that ATP-dependent chromatin remodelers play in self-renewal, differentiation processes, and DNA repair, their involvement in neoplastic pathologies is not surprising. In particular, members of the CHD and SWI/SNF families appear important in this context.

Several CHD family subunits are thus mutated or misexpressed in tumor cells. For example CHD1 has been identified as a tumor suppressor gene, located in a genomic region often deleted in prostate cancer [112–114].

An even more prominent role has been observed for the NuRD and NuRD-like complexes (Mi-2 subfamily). Recent studies found their ATPase subunits to be frequently mutated in a variety of cancers. For example, Chd3 was frequently mutated in lymphomas [33], Chd4 in endometrial tumors [115], or Chd5 in several cancers including neuroblastomas [116]. Moreover, the NuRD accessory subunit MeTastasis-Associated protein 3 (MTA3) has been shown to inhibit epithelial-to-mesenchymal transition in breast cancer, implicating MTA3 in the prevention of a necessary step for metastasis formation and for cancer progression [117]. Given these findings, the NuRD complexes could be considered to function as tumor suppressors in breast cancer. However, their role in tumor development is more complex, as there is also contrasting data available, most notably for the NuRD subunit MTA1. MTA1 is thus a direct target of c-MYC in human cancer cell lines and promotes the transformation potential of c-MYC [118,119]. In line with this observation, elevated levels of MTA1 have been detected in many types of tumors, which often positively correlates with their severity and progression to metastatic stages [118,119]. These tumor-promoting roles of MTA1 clearly differ from the tumor-suppressive roles of other NuRD subunits, making it difficult to dissect the exact role of NuRD complexes in cancer. Future studies will have to clarify how these different phenotypes can arise.

Also the SWI/SNF family of chromatin remodelers has been linked to neoplastic diseases. Recent sequencing efforts on numerous cancer genomes have shown that mutations affecting SWI/SNF complex subunits are present in 20% of all human tumors [120]. Several studies have indicated that both of its ATPase subunits, BRG1 and BRM, are necessary for proper regulation of cell proliferation. Both are downregulated in many tumor cell lines, and, more importantly, proliferation can be slowed down or arrested by reintroducing their expression [121–124]. Their tumor suppressor function may partly be explained by their ability to interact with retinoblastoma protein to jointly repress E2F1, thereby inhibiting the cell cycle [121,125]. Also the accessory subunits of SWI/SNF can have strong tumor-suppressive roles. For example, loss of BAF47 (also known as INI1 or SNF5) causes rhabdoid tumors [126,127], a tumor-suppressive capability that seems to derive from SWI/SNF's role in functionally counteracting the gene silencing activity of the Polycomb complexes. While in normal cells SWI/SNF complexes regulate Polycomb localization and function, this regulation appears to be lost in cancer cells. Particularly when BAF47 is mutated, SWI/SNF fails to remove Polycomb from the Ink4a locus, which results in sustained cell proliferation [128]. Another subunit frequently mutated in cancers is the BAF signature subunit BAF250. Loss-of-function mutations in BAF250 have been detected in a variety of cancers, most notably in neuroblastomas and ovarian carcinomas [9,129,130]. Moreover, lack of BAF250a expression correlates with malignancy in numerous cancer types [131–133]. Taken together, SWI/SNF complexes predominantly function as tumor suppressors, although also here exceptions exist. For example, a translocation affecting the BAF subunit SS18 leads to the expression of a fusion protein that drives the proliferation of synovial sarcoma and relies on SWI/SNF activity [134].

Finally, the INO80 family of chromatin remodelers shows several links to cancer, too. For example, haploinsufficiency of mammalian TIP60 (part of the TIP60–p400 complex) has been found to accelerate oncogene-induced tumorigenesis in mice [135]. On the other hand, the INO80 family accessory subunits RUVBL1 and 2 are often overexpressed in cancers [136].

Interestingly though, while most cancer-related roles of ATP-dependent chromatin remodelers derive from their influence on transcription, the INO80 family contributes to cancer development also with transcription-independent functions. The most prominent is their role in the repair of DNA damage. The INO80 complex is recruited to sites of DNA double-strand breaks by its ability to directly bind phosphorylated H2AX (γ-H2AX), an early marker of DNA double-strand breaks, and is thought to facilitate the later repair events by homologous recombination [137,138]. Also the SRCAP complex may support DNA repair, but by nonhomologous end joining [12]. Consistently, the INO80 family of chromatin remodelers promote cell survival under genotoxic stress [137–141]. Finally, INO80 family remodelers are found at telomeres and are implicated in telomere maintenance [136,142].

4.6 THE ROLE OF CHROMATIN REMODELING COMPLEXES DURING AGING

Recent studies have found that ATP-dependent chromatin remodeling also plays a role in the regulation of aging, a major determinant of human health and longevity. The first link was provided by a study in humans, showing that multiple NuRD subunits are downregulated in aged individuals, as well as, in patients with Hutchinson–Gilford Progeria Syndrome (an accelerated aging syndrome). This decline in NuRD complex function appears to contribute to aging-associated chromatin defects and DNA damage accumulation [143].

Another study in *Caenorhabditis elegans* has demonstrated that the SWI/SNF complex participates in the promotion of stress resistance and longevity, because it functions as a coactivator of the transcription factor DAF-16/FOXO, a well-known driver of these processes [21,144]. Activated DAF-16/FOXO recruits a BAF-like SWI/SNF complex to target promoters, where BAF presumably initiates chromatin rearrangements to facilitate expression of the downstream stress resistance and longevity-promoting genes [21].

Finally, one of the catalytic subunits of ISWI complexes in yeast, Isw2, has been identified in a screen for lifespan changes. Deletion of *ISW2* extends replicative lifespan as a result of stress response gene derepression and DNA damage repair pathway enhancement [145]. In addition, the authors observed increased longevity in *C. elegans* after knockdown of the ISWI accessory subunit and ACF1 ortholog *athp-2*.

All these studies indicate that ATP-dependent chromatin remodelers can affect stress resistance, aging, and eventually longevity. However, this research area is still in its infancy, and thus further studies will be required to fully understand the involvement of ATP-dependent chromatin remodelers in this important physiological context.

4.7 CONCLUSIONS

Many years of research investigating ATP-dependent chromatin remodelers and their functions have demonstrated that these complexes are essential regulators of chromatin assembly and accessibility and that they play a crucial role in diverse cellular processes, such as, transcription, DNA replication, recombination, and repair. In this book chapter, we have discussed the different chromatin remodelers of the Snf2 family of helicase-like proteins, their subfamilies, and a multitude of catalytic and accessory subunits. Different subfamilies, different expression patterns of their subunits, and the combinatorial assembly of these subunits into chromatin remodeling complexes provide an impressive portfolio of remodeling activities and specificities, some of which are present only in particular tissues and physiological contexts.

Many in vitro and also some in vivo studies have attempted to understand the mechanisms by which ATP-dependent chromatin remodelers are targeted to certain genomic regions, and how they catalyze their remodeling reactions. We presented examples of these studies in Section 4.3, however, a complete understanding of how targeting and remodeling occurs is still lacking. In addition, some of the remodeling complexes can fulfill functions that go beyond the simple repositioning of nucleosomes, which often are even less understood. Future studies will have to provide this missing mechanistic insight.

In metazoans, most of the phenotypes caused by loss- or gain-of-function mutations in ATP-dependent chromatin remodelers have been observed during development. This is not surprising, given that these enzymes are so crucial for key cellular processes, such as, transcription or DNA replication. However, more recent research has demonstrated ATP-dependent chromatin remodeler functions in many other physiological and pathological contexts, too. The discovery of their frequent roles in cancer, predominantly as tumor suppressors but sometimes also as oncogenes, has identified ATP-dependent chromatin remodelers and the processes they catalyze as interesting therapeutic targets, with the search for small molecules that influence their actions being well underway [146]. Moreover, the involvement of ATP-dependent chromatin remodelers in the regulation of aging indicates that understanding how these complexes function and how we can manipulate their actions could represent a promising route, not only to resolve pathological conditions, but also to improve human health and longevity in general.

Abbreviations

ACF	ATP-utilizing Chromatin assembly and remodeling Factor
BAF	BRG1-Associated Factor
BRK	BRahma and Kismet
CERF	CERC2-containing Remodeling Factor
CHD	Chromodomain Helicase DNA-binding
CHRAC	CHRomatin Accessibility Complex
ESCs	Embryonic Stem Cells
HSA	Helicase/SANT-Associated
γ-H2AX	Phosphorylated H2AX
HELICc	HELICase superfamily c-terminal
INO80	INOsitol-requiring 80
ISWI	Imitation SWItch
MBD	Methyl CpG–Binding Domain
MTA	MeTastasis-Associated
NoRC	Nucleolar Remodeling Complex
NuRD	Nucleosome Remodeling and Deacetylase
NURF	NUcleosome Remodeling Factor
PBAF	Polybromo-containing BAF
PHD	Plant HomeoDomain
RbAP	Retinoblastoma-Associated Protein
RSF	Remodeling and Spacing Factor
SANT	Switch-defective protein 3 (Swi3), Adaptor 2 (Ada2), Nuclear receptor CoRepressor (N-CoR), Transcription factor (TF)IIIB
SLIDE	SANT-Like ISWI
SRCAP	Snf2-Related CBP Activator Protein
SWI/SNF	SWItch/Sucrose Non-Fermenting
ToRC	Toutatis-containing chromatin Remodeling Complex
WAC	WSTF/Acf1/Cbp146
WICH	WSTF–ISWI CHromatin remodeling complex

Acknowledgments

We thank Karim Bouazoune and Lluís Millán Ariño for the critical reading of the manuscript.

References

[1] Kornberg RD. Chromatin structure: a repeating unit of histones and DNA. Science 1974;184(4139):868–71.

[2] Zaret KS, Carroll JS. Pioneer transcription factors: establishing competence for gene expression. Genes Dev 2011;25(21):2227–41.

[3] Strahl BD, Allis CD. The language of covalent histone modifications. Nature 2000;403(6765):41–5.

[4] Cedar H, Bergman Y. Programming of DNA methylation patterns. Annu Rev Biochem 2012;81:97–117.

[5] Holoch D, Moazed D. RNA-mediated epigenetic regulation of gene expression. Nat Rev Genet 2015;16(2):71–84.

[6] Martienssen R, Moazed D. RNAi and heterochromatin assembly. Cold Spring Harb Perspect Biol 2015;7(8):a019323.

[7] Clapier CR, Cairns BR. The biology of chromatin remodeling complexes. Annu Rev Biochem 2009;78:273–304.

[8] Taverna SD, Li H, Ruthenburg AJ, Allis CD, Patel DJ. How chromatin-binding modules interpret histone modifications: lessons from professional pocket pickers. Nat Struct Mol Biol 2007;14(11):1025–40.

[9] Längst G, Manelyte L. Chromatin remodelers: from function to dysfunction. Genes 2015;6(2):299–324.

[10] Eisen JA, Sweder KS, Hanawalt PC. Evolution of the SNF2 family of proteins: subfamilies with distinct sequences and functions. Nucleic Acids Res 1995;23(14):2715–23.

[11] Flaus A, Martin DMA, Barton GJ, Owen-Hughes T. Identification of multiple distinct Snf2 subfamilies with conserved structural motifs. Nucleic Acids Res 2006;34(10):2887–905.

[12] Morrison AJ, Shen X. Chromatin remodelling beyond transcription: the INO80 and SWR1 complexes. Nat Rev Mol Cell Biol 2009;10(6):373–84.

[13] Szerlong H, Hinata K, Viswanathan R, Erdjument-Bromage H, Tempst P, Cairns BR. The HAS domain binds nuclear actin-related proteins to regulate chromatin-remodeling ATPases. Nat Struct Mol Biol 2008;15(5):469–76.

[14] Filippakopoulos P, Knapp S. The bromodomain interaction module. FEBS Lett 2012;586(17):2692–704.

[15] Wang W, Xue Y, Zhou S, Kuo A, Cairns BR, Crabtree GR. Diversity and specialization of mammalian SWI/SNF complexes. Genes Dev 1996;10(17):2117–30.

[16] Wang W, Côté J, Xue Y, Zhou S, Khavari PA, Biggar SR, et al. Purification and biochemical heterogeneity of the mammalian SWI–SNF complex. EMBO J 1996;15(19):5370–82.

[17] Lemon B, Inouye C, King DS, Tjian R. Selectivity of chromatin-remodelling cofactors for ligand-activated transcription. Nature 2001;414(6866):924–8.

[18] Yan Z, Cui K, Murray DM, Ling C, Xue Y, Gerstein A, et al. PBAF chromatin-remodeling complex requires a novel specificity subunit, BAF200, to regulate expression of selective interferon-responsive genes. Genes Dev 2005;19(14):1662–7.

[19] Ho L, Crabtree GR. Chromatin remodelling during development. Nature 2010;463(7280):474–84.

[20] Yudkovsky N, Logie C, Hahn S, Peterson CL. Recruitment of the SWI/SNF chromatin remodeling complex by transcriptional activators. Genes Dev 1999;13(18):2369–74.

[21] Riedel CG, Dowen RH, Lourenco GF, Kirienko NV, Heimbucher T, West JA, et al. DAF-16 employs the chromatin remodeller SWI/SNF to promote stress resistance and longevity. Nat Cell Biol 2013;15(5):491–501.

[22] Grüne T, Brzeski J, Eberharter A, Clapier CR, Corona DFV, Becker PB, et al. Crystal structure and functional analysis of a nucleosome recognition module of the remodeling factor ISWI. Mol Cell 2003;12(2):449–60.

[23] Banting GS, Barak O, Ames TM, Burnham AC, Kardel MD, Cooch NS, et al. CECR2, a protein involved in neurulation, forms a novel chromatin remodeling complex with SNF2L. Hum Mol Genet 2005;14(4):513–24.

[24] Varga-Weisz PD, Wilm M, Bonte E, Dumas K, Mann M, Becker PB. Chromatin-remodelling factor CHRAC contains the ATPases ISWI and topoisomerase II. Nature 1997;388(6642):598–602.

[25] Ito T, Levenstein ME, Fyodorov DV, Kutach AK, Kobayashi R, Kadonaga JT. ACF consists of two subunits, Acf1 and ISWI, that function cooperatively in the ATP-dependent catalysis of chromatin assembly. Genes Dev 1999;13(12):1529–39.

[26] Fyodorov DV, Kadonaga JT. Binding of Acf1 to DNA involves a WAC motif and is important for ACF-mediated chromatin assembly. Mol Cell Biol 2002;22(18):6344–53.

[27] Xiao H, Sandaltzopoulos R, Wang HM, Hamiche A, Ranallo R, Lee KM, et al. Dual functions of largest NURF subunit NURF301 in nucleosome sliding and transcription factor interactions. Mol Cell 2001;8(3):531–43.

[27a] Li X1, Wang S, Li Y, Deng C, Steiner LA, Xiao H, Wu C, Bungert J, Gallagher PG, Felsenfeld G, Qiu Y, Huang S. Chromatin boundaries require functional collaboration between the hSET1 and NURF complexes. Blood 2011;118(5):1386–94.

[28] Dirscherl SS, Krebs JE. Functional diversity of ISWI complexes. Biochem Cell Biol 2004;82(4):482–9.

[29] Fyodorov DV, Blower MD, Karpen GH, Kadonaga JT. Acf1 confers unique activities to ACF/CHRAC and promotes the formation rather than disruption of chromatin in vivo. Genes Dev 2004;18(2):170–83.

[30] Santoro R, Li J, Grummt I. The nucleolar remodeling complex NoRC mediates heterochromatin formation and silencing of ribosomal gene transcription. Nat Genet 2002;32(3):393–6.

[31] Guetg C, Lienemann P, Sirri V, Grummt I, Hernandez-Verdun D, Hottiger MO, et al. The NoRC complex mediates the heterochromatin formation and stability of silent rRNA genes and centromeric repeats. EMBO J 2010;29(13):2135–46.

[32] Alkhatib SG, Landry JW. The nucleosome remodeling factor. FEBS Lett 2011;585(20):3197–207.

[33] Marfella CGA, Imbalzano AN. The Chd family of chromatin remodelers. Mutat Res 2007;618(1–2):30–40.

[34] Stokes DG, Perry RP. DNA-binding and chromatin localization properties of CHD1. Mol Cell Biol 1995;15(5):2745–53.

[35] Ryan DP, Sundaramoorthy R, Martin D, Singh V, Owen-Hughes T. The DNA-binding domain of the Chd1 chromatin-remodelling enzyme contains SANT and SLIDE domains. EMBO J 2011;30(13):2596–609.

[36] Paul S, Kuo A, Schalch T, Vogel H, Joshua-Tor L, McCombie WR, et al. Chd5 requires PHD-mediated histone 3 binding for tumor suppression. Cell Rep 2013;3(1):92–102.

[37] Pointner J, Persson J, Prasad P, Norman-Axelsson U, Strålfors A, Khorosjutina O, et al. CHD1 remodelers regulate nucleosome spacing in vitro and align nucleosomal arrays over gene coding regions in S. pombe. EMBO J 2012;31(23):4388–403.

[38] Lin JJ, Lehmann LW, Bonora G, Sridharan R, Vashisht AA, Tran N, et al. Mediator coordinates PIC assembly with recruitment of CHD1. Genes Dev 2011;25(20):2198–209.

[39] Skene PJ, Hernandez AE, Groudine M, Henikoff S. The nucleosomal barrier to promoter escape by RNA polymerase II is overcome by the chromatin remodeler Chd1. Elife 2014;3(3):e02042.

[40] Feng Q, Zhang Y. The MeCP1 complex represses transcription through preferential binding, remodeling, and deacetylating methylated nucleosomes. Genes Dev 2001;15(7):827–32.

[41] Bowen NJ, Fujita N, Kajita M, Wade PA. Mi-2/NuRD: multiple complexes for many purposes. Biochim Biophys Acta 2004;1677(1–3):52–7.

[42] Kolla V, Naraparaju K, Zhuang T, Higashi M, Kolla S, Blobel GA, et al. The tumour suppressor CHD5 forms a NuRD-type chromatin remodelling complex. Biochem J 2015;468(2):345–52.

[43] Spruijt CG, Bartels SJJ, Brinkman AB, Tjeertes JV, Poser I, Stunnenberg HG, et al. CDK2AP1/DOC-1 is a bona fide subunit of the Mi-2/NuRD complex. Mol Biosyst 2010;6(9):1700–6.

[44] Zhang Y, LeRoy G, Seelig HP, Lane WS, Reinberg D. The dermatomyositis-specific autoantigen Mi2 is a component of a complex containing histone deacetylase and nucleosome remodeling activities. Cell 1998;95(2):279–89.

[45] Wade PA, Jones PL, Vermaak D, Wolffe AP. A multiple subunit Mi-2 histone deacetylase from Xenopus laevis cofractionates with an associated Snf2 superfamily ATPase. Curr Biol 1998;8(14):843–6.

[46] Shen X, Mizuguchi G, Hamiche A, Wu C. A chromatin remodelling complex involved in transcription and DNA processing. Nature 2000;406(6795):541–4.

[47] Papamichos-Chronakis M, Watanabe S, Rando OJ, Peterson CL. Global regulation of H2A.Z. localization by the INO80 chromatin-remodeling enzyme is essential for genome integrity. Cell 2011;144(2):200–13.

[48] Alatwi HE, Downs JA. Removal of H2A.Z. by INO80 promotes homologous recombination. EMBO Rep 2015;16(8):986–94.

[49] Bargaje R, Alam MP, Patowary A, Sarkar M, Ali T, Gupta S, et al. Proximity of H2A.Z. containing nucleosome to the transcription start site influences gene expression levels in the mammalian liver and brain. Nucleic Acids Res 2012;40(18):8965–78.

[50] Narlikar GJ, Sundaramoorthy R, Owen-Hughes T. Mechanisms and functions of ATP-dependent chromatin-remodeling enzymes. Cell 2013;154(3):490–503.

[51] Becker PB, Workman JL. Nucleosome remodeling and epigenetics. Cold Spring Harb Perspect Biol 2013;5(9.):a017905.

[52] Kwon H, Imbalzano AN, Khavari PA, Kingston RE, Green MR. Nucleosome disruption and enhancement of activator binding by a human SW1/SNF complex. Nature 1994;370(6489):477–81.

[53] Brehm A, Längst G, Kehle J, Clapier CR, Imhof A, Eberharter A, et al. dMi-2 and ISWI chromatin remodelling factors have distinct nucleosome binding and mobilization properties. EMBO J 2000;19(16):4332–41.

[54] Gangaraju VK, Bartholomew B. Dependency of ISW1a chromatin remodeling on extranucleosomal DNA. Mol Cell Biol 2007;27(8):3217–25.

[55] Whitehouse I, Stockdale C, Flaus A, Szczelkun MD, Owen-Hughes T. Evidence for DNA translocation by the ISWI chromatin-remodeling enzyme. Mol Cell Biol 2003;23(6):1935–45.

[56] Yang JG, Madrid TS, Sevastopoulos E, Narlikar GJ. The chromatin-remodeling enzyme ACF is an ATP-dependent DNA length sensor that regulates nucleosome spacing. Nat Struct Mol Biol 2006;13(12):1078–83.

[57] McKnight JN, Jenkins KR, Nodelman IM, Escobar T, Bowman GD. Extranucleosomal DNA binding directs nucleosome sliding by Chd1. Mol Cell Biol 2011;31(23):4746–59.

[58] Bouazoune K, Kingston RE. Chromatin remodeling by the CHD7 protein is impaired by mutations that cause human developmental disorders. Proc Natl Acad Sci USA 2012;109(47):19238–43.

[59] Chaban Y, Ezeokonkwo C, Chung W-H, Zhang F, Kornberg RD, Maier-Davis B, et al. Structure of a RSC-nucleosome complex and insights into chromatin remodeling. Nat Struct Mol Biol 2008;15(12):1272–7.

[60] Dechassa ML, Zhang B, Horowitz-Scherer R, Persinger J, Woodcock CL, Peterson CL, et al. Architecture of the SWI/SNF-nucleosome complex. Mol Cell Biol 2008;28(19):6010–21.

[61] Yamada K, Frouws TD, Angst B, Fitzgerald DJ, DeLuca C, Schimmele K, et al. Structure and mechanism of the chromatin remodelling factor ISW1a. Nature 2011;472(7344):448–53.

[62] Zofall M, Persinger J, Kassabov SR, Bartholomew B. Chromatin remodeling by ISW2 and SWI/SNF requires DNA translocation inside the nucleosome. Nat Struct Mol Biol 2006;13(4):339–46.

[63] Schwanbeck R, Xiao H, Wu C. Spatial contacts and nucleosome step movements induced by the NURF chromatin remodeling complex. J Biol Chem 2004;279(38):39933–41.

[64] Saha A, Wittmeyer J, Cairns BR. Chromatin remodeling through directional DNA translocation from an internal nucleosomal site. Nat Struct Mol Biol 2005;12(9):747–55.

[65] Strohner R, Wachsmuth M, Dachauer K, Mazurkiewicz J, Hochstatter J, Rippe K, et al. A "loop recapture" mechanism for ACF-dependent nucleosome remodeling. Nat Struct Mol Biol 2005;12(8):683–90.

[66] Deindl S, Hwang WL, Hota SK, Blosser TR, Prasad P, Bartholomew B, et al. ISWI remodelers slide nucleosomes with coordinated multi-base-pair entry steps and single-base-pair exit steps. Cell 2013;152(3):442–52.

[67] Saha A, Wittmeyer J, Cairns BR. Chromatin remodeling by RSC involves ATP-dependent DNA translocation. Genes Dev 2002;16(16):2120–34.

[68] Lia G, Praly E, Ferreira H, Stockdale C, Tse-Dinh YC, Dunlap D, et al. Direct observation of DNA distortion by the RSC complex. Mol Cell 2006;21(3):417–25.

[69] Blosser TR, Yang JG, Stone MD, Narlikar GJ, Zhuang X. Dynamics of nucleosome remodelling by individual ACF complexes. Nature 2009;462(7276):1022–7.

[70] Sirinakis G, Clapier CR, Gao Y, Viswanathan R, Cairns BR, Zhang Y. The RSC chromatin remodelling ATPase translocates DNA with high force and small step size. EMBO J 2011;30(12):2364–72.

[71] Erdel F, Krug J, Längst G, Rippe K. Targeting chromatin remodelers: signals and search mechanisms. Biochim Biophys Acta 2011;1809(9):497–508.

[72] Yoshinaga SK, Peterson CL, Herskowitz I, Yamamoto KR. Roles of SWI1, SWI2, and SWI3 proteins for transcriptional enhancement by steroid receptors. Science 1992;258(5088):1598–604.

[73] Hong W, Nakazawa M, Chen Y-Y, Kori R, Vakoc CR, Rakowski C, et al. FOG-1 recruits the NuRD repressor complex to mediate transcriptional repression by GATA-1. EMBO J 2005;24(13):2367–78.

[74] Rowbotham SP, Barki L, Neves-Costa A, Santos F, Dean W, Hawkes N, et al. Maintenance of silent chromatin through replication requires SWI/SNF-like chromatin remodeler SMARCAD1. Mol Cell 2011;42(3):285–96.

[75] Tyler JK, Collins KA, Prasad-Sinha J, Amiott E, Bulger M, Harte PJ, et al. Interaction between the *Drosophila* CAF-1 and ASF1 chromatin assembly factors. Mol Cell Biol 2001;21(19):6574–84.

[76] Swygert SG, Peterson CL. Chromatin dynamics: interplay between remodeling enzymes and histone modifications. Biochim Biophys Acta 2014;1839(8):728–36.

[77] Denslow Sa, Wade Pa. The human Mi-2/NuRD complex and gene regulation. Oncogene 2007;26(37):5433–8.

[78] Ranjan A, Mizuguchi G, FitzGerald PC, Wei D, Wang F, Huang Y, et al. Nucleosome-free region dominates histone acetylation in targeting SWR1 to promoters for H2A.Z. replacement. Cell 2013;154(6):1232–45.

[79] Cajigas I, Leib DE, Cochrane J, Luo H, Swyter KR, Chen S, et al. Evf2 lncRNA/BRG1/DLX1 interactions reveal RNA-dependent inhibition of chromatin remodeling. Development 2015;142(15):2641–52.

[80] Kennison Ja, Tamkun JW. Dosage-dependent modifiers of polycomb and antennapedia mutations in *Drosophila*. Proc Natl Acad Sci USA 1988;85(21):8136–40.

[81] Elfring LK, Daniel C, Papoulas O, Deuring R, Sarte M, Moseley S, et al. Genetic analysis of brahma: the *Drosophila* homolog of the yeast chromatin remodeling factor SWI2/SNF2. Genetics 1998;148(1):251–65.

[82] Bultman S, Gebuhr T, Yee D, La Mantia C, Nicholson J, Gilliam A, et al. A Brg1 null mutation in the mouse reveals functional differences among mammalian SWI/SNF complexes. Mol Cell 2000;6(6):1287–95.

[83] Bultman SJ, Gebuhr TC, Pan H, Svoboda P, Schultz RM, Magnuson T. Maternal BRG1 regulates zygotic genome activation in the mouse. Genes Dev 2006;20(13):1744–54.

[84] Reyes JC, Barra J, Muchardt C, Camus A, Babinet C, Yaniv M. Altered control of cellular proliferation in the absence of mammalian brahma (SNF2alpha). EMBO J 1998;17(23):6979–91.

[85] Yan Z, Wang Z, Sharova L, Sharov AA, Ling C, Piao Y, et al. BAF250B-associated SWI/SNF chromatin-remodeling complex is required to maintain undifferentiated mouse embryonic stem cells. Stem Cells 2008;26(5):1155–65.

[86] Ho L, Jothi R, Ronan JL, Cui K, Zhao K, Crabtree GR. An embryonic stem cell chromatin remodeling complex, esBAF, is an essential component of the core pluripotency transcriptional network. Proc Natl Acad Sci USA 2009;106(13):5187–91.

[87] Gao X, Tate P, Hu P, Tjian R, Skarnes WC, Wang Z. ES cell pluripotency and germ-layer formation require the SWI/SNF chromatin remodeling component BAF250a. Proc Natl Acad Sci USA 2008;105(18):6656–61.

[88] Kidder BL, Palmer S, Knott JG. SWI/SNF-Brg1 regulates self-renewal and occupies core pluripotency-related genes in embryonic stem cells. Stem Cells 2009;27(2):317–28.

[89] Lessard J, Wu JI, Ranish JA, Wan M, Winslow MM, Staahl BT, et al. An essential switch in subunit composition of a chromatin remodeling complex during neural development. Neuron 2007;55(2):201–15.

[90] Takeuchi JK, Bruneau BG. Directed transdifferentiation of mouse mesoderm to heart tissue by defined factors. Nature 2009;459(7247):708–11.

[91] Wang Z, Zhai W, Richardson JA, Olson EN, Meneses JJ, Firpo MT, et al. Polybromo protein BAF180 functions in mammalian cardiac chamber maturation. Genes Dev 2004;18(24): 3106–16.

[92] Huang X, Gao X, Diaz-Trelles R, Ruiz-Lozano P, Wang Z. Coronary development is regulated by ATP-dependent SWI/SNF chromatin remodeling component BAF180. Dev Biol 2008;319(2):258–66.

[93] Bramswig NC, Lüdecke H-J, Alanay Y, Albrecht B, Barthelmie A, Boduroglu K, et al. Exome sequencing unravels unexpected differential diagnoses in individuals with the tentative diagnosis of Coffin–Siris and Nicolaides–Baraitser syndromes. Hum Genet 2015;134(6):553–68.

[94] Deuring R, Fanti L, Armstrong Ja, Sarte M, Papoulas O, Prestel M, et al. The ISWI chromatin-remodeling protein is required for gene expression and the maintenance of higher order chromatin structure in vivo. Mol Cell 2000;5(2):355–65.

[95] Corona DFV, Siriaco G, Armstrong JA, Snarskaya N, McClymont SA, Scott MP, et al. ISWI regulates higher-order chromatin structure and histone H1 assembly in vivo. PLoS Biol 2007;5(9):e232.

[96] Landry J, Sharov AA, Piao Y, Sharova LV, Xiao H, Southon E, et al. Essential role of chromatin remodeling protein Bptf in early mouse embryos and embryonic stem cells. PLoS Genet 2008;4(10):e1000241.

[97] Stopka T, Skoultchi AI. The ISWI ATPase Snf2h is required for early mouse development. Proc Natl Acad Sci USA 2003;100(24):14097–102.

[98] Francke U. Williams–Beuren syndrome: genes and mechanisms. Hum Mol Genet 1999;8(10):1947–54.

[99] Konev AY, Tribus M, Park SY, Podhraski V, Lim CY, Emelyanov AV, et al. CHD1 motor protein is required for deposition of histone variant H3.3 into chromatin in vivo. Science 2007;317(5841):1087–90.

[100] Hödl M, Basler K. Transcription in the absence of histone H3.3. Curr Biol 2009;19(14): 1221–36.

[101] Gaspar-Maia A, Alajem A, Polesso F, Sridharan R, Mason MJ, Heidersbach A, et al. Chd1 regulates open chromatin and pluripotency of embryonic stem cells. Nature 2009;460(7257):863–8.

[102] Kaji K, Caballero IM, MacLeod R, Nichols J, Wilson Va, Hendrich B. The NuRD component Mbd3 is required for pluripotency of embryonic stem cells. Nat Cell Biol 2006;8(3):285–92.

[103] Kaji K, Nichols J, Hendrich B. Mbd3, a component of the NuRD co-repressor complex, is required for development of pluripotent cells. Development 2007;134(6):1123–32.

[104] Schnetz MP, Bartels CF, Shastri K, Balasubramanian D, Zentner GE, Balaji R, et al. Genomic distribution of CHD7 on chromatin tracks H3K4 methylation patterns. Genome Res 2009;19(4):590–601.

[105] Schnetz MP, Handoko L, Akhtar-Zaidi B, Bartels CF, Pereira CF, Fisher AG, et al. CHD7 targets active gene enhancer elements to modulate ES cell-specific gene expression. PLoS Genet 2010;6(7):e1001023.

[106] He D, Marie C, Zhao C, Kim B, Wang J, Deng Y, et al. Chd7 cooperates with Sox10 and regulates the onset of CNS myelination and remyelination. Nat Neurosci 2016;19(5):678–89.

[107] Vissers LELM, van Ravenswaaij CMA, Admiraal R, Hurst JA, de Vries BBA, Janssen IM, et al. Mutations in a new member of the chromodomain gene family cause CHARGE syndrome. Nat Genet 2004;36(9):955–7.

[108] Engelen E, Akinci U, Bryne JC, Hou J, Gontan C, Moen M, et al. Sox2 cooperates with Chd7 to regulate genes that are mutated in human syndromes. Nat Genet 2011;43(6):607–11.

[109] Nishiyama M, Oshikawa K, Tsukada Y, Nakagawa T, Iemura S, Natsume T, et al. CHD8 suppresses p53-mediated apoptosis through histone H1 recruitment during early embryogenesis. Nat Cell Biol 2009;11(2):172–82.

[110] Fazzio TG, Huff JT, Panning B. An RNAi screen of chromatin proteins identifies Tip60–p400 as a regulator of embryonic stem cell identity. Cell 2008;134(1):162–74.

[111] Wang L, Du Y, Ward JM, Shimbo T, Lackford B, Zheng X, et al. INO80 facilitates pluripotency gene activation in embryonic stem cell self-renewal, reprogramming, and blastocyst development. Cell Stem Cell 2014;14(5):575–91.

[112] Huang S, Gulzar ZG, Salari K, Lapointe J, Brooks JD, Pollack JR. Recurrent deletion of CHD1 in prostate cancer with relevance to cell invasiveness. Oncogene 2012;31(37):4164–70.

[113] Liu W, Lindberg J, Sui G, Luo J, Egevad L, Li T, et al. Identification of novel CHD1-associated collaborative alterations of genomic structure and functional assessment of CHD1 in prostate cancer. Oncogene 2012;31(35):3939–48.

[114] Burkhardt L, Fuchs S, Krohn A, Masser S, Mader M, Kluth M, et al. CHD1 is a 5q21 tumor suppressor required for ERG rearrangement in prostate cancer. Cancer Res 2013;73(9): 2795–805.

[115] Le Gallo M, O'Hara AJ, Rudd ML, Urick ME, Hansen NF, O'Neil NJ, et al. Exome sequencing of serous endometrial tumors identifies recurrent somatic mutations in chromatin-remodeling and ubiquitin ligase complex genes. Nat Genet 2012;44(12):1310–5.

[116] Kolla V, Zhuang T, Higashi M, Naraparaju K, Brodeur GM. Role of CHD5 in human cancers: 10 years later. Cancer Res 2014;74(3):652–8.

[117] Fujita N, Jaye DL, Kajita M, Geigerman C, Moreno CS, Wade PA. MTA3, a Mi-2/NuRD complex subunit, regulates an invasive growth pathway in breast cancer. Cell 2003;113(2): 207–19.

[118] Nicolson GL, Nawa A, Toh Y, Taniguchi S, Nishimori K, Moustafa A. Tumor metastasis-associated human MTA1 gene and its MTA1 protein product: role in epithelial cancer cell invasion, proliferation and nuclear regulation. Clin Exp Metastasis 2003;20(1):19–24.

[119] Zhang X-Y, DeSalle LM, Patel JH, Capobianco AJ, Yu D, Thomas-Tikhonenko A, et al. Metastasis-associated protein 1 (MTA1) is an essential downstream effector of the c-MYC oncoprotein. Proc Natl Acad Sci USA 2005;102(39):13968–73.

[120] Gonzalez-Perez A, Jene-Sanz A, Lopez-Bigas N. The mutational landscape of chromatin regulatory factors across 4,623 tumor samples. Genome Biol 2013;14(9):r106.

[121] Dunaief JL, Strober BE, Guha S, Khavari PA, Alin K, Luban J, et al. The retinoblastoma protein and BRG1 form a complex and cooperate to induce cell cycle arrest. Cell 1994;79(1):119–30.

[122] Reisman DN, Strobeck MW, Betz BL, Sciariotta J, Funkhouser W, Murchardt C, et al. Concomitant down-regulation of BRM and BRG1 in human tumor cell lines: differential effects on RB-mediated growth arrest vs CD44 expression. Oncogene 2002;21(8):1196–207.

[123] Bourachot B, Yaniv M, Muchardt C. Growth inhibition by the mammalian SWI-SNF subunit Brm is regulated by acetylation. EMBO J 2003;22(24):6505–15.

[124] Bartlett C, Orvis TJ, Rosson GS, Weissman BE. BRG1 mutations found in human cancer cell lines inactivate Rb-mediated cell-cycle arrest. J Cell Physiol 2011;226(8):1989–97.

[125] Trouche D, Le Chalony C, Muchardt C, Yaniv M, Kouzarides T. RB and hbrm cooperate to repress the activation functions of E2F1. Proc Natl Acad Sci USA 1997;94(21):11268–73.

[126] Versteege I, Sévenet N, Lange J, Rousseau-Merck MF, Ambros P, Handgretinger R, et al. Truncating mutations of hSNF5/INI1 in aggressive paediatric cancer. Nature 1998;394(6689):203–6.

[127] Roberts CW, Galusha SA, McMenamin ME, Fletcher CD, Orkin SH. Haploinsufficiency of Snf5 (integrase interactor 1) predisposes to malignant rhabdoid tumors in mice. Proc Natl Acad Sci USA 2000;97(25):13796–800.

[128] Wilson BG, Wang X, Shen X, McKenna ES, Lemieux ME, Cho Y-J, et al. Epigenetic antagonism between polycomb and SWI/SNF complexes during oncogenic transformation. Cancer Cell 2010;18(4):316–28.

[129] Sausen M, Leary RJ, Jones S, Wu J, Reynolds CP, Liu X, et al. Integrated genomic analyses identify ARID1A and ARID1B alterations in the childhood cancer neuroblastoma. Nat Genet 2013;45(1):12–7.

[130] Wiegand KC, Shah SP, Al-Agha OM, Zhao Y, Tse K, Zeng T, et al. ARID1A mutations in endometriosis-associated ovarian carcinomas. N Engl J Med 2010;363(16):1532–43.

[131] Itamochi H, Oumi N, Oishi T, Shoji T, Fujiwara H, Sugiyama T, et al. Loss of ARID1A expression is associated with poor prognosis in patients with stage I/II clear cell carcinoma of the ovary. Int J Clin Oncol 2015;20(5):967–73.

[132] Kim MJ, Gu MJ, Chang H-K, Yu E. Loss of ARID1A expression is associated with poor prognosis in small intestinal carcinoma. Histopathology 2015;66(4):508–16.

[133] Park JH, Lee C, Suh JH, Chae JY, Kim HW, Moon KC. Decreased ARID1A expression correlates with poor prognosis of clear cell renal cell carcinoma. Hum Pathol 2015;46(3):454–60.

[134] Kadoch C, Crabtree GR. Reversible disruption of mSWI/SNF (BAF) complexes by the SS18–SSX oncogenic fusion in synovial sarcoma. Cell 2013;153(1):71–85.

[135] Gorrini C, Squatrito M, Luise C, Syed N, Perna D, Wark L, et al. Tip60 is a haplo-insufficient tumour suppressor required for an oncogene-induced DNA damage response. Nature 2007;448(7157):1063–7.

[136] Huber O, Ménard L, Haurie V, Nicou A, Taras D, Rosenbaum J. Pontin and reptin, two related ATPases with multiple roles in cancer. Cancer Res 2008;68(17):6873–6.

[137] Morrison AJ, Highland J, Krogan NJ, Arbel-Eden A, Greenblatt JF, Haber JE, et al. INO80 and gamma-H2AX interaction links ATP-dependent chromatin remodeling to DNA damage repair. Cell 2004;119(6):767–75.

[138] Downs JA, Allard S, Jobin-Robitaille O, Javaheri A, Auger A, Bouchard N, et al. Binding of chromatin-modifying activities to phosphorylated histone H2A at DNA damage sites. Mol Cell 2004;16(6):979–90.

[139] Ikura T, Ogryzko VV, Grigoriev M, Groisman R, Wang J, Horikoshi M, et al. Involvement of the TIP60 histone acetylase complex in DNA repair and apoptosis. Cell 2000;102(4):463–73.

[140] van Attikum H, Fritsch O, Hohn B, Gasser SM. Recruitment of the INO80 complex by H2A phosphorylation links ATP-dependent chromatin remodeling with DNA double-strand break repair. Cell 2004;119(6):777–88.

[141] Wu S, Shi Y, Mulligan P, Gay F, Landry J, Liu H, et al. A YY1-INO80 complex regulates genomic stability through homologous recombination-based repair. Nat Struct Mol Biol 2007;14(12):1165–72.

[142] Yu EY, Steinberg-Neifach O, Dandjinou AT, Kang F, Morrison AJ, Shen X, et al. Regulation of telomere structure and functions by subunits of the INO80 chromatin remodeling complex. Mol Cell Biol 2007;27(16):5639–49.

[143] Pegoraro G, Kubben N, Wickert U, Göhler H, Hoffmann K, Misteli T. Ageing-related chromatin defects through loss of the NURD complex. Nat Cell Biol 2009;11(10):1261–7.

[144] Kenyon C. The plasticity of aging: insights from long-lived mutants. Cell 2005;120(4):449–60.

[145] Dang W, Sutphin GL, Dorsey JA, Otte GL, Cao K, Perry RM, et al. Inactivation of yeast Isw2 chromatin remodeling enzyme mimics longevity effect of calorie restriction via induction of genotoxic stress response. Cell Metab 2014;19(6):952–66.

[146] Finley A, Copeland RA. Small molecule control of chromatin remodeling. Chem Biol 2014;21(9):1196–210.

CHAPTER 5

Chromatin Dynamics During the Cell Cycle

D. Doenecke

Institute for Molecular Biology, University of Göttingen, Göttingen, Lower Saxony, Germany

5.1 INTRODUCTION

The organization of eukaryotic chromosomes undergoes phase-specific extensive changes during the cell division cycle. It thereby reflects the consecutive functional transitions between phases of high transcriptional activity during the G1 phase, the coordinated synthesis of DNA and chromosomal proteins during the S phase, and the sequence of individual steps culminating in mitotic cell division. In morphological terms, functional transitions are represented by just two histologically different states, that is, the apparently loosely packed chromatin during interphase contrasts with the highly condensed metaphase chromosomes. Morphological differences in the state of condensation of interphase chromatin led to the differentiation between euchromatin and heterochromatin, the latter being tentatively defined as the transcriptionally less-active chromatin portion during interphase [1,2]. This may be an oversimplification, but helps to correlate structural and functional features of chromatin.

The major constituents of chromatin are, in quantitative terms, DNA and histones in a 1:1 ratio. Histones are extremely conserved basic proteins, their primary structures vary at just few positions, and their conserved tertiary structures are the basis for the formation of the nucleosome as the repeat and basal packaging unit of chromatin in all eukaryotes [3]. This implies that modifications of amino acid side chains of histone proteins, replacement of one subtype of histone by a variant isoform, or even eviction of histones have a major impact on the local or general organization of chromatin with the respective functional consequences, as will be discussed in this chapter.

Functional and structural needs during any chromatin-related nuclear activity will depend besides histones on further specific proteins, such as factors involved in the regulation of transcription, in the coordination and execution of individual steps during DNA replication or in the organization of the individual steps of mitosis. The regulation of the individual steps of the cell

CONTENTS

117

cycle through activation and inactivation, de novo synthesis and degradation, deposition, and eviction of enzymes and structural components is based on a network of covalent modifications of proteins including histones as major constituents of chromatin.

The structure of chromatin is organized on several levels. As mentioned earlier, the basal repeat subunit is the nucleosome. One of the histones, H1, which is not part of the nucleosomal core particle, but interacts with the DNA linking the core particles, is involved in the formation of higher order structures by the chain of nucleosome particles [4,5]. Even during interphase, when chromatin is apparently not condensed and no distinct chromosomes can be morphologically discriminated, territories occupied by individual chromosomes persist [6]. Within these territories, folding of the genome in topologically associated, megabase-sized domains is achieved through protein–protein interactions [7]. Within these domains, the local chromatin organization at defined subregions controls cell type-specific patterns of transcriptional activity. This general multilevel organization of chromatin and its cell type-specific suborganization is the basis for the identity of individual cell types. Hence, the mechanics of the cell cycle must guarantee that cell type identity is maintained and that cellular programs are memorized during the cell division cycle.

5.2   CHROMATIN ORGANIZATION: THE CHAIN OF NUCLEOSOMES FORMS HIGHER ORDER STRUCTURES

The basal chromatin structure is the chain of nucleosomes [3]. The core nucleosome is formed by 147 nucleotide pairs of DNA which are wrapped in 1.75 turns around the surface of a histone octamer consisting of the histones H2A, H2B, H3, and H4. The basis for the formation of this core histone octamer is the particular three domain structure of these proteins. Each of these has a central globular domain consisting of three alpha helices which are structured in a way that promotes the interaction of H3 with H4 and of H2A with H2B. Based on this principle, two H3-H4 dimers form a central tetramer, to which two HA-H2B dimers are attached in completing the octamer [8]. Of particular interest are the N-terminal portions of the core histones. They extrude toward the outside of the octamer and interact with DNA and further chromatin components. These N-terminal tails are enriched in basic amino acids and are the prime targets of posttranslational modifications, which consequently influence the local functional chromatin state. In the electron microscope, the basal string of nucleosomes appears as a 10-nm fiber. Linker histones contribute to the assembly of nucleosomes into a next level of compaction, the 30-nm fiber [5].

H1 is not a component of the nucleosomal core particle, but completes together with linker DNA the nucleosome as the chromatin repeat unit, as it was initially defined [9,10]. H1 histones consist of three domains. The N-terminal extended region of about 45 residues is enriched in basic amino acids, the highly conserved central globular domain (about 75 amino acids) is organized as a winged helix motif which before had been similarly observed in other DNA-binding proteins [11]. The determination of the binding site of H1 at the linker and core particle surface is still a matter of controversy. FRAP analysis and computer modeling [12] were the basis for a model in which the globular H1 domain interacts with DNA about one helical turn away from the dyad axis of the nucleosome and at a second site on the linker DNA near the exit site of the linker DNA outside the core particle.

Compaction of the chain of nucleosomes into 30-nm fibers in the presence of H1 is based on a left-handed twist of repeating tetranucleosomal structural units. Within these units, the nucleosomes zigzag back and forth with straight linker DNA in between. H1 is positioned in a way that allows interactions with both the dyad and the exiting and entering DNA at the nucleosomal core particle [12]. The detailed organization of the 30-nm fiber had long been discussed after its original description as a regularly coiled solenoid with about 7 nucleosomes per turn and the linker histones oriented toward the interior of the fiber [13]. However, cryoelectron microscopy studies [14] of reconstituted 30-nm chromatin fibers agreed with the results of X-ray investigations demonstrating a zigzag organization within a tetranucleosome particle [15]. A regular compaction of chromatin in a 30-nm fiber as obtained by reconstitution with core and linker histones excludes the interaction with any factors that may locally mediate chromatin function. Thus, it may represent a silent state between the "open" 10-nm fiber and higher order structures, and the description of the complex patterns of dynamic transitions in chromatin organization may just be based on the 10-nm fiber [16].

Interphase chromosomes occupy within the cell nucleus distinct territories. The three-dimensional organization of chromatin within these territories depends on long-range intrachromosomal interactions which define topologically associated domains (TADs) with an average size of about 0.8 Mb [17]. The borders of TADs are defined as regions with comparatively few protein-mediated DNA–DNA interactions. The next level in the substructure of TADs is organized as gene loops which reflect interactions between chromatin architectural proteins. The main protein involved in the formation of loops in vertebrates is the CCCTC-binding factor CTCF. Chromatin loops often form via interactions between two convergent CTCF binding sites at the base of the loop [18] in the presence or absence of cohesin [19,20]. Such sites of interaction are maintained in several cell types, whereas shorter loops anchored just by cohesin and regulators of transcription appear to be cell type specific [21]. Regular positioning and maintenance of the system of insulated neighborhoods is the basis for cell type specificity. It must be maintained during the dynamic changes in chromatin organization during the cell division cycle.

5.3 THE TWO STATES OF CHROMATIN DURING INTERPHASE: EUCHROMATIN VERSUS HETEROCHROMATIN

Two different states of interphase chromatin were initially defined on the basis of different staining characteristics: euchromatin and heterochromatin (discussed in detail in Chapters 1–3). Euchromatin appeared as loosely packed material in contrast to the highly condensed heterochromatic nuclear subcompartment. The open state of euchromatin functionally correlates with high transcriptional activity, whereas heterochromatin appears to represent the inactivated portion of the genome (for review, see Ref. [22]). As a consequence of this varied chromatin compaction, access for the transcription machinery or other factors mediating chromatin-based processes is facilitated in euchromatin and appears to be blocked in heterochromatic portions of the genome. This differential compaction of chromatin is based on specific patterns of DNA and histone modifications or histone replacement by variant histones and on differential patterns of factors interacting with these modified sites in DNA or histones.

Heterochromatin occurs in two different categories. Facultative heterochromatin refers to chromatin harboring genes that are specifically inactivated, but may be active in other cell types or at specific developmental stages. In contrast, constitutive heterochromatin is essentially devoid of genes and is permanently inactivated in any cell type. This is the characteristic state of chromatin at centromeric and telomeric regions of chromatin (for review, see Ref. [2]). Euchromatin is demarcated from facultative heterochromatin by protein components, such as the CCCTC binding factor CTCF, which has been mentioned earlier as the DNA binding protein at the basis of gene loops within TADs. CTCF exhibits a broad spectrum of functions, which are dependent on the combinatorial usage of its zinc fingers and the chromatin context of CTCF binding sites [19,23]. Besides forming gene loops, CTCF can also activate or inhibit transcription. Well-documented binding partners of CTCF include transcription factors, the mediator complex, RNA polymerase II, and transcriptional repressors. At the boundaries between eu- and facultative heterochromatin, it demarcates the transcriptionally active domains from inactive zones [24]. It depends on the genomic site and on the pattern of interacting factors, which of the different functions of CTCF is actually predominant (for review, see Ref. [25]).

As will be discussed in detail, modifications of histones play an essential role in the local organization of chromatin as docking sites for proteins that mediate the respective function. Typical histone modifications within heterochromatin states are, for example, the hypoacetylation of lysine side chains in the N-terminal tails of histones, di- [26] and trimethylation of lysine 9 in histone H3 (constitutive heterochromatin) or of lysine 27 in H3 (facultative, repressive heterochromatin). These modified sites in turn serve as docking sites for protein complexes that mediate the differential states of compaction (for review, see Ref. [2]). In the regulation of developmental target genes, trimethylation of H3-K27 by the polycomb group protein PRC2 generates the binding site for PRC1 for the formation of repressive chromatin [27].

5.4 POSTTRANSLATIONAL HISTONE MODIFICATIONS: WRITERS, READERS, AND ERASERS

As discussed in Chapter 2, covalent modification of histones at specific positions is a major tool for the functional and structural organization of chromatin. Modification of histone side chains may either directly alter the chromatin organization by interfering with protein–DNA or protein–protein interactions, or the modified site may be recognized by other proteins as sites for binding with the respective consequences for local or general chromatin organization [28]. The roles of phosphorylation of serine or threonine, acetylation of lysine, and methylation of lysine and arginine side chains are the most intensely studied histone modifications. In contrast to acetylation of the peripheral amino group of lysine, the methylation of this group does not alter the local charge. Moreover, lysine may be mono-, di-, or trimethylated. Similarly, arginine methylation at the guanidine group of the side chain can occur in three versions, that is, as monomethylation and as symmetrical versus asymmetrical dimethylation. Further modifications are ubiquitination and ADP-ribosylation of specific lysines [29], arginine deimination [30], and proline isomerization [31]. In each case, the positions to be modified must be recognized by the corresponding site-specific enzymes (the "writers"), the type, and position of modification has to be recognized and translated into a specific function upon binding of a specialized interaction partner (the "reader"). Reversal of this effect elicited upon binding of the respective regulatory protein or protein complex requires the specific and controlled activity of enzymes that mediate removal of the respective groups, that is, enzymes that deacetylate, demethylate, or dephosphorylate, respectively (the "erasers").

Typical patterns of histone modifications have been described for most chromatin-related nuclear local or more general processes, such as transcriptional activity, DNA replication, DNA repair, and chromatin condensation. Hence, phase-specific changes of histone modification patterns are essential components of the systems regulating the dynamic changes of chromatin structure during the cell cycle.

Table 5.1 Nonallelic Variants of Core Histones in Mammals

H2A[a]
H2A.Z (subtypes H2A.Z.1, H2A.Z.2, H2A.Z.3): enriched at transcription start sites
H2A.X: involved in DNA repair function, phosphorylated H2A.X recruits DNA repair factors
H2A.B (H2A.B.1, H2A.B.2, H2A.B.3): involved in male germ cell development
macroH2A (mH2A.1.1, mH2A.2.1, mH2A.2.2): involved in transcriptional silencing
H2B[b]
TS H2B: male germ cell development
H3
H3.3: enriched at transcription start sites[c]
H3.4: (same as H3t) involved in differentiation of male germ cells[d]
CENP-A (same as CenH3): replaces part of main type H3 in centromeres[e]

[a]Bönisch C, Hake S. Histone H2A variants in nucleosomes and chromatin: more or less stable? Nucleic Acids Res 2012;40:10719–41.
[b]Shinagawa T, et al. Disruption of Th2a and Th2b genes causes defects in spermatogenesis. Development 2015;142:1287–92.
[c]Ahmad K, Henikoff S. The histone variant H3.3 marks active chromatin by replication-independent nucleosome assembly. Mol Cell 2002;9:1191–1200.
[d]Witt O, et al. Testis-specific expression of a novel human H3 histone gene. Exp Cell Res 1996;229:301–6.
[e]De Rop V, et al. CENP-A: the key player behind centromere identity, propagation, and kinetochore assembly. Chromosoma 2012;121:527–38.

In addition to the modification of core and linker histones, replacement of canonical histones by nonallelic variants mediates changes in the local chromatin organization. These variant histones are are also subject to covalent modification and may provide binding sites for regulatory factors or they may influence protein-protein or protein-DNA interactions within or between nucleosomes (Table 5.1). In the context of this chapter we will in particular refer to the H3 variants CENP-A and H3.3.

5.5 DIFFERENTIAL PATTERNS OF H1 AND CORE HISTONE MODIFICATIONS DURING THE CELL CYCLE

Modifications and demodifications of specific amino acid side chains at defined positions with the consequence of binding or eviction of phase-specific partner proteins are key events in the regulation of the cell cycle (summarized in Fig. 5.1). One of the major regulatory principles of the cell cycle is the phosphorylation and dephosphorylation of regulatory proteins. In respect to chromatin, this has been described already in 1973 by Bradbury et al. [32,33] and Gurley et al. [34] without differentiating between H1 variants. Talasz et al. [35] could then show that individual variants of H1 have varied phosphorylation patterns during the cell cycle with a dynamic, limited phosphorylation process during the S phase, during which both phosphorylated and unphosphorylated H1 is observed, and phosphorylated H1 colocalizes with replicating DNA [36]. A rapid increase of H1 phosphorylation occurs at the G2/M transition, when

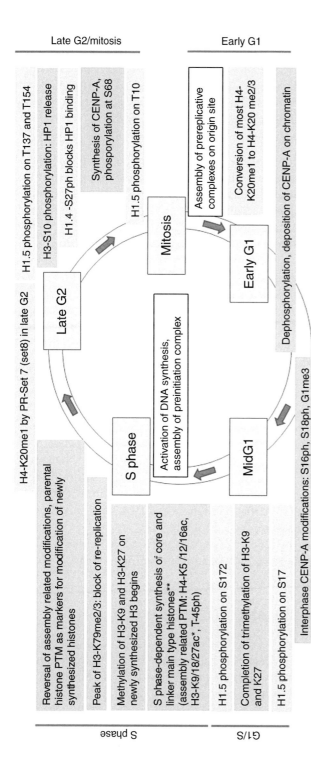

FIGURE 5.1 Cell cycle-regulated posttranslational modifications of histone proteins.

Examples are given for just acetylation, methylation and phosphorylation. Modifications that are related to gene regulation during interphase are mostly omitted. H1.5 and H1.4 phosphorylations are presented as examples for H1 modifications (for other subtypes, see Refs. [40,41]). H3 lysine methylation data as mentioned in the text and in Ref. [42]. *PTM*, posttranslational modification; *K*, lysine; *S*, serine; *T*, threonine; *ac*, acetylation; *me*, methylation; *ph*, phosphorylation. Most examples are based on data from mammalian cells. *Mouse data differ from results obtained with human cells; **nonreplication-dependent histones include H3.3, H1.0, H1.X (H1.10).

chromatin condensation is linked to H1 phosphorylation [37]. Participation of H1 in mitotis has also been demonstrated in respect to alignment and segregation of chromatids [38]. A second level of regulation in H1 patterns during the cell cycle is achieved by the differential expression of H1 nonallelic variant genes. With minor differences the main type variant genes H1.1–H1.5 are transcribed during the S phase, whereas histones H1.0 and H1.X (i.e., H1.10 in unified nomenclature) are synthesized throughout the cell cycle [39].

Modification patterns of core histones will be discussed later in detail in the context of establishing origins of replication, the assembly of new chromatin, and the steps from the end of S phase until the end of mitosis. In short, the state of methylation of lysine 20 in histone H4 [43] and acetylation of lysines 5 and 12 in H4 [44] are essential modifications for the formation of prereplication complexes and licensing of replication origins. Thereafter, during the process of DNA replication, newly synthesized H3 and H4 are acetylated before deposition. Histone H4 carries acetyl groups at lysines 5, 12, and 56 [45,46], whereas varied acetylation patterns in newly synthesized H3 have been observed. Transient acetylation of lysines 9, 18, and 27 have been described in murine cells, but this patterns differs from S phase HeLa cells [47]. Furthermore, histone H3 acetylation on lysine 56 and phosphorylation on threonine 45 of histone H3 are associated with chromatin assembly [48]. Methylation of H3 on lysine 79 during S phase (H3-K79me2) is considered to mark replicated chromatin in order to block rereplication [49]. After chromatin assembly, histone deacetylation by deacetylases, histone deacetylases (HDAC1 and HDAC2) [50] is essential in order to facilitate formation of condensed chromatin. In terms of histone modification, heterochromatin formation depends on H3 trimethylation at lysine 9 (H3-K9me) creating a binding site for heterochromatin protein 1 (HP1) [51]. During mitosis, the phosphorylation of the S10 residue in H3 (H3-S10ph, Ref. [52]) neighboring H3K9 causes the ejection of HP1 from chromatin despite the maintenance of H3K9me3 modifications. Finally, chromatin condensation at centromeres and kinetochore formation require incorporation of a centromeric H3 variant (CENP-A/CenH3). This centromere-specific H3 is subject to modification on serines and the N-terminal glycine (S16ph, S18ph, G1me3, see Ref. [53]).

5.6   CHROMATIN ORGANIZATION AND DNA REPLICATION

The origin of replication is already established at the transition from mitosis to the G1 phase, when origins of replication are "licensed" for the ensuing S phase by assembling prereplicative complexes on the origin site. The site of origin is not specified by a particular DNA sequence, but by the context of local chromatin organization and DNA topology. An exception is the genome of the yeast *Saccharomyces cerevisiae*, which contains origin consensus sites termed autonomously replicating sequences. In contrast, origins of replication in higher eukaryotes occur just in preferential zones comprising more than 10,000 nucleotide pairs [54], and they are more frequent in transcriptionally active than in silent chromatin [55]. The licensing step consists of the deposition of licensing factors onto a basal origin of replication complex (ORC). The final activation of the origin in the S phase then affords the acquisition of factors directly involved in the replication process [helicase, polymerases, proliferating cell nuclear antigen (PCNA)]. At the stage of origin licensing, lysine 20 of histone H4 is monomethylated (H4-K20me1). Depending on the structural context, this H4 lysine 20 methylation is involved in several chromatin-related processes, such as DNA damage response, chromatin condensation, and regulation of transcription [43,56]. Knockout of the gene encoding the methyltransferase PR-Set7 that mediates this monomethylation leads to a reduced loading of the ORC [57]. H4-K20me1 also contributes to the activation of the acetyltransferase HBO1 which acetylates histone H4 at lysines K5 and K16 and thus contributes to DNA replication licensing [44]. This process is counteracted by a protein termed Geminin that blocks reinitiation of DNA replication [44]. Further methylations of the monomethylated H4-K20 in early G1 yielding H4-K20me2 and -K20me3 help to recruit components of the ORC [58]. The state of histone H4 methylation during the cell cycle is controlled by three methyltransferases: PR-Set7 (SET8) is the monomethyltransferase, whereas Suv4-20h1 and Suv4-20h2 further methylate H4K20me to H4K20me2 and -me3. One demethylase (PHF8) has been described that demethylates H4K20me1 [59]. Each of the different states of H4K20 methylation is recognized by distinct binding proteins depending on their genomic distribution and on the progression of the cell cycle. PR-Set7 (SET8) is maximally active in late G2, and then becomes less active during mitosis and reaches minimal activity during the S phase [60]. In parallel, monomethylated H4-K20me1 at specific promoters participates in the repression of E2F target genes that regulate entry into S phase, and is demethylated by PHF8 at the G1-S transition to enable cell cycle progression past the G1-S transition. The genome-wide removal of PHF8 during mitotic prophase again allows the accumulation of H4-K20me1 in the context of chromatin condensation and promotes condensin II deposition [59]. After mitosis, when heterochromatin is reestablished in early G1, H4-K20me1 is converted to H4-K20me3 by the enzymes Suv4-20h1/h2 [60]. The methyl transferase Suv4-20h2 depositing H4-K20me3 is targeted to heterochromatic regions by the HP1 or LncRNA and is involved in chromatin condensation and transcriptional repression [61,62].

Nucleosomes must be disassembled during replication ahead of the replication machinery and the parental histones are reformed on the replicated DNA double helices. In parallel, de novo synthesis of histones is needed to provide the full complement of histones for the assembly of new nucleosomes on the replicated DNA, and thereby the correct nucleosome density is established. This recycling of parental histones and addition of newly synthesized histones implies that part of the posttranslational modifications potentially are maintained in replicated chromatin, whereas newly deposited histones need to be modified. It has been shown that modifications like H3-K9me3 and H3-K27me3 can serve as markers for recruitment of the respective enzymes and thus make sure that these marks are introduced onto the newly synthesized histones at the correct sites in the genome [63]. Other marks may be erased and gradually reestablished as has been demonstrated in *Drosophila* embryos [64]. Alabert et al. [65] have analyzed in human cells the propagation of posttranslational histone modifications by a combination of the nascent chromatin capture technique with stable isotope labeling with amino acids in cell culture and could demonstrate that no obvious erasure of modifications occurred, and histone marks like diacetylation of H4 and acetylation of H3 (H3-K14ac, H3-K18/K23ac) as well as H3-K9me1 are efficiently maintained upon DNA replication. In addition, it was shown that H3 de novo methylations (K27me1, K27me2, K36me1) take place on nascent chromatin. Analysis of histone variants during DNA replication revealed that H3.3 and H2A.X are not lost at the replication fork whereas H2A.Z appeared to be lost in part and inefficiently supplied during replication, but gradually replenished thereafter. The authors conclude that posttranslational modifications are maintained by two modes: in most cases old histones maintain their state of modification and new histones acquire modifications to reproduce the pre-replicative state, whereas in other cases histone modifications are continuously established on newly deposited and old histones [65].

The assembly of new nucleosomes requires the synthesis of a vast amount of histones during the S phase of the cell cycle. Since chromatin roughly consists of equal amounts of histones and DNA, the amount of de novo synthesized histones equals the mass of newly synthesized DNA. This requires a coordination of DNA and histone synthesis, and the need for a high quantity of histones to be synthesized during the S phase is met by transcription of multiple copies of histone genes arranged in clusters. In addition to the coordinated transcription of multiple copies of histone genes and efficient translation of the respective mRNAs, an efficient transport of the newly synthesized histones through the nuclear pore is needed, before histone chaperones help in the formation of new nucleosomes (for review, see Ref. [66]). Histone chaperones are needed for guiding newly synthesized histones onto replicated DNA. The chromatin assembly factor CAF-1 I mediates deposition of H3–H4 dimers [67], it is targeted to replication forks by PCNA which is a processivity factor for DNA polymerases. Subsequent H2A–H2B dimer deposition is mediated by the nucleosome assembly protein NAP1 [68]. As for H1 histones, the protein nuclear autoantigenic sperm protein has been considered as an H1 chaperone, but it also is involved in H3–H4 deposition [69]. This faithful restoration of chromatin organization in coordination with the replication complex requires remodeling of nucleosomes (for review, see [70]). For nuclear transport and assembly of nucleosomes, newly synthesized histones must be modified. The acetyl transferase Hat1 specifically modifies free (in contrast to nucleosome bound) histones. It acetylates H4 at lysines 5 and 12 [45,46]. In addition to the acetylation of lysines 5 and 12, Hat1 is involved in maintaining the acetylation state of H3 at lysine positions 9, 18, and 27 [47] and in transferring the acetylated H3/H4 dimers to the replicating chromatin [71].

After replication, the general organization of the chromatin landscape must be reestablished. This requires removal of the S phase-specific acetylation pattern before regions of silenced chromatin can be reestablished, such as the constitutive heterochromatin at telomeres and in the pericentromeric region. As mentioned, earlier, deposition of HP1 depends on the di- or trimethylation of lysine 9 in H3, a step which can only take place after deacetylation of the respective acetyl group. The enzymes responsible for this deacetylation are HDAC1 and HDAC2, which are recruited to sites of DNA replication via an interaction with a PCNA-bound chromatin remodeler complex SMARCAD1a [50] that also recruits transcriptional repressors and H3K9 histone methyltransferases to reestablish heterochromatin.

5.7   THE HISTONE H3 VARIANT CENP-A ESTABLISHES THE SITE FOR KINETOCHORE ASSEMBLY

The site of kinetochore assembly is not dictated by specific DNA sequences, but by the characteristic chromatin structure of centromeres. A constitutive centromere-associated network is maintained throughout the cell cycle [72] and forms the basis for kinetochore assembly as a prerequisite for the correct segregation of chromatids in mitosis. The epigenetic mark that activates centromeric chromatin is the histone H3 variant CENP-A [73]. This is a bona fide H3 histone with the characteristic histone fold consisting of three α-helices. Within the histone fold, human CENP-A is 56% identical to human H3.1, but just 24% in the N-terminal domain [53]. In centromere-specific nucleosomes, it replaces histone H3.1. CENP-A was discovered as one of several centromere-derived antigens (other antigens were e.g., CENP-B and CENP-C) in a human autoimmune disease (for review, see [74]) and was later identified as a histone H3 variant. In a more systematic nomenclature based on phylogenetic criteria CENP-A (and also its yeast ortholog Cse4) is termed CenH3 [75].

CENP-A is synthesized during late G2 [76] and is recruited to chromatin during G1 together with its chaperone Holliday junction recognition protein (HJURP) by the Mis 18 (Mis 18α/β) complex [77] which is, in turn, recruited to centromeres via Mis 18BP1 binding to the centromere protein CENP-C [78]. Cyclin-dependent kinase (Cdk) 1-mediated phosphorylation of serine 68 in CENP-A blocks interaction of CENP-A and HJURP and thereby prevents CENP-A incorporation into nucleosomes prior to the exit of mitosis when serine 68 is dephosphorylated [79]. CENP-A containing nucleosomes then are interspersed between H3-containing nucleosomes (about 1 in 6–8 nucleosomes, see Ref. [64]), and during S phase CENP-A nucleosomes are equally distributed between daughter chromatids. Since no new CENP-A is synthesized during the S phase, the resulting gaps may be filled with other H3 variants as placeholders, such as H3.3 [80] or H3.1. The short C-terminal tail of CENP-A differs from main type H3 in sequence and length and is involved in recruiting kinetochore components [81]. The N-terminal domain of CENP-A is phosphorylated at serines 16 and 18, and the N-terminal glycine amino group is trimethylated as described in human centromeric nucleosomes [53]. Thus it appears that a combination of posttranslational CENP-A modifications and differences between nucleosomes containing canonical H3 and nucleosomes marked by CENP-A provide the platform for the assembly of kinetochores. In addition to the state of the H3 moiety of centromeric nucleosomes, monomethylation of lysine 20 in histone 4 of CENP-A-containing nucleosomes is essential for kinetochore assembly [82].

Pericentromeric heterochromatin is hypoacetylated and is enriched in methylated histone H3 (H3-K9me3). Methylated H3 (H3-K9me3) is essential for binding of the HP1 [46]. Interestingly, de novo assembly of ectopic centromeres is supported by the organization of chromatin at the boundaries of pericentromeric heterochromatin and euchromatin [83], supporting a role for chromatin states in addition to CENP-A in the establishment of centromere identity.

5.8   MITOTIC CONDENSATION OF CHROMATIN

The change in chromatin organization from the interphase state, when no individual chromosomes are microscopically visible, to the densely packed mitotic chromosomes consisting of two chromatids that are attached to each other requires an intricate system of protein modifications and protein–protein interactions. This structural reorganization provides the basis for a proper segregation of the duplicated chromatids during metaphase. The condensation of the replicated chromatids and their attachment to each other until the onset of anaphase allows the accurate interaction with the mitotic spindle apparatus, the proper arrangement of individual chromosomes during metaphase and the correct partitioning of the individual chromatids after cleavage of the cohesin complex [84]. As discussed in the preceding paragraph, proper interaction of the spindle apparatus with individual chromosomes during the mitotic prophase takes place at the kinetochore complex which is assembled at centromeric chromatin.

Condensation of chromosomes largely depends on a protein complex termed structural-maintenance-of chromosome complex (SMC). This is composed of condensin, cohesin, topoisomerase II, and several other nonhistone proteins (for review, see Ref. [85]). In addition to the SMC, histone modifications mediate major steps of chromatin condensation by creating sites for interaction with factors involved in structural reorganization of chromatin.

Variations in the state of phosphorylation of histones, nonhistone proteins, and several stage-specific factors are characteristic for the different phases of the cell cycle. As mentioned earlier, a correlation between the phosphorylation of histone H1 and of H3 has been described already in 1975 [34]. The phosphorylation sites in the N-terminal domain of histone H3.1 during mitosis are serines 10 and 28 and threonines 3 and 6 [86]. In addition, phosphorylation within the globular part of H3, affecting the histone fold domain, has been observed at threonine 80 during early (prophase) and late (anaphase, telophase) stages of mitosis [87]. Histone H3.3 is a nonreplication-dependent variant subtype of H3, which is associated with transcriptionally active chromatin and may also help to fill the gaps that result upon replication of CENP-A containing nucleosomes during S phase, when no de novo synthesis of CENP-A occurs [80]. Phosphorylation of this H3 isoform that differs from main type H3 at just five positions, has been observed in the vicinity of centromeres just during the mitotic prophase [88].

Protein phosphorylation during prometaphase and metaphase is due to an increase in CDK and chromosome passenger kinase activity [89]. Phosphorylation of serine 10, just adjacent to lysine 10, which is trimethylated during mitosis, is done by Aurora B kinase [90] which is part of the chromosome passenger complex. This multiprotein complex interacts with chromosomes during the mitotic prophase and is mainly found at centromeres during metaphase [90]. Wilkins et al. [91] have elucidated one of the effects of H3-S10 phosphorylation, in the yeast *S. cerevisiae*. They have used a system of genetically encoded, UV-light-inducible cross-linkers to monitor the impact of H3-S10-phosphorylation on other parameters of chromatin condensation. One of these is the interaction of an acidic patch in the H2A/H2B-dimer with the N-terminal tail of H4 in a neighboring nucleosome [92,93]. This condensation-specific contact is blocked by acetylation of lysine 16 in H4 (H4-K16ac), and this inhibition is removed upon deacetylation of H4-K16. As demonstrated by Wilkins et al. [91], recruitment of the respective deacetylase depends on the phosphorylation of serine 10 in H3.

The Aurora-B-mediated phosphorylation of serine 10 in H3 has another profound effect on chromatin dynamics during mitosis. The HP1 is a characteristic component of heterochromatin, and its deposition depends on the trimethylation of lysine 9 in H3. This state of modification persists during mitosis, but during prophase HP1α, -β, and -γ are released from chromatin. This is due to the Aurora-B-mediated phosphorylation of the neighboring serine 10 [52,94]. This demonstrates the importance of varied patterns of different modifications which result in changes of interaction partners that each characterize a specific functional state.

As mentioned earlier, the phosphorylation of threonine 3 in H3 is another hallmark of mitotic chromatin. The kinase responsible for this modification is haspin (haploid-sperm-cell-specific nuclear protein kinase, Ref. [95]).

Although it initially had been detected in germ cells, this kinase also acts in somatic cells, where it phosphorylates threonine 3 in histone H3 (H3-T3ph) during the mitotic prophase [96]. H3-T3ph can be detected with the respective antibodies during prometaphase and metaphase and appears to be particularly important for proper alignment of chromosomes during metaphase [96–98].

The phosphorylation of histone H1 is known since a long time as a hallmark of mitosis. In fact, H1 is often used as a substrate for in vitro phosphorylation tests. The participation of H1 in the formation of higher order chromatin structures is well established, and it therefore appears reasonable that H1 should be involved in chromosomal condensation and decondensation events (for a review, see Ref. [4]). However, investigations on histone modifications, the state of chromatin packaging and patterns of protein interactions with modified histones have until now concentrated on core histones. Mitotic H1 phosphorylation has been described at five to six phosphorylation sites [35]. Most of these sites correspond to the consensus motif S/TPXK, which is characteristic for CDK-mediated phosphorylation. However, it appears that the different H1 isoforms are differentially phosphorylated, and additional modification sites have been described. Happel et al. [99] have shown that the subtype H1.5 is phosphorylated (in addition to the CDK-specific modifications) at threonine 10 by glycogen synthase 3. This modification is first detected during prometaphase and persists until telophase. This modified H1.5 remains associated with chromosomes throughout mitosis, whereas other phosphorylated H1 subtypes (modified by CDK) are partially released from mitotic chromosomes. This differential association of different H1 histones during mitosis may reflect differential needs to interact with other factors involved in controlling the structural transitions during mitosis.

The differential methylation of lysine 20 in histone H4 contributes to several functional transitions during the cell cycle. As mentioned earlier, monomethylation at this position (H4-K20me1) at the end of G2 contributes to the establishment of origins of replication for the next S phase. The G2-specific monomethylation is involved in binding of the condensin II complex [59,100]. As shown in yeast, acetylation of lysine 56 in H3 (H3-K56ac) appears to block this condensin II recruitment during the S phase. After the S phase this block is relieved by deacetylation mediated by a Sir2-related deacetylase prior to the H4K20me1-mediated loading of condensin II [101].

5.9   REESTABLISHMENT OF INTERPHASE CHROMATIN

The end of mitosis and restoration of nuclei in daughter cells require a reversal of mitosis-specific patterns of chromatin-associated protein complexes and assemblies. The organization of interphase chromatin must be established again in order to allow the transcriptional activity of the newly formed daughter cells. This implies a restoration of heterochromatic and euchromatic regions within the context of chromatin territories and the maintenance of topologically associated chromatin domains and their substructures. Dephosphorylation at mitosis-specific phosphorylation sites in histones and other mitotic control proteins is needed for decondensation of chromatin. The chromosome passenger complex dissociates from chromatin, whereas dephosphorylated core histones are retained in nucleosomes. The phosphorylated serine 10 in H3, which causes the removal of HP1 during mitosis, is released and this allows renewed HP1 binding to H3-K9me3 leading to reestablishment of heterochromatin.

The reversal of phosphorylation patterns of components and regulators of mitotic chromatin organization is an essential part of the transition from mitosis to G1. A key protein phosphatase in this context is Repo-Man [102], which is recruited to chromosomes during anaphase. It is a subunit of protein phosphatase γ [103] and dephosphorylates H3 at positions 3, 10, and 28 with the consequence of dissociation of the chromosome passenger complex and binding of HP1 after removal of the inhibitory phosphate group at H3-S10. In addition to its role in allowing the restoration of heterochromatin, Repo-Man interacts with importin β, which is an essential step of the reassembly of the nuclear envelope toward the end of mitosis [104].

5.10 RETENTION OF BASIC CHROMATIN ORGANIZATION DURING MITOSIS

Mitotic chromatin is almost completely silent in terms of gene expression. This may in part be a consequence of the varied patterns of protein modifications and interactions due to the transition from interphase chromatin to the condensed mitotic chromosomes, but it is mainly due to the eviction of the components of the transcriptional control machinery, that is, transcription factors and their coactivators. However, the characteristics of a given cell type must be maintained and specific transcription patterns must be reestablished after mitosis. This is achieved by maintaining cell-type-specific local patterns of histone modifications and DNA-binding proteins in addition to DNA methylation patterns, which may help to retrieve the components that are characteristic for transcriptional activation or heterochromatin formation, respectively. This memorizing of cellular programs based on patterns of histone modifications and nonhistone DNA binding proteins has been defined as mitotic bookmarking [105]. Eviction of the respective factors during mitosis may be a direct effect of their modification on their chromatin binding or the modification indirectly may cause the binding or eviction of factors that are relevant for the local mitotic or postmitotic state of chromatin [106]. As described earlier, the HP1 is bound to di- or trimethylated lysine 9 in H3 and is displaced in mitosis by the Aurora-B-dependent phosphorylation of the adjacent serine 10 (H3-S10ph). Dephosphorylation of this site during anaphase reestablishes the heterochromatic interphase state. Similarly, acetylation of lysine 16 in H4 blocks condensation of chromatin, its deacetylation allows the interaction of the acidic patch in H2A/H2B dimers with H4 as one step of condensation [91], and it is reasonable to conclude that reacetylation of the respective lysine residues may be a means to locally decondense chromatin upon transition from mitosis to G1.

In contrast to these two examples of inhibitory modifications that are needed for mitotic chromatin condensation and may be reversed at the transition to interphase organization, centromeres provide an example for the maintenance of a local chromatin organization that has to be maintained throughout several cell cycles. This centromeric chromatin is characterized by the inner centromere with its CENP-A-containing nucleosomes and the pericentromeric chromatin. Both are landmarks to be recognized by the machinery for kinetochore assembly in preparation of the next mitosis [107]. In vertebrates, this type of organization does not depend on a specific DNA sequence, but on the maintenance of a local chromatin organization.

The reestablishment of the specific transcription pattern of a given cell type requires the maintenance of specific local modification patterns of histones and regulatory factors at specific genomic loci. One example for gene bookmarking that mediates the postmitotic rerecruitment of RNA polymerase II is the acetylation at lysine 5 of histone H4 (H4-K5ac). This is a binding site for the transcriptional coactivator BRD4 that remains associated with chromatin at certain sites throughout mitosis [108,109]. BRD4 binding further increases at the onset of telophase to ensure their prompt transcriptional activation in the daughter cells immediately after mitosis [110]. In addition to bookmarking at specific loci, the topological organization of chromatin at a higher level also must reflect the transcriptional program of a given cell type. As mentioned in the introduction, TADs are building blocks of the genome containing loops and other elements that reflect the local transcriptional state. The boundaries of transcriptionally activated gene loops are enriched in binding sites for the multifunctional factor CTCF, and it has been shown in *Drosophila melanogaster* that CTCF contributes to maintenance of local states of chromatin activity throughout the cell cycle [111]. On the same lines, sequence-specific binding of CTCF to mitotic chromosomes has been demonstrated at the differentially methylated Igf2/H19 locus in mammalian cells [112]. Similarly, the Notch signaling effector RBPJ, which can bind to CTCF, is retained on chromatin throughout mitosis [113] and thus appears to be another example for mitotic bookmarking.

5.11   CONCLUSIONS

The state of histone modification patterns and of distribution of histone variants contributes to phase-specific interactions of chromatin with protein complexes that are essential for the respective phase-specific function. This applies to the assembly and licensing of the origins of replication as well as to factors involved in DNA replication, deposition of histones, establishment of eu- and heterochromatin both after the S phase and particularly after cell division when cell-type-specific patterns of gene expression must be reinstalled on the chromatin level. The binding of heterochromatin-specific HP1 protein depends on the trimethylation of a specific lysine residue, whereas phosphorylation of an adjacent serine leads to release of HP1 and thus nicely demonstrates the mutual dependence of histone modification and its "reading" by a specific protein. Incorporation of the centromere-specific H3 variant CENP-A is a prerequisite for the assembly of kinetochores and depends on a specific phosphorylation state of this variant. In analogy to these two examples, several cell cycle phase-specific types of modification and binding of phase-specific proteins have been described. Future work will concentrate on deciphering the network of modifications and structural transitions that are the basis for the dynamics of chromatin during the cell cycle, and are guided by the dynamics of protein–protein and protein–nucleic acid interactions, including the roles of noncoding RNAs. Detailed studies of chromatin proteomics at different stages of the cell cycle and comprehensive bioinformatics analysis may be the next step in understanding the transitions of chromatin structure along the cell cycle. This will be the basis for studies on stage-specific interactions between side chains of modified main type or variant histones with effectors of signaling in cell division as well as regulators of gene expression. New microscopic tools and biophysical as well as biochemical methods will focus on structural aspects of the three-dimensional subnuclear organization and its functional implications.

Glossary

Centromere The centromere is the region in the condensed mitotic chromosome where the kinetochore is assembled. At this complex the spindle apparatus interacts with the duplicated chromatin threads (the chromatids) and executes the separation of the chromatids which thereafter become the chromosomes of the daughter cells.

Chromatin Deoxyribonucleoprotein complex consisting of DNA and histones in about a 1:1 ratio, plus nonhistone proteins, which altogether form the chromosomes of any eukaryotic organism. During the interphase of the cell cycle, chromatin is loosely packed, for cell division, chromosomes must be densely packed (condensed) in order to allow their proper separation into daughter cells.

Histones A family of proteins with a high content of basic amino acids, that is, lysine and arginine. Histones interact with DNA. The histone group of proteins consists of five classes termed H1, H2A, H2B, H3, and H4. Histones form together with DNA the nucleosome particles as repeat units of chromatin suborganization.

Nucleosomes Nucleosomes are the repeat units that form the first level of DNA compaction in chromatin. As initially defined, a nucleosome is composed by a core particle, that consists of a histone octamer (two times the histones H2A, H2B, H3, and H4) around which 147 base pairs of DNA are wrapped plus a stretch of linker DNA that connects core particles with each other, and one molecule of histone H1 which interacts with linker DNA at its exit/entry site at the core particle. In newer articles, authors may use the term "nucleosome" for just the core particle.

Phases of the cell cycle The cell division cycle is divided into four phases. The S phase is the period during which DNA is replicated. At the same time, the respective amount of histones has to be synthesized in order to achieve the complete reduplication of chromatin. The second key process of cell division is mitosis (the M phase) during which the cell nucleus is divided into two genetically identical daughter cell nuclei. Mitosis is completed by cytokinesis, which is the division of the cytoplasm and final generation of daughter cells. The phases in between S phases and mitosis are the gap-1 phase (G1 phase) which is characterized by cell growth before initiation of the S phase and the G2 phase, which after DNA replication sets the stage for mitosis.

Abbreviations

Caf-1	Chromatin assembly factor
CENP	Centromeric protein
CTCF	CCCTC-binding factor
CDK	Cyclin-dependent kinase
Haspin	Haploid germ cell-specific nuclear protein kinase
HBO1	Human acetylase binding to ORC1
HDAC	Histone deacetylase
HP	Heterochromatin protein
PCNA	Proliferating cell nuclear antigen
PTM	Posttranslational modification
Smarcad	SWI/SNF-related, matrix-associated actin dependent regulator of chromatin
TAD	Topologically associated domain

References

[1] Huisinga KL, et al. The contradictory definitions of heterochromatin: transcription and silencing. Chromosoma 2006;115:110–22.

[2] Saksouk N, et al. Constitutive heterochromatin formation and transcription in mammals. Epigenet Chromatin 2015;8:3.

[3] Luger K, et al. New insights into nucleosome and chromatin structure: an ordered state or a disordered affair? Nat Rev Mol Cell Biol 2012;13. 436-447.15.

[4] Happel N, Doenecke D. Histone H1 and its isoforms: contribution to chromatin structure and function. Gene 2009;431:1–12.

[5] Bian Q, Belmont AS. Revisiting higher order and large-scale chromatin organization. Curr Opin Cell Biol 2012;24:359–66.

[6] Gibcus JH, Dekker J. The hierarchy of the 3D genome. Mol Cell 2013;49:773–82.

[7] Dixon JR, et al. Topological domains in mammalian genomes identified by analysis of chromatin interactions. Nature 2012;485:376–80.

[8] Luger K, et al. Crystal structure of the nucleosome core particle at 2.8 A resolution. Nature 1997;389:251–60.

[9] Oudet P, et al. Electron microscopic and biochemical evidence that chromatin structure is a repeating unit. Cell 1975;4:281–300.

[10] Oudet P, Chambon P. Seeing is believing. Cell 2004;S116:S79–80.

[11] Clark KL, et al. Co-crystal structure of the HNF-3/fork head DNA recognition motif resembles histone H5. Nature 1993;264:412–20.

[12] Brown DT, et al. Mapping the interaction surface of linker histone H1(0) with the nucleosome of native chromatin in vivo. Nat Struct Mol Biol 2006;13:250–5.

[13] Thoma F, et al. Involvement of histone H1 in the organization of the nucleosome and of the salt-dependent superstructures of chromatin. J Cell Biol 1979;83:403–27.

[14] Song F, et al. Cryo-EM study of the chromatin fiber reveals a double helix twisted by tetranucleosomal units. Science 2014;344:376–80.

[15] Schalch T, et al. X-ray structure of a tetranucleosome and its implications for the chromatin fibre. Nature 2005;436:138–41.

[16] Fussner E, et al. Living without 30 nm chromatin fibers. Trends Biochem Sci 2011;36:1–6.

[17] Ciabrelli F, Cavalli G. Chromatin-driven behavior of topologically associating domains. J Mol Biol 2015;427:608–25.

[18] Rao SS, et al. A 3D map of the human genome at kilobase resolution reveals principles of chromatin looping. Cell 2014;159:1665–80.

[19] Ong CT, Corces VG. CTCF: an architectural protein bridging genome topology and function. Nat Rev Genet 2014;15:234–46.

[20] Dowen JM, et al. Control of cell identity genes occurs in insulated neighborhoods in mammalian chromosomes. Cell 2014;159:374–87.

[21] Baranello L, et al. CTCF and cohesin cooperate to organize the 3D structure of the mammalian genome. Proc Natl Acad Sci USA 2014;111:889–90.

[22] Filion GJ, et al. Systematic protein location mapping reveals five principal chromatin types in *Drosophila* cells. Cell 2010;143:212–4.

[23] Ohlsson R, et al. CTCF is a uniquely versatile transcription regulator linked to epigenetics and disease. Trends Genet 2001;17:520–7.

[24] Cuddapah S, et al. Global analysis of the insulator protein CTCF in chromatin barrier regions reveals demarcation of active and repressive domains. Genome Res 2009;19:24–32.

[25] Holwerda SJB, de Laat W. CTCF: the protein, the binding partners, the binding sites and their chromatin loops. Phil Trans Royal Society 2013;368:20120369.

[26] Wen B, et al. Large histone H3 lysine 9 dimethylated chromatin blocks distinguish differentiated from embryonic stem cells. Nature Genet 2009;41:246–50.

[27] Aoto T, et al. Polycomb group protein-associated chromatin is reproduced in post-mitotic G1 phase and is required for S phase progression. J Biol Chem 2008;283:18905–15.

[28] Zentner GE, Henikoff S. Regulation of nucleosome dynamics by histone modifications. Nat Struct Mol Biol 2013;20:259–66.

[29] Messner S, Hottiger MO. Histone ADP-ribosylation in DNA repair, replication and transcription. Trends Cell Biol 2011;21:534–42.

[30] Cuthbert G, et al. Histone deimination antagonizes arginine methylation. Cell 2004;118:545–53.

[31] Nelson CJ, et al. Proline isomerization of histone H3 regulates lysine methylation and gene expression. Cell 2006;126:905–16.

[32] Bradbury EM, et al. Phosporylation of very-lysine-rich histone in *Physarum polycephalum*. Correlation with chromosome condensation. Eur J Biochem 1973;33:131–9.

[33] Bradbury EM, et al. Control of cell division by very lysine rich histone (F1) phosphorylation. Nature 1974;247:257–61.

[34] Gurley LR, et al. Sequential phosphorylation of histone subfractions in the Chinese hamster cell cycle. J Biol Chem 1975;250:3936–44.

[35] Talasz H, et al. Site-specifically phosphorylated forms of H1.5 and H1.2 localized at distinct regions of the nucleus are related to different processes during the cell cycle. Chromosoma 2009;118:693–709.

[36] Alexandrow MG, Hamlin JL. Chromatin decondensation in S-phase involves recruitment of Cdk2 by Cdc45 and histone H1 phosphorylation. J Cell Biol 2005;168:875–86.

[37] Baatout S, Derradji H. About histone H1 phosphorylation during mitosis. Cell Biochem Funct 2006;24:93–4.

[38] Maresca TJ, Heald R. The long and the short of it: linker histone H1 is required for metaphase chromosome compaction. Cell Cycle 2006;5:589–91.

[39] Happel N, et al. H1 subtype expression during cell proliferation and growth arrest. Cell Cycle 2009;8:2226–32.

[40] Hergeth SP, et al. Isoform-specific phosphorylation of human linker histone H1.4 in mitosis by the kinase Aurora B. J Cell Sci 2011;124:1623–8.

[41] Hergeth SP, Schneider R. The H1 linker histones: multifunctional proteins beyond the nucleosomal core particle. EMBO Rep 2015;16:1439–53.

[42] Zee BM, et al. Origins and formation of histone methylation across the human cell cycle. Mol Cell Biol 2012;32:2503–14.

[43] Beck DB, et al. PR-Set7 and H4K20me1: at the crossroads of genome integrity, cell cycle, chromosome condensation, and transcription. Genes Dev 2012;26:325–37.

[44] Miotto B, Struhl K. HBO1 histone acetylase activity is essential for DNA replication licensing and inhibited by Geminin. Mol Cell 2010;37:57–66.

[45] Parthun MR. Histone acetyltransferase 1: more than just an enzyme? Biochim Biophys Acta 2013;1819:256–63.

[46] Brownell JE, Allis CD. Special HATs for special occasions: linking histone acetylation to chromatin assembly and gene activation. Curr Opin Genet Dev 1996;6:176–84.

[47] Nagarajan P, et al. Histone acetyl transferase 1 is essential for mammalian development, genome stability, and the processing of newly synthesized histones H3 and H4. PLoS Genet 2013;9:e1003518.

[48] Darieva Z, et al. Protein kinase C coordinates histone H3 phosphorylation and acetylation. Elife 2015;. e09886.

[49] Fu H, et al. Methylation of histone H3 on lysine 79 associates with a group of replication origins and helps limit DNA replication once per cell cycle. PLoS Genet 2013;9:e1003542.

[50] Rowbotham SP, et al. Maintenance of silent chromatin through replication requires SWI/SNF-like chromatin remodeler SMARCAD1. Mol Cell 2011;42:285–96.

[51] Stewart MD, et al. Relationship between histone H3 lysine methylation, transcription repression, and heterochromatin protein 1 recruitment. Mol Cell Biol 2005;25:2525–38.

[52] Fischle W, et al. Regulation of HP-1-chromatin binding by histone H3 methylation and phosphorylation. Nature 2005;438:1116–22.

[53] Bailey AO, et al. Posttranslational modification of CENP-A influences the conformation of centromeric chromatin. Proc Natl Acad Sci USA 2013;110:11827–32.

[54] Mesner LD, et al. Bubble-chip analysis of human origin distributions demonstrates on a genomic scale significant clustering into zones and significant association with transcription. Genome Res 2011;21:377–89.

[55] Mesner LD, et al. Bubble-seq analysis oft the human genome reveals distinct chromatin-mediated mechanisms for regulating early- and late-firing origins. Genome Res 2013;23:1774–88.

[56] Sherstyuk VV, et al. Epigenetic landscape for initiation of DNA replication. Chromosoma 2014;123:183–99.

[57] Tardat M, et al. The histone H4 Lys 20 methyltransferase PR-Set7 regulates replication origins in mammalian cells. Nat Cell Biol 2010;12:1086–93.

[58] Beck DB, et al. The role of PR-Set7 in replication licensing depends on Suv4-20h. Genes Dev 2012;26:2580–9.

[59] Liu W, et al. PHF8 mediates histone H4 lysine 20 demethylation events involved in cell cycle progression. Nature 2010;466:508–12.

[60] Jorgensen S, et al. Histone H4 lysine methylation: key player in epigenetic regulation of genomic integrity. Nucleic Acids Res 2013;41:2797–806.

[61] Bierhoff H, et al. Quiescence-induced LncRNAs trigger H4K20 trimethylation and transcriptional silencing. Mol Cell 2014;54:675–82. 2014.

[62] Kapoor-Vazirani P, Vertino PM. A dual role for the histone methyltransferase PR-SET7/SETD8 and histone H4 lysine monomethylation in the local regulation of RNA polymerase II pausing. J Biol Chem 2014;289:7425–37.

[63] Margueron R, Reinberg D. Chromatin structure and the inheritance of epigenetic information. Nat Rev Genet 2010;11:285–96.

[64] Petruk S, et al. Stepwise histone modifications are mediated by multiple enzymes that rapidly associate with nascent DNA during replication. Nat Commun 2013;4:2841.

[65] Alabert C, et al. Two distinct modes for propagation of histone PTMs across the cell cycle. Genes Dev 2015;29:585–90.

[66] Keck KM, Pemberton LF. Histone chaperones link histone nuclear import and chromatin assembly. Biochim Biophys Acta 2012;1819:277–89.

[67] Verreault A, et al. Nucleosome assembly by a complex of CAF-1 and acetylated histones H3/H4. Cell 1996;87:95–104.

[68] Zlatanova J, et al. Nap1: taking a closer look at a juggler protein of extraordinary skills. FASEB J 2007;21:1294–310.

[69] Osakabe A, et al. Nucleosome formation activity of human somatic nuclear autoantigenic sperm protein (sNASP). J Biol Chem 2010;285:11913–21.

[70] Mermoud JE, et al. Keeping chromatin quiet: how nucleosome remodeling restores heterochromatin after replication. Cell Cycle 2011;10:4017–25.

[71] Ejlassi-Lassallette A, et al. H4 replication-dependent diacetylation and HAT1 promote S-phase chromatin assembly in vivo. Mol Biol Cell 2011;22.245–55.

[72] McAinsh AD, Meraldi P. The CCAN complex: linking centromere specification to control of kinetochore-microtubule dynamics. Semin Cell Dev Biol 2011;22:946–52.

[73] Earnshaw WC, et al. Esperanto for histones: CENP-A, not CenH3, is the centromeric histone H3 variant. Chromosome Res 2013;21:101–6.

[74] Fukagawa T, Earnshaw WC. The centromere: chromatin foundation for the kinetochore machinery. Dev Cell 2014;30:496–508.

[75] Talbert PB, et al. A unified phylogeny-based nomenclature for histone variants. Epigenet Chromatin 2012;5:7.

[76] Shelby RD, et al. Chromatin assembly at kinetochores is uncoupled from DNA replication. J Cell Biol 2000;151:1113–8.

[77] Fujita Y, et al. Priming of centromere for CENP-A recruitment by human hMis18lpha, hMis18beta, and M18BP1. Dev Cell 2007;12:17–30.

[78] Moree B, et al. CENP-C recruits M18BP1 to centromeres to promote CENP-A chromatin assembly. J Cell Biol 2011;194:855–71.

[79] Yu Z, et al. Dynamic phosphorylation of CENP-A at Ser68 orchestrates its cell-cycle-dependent deposition at centromeres. Dev Cell 2015;32:68–81.

[80] Dunleavy EM, et al. H3.3 is deposited at centromeres in S phase as a placeholder for newly assembled CENP-A in G1 phase. Nucleus 2011;2:146–57.

[81] Guse A, et al. In vitro centromere and kinetochore assembly on defined chromatin templates. Nature 2011;477:354–8.

[82] Hori T, et al. Histone H4 Lys 20 monomethylation of the CENP-A nucleosome is essential for kinetochore assembly. Dev Cell 2014;29:740–9.

[83] Olszak AM, et al. Heterochromatin boundaries are hotspots for de novo kinetochore formation. Nature Cell Biol 2011;13:799–809.

[84] Meadows J, Millar JB. Sharpening the anaphase switch. Biochem Soc Trans 2015;43:19–22.

[85] Aragon E, et al. Condensin, cohesin and the control of chromatin states. Curr Opin Genet Dev 2013;23:204–11.

[86] Sawicka A, Seiser C. Histone H3 phosphorylation—a versatile chromatin modification for different occasions. Biochimie 2012;94:2193–201.

[87] Hammond SL, et al. Mitotic phosphorylation of histone H3 threonine 80. Cell Cycle 2014;13:440–52.

[88] Hake SB, et al. Serine 31 phosphorylation of histone variant H3.3 is specific to regions bordering centromeres in metaphase chromosomes. Proc Natl Acad Sci USA 2005;102:6344–9.

[89] Carmena M, et al. The chromosomal passenger complex (CPC): from easy rider to the godfather of mitosis. Nat Rev Mol Cell Biol 2012;13:789–803.

[90] Giet R, Glover DM. Drosophila aurora B kinase is required for histone H3 phosphorylation and condensing recruitment during chromosome condensation and to organize the central spindle during cytokinesis. J Cell Biol 2001;152:669–82.

[91] Wilkins BJ, et al. A cascade of histone modifications induces chromatin condensation in mitosis. Science 2014;343:77–80.

[92] Robinson PJ, et al. 30 nm chromatin fibre decompaction requires both H4-K16 acetylation and linker histone eviction. J Mol Biol 2008;381:816–25.

[93] Dorigo B, et al. Chromatin fiber folding: requirement for the histone H4 N-terminal tail. J Mol Biol 2003;327:85–96.

[94] McManus KJ, Hendzel MJ. The relationship between histone H3 phosphorylation and acetylation throughout the mammalian cell cycle. Biochem Cell Biol 2006;84:640–57.

[95] Tanaka H, et al. Identification and characterization of a haploid germ cell-specific nuclear protein kinase (Haspin) in spermatid nuclei and its effect on somatic cells. J Biol Chem 1999;274:17049–57.

[96] Dai J, et al. The kinase haspin is required for mitotic histone H3 Thr 3 phosphorylation and normal metaphase chromosome alignment. Genes Dev 2005;19:472–88.

[97] Higgins JMG. Haspin: a newly discovered regulator of mitotic chromosome behavior. Chromosoma 2010;119:137–47.

[98] Higgins JMG, Herbert M. Nucleosome assembly proteins get set to defeat the guardian of chromosome cohesion. PLoS Genet 2013;9:e1003829.

[99] Happel N, et al. M phase-specific phosphorylation of histone H1.5 at threonine 10 by GSK-3. J Mol Biol 2009;386:339–50.

[100] Tanaka A, et al. Epigenetic regulation of condensin-mediated genome organization during the cell cycle and upon DNA damage through histone H3 lysine 56 acetylation. Mol Cell 2012;48:532–46.

[101] Bilodeau S, Cote J. A chromatin switch for chromatin condensation. Dev Cell 2012;23:1127–8.

[102] Vagnarelli P, Earnshaw WC. Repo-Man-PP1. A link between chromatin remodeling and nuclear envelope assembly. Nucleus 2012;3:138–42.

[103] Trinkle-Mulcahy L, et al. Repo-Man recruits PP1 gamma to chromatin and is essential for cell viability. J Cell Biol 2006;172:679–92.

[104] Vagnarelli P, et al. Repo-Man coordinates chromosomal reorganization with nuclear envelope reassembly during mitotic exit. Dev Cell 2011;21:328–42.

[105] Kadauke S, Blobel GA. Mitotic bookmarking by transcription factors. Epigenet Chromatin 2013;6:6.

[106] Wang F, Higgins JMG. Histone modifications and mitosis: countermarks, landmarks, and bookmarks. Trends Cell Biol 2012;23:175–84.

[107] Probst AV, et al. Epigenetic inheritance during the cell cycle. Nat Rev Mol Cell Biol 2009;10:192–206.

[108] Zhao R, et al. Gene bookmarking accelerates the kinetics of post-mitotic transcriptional re-activation. Nat Cell Biol 2011;13:1295–304.

[109] Voigt P, Reinberg D. BRD4 jump-starts transcription after mitotic silencing. Genome Biol 2011;12:133.

[110] Dey A, et al. Brd4 marks select genes on mitotic chromatin and directs postmitotic transcription. Mol Biol Cell 2009;20:4899–909.

[111] Shen W, et al. A possible role of *Drosophila* CTCF in mitotic bookmarking and maintaining chromatin domains during the cell cycle. Biol Res 2015;58:27.

[112] Burke LJ. CTCF binding and higher order chromatin structure of the H19 locus are maintained in mitotic chromatin. EMBO J 2005;24:3291–300.

[113] Lake RJ, et al. RBPJ, the major transcriptional effector of Notch signaling, remains associated with chromatin throughout mitosis, suggesting a role in mitotic bookmarking. PLoS Genet 2014;10:e1004204.

CHAPTER 6

Epigenetic Regulation of Replication and Replication Timing

C. Brossas*, A-.L. Valton*,**, B. Duriez*, M-.N. Prioleau*

*Institut Jacques–Monod, CNRS, Paris Diderot University, Paris, France
**Program in Systems Biology, Department of Biochemistry and
Molecular Pharmacology, University of Massachusetts Medical School,
University of Massachusetts, Worcester, MA, United States

DNA replication is a highly regulated process by which the whole genome is precisely duplicated at each cell division. It requires the establishment of complex regulatory mechanisms operating throughout the cell cycle and therefore strongly affecting and interacting with other DNA processes, such as, transcription, chromatin formation and dynamics, and DNA repair. Two phases of the cell cycle are particularly important for the regulation of DNA replication. During G1, the chromatin is loaded with prereplicative complexes (pre-RCs), but the absence of specific kinases prevents the activation of these complexes at this time. This phase corresponds to the licensing period. On entry into S phase, the activation of specific S-phase kinases leads to the firing of pre-RCs, which are then released from the chromatin with no possibility of refolding onto it, thereby preventing rereplication. Genome-wide approaches for studying the replication timing (RT) program have revealed that complex genomes are organized into chromosomal domains that replicate according to a precise temporal pattern during S phase. This property is universal in eukaryotes, suggesting an important function that has nevertheless remained largely enigmatic. In eukaryotes, the number of replication start sites (RSSs), or replication origins, is strongly related to genome size, and ranges from several hundred in yeasts to over 50,000 in the human genome. The identification of replication origins in several model systems was a critical step toward understanding the mechanisms involved in the global regulation of the temporal program. We discuss here recent advances in our understanding of the regulation of the spatiotemporal program of DNA replication. We focus mostly on studies in yeasts and vertebrate cells, for the sake of clarity. Many other excellent, exhaustive reviews have been published recently and provide more detailed discussions [1–5]. We will begin by describing the characteristics of the *cis*-elements found at RSSs in yeasts and vertebrate cells. We will then describe the

CONTENTS

Chromatin Regulation and Dynamics. http://dx.doi.org/10.1016/B978-0-12-803395-1.00006-X

key *trans*-factors involved in origin firing and how their presence in limiting numbers may contribute to the establishment of a temporal program. We will then consider the mapping of RT domains in vertebrate cells and their connection with the three-dimensional (3D) organization of chromosomes in the nucleus. A more detailed consideration of the specific marks associated with replication origins in early- or late-replicating domains will then enable us to propose specific mechanisms of regulation, for which experimental validation is required. Finally, we will describe the recently identified general *trans*-factors involved in the control of RT. The identification of these factors has provided support for the hypothesis that the 3D organization of the nucleus plays a crucial role in controlling the spatiotemporal program of DNA replication.

6.1   THE NATURE OF REPLICATION START SITES, INITIATOR FACTORS, AND THEIR POTENTIAL REGULATION

6.1.1   Characteristics of Replication Origins

The definition of replication origins, the targets of the firing factors, was a crucial step in the development of our understanding of the regulation of the temporal program of replication. Replication origins have two properties: the timing of their firing during S phase and their efficiency, corresponding to the percentage of the cell cycle during which a specific origin is active. In budding yeast, DNA replication is initiated at AT-rich consensus elements called autonomously replicating sequence (ARS) consensus sequences (ACSs). These sequences contain an 11–17 bp consensus element that is recognized by the origin recognition complex (ORC), the landing platform for the pre-RC [6,7]. However, only a small portion of these ARS elements (~400 of the 10,000 present) are used as replication origins, implying the involvement of another important regulation mechanism in the selection of these specific sites. Pioneering experiments on the ARS1 replication origin demonstrated that forcing the positioning of a nucleosome over the ACS-impaired origin activity [8]. This suggested a role for nucleosomes in controlling the accessibility of the ACS to the ORC. More recent genome-wide nucleosome-positioning maps have shown that the presence of well-positioned nucleosomes flanking the nucleosome-depleted ACS are a conserved feature of replication origins throughout the genome [9,10]. It has been suggested that the initial nucleosome-depleted region (NDR) creates a chromatin environment permissive for ORC binding to the ACS. The binding of the ORC and the recruitment of chromatin remodelers therefore probably specify the position of the +1 and −1 nucleosomes, resulting in arrays of phased nucleosomes on either side of the ACS.

In fission yeast, origin positioning is mostly dependent on the presence of sequences with a high adenine and thymine content (A+T-rich sequences) but without consensus elements [11–13]. The information is recognized by the Orc4 subunit of the ORC, which is unique among eukaryotes in that it contains a large domain harboring AT-hook subdomains that target the ORC to A+T-rich sequences [14]. The chromatin organization of replication origins in *Schizosaccharomyces pombe* was analyzed to determine whether NDRs were favorable sites for the initiation of replication, as in *Saccharomyces cerevisiae*. Many replication origins are located in large intergenic regions, possibly because these regions tend to be more A+T-rich (70%) than the rest of the genome, increasing the likelihood of ORC-binding sites being present. Most (86%) Orc1/Orc4-binding sites, however, do not overlap with NDRs in *S. pombe* [15]. This finding suggests that care is required in the use of nucleosome positioning models based on observations in *S. cerevisiae* for the prediction of origin organization in other organisms, and raises the question whether or not the unique structure of the Orc4 protein would compensate for the need for NDRs.

The recent development of genome-wide mapping for the replication origins of vertebrate cells has provided important new insight into the potential mechanisms involved in origin firing. The first large-scale studies covered about 1% of the human and mouse genomes and suggested that CpG islands (CGIs) played a key role in the control of origin positioning [16,17]. This association was in agreement with the few well-characterized origins and a pioneering study identifying CGIs as preferential sites of initiation [18]. These studies also made an important new discovery: they identified many non-CGI origins as another important class of replication origins, although CGI origins were found to be the most efficient. Finally, these initial studies also established a clear correlation between origin density and GC richness and identified large regions devoid of strong replication initiation sites [16]. This result will be discussed further, as GC richness is also highly correlated with RT. These analyses laid the foundations for genome-wide studies, which confirmed observations made at a smaller scale and provided evidence for strong enrichment in a specific G-rich motif capable of forming a G-quadruplex (G4) within replication origins in mouse- and human cells [19,20]. G4 is a four-stranded helical DNA structure. Its basic structural unit is the G-quartet, a square planar assembly of four Hoogsteen-bonded guanine bases. The planar G-quartets stack on top of one another, forming four-stranded helical structures. Origin efficiency also correlated with G4 density at the site of initiation, suggesting a connection between the probability of G4 formation and origin efficiency. The signature motif predictive of G4 formation is four tracts of at least three guanines separated by other bases that can fold to form the secondary structure of the G4 [21]. Computational analyses have indicated that there are more than 370,000 G4 motifs in the human genome [22]. A recent genome-wide study detected ~80,000 origins of replication in five human cell lines and showed that not all the G4 structures overlapped with a replication origin, thus suggesting that

G4 may be important determinants for origin specification but they are not sufficient in themselves. A recent genetic study on the chicken DT40 cell line demonstrated that G4s were required for the activity of two model origins [23]. Precise point mutations at one of these origins revealed the existence of a strong correlation between G4 stability and origin efficiency, convincingly suggesting the involvement of a folded G4 in origin regulation. Moreover, G4 motif orientation determines the positioning of the replication start site. This finding is consistent with genome-wide studies in mice and *Drosophila* showing a specific orientation of initiation profiles with respect to the orientation of G4 motifs [19,24]. Finally, as suspected, it was found that the G4 motif of the model origin needed to cooperate with a 250 bp module to yield a functional origin. This element contains several binding sites for transcription factors potentially involved in origin selection. These findings raise a new key question concerning the way in which G4s affect origin function. G4s may thus exclude nucleosomes, thereby favoring pre-RC formation, as reported in *S. cerevisiae* in which NDR ACS are the best substrates for pre-RC formation. Alternatively, G4 formation may facilitate DNA unwinding and the initiation of replication, or it may be recognized by specific factors involved in the formation of a functional origin. Support for this last hypothesis has recently been provided by the observation that the ORC binds preferentially to G4s formed on RNA or single-stranded DNA [25]. The N-terminal part of human RecQL4, the ortholog of the essential initiating factor Sld2 of *S. cerevisiae* also binds G4 structures with high affinity [26].

6.1.2   Regulation of Origin Firing by Competition for Limiting Factors

The process of replication initiation is divided into two periods to ensure the initiation of replication at each origin once, and only once, in each cell cycle. During G1 phase, two copies of the MCM2-7 ATPases are loaded onto replication origins to form a pre-RC (Fig. 6.1). This step requires three "licensing" factors, the hexameric ORC, Cdc6, and Cdt1. Pre-RC formation can occur only in late mitosis and G1 phase because cyclin-dependent kinase (CDK), a potent inhibitor of pre-RC formation, is inhibited by the APC/C and cyclin-dependent kinase inhibitors (CKIs) during these phases of the cell cycle [27]. Many more origins than are actually used during S phase are licensed during this period, thereby ensuring sufficient loading of the DNA with pre-RCs before the start of S phase. These "dormant" origins provide a backup in case of fork stalling and may render the replication program more flexible and robust against perturbations [28,29]. At the transition between the G1- and S phases, inactivation of the APC/C complex and CKIs allows progression to the initiation of S phase through the cumulative effects of two kinases, S-CDK and Dbf4-dependent kinase composed of Cdc7 kinase and its activator (DDK) required for pre-RC activation (Fig. 6.1). However, not all replication origins fire at the

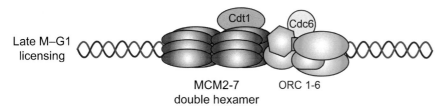

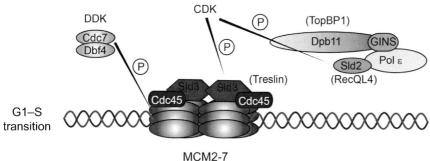

FIGURE 6.1 Formation and activation of a DNA replication origin.
Budding yeast protein names are used and, when different, the names of the human orthologs are indicated in brackets. During G1, the eukaryotic replicative helicase *MCM2-7* is loaded onto double-stranded DNA in an inactive state. The hexameric origin recognition complex *(ORC)* is recruited to replication origins. The binding of *Cdc6* and *Cdt1* then allows the recruitment of the six subunits, *MCM2-7*. At the G1–S transition, two kinases are required to activate the helicase: Dbf4-dependent kinase *(DDK)* and cyclin-dependent kinase *(CDK)*. DDK directly phosphorylates subunits of the loaded *MCM2-7* double hexamer, thus allowing the recruitment of *Sld3* and *Cdc45*. The phosphorylation of *Sld2* and *Sld3* by *CDK* is required for interaction with *Dpb11*. Initiation on the leading strand requires assembly of the active form of the replicative helicase, the *Cdc45–MCM2-7–GINS* (CMG) complex, in addition to the polymerase *(Pol ε)*. Five limiting initiating factors have been identified in yeast: *Sld2, Sld3, Dpb11, Dbf4,* and *Cdc45*.

same time. Instead, a precise temporal program specific to each cell type regulates the coordinate firing of groups of origins, defining RT domains displaying homogeneous timing. Several genetic studies in budding yeast have suggested that the essential role of DDK is to activate the MCM2-7 helicase by phosphorylating several of its subunits. CDK phosphorylates Sld2 and Sld3, allowing the recruitment of Dpb11. These events lead to the final formation of the replicative helicase, the CMG complex (Cdc45, MCM2-7, and GINS), and to the loading of DNA polymerase onto the pre-RCs (Fig. 6.1) [30]. The demonstration that budding yeast contains limiting amounts of initiation proteins, including CDK targets (Sld2 and Sld3); their partner Dpb11, the Dbf4 subunit of DDK;

and Cdc45, was a key discovery for our understanding of the regulation of RT. The overexpression of several of these factors together thus induces the abnormal early firing of late-replicated origins [31,32]. These important findings inspired the hypothesis that suggests a general mechanism for the correct temporal activation of origins during S phase. In early S phase, only early-firing origins can be activated, due to their better "accessibility" to firing factors. After firing, the limiting factors are released and can bind to and activate less accessible origins, and so on. This discovery led to studies aiming to identify the factors regulating the accessibility of these limiting factors to specific pre-RCs.

6.2 COMPLEX GENOMES ARE STRUCTURED INTO REPLICATION TIMING DOMAINS

6.2.1 Establishment of the Replication Timing Program in G1

It has been clearly established that the replication of eukaryotic genomes is a highly organized process. Specific regions of the genome are reproducibly replicated during precise windows of the S phase. This pattern of replication was observed in pioneering studies based on the labeling of replicating DNA during different time windows and the visualization of metaphase chromosomes [33]. These studies also provided the first evidence to suggest that the specific RT of a genomic region is tightly connected to gene activation within that region because Giemsa-light (R) bands containing transcriptionally active chromosomal domains replicate earlier in S phase than Giemsa-dark (G) bands comprising transcriptionally silent heterochromatin. The mechanism by which these replicating domains are established has been addressed in studies involving the transfer of mammalian G1 nuclei into *Xenopus* egg extracts, a system allowing the rapid and efficient replication of the transferred nuclei [34]. By differentially labeling early- and late-replicating domains of CHO cell chromosomes, the authors were able to follow the pattern of duplication in the *Xenopus* system. They showed that there was a transition point in early G1 (about 2 h after mitosis) at which the nuclei acquired information about the correct progression of the timing program. This acquisition coincided with the nuclear repositioning of specific early- and late regions within the nucleus, suggesting an important role for nuclear positioning in RT. However, the molecular events occurring at this transition, known as the "timing decision point," remain elusive. As a first key step toward understanding these events, we need to identify at the molecular level, the regulatory mechanisms involved in the timing program.

6.2.2 Characteristics of Replication Timing Domains

The recent development of genome-scale analysis of the temporal program of replication has provided a more refined vision of the specific properties of late- and early domains of replication. RT is universal in eukaryotes [2,35,36]. In vertebrate cells, there are large constant timing regions (CTRs, from 200 kb to 2 Mb in size) replicated in early-, late-, or mid-S phase, punctuated by slopes of progressive change in RT known as timing transition regions (TTRs). Early-replicated regions are GC-rich and gene-rich, whereas late-replicated domains are AT-rich and gene-poor (Fig. 6.2). Regions of intermediate composition have a RT that may differ between cell types, and these regions

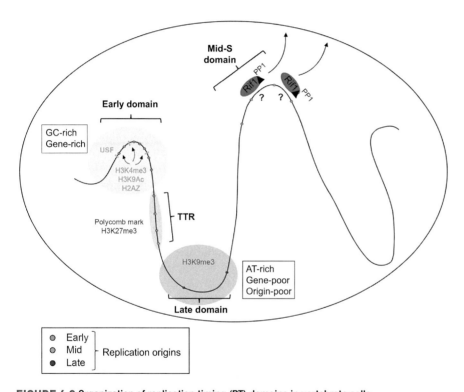

FIGURE 6.2 Organization of replication timing (RT) domains in vertebrate cells.
Active early-, mid-S phase-, and late replication origins are represented as small *green*, *blue*, and *red* dots, respectively. Early-replicated domains are GC-, gene-, and origin-rich. They are located toward the interior of the nucleus and origins found in these domains are strongly associated with open chromatin marks, such as *H3K4me3*, *H2AZ*, and *H3K9Ac*. The upstream factor, *USF*, previously identified as a barrier element [42] can also organize early domains of replication flanking active origins. Timing transition regions (TTRs) are origin-rich and their origins are strongly associated with the *polycomb mark H3K27me3*. Finally, *Rif1* is involved in the control of mid-S-replicating regions, probably through direct interaction with these regions. This leads to the formation of a microenvironment at the periphery of the nucleus or around the nucleolus, enriched in protein phosphatase 1, *PP1*, which counteracts *DDK* activity.

account for ~50% of the human genome. As mentioned earlier, the genome-wide mapping of replication origins has demonstrated the existence of a strong correlation between GC richness and origin density. Thus, according to the current model, early CTRs result from the essentially synchronous activation of clusters of origins. The RT of TTRs can be explained by two models: unidirectional forks emanating from the early CTRs toward the late CTRs or sequential firing of origins along the transition zone. The genome-wide mapping of replication origins has shown that many of these regions contain large numbers of strong initiation sites, consistent with the sequential firing hypothesis, which seems to hold for most TTRs [16,37,38]. Moreover, multi-scale analyses of genome-wide RT have confirmed that fork velocity is consistent with the firing of multiple origins along the length of the TTRs [39]. One TTR has been shown to replicate predominantly as a single long unidirectional fork, suggesting that some TTRs may be origin-poor [40]. The genome-wide mapping of replication origins has confirmed that late CTRs are origin-poor, suggesting that their replication results from the progression of forks initiated in flanking regions (merging TTRs), the late activation of inefficient origins along these domains, or both. Some late CTRs containing large transcribed genes correspond to common fragile site (CFSs). CFSs are loci where breaks and constrictions appear recurrently in metaphase chromosomes from cells grown under conditions of replication stress. One of the most active CFS (FRA3B) results from a pattern of late replication and a paucity of initiation events, leading to the persistence of unreplicated regions at the end of the S phase [41]. The mechanisms involved in the control of early- and late CTRs therefore differ considerably, depending on whether strong initiation sites are present or absent.

Genome-wide maps of 3D chromatin interactions have been shown to be highly predictive of the RT of specific regions [43]. It was initially shown, at low resolution, that active chromatin was predominantly associated with other active regions, whereas repressed chromatin was associated with other silent regions, with little mixing of these two types of regions [44]. This division of the genome into two spatially separate compartments, one corresponding to open chromatin and of the other to less accessible chromatin, was strongly correlated with the division of the genome into early- and late-replicating domains, respectively. More recent high-resolution interaction maps have revealed that metazoan genomes fold into distinct modules called topologically associating domains (TADs), within which genomic interactions are strong, with weaker interactions observed between TADs [45–47]. Again, these high-resolution maps are well correlated with RT domains, suggesting a functional link between the two [48]. However, the mechanisms of TAD establishment and maintenance are largely unknown. It remains to be determined whether they result from the overall dynamics of transcriptional regulation within these domains or whether they are critical architectural elements for the correct regulation of complex genomes.

6.2.3 The Specific Case of the Inactive X Chromosome: Random Replication Versus Structured Replication

Genome-wide studies have confirmed the strong positive correlation between early replication and transcriptional activity in multicellular organisms. No such correlation is observed in fission and budding yeasts, except for the late-replicating heterochromatin present at the telomeres. A clear example of the tight connection between RT and gene activity in mammals is provided by the asynchrony observed for the X chromosome of females. One of the two X chromosomes becomes inactivated during development, to compensate for the double dose of this chromosome in female cells. The inactive X (Xi) chromosome is selected at random and clonally maintained by a series of events including the coating of the chromosome with the XIST noncoding RNA and repressive chromatin modifications. These modifications delay the replication of the Xi chromosome with respect to the active early-replicated X chromosome (Xa). A recent analysis based on copy number variation distinguished between the RT patterns of Xa and Xi on the basis of the SNP characteristics of the father or the mother, thereby providing the first detailed patterns of replication for these chromosomes [49]. The cell lines used have XCI (X-chromosome inactivation) skew, with the same copy of the X chromosome inactivated in >90% of the cells in culture, the paternally-derived X chromosome being the clonally inactive copy. Consistent with previous studies, Xi was found to replicate much later than Xa. Surprisingly, Xi did not follow a structured replication profile. Autocorrelation analyses focusing on the correlation between RT at different sites as a function of the physical distance between them showed that the Xi replication pattern appeared to be random, unlike the structured pattern of replication observed for the autosomal chromosomes and Xa. The only exception was an 8-Mb region on the distal short arm of the X chromosome, a region also containing a cluster of genes that escape X inactivation. These results provided the first demonstration of random replication in somatic human cells. Xi replication was found to be rapid, random, and associated with transcriptional quiescence, reproducing all the properties of early embryonic replication in frogs and flies. A previous survey of the positioning of replication origins at several specific CGI origins on Xi and Xa showed that the same sets of origins were used with similar efficiency, suggesting that the strong origins found at CGIs continued to be active on Xi [50]. This is a unique situation for late-replicating regions, most of which are AT-rich regions lacking strong initiation sites. It remains unclear whether this unusual property is a key determinant of the random and rapid replication of Xi.

6.3 REGULATION OF REPLICATION ORIGINS BY EPIGENETIC MARKS

6.3.1 What Can We Learn From Correlative Studies Based on Genome-Wide Analyses?

The recent development of genome-wide studies of RT associated with maps of diverse histone modifications has made it possible to search for specific enrichments associated with a specific pattern of RT. As expected, given the correlation with transcriptional activity, early-replicated domains have been shown to be correlated with general nuclease accessibility and open chromatin marks, such as H3K4me3 and H3K36me3. However, few associations, if any, were detected between early replicating domains and marks characteristics of facultative or constitutive heterochromatin, such as H3K27me3, H3K9me3, or H4K20me3 [2]. The recent genome-wide mapping of replication origins has made it possible to investigate the relationship between the spatiotemporal program and the landscape of chromatin modifications at a fine scale [38]. Several studies have shown that PR-set7, which is involved in depositing the histone mark H4K20me1, is involved in the control of origin firing [51]. This mark was also shown to be associated with 50% of origins, particularly those activated in early- and mid-S phase [38]. Open chromatin marks, such as, H3K9ac, H3K4me3, and H2AZ, were found to be strongly associated with origins replicated early in S phase, but much less associated with origins activated in the second part of S phase. Moreover, these marks tended to be absent from late-activated origins. Overall, 64% of origins carried none of these three open chromatin mark, indicating that many origins are probably not directly driven by the presence of these open chromatin marks, in agreement with previous suggestions based on the analysis of specific loci [52,53]. An investigation of heterochromatin marks showed that early-replicated origins displayed a depletion of H3K9me3, whereas late origins were characterized by a significant enrichment in this mark. By contrast, origins activated in early- and mid-S phase were enriched in H3K27me3, which overlapped with 40% of the origins. A discriminant analysis—aiming at pinpointing combinations of chromatin marks that characterize early-, mid-, and late-firing origins—showed that early origins were characterized by proximity to the combination of open chromatin marks (H3K9ac, H3K4me3, and H2AZ) and H4K20me1. The physical distance between early origins and open chromatin marks increased with the progression of replication. However, mid-S phase origins remained strongly associated with H4K20me1. Moreover, the coupling of H3K27me3 and H4K20me1 with the absence of other marks constituted a strong characteristic of mid-S phase origins. Assays measuring origin efficiency revealed that H3K27me3 and H4K20me1 were also potential positive regulators of origin firing. Finally, H3K9me3 was shown to be associated with late origins [38].

Analyses of nucleosome architecture at several well-characterized mouse origins showed that RSSs located upstream of the TSS of active promoters occurred at positions of high nucleosome occupancy [54]. These results were validated by the genome-wide mapping of replication origins in HeLa cells. A fine-scale analysis of replication initiation and ORC-binding sites showed that the ORC, on the other hand, occupied the NDR adjacent to the origin (RSS), at the position of unstable nucleosomes containing H3.3/H2AZ histone variants. These findings suggest that, as in *S. cerevisiae*, the ORC localizes preferentially to NDRs at more efficient ORIs. A recent study in two *Drosophila* cell lines reached similar conclusions: RSSs are characterized by a well-positioned nucleosome. Moreover, as observed in human cells, enrichment in the H4K20me1 histone mark was also observed at such positions [24].

6.3.2 Open Chromatin Marks and the Regulation of Origin Firing

Genome-wide studies in human cells showed a specific enrichment in open chromatin marks close to early-replicated origins and a depletion of these marks at late-replicating origins, suggesting an involvement of these marks in the regulation of early firing [38]. Pioneering studies in the yeast *S. cerevisiae* showed that use of the histone acetyl transferase, Gcn5p, for the targeted histone acetylation of a late origin caused the earlier activation of this origin, demonstrating a direct role for open chromatin marks in early firing [55]. The same study showed that the deletion of Rpd3, a canonical histone deacetylase (HDAC), globally advanced the timing of replication. This conclusion was recently revisited in a more systematic analysis based on the genome-wide mapping of replication profiles and the deletion of several HDACs [56]. This analysis revealed that two HDACs, Rpd3 and Sir2, controlled RT in opposite ways. In *rpd3Δ* cells, a large proportion of the late origins fired prematurely, suggesting that Rpd3 was necessary for delayed initiation at late origins. However, surprisingly, a substantial fraction of early origins did not fire at all in *sir2Δ* cells, indicating that Sir2 was promoting the activation of these early origins. At the same time, this study also uncovered that Sir2 represses initiation at rDNA origins, whereas Rpd3 counteracts this effect. Consequently, deletion of SIR2 restored normal replication in *rpd3Δ* cells by reactivating rDNA origins. As in *S. cerevisiae*, one-third of replication origins are located in repetitive sequences, these findings indicate that HDACs might control the RT program in budding yeast by modulating the ability of rDNA repeats to compete with single-copy origins for limiting firing factors. In parallel, Rpd3 might also act locally on single origins. This important study highlights the caution required when interpreting data based on the deletion of *trans* factors. Indeed, an unpredicted indirect effect might radically change the interpretation. It also highlights the considerable power of specifically altering a single locus and testing its impact locally. As opposed to yeast, the disruption of histone modifications in *trans* following the genetic knockout of histone-modifying enzymes was found to result in minor changes to RT in vertebrate cells [57]. Only a minority (5/23) of 23 single-copy loci displayed a weak shift in RT in a panel of mutant mouse ES cell lines in which histone deacetylation; H3K4, H3K9, and H3K27 methylation; or DNA methylation was disrupted. Furthermore, H3K9me2 depletion had no effect on RT genome-wide, except for a small effect on major satellite repeats [58].

Similarly, the *cis* targeting of histone acetyltransferase activity adjacent to the human β-globin origin has been shown to induce a slight shift towards earlier replication, from late to mid–late S phase in lymphocytes, suggesting that this signal is insufficient to organize the early domain of replication present in erythroid cells [59]. Recent studies on the avian DT40 cell line have shown that the strong β^A-globin replication origin flanked by upstream stimulating factor (USF) protein–binding sites is sufficient to advance the timing of replication significantly [60]. The observation that USF-binding sites flanking the origin were required to drive a shift in RT suggests the likely involvement of microdomain formation. The recruitment of enzymes depositing open chromatin marks by USF also suggests that these marks may be involved in the control of RT. Finally, a recent study showed that recruitment of the strong VP16-activator domain close to a normally inactive promoter resulted in its earlier replication, again through a process associated with the deposition of open chromatin marks and/or transcriptional activation [61]. These results are consistent with a complex role for open chromatin marks in the establishment of early replication.

6.3.3 *Schizosaccharomyces pombe* Centromere and the Mat Locus: Special Cases in Which Heterochromatin Controls Early Replication

In *S. pombe*, the pericentromeric region and the silent mating-type locus replicate in early S phase, whereas the subtelomeric region is replicated late (Fig. 6.3). These findings suggest that complex mechanisms regulate RT in the heterochromatic regions of fission yeasts. Swi6, the *S. pombe* counterpart of heterochromatin protein 1 (HP1), is required for early replication of the pericentromeric region and the *mat* locus [62]. Moreover, an HP1-binding motif within Dfp1 (the Dbf4-dependent kinase Cdc7 or DDK) is also required. The tethering of Dfp1 to the pericentromeric region and the *mat* locus in *swi6*-deficient cells restored the early replication of these loci. Thus the heterochromatin protein, Swi6, regulates the early firing of origins by interacting with Dfp1, which in turn activates a kinase involved in the firing of origins in S phase and is limiting for this process (Fig. 6.3). The HP1-binding motif of Dfp1 does not seem to be conserved in Dbf4 homologs in other eukaryotes,

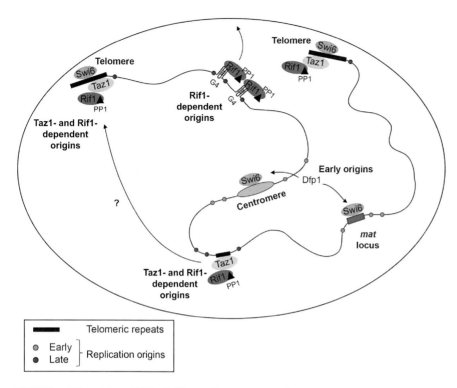

FIGURE 6.3 Regulation of RT in *Schizosaccharomyces pombe*.
Early- and late-replicating origins are indicated by *green* and *red* dots, respectively. The centromeric and *mat* loci are heterochromatic regions replicated in early S phase. This control is achieved through the recruitment of *Dfp1* (the homolog of Dbf4) by *Swi6* (the homolog of HP1), through direct protein–protein interactions. Telomeric regions and late-replicated regions internal to the chromosome recruit *Taz1* by direct binding to telomeric repeats. The *question mark* suggests that *Taz1* internal origins might also cluster together with telomeric regions at the periphery of the nucleus. The timing of replication in these regions is also under the control of *Rif1*. *Taz1*-independent late-replicating regions are organized around G4-like structures formed at specific intergenic regions (IGRs), and are recognized by *Rif1*. These regions also cluster toward the periphery of the nucleus.

suggesting that this regulation is specific to S. *pombe*. Subtelomeric regions containing Swi6 are replicated in late S phase, suggesting that another mechanism prevents origin activation in these regions during early S phase. This aspect will be dealt with further, in the discussion of the role of Taz1 in the control of RT. The centromeric and pericentromeric regions of S. *cerevisiae* are not associated with a specific counterpart of HP1, but they are nevertheless replicated early. As in S. *pombe*, the Dbf4 subunit of DDK is specifically recruited to centromeres by an alternative pathway involving the Ctf19 kinetochore complex (Fig. 6.4) [63]. These results highlight the key regulatory role of the sequestration of S-phase kinases at specific loci to drive their early replication.

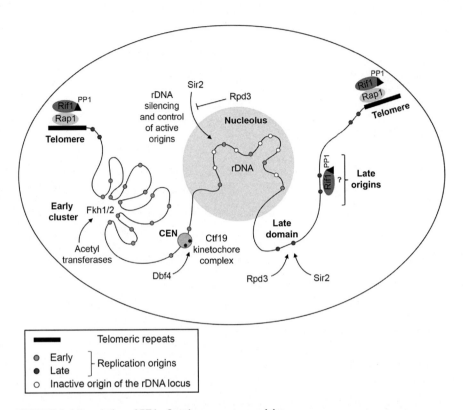

FIGURE 6.4 Regulation of RT in *Saccharomyces cerevisiae*.
Early- and late-replicating origins are indicated by *green* and *red* dots, respectively. Inactive dormant origins within the rDNA locus are indicated by *white* dots. Early-replicated origins are clustered by the *Fkh1/2* factors, favoring their activation at the beginning of S phase. *Acetyl transferases* may also be involved in controlling some early origins. The recruitment of *Dbf4* by the *kinetochore complex* also leads to the early firing of centromeric regions. Late-firing origins at the telomeres are controlled by *Rif1*, which also influences some internal origins, through an as yet unknown mechanism. The histone deacetylases (HDACs) *Sir2* and *Rpd3* control the timing of replication throughout the whole genome, by limiting the numbers of active origins within rDNA repeats. They may also delay the timing of replication for other isolated origins.

6.4 GLOBAL REGULATORS OF REPLICATION TIMING ACT THROUGH 3D NUCLEAR ORGANIZATION

Pulse labeling with modified dNTPs in different cell types and organisms have revealed a punctuate appearance of the labeled sites of DNA synthesis named replication foci [64–66]. These foci result from the more or less coordinate activation of several replicons, which fire in spatially and temporally well-defined clusters [67]. The nature and the formation of these "replication factories" within the 3D nuclear architecture remain poorly characterized. In mammals based on the number of replication foci and their lifetime (~60 min), it is estimated that each contains 0.5–1 Mb of DNA, close to the size of RT domains.

6.4.1 Forkhead Proteins

Studies in the yeast *S. cerevisiae* showed the Fkh1 and FKh2, G2/M-phase specific transcription factors regulated the timing of replication initiation at most of the earliest origins, through a mechanism involving origin clustering in G1 phase (Fig. 6.4) [68]. The deletion of both Fkh1 and -2 thus delays the activation of many origins, whereas that of another subset of origins is advanced. These two classes of affected origins tend to cluster together along the chromosomes, suggesting that Fkh1/2 establish domains of RT. Consistent with a direct role for Fkh proteins in early-origin firing, Fkh1 and Fkh2 binding sites are highly abundant near Fkh-activated origins and depleted near Fkh-repressed origins. Moreover, the mutation of two Fkh1/2-binding sites near ARS305 has been shown to delay its activation, demonstrating the direct role, in *cis*, of Fkh proteins. A global analysis of intra- and interchromosomal interactions in the yeast genome suggested that early origins cluster in G1 phase [69] and that most of these interactions are lost in mutants with deletions of Fkh1 and Fkh2, highlighting the important role of Fkh1/2 in determining long-range chromatin contacts. Finally, a direct or indirect physical interaction between Fkh1/2 and ORC has been detected in coimmunoprecipitation experiments, suggesting a model in which Fkh1/2 located close to ACSs can stabilize the ORC at early-firing origins. The clustering of origins and initiation factors is probably kinetically advantageous for the assembly of the factors required for replication initiation on entry into S phase, which transforms these origin clusters into early replication factories in a coordinated manner.

6.4.2 Rif1

Rap1-interacting factor 1 (Rif1) was originally discovered in budding yeast as a telomere-binding protein required for telomere length regulation. Its orthologs have been identified in fission yeast and in humans. In fission yeast, Rif1 binds to telomeres through interaction with Taz1, an ortholog of human TRF1/TRF2 (Fig. 6.3). The first evidence that Rif1 also plays a critical role in the regulation of the temporal program of DNA replication came from studies in *S. pombe* [70]. The replication pattern is thus profoundly affected in *rif1Δ* cells, with replication being advanced not only for telomeric regions, but also for origins internal to the chromosome that normally fire late. Moreover, some active early-firing origins are suppressed in *rif1Δ* cells. Rif1 was subsequently implicated in the genome-wide control of DNA replication in human and mouse cells and in budding yeast [70–73].

In human cells, *Rif1* depletion leads to the loss of mid-S replication foci profile. As Rif1 partially overlaps mid-S replication foci during unperturbed S phase, these results are consistent with the notion that Rif1 directs the formation of nuclear structures involved in the regulation of RT of mid-S domains (Fig. 6.2). Moreover, absence of Rif1 leads to major changes in RT domain structures, which occur over scales of several hundred kilobases to megabases. A global analysis of RT changes thus showed a shift to mid-S phase for the replication of both early- and late-replicating regions in cells lacking Rif1. This protein was found to bind to DNaseI-insoluble nuclear structures at late-M/early-G1 phase, and to colocalize with lamin B1. Interestingly, the depletion of Rif1 increases chromatin loop sizes, as shown in nuclear halo assays, suggesting the Rif1 might affect RT by coordinating chromatin fiber interactions in 3D.

Both in the fission yeast and in human cells, Rif1 depletion has no impact on the recruitment of pre-RC complexes, highlighting the lack of its involvement in pre-RC formation [70,72]. However, studies in S. pombe have shown that advanced origins bind Cdc45 earlier, suggesting a probable effect of Rif1 on steps regulating Cdc45 loading. Similar conclusions can be drawn from the increases in Cdc7-dependent MCM protein phosphorylation and Cdc45 loading in early S phase in human cells lacking Rif1. Molecular events regulated by Rif1 have recently been dissected in the budding yeast [73]. Rif1 was known to interact with protein phosphatase 1 (PP1). PP1 interaction motifs within the Rif1 N-terminal domain are thus crucial for the repression of replication. As PP1 activity prevents premature Mcm4 phosphorylation by the DDK, Rif1 functions as a PP1 substrate-targeting subunit that counteracts the DDK-mediated phosphorylation involved in origin firing during S phase. It is therefore essential to determine how Rif1 is specifically targeted to regions replicated in mid- or late S phase.

The specific effects of Rif1 on a subset of origins were investigated by mapping Rif1-binding sites in S. pombe. Chromatin immunoprecipitation experiments led to the identification of 35 high-affinity Rif1-binding sites that tended to be located near dormant origins and to contain at least two copies of a conserved motif containing a track of five Gs [74]. Specific mutations at these sites led to a loss of Rif1 binding and to the concomitant activation of late-firing or dormant origins located up to 50 kb away. Rif1-binding sites can adopt a G-quadruplex structure in vitro and Rif1 has been shown to bind G4 preferentially in vitro. These results suggest that Rif1 associates with G-quadruplex structures at specific sites in vivo and organizes specific nucleoprotein structures at the nuclear membrane (Fig. 6.3). Rif1 can form oligomers in vitro and may therefore group G-quadruplex-like structures together in a nuclear compartment, thereby preventing pre-RC activation through the recruitment of PP1.

6.4.3 Taz1

The telomere-binding protein Taz1 also plays an essential role in the control of RT in fission yeast. Taz1, an ortholog of mammalian TRF1 and TRF2, binds to fission yeast telomeres and is required for their protection and maintenance. In a *taz1Δ* strain, about half of late origins, including those in the subtelomeres, fire in early S phase [75]. Telomeric repeats were associated with a subset of these taz1-regulated origins, suggesting a critical role of this sequence and the binding of Taz1. Furthermore, the insertion of a DNA fragment containing four telomeric repeats at ectopic sites repressed firing from neighboring early origins. These results demonstrate the critical role of Taz1 in RT control in the fission yeast (Fig. 6.3). As mentioned earlier, Rif1, another telomere-binding protein in fission yeast, is also involved in the global control of RT [70]. A genome-wide analysis of replication in *taz1Δ* and in *rif1Δ* cells revealed that all of the late origins firing in *taz1Δ* were also activated in *rif1Δ*. Moreover, late origins not firing in *taz1Δ* were activated in *rif1Δ* [75]. These observations suggest that there are Taz1-dependent and Taz1-independent pathways for timing control and that both require Rif1 (Fig. 6.3). It therefore seems likely that Rif1 acts downstream from Taz1. The clustering of Taz1-dependent origins at telomeres may regulate this subset of origins allowing coordination with the replication of telomeric regions (Fig. 6.3, *question mark*).

6.5 INTERPLAY BETWEEN INHERITANCE OF CHROMATIN MARKS AND REPLICATION

The organization of genomes into RT domains is universal in eukaryotes, suggesting an important but as yet largely undetermined function of this organization. DNA replication has been considered as a window of opportunity for changing the chromatin landscape because the chromatin is destabilized and reassembled on the two daughter strands behind the replication forks. It has been suggested that early replication favors the assembly of a more open chromatin structure. This hypothesis is supported by only one observation for plasmids, showing that the time at which a plasmid is injected into a cell during S phase influences its chromatin structure and its expression level [76]. It will, therefore, now be important to address this key question in a physiological chromosomal context. An attractive feature of this model is its self-reinforcing nature. As described here, an open chromatin structure regulates early replication and the early timing of replication favors the assembly of a more open chromatin structure. The de novo construction of a domain with advanced RT from a region in which replication naturally occurs late should provide a clear answer to this important question. Recently, newly synthesized histones were tracked by quantitative mass spectrometry after their deposition behind the replication fork using nascent chromatin capture [77]. In this method, biotin–dUTP labeling of newly replicated DNA and cross-linking of associated proteins allow the isolation of nascent chromatin. Histones post-translational modifications in nascent chromatin were largely similar in early-, mid-, and late S phase thus supporting the view that globally, RT does not have a strong impact on the newly assembled chromatin. However, a small increase in H3K27me1 and H3K36me1 and reduction of H4 diacetylation on new histones incorporated in late S phase compared to early S was observed, suggesting a faster establishment of repressive chromatin in late S phase.

It has been shown that defective replication has a major impact on the maintenance of the epigenetic status of specific loci. For example, in chicken DT40 cells, fork stalling due to the presence of unresolved secondary structures, such as G4, in the absence of the translesion polymerase REV1 or the helicases FANCJ, RecQ helicase-like Werner (WRN), and Bloom (BLM), leads to the formation of gaps in the DNA duplex, which are filled in after bulk replication [78–80]. This uncouples DNA synthesis and chromatin reassembly from the displacement of the parental histones by the replicative helicase. Thus, the chromatin deposited during gap filling has only newly synthesized histones and lacks the histone marks present on the parental nucleosomes. This leads to a failure to maintain the chromatin and expression states of genes associated with G4-forming DNA. This highlights the importance of regulating the normal progression of replication forks for the maintenance of cell type–specific gene expression patterns.

6.6   CONCLUDING REMARKS

The organization of mammalian chromosomes into TADs is strongly correlated with RT units, suggesting common mechanisms of regulation and/or causal links between the two processes. These basic units of higher-order chromosomal structures are probably constructed from key stabilizing interactions between regulatory elements, allowing coordinated transcription and/or replication origin control. Studies in yeasts have suggested important mechanisms involving the clustering of replication origins such that a similar microenvironment with controlled DDK- and CDK activities mediates the coordinated firing of origins embedded in these structures [68,74,75]. Similar processes probably operate in vertebrate cells. However, the higher level of complexity of this organization renders the analysis much more complex. The development of genome engineering tools, such as CRISPR/Cas9, should lead to major progress. Studies of specific deletions of TAD boundaries should provide insight into the role of these regions (or lack of it) in the regulation of the temporal program. Similarly, the de novo creation of replicons with advanced, early replication in regions of naturally late replication should provide information about the processes involved and the factors required for this regulation [60,61]. These genetic approaches, based on insertion of regulatory *cis*-elements by homologous recombination, should establish mechanistic links that are impossible to obtain through purely correlative studies. The observation that Rif1 deletion in human and mouse cells leads to profound disturbances of the temporal replication program clearly identified Rif1 as a key factor controlling RT. However, as mentioned earlier, caution is required when interpreting the findings for the deletion of this *trans*-acting factor because Rif1 is a general factor also involved in transcriptional regulation and DNA repair, and transcription may be directly involved in the control of RT. It is therefore important to find new ways to test the role of Rif1 in the regulation of the temporal program of DNA replication directly in vertebrate cells.

Another new finding to have emerged from recent studies is that G4 motifs appear to be important *cis*-regulatory elements involved in controlling not only fork progression, but also origin selection and firing. Several critical replication factors have been found to interact more specifically with G4 structures in vitro. These factors include Rif1, ORC, and RecQL4, suggesting new hypotheses concerning the role of G4 in the control of DNA replication. Unfortunately, it remains impossible to track the formation of G4 structures exhaustively in vivo. Thus, to date, only several indirect lines of evidence based on series of mutations affecting G4 stability to different extents have been obtained. However, several studies have provided clear evidence that these structures are indeed formed in vivo and that their abundance increases during S phase [21]. Future studies should try to identify the specific factors involved in the dynamics and specific recognition of these particular DNA structures to improve our understanding of the regulation of DNA replication by G4.

References

[1] Fragkos M, Ganier O, Coulombe P, Méchali M. DNA replication origin activation in space and time. Nat Rev Mol Cell Biol 2015;16(6):360–74.

[2] Rhind N, Gilbert DM. DNA replication timing. Cold Spring Harb Perspect Med 2013;3(7):1–26.

[3] Méchali M, Yoshida K, Coulombe P, Pasero P. Genetic and epigenetic determinants of DNA replication origins, position and activation. Curr Opin Genet Dev 2013;23(2):124–31.

[4] Renard-Guillet C, Kanoh Y, Shirahige K, Masai H. Temporal and spatial regulation of eukaryotic DNA replication: from regulated initiation to genome-scale timing program. Semin Cell Dev Biol 2014;30:110–20.

[5] MacAlpine DM, Almouzni G. Chromatin and DNA replication. Cold Spring Harb Perspect Biol 2013;5(8):a010207.

[6] Bell SP, Stillman B. ATP-dependent recognition of eukaryotic origins of DNA replication by a multiprotein complex. Nature 1992;357(6374):128–34.

[7] Marahrens Y, Stillman B. A yeast chromosomal origin of DNA replication defined by multiple functional elements. Science 1992;255(5046):817–23.

[8] Simpson RT. Nucleosome positioning can affect the function of a cis-acting DNA element in vivo. Nature 1990;343(6256):387–9.

[9] Eaton ML, Galani K, Kang S, Bell SP, MacAlpine DM. Conserved nucleosome positioning defines replication origins. Genes Dev 2010;24(8):748–53.

[10] Berbenetz NM, Nislow C, Brown GW. Diversity of eukaryotic DNA replication origins revealed by genome-wide analysis of chromatin structure. PLoS Genet 2010;6(9):e1001092.

[11] Segurado M, de Luis A, Antequera F. Genome-wide distribution of DNA replication origins at A+T-rich islands in *Schizosaccharomyces pombe*. EMBO Rep 2003;4(11):1048–53.

[12] Heichinger C, Penkett CJ, Bähler J, Nurse P. Genome-wide characterization of fission yeast DNA replication origins. EMBO J 2006;25(21):5171–9.

[13] Hayashi M, Katou Y, Itoh T, Tazumi A, Tazumi M, Yamada Y, et al. Genome-wide localization of pre-RC sites and identification of replication origins in fission yeast. EMBO J 2007;26(5):1327–39.

[14] Chuang RY, Kelly TJ. The fission yeast homologue of Orc4p binds to replication origin DNA via multiple AT-hooks. Proc Natl Acad Sci USA 1999;96(6):2656–61.

[15] de Castro E, Soriano I, Marín L, Serrano R, Quintales L, Antequera F. Nucleosomal organization of replication origins and meiotic recombination hotspots in fission yeast: nucleosomal organization of ORIs and DSBs in *S. pombe*. EMBO J 2012;31(1):124–37.

[16] Cadoret J-C, Meisch F, Hassan-Zadeh V, Luyten I, Guillet C, Duret L, et al. Genome-wide studies highlight indirect links between human replication origins and gene regulation. Proc Natl Acad Sci USA 2008;105(41):15837–42.

[17] Sequeira-Mendes J, Díaz-Uriarte R, Apedaile A, Huntley D, Brockdorff N, Gómez M. Transcription initiation activity sets replication origin efficiency in mammalian cells. PLoS Genet 2009;5(4):e1000446.

[18] Delgado S, Gómez M, Bird A, Antequera F. Initiation of DNA replication at CpG islands in mammalian chromosomes. EMBO J 1998;17(8):2426–35.

[19] Cayrou C, Coulombe P, Puy A, Rialle S, Kaplan N, Segal E, et al. New insights into replication origin characteristics in metazoans. Cell Cycle 2012;11(4):658–67.

[20] Besnard E, Babled A, Lapasset L, Milhavet O, Parrinello H, Dantec C, et al. Unraveling cell type-specific and reprogrammable human replication origin signatures associated with G-quadruplex consensus motifs. Nat Struct Mol Biol 2012;19(8):837–44.

[21] Rhodes D, Lipps HJ. G-quadruplexes and their regulatory roles in biology. Nucleic Acids Res 2015;43(18):8627–37.

[22] Huppert JL, Balasubramanian S. Prevalence of quadruplexes in the human genome. Nucleic Acids Res 2005;33(9):2908–16.

[23] Valton A-L, Hassan-Zadeh V, Lema I, Boggetto N, Alberti P, Saintomé C, et al. G4 motifs affect origin positioning and efficiency in two vertebrate replicators. EMBO J 2014;33(7):732–46.

[24] Comoglio F, Schlumpf T, Schmid V, Rohs R, Beisel C, Paro R. High-resolution profiling of *Drosophila* replication start sites reveals a DNA shape and chromatin signature of metazoan origins. Cell Rep 2015;11(5):821–34.

[25] Hoshina S, Yura K, Teranishi H, Kiyasu N, Tominaga A, Kadoma H, et al. Human origin recognition complex binds preferentially to G-quadruplex-preferable RNA and single-stranded DNA. J Biol Chem 2013;288(42):30161–71.

[26] Keller H, Kiosze K, Sachsenweger J, Haumann S, Ohlenschläger O, Nuutinen T, et al. The intrinsically disordered amino-terminal region of human RecQL4: multiple DNA-binding domains confer annealing, strand exchange and G4 DNA binding. Nucleic Acids Res 2014;42(20):12614–27.

[27] Diffley JFX. Regulation of early events in chromosome replication. Curr Biol 2004;14(18):R778–86.

[28] Ibarra A, Schwob E, Méndez J. Excess MCM proteins protect human cells from replicative stress by licensing backup origins of replication. Proc Natl Acad Sci USA 2008;105(26):8956–61.

[29] Ge XQ, Jackson DA, Blow JJ. Dormant origins licensed by excess Mcm2-7 are required for human cells to survive replicative stress. Genes Dev 2007;21(24):3331–41.

[30] Zegerman P. Evolutionary conservation of the CDK targets in eukaryotic DNA replication initiation. Chromosoma 2015;124(3):309–21.

[31] Tanaka S, Nakato R, Katou Y, Shirahige K, Araki H. Origin association of Sld3, Sld7, and Cdc45 proteins is a key step for determination of origin-firing timing. Curr Biol 2011;21(24): 2055–63.

[32] Mantiero D, Mackenzie A, Donaldson A, Zegerman P. Limiting replication initiation factors execute the temporal programme of origin firing in budding yeast. EMBO J 2011;30(23): 4805–14.

[33] Gilbert DM. Replication timing and transcriptional control: beyond cause and effect. Curr Opin Cell Biol 2002;14(3):377–83.

[34] Dimitrova DS, Gilbert DM. The spatial position and replication timing of chromosomal domains are both established in early G1 phase. Mol Cell 1999;4(6):983–93.

[35] MacAlpine DM, Rodríguez HK, Bell SP. Coordination of replication and transcription along a *Drosophila* chromosome. Genes Dev 2004;18(24):3094–105.

[36] Hiratani I, Ryba T, Itoh M, Yokochi T, Schwaiger M, Chang C-W, et al. Global reorganization of replication domains during embryonic stem cell differentiation. PLoS Biol 2008;6(10):e245.

[37] Cayrou C, Coulombe P, Vigneron A, Stanojcic S, Ganier O, Peiffer I, et al. Genome-scale analysis of metazoan replication origins reveals their organization in specific but flexible sites defined by conserved features. Genome Res 2011;21(9):1438–49.

[38] Picard F, Cadoret J-C, Audit B, Arneodo A, Alberti A, Battail C, et al. The spatiotemporal program of DNA replication is associated with specific combinations of chromatin marks in human cells. PLoS Genet 2014;10(5):e1004282.

[39] Guilbaud G, Rappailles A, Baker A, Chen C-L, Arneodo A, Goldar A, et al. Evidence for sequential and increasing activation of replication origins along replication timing gradients in the human genome. PLoS Comput Biol 2011;7(12):e1002322.

[40] Norio P, Kosiyatrakul S, Yang Q, Guan Z, Brown NM, Thomas S, et al. Progressive activation of DNA replication initiation in large domains of the immunoglobulin heavy chain locus during B cell development. Mol Cell 2005;20(4):575–87.

[41] Letessier A, Millot GA, Koundrioukoff S, Lachagès A-M, Vogt N, Hansen RS, et al. Cell-type-specific replication initiation programs set fragility of the FRA3B fragile site. Nature 2011;470(7332):120–3.

[42] West AG, Huang S, Gaszner M, Litt MD, Felsenfeld G. Recruitment of histone modifications by USF proteins at a vertebrate barrier element. Mol Cell 2004;16(3):453–63.

[43] Ryba T, Hiratani I, Lu J, Itoh M, Kulik M, Zhang J, et al. Evolutionarily conserved replication timing profiles predict long-range chromatin interactions and distinguish closely related cell types. Genome Res 2010;20(6):761–70.

[44] Lieberman-Aiden E, van Berkum NL, Williams L, Imakaev M, Ragoczy T, Telling A, et al. Comprehensive mapping of long-range interactions reveals folding principles of the human genome. Science 2009;326(5950):289–93.

[45] Dixon JR, Selvaraj S, Yue F, Kim A, Li Y, Shen Y, et al. Topological domains in mammalian genomes identified by analysis of chromatin interactions. Nature 2012;485(7398):376–80.

[46] Nora EP, Lajoie BR, Schulz EG, Giorgetti L, Okamoto I, Servant N, et al. Spatial partitioning of the regulatory landscape of the X-inactivation centre. Nature 2012;485(7398):381–5.

[47] Sexton T, Cavalli G. The role of chromosome domains in shaping the functional genome. Cell 2015;160(6):1049–59.

[48] Pope BD, Ryba T, Dileep V, Yue F, Wu W, Denas O, et al. Topologically associating domains are stable units of replication-timing regulation. Nature 2014;515(7527):402–5.

[49] Koren A, McCarroll SA. Random replication of the inactive X chromosome. Genome Res 2014;24(1):64–9.

[50] Gómez M, Brockdorff N. Heterochromatin on the inactive X chromosome delays replication timing without affecting origin usage. Proc Natl Acad Sci USA 2004;101(18):6923–8.

[51] Tardat M, Brustel J, Kirsh O, Lefevbre C, Callanan M, Sardet C, et al. The histone H4 Lys 20 methyltransferase PR-Set7 regulates replication origins in mammalian cells. Nat Cell Biol 2010;12(11):1086–93.

[52] Prioleau M-N, Gendron M-C, Hyrien O. Replication of the chicken beta-globin locus: early-firing origins at the 5′ HS4 insulator and the rho- and betaA-globin genes show opposite epigenetic modifications. Mol Cell Biol 2003;23(10):3536–49.

[53] Gay S, Lachages A-M, Millot GA, Courbet S, Letessier A, Debatisse M, et al. Nucleotide supply, not local histone acetylation, sets replication origin usage in transcribed regions. EMBO Rep 2010;11(9):698–704.

[54] Lombraña R, Almeida R, Revuelta I, Madeira S, Herranz G, Saiz N, et al. High-resolution analysis of DNA synthesis start sites and nucleosome architecture at efficient mammalian replication origins. EMBO J 2013;32(19):2631–44.

[55] Vogelauer M, Rubbi L, Lucas I, Brewer BJ, Grunstein M. Histone acetylation regulates the time of replication origin firing. Mol Cell 2002;10(5):1223–33.

[56] Yoshida K, Bacal J, Desmarais D, Padioleau I, Tsaponina O, Chabes A, et al. The histone deacetylases sir2 and rpd3 act on ribosomal DNA to control the replication program in budding yeast. Mol Cell 2014;54(4):691–7.

[57] Jørgensen HF, Azuara V, Amoils S, Spivakov M, Terry A, Nesterova T, et al. The impact of chromatin modifiers on the timing of locus replication in mouse embryonic stem cells. Genome Biol 2007;8(8):R169.

[58] Yokochi T, Poduch K, Ryba T, Lu J, Hiratani I, Tachibana M, et al. G9a selectively represses a class of late-replicating genes at the nuclear periphery. Proc Natl Acad Sci USA 2009;106(46):19363–8.

[59] Goren A, Tabib A, Hecht M, Cedar H. DNA replication timing of the human β-globin domain is controlled by histone modification at the origin. Genes Dev 2008;22(10):1319–24.

[60] Hassan-Zadeh V, Chilaka S, Cadoret J-C, Ma MK-W, Boggetto N, West AG, et al. USF binding sequences from the HS4 insulator element impose early replication timing on a vertebrate replicator. PLoS Biol 2012;10(3):e1001277.

[61] Therizols P, Illingworth RS, Courilleau C, Boyle S, Wood AJ, Bickmore WA. Chromatin decondensation is sufficient to alter nuclear organization in embryonic stem cells. Science 2014;346(6214):1238–42.

[62] Hayashi MT, Takahashi TS, Nakagawa T, Nakayama J, Masukata H. The heterochromatin protein Swi6/HP1 activates replication origins at the pericentromeric region and silent mating-type locus. Nat Cell Biol 2009;11(3):357–62.

[63] Natsume T, Müller CA, Katou Y, Retkute R, Gierliński M, Araki H, et al. Kinetochores coordinate pericentromeric cohesion and early DNA replication by Cdc7-Dbf4 kinase recruitment. Mol Cell 2013;50(5):661–74.

[64] Nakamura H, Morita T, Sato C. Structural organizations of replicon domains during DNA synthetic phase in the mammalian nucleus. Exp Cell Res 1986;165(2):291–7.

[65] O'Keefe RT, Henderson SC, Spector DL. Dynamic organization of DNA replication in mammalian cell nuclei: spatially and temporally defined replication of chromosome-specific alpha-satellite DNA sequences. J Cell Biol 1992;116(5):1095–110.

[66] Jackson DA, Pombo A. Replicon clusters are stable units of chromosome structure: evidence that nuclear organization contributes to the efficient activation and propagation of S phase in human cells. J Cell Biol 1998;140(6):1285–95.

[67] Berezney R, Dubey DD, Huberman JA. Heterogeneity of eukaryotic replicons, replicon clusters, and replication foci. Chromosoma 2000;108(8):471–84.

[68] Knott SRV, Peace JM, Ostrow AZ, Gan Y, Rex AE, Viggiani CJ, et al. Forkhead transcription factors establish origin timing and long-range clustering in *S. cerevisiae*. Cell 2012;148(1–2):99–111.

[69] Duan Z, Andronescu M, Schutz K, McIlwain S, Kim YJ, Lee C, et al. A three-dimensional model of the yeast genome. Nature 2010;465(7296):363–7.

[70] Hayano M, Kanoh Y, Matsumoto S, Renard-Guillet C, Shirahige K, Masai H. Rif1 is a global regulator of timing of replication origin firing in fission yeast. Genes Dev 2012;26(2):137–50.

[71] Cornacchia D, Dileep V, Quivy J-P, Foti R, Tili F, Santarella-Mellwig R, et al. Mouse Rif1 is a key regulator of the replication-timing programme in mammalian cells. EMBO J 2012;31(18):3678–90.

[72] Yamazaki S, Ishii A, Kanoh Y, Oda M, Nishito Y, Masai H. Rif1 regulates the replication timing domains on the human genome. EMBO J 2012;31(18):3667–77.

[73] Hiraga S-I, Alvino GM, Chang F, Lian H-Y, Sridhar A, Kubota T, et al. Rif1 controls DNA replication by directing Protein Phosphatase 1 to reverse Cdc7-mediated phosphorylation of the MCM complex. Genes Dev 2014;28(4):372–83.

[74] Kanoh Y, Matsumoto S, Fukatsu R, Kakusho N, Kono N, Renard-Guillet C, et al. Rif1 binds to G quadruplexes and suppresses replication over long distances. Nat Struct Mol Biol 2015;22(11):888–97.

[75] Tazumi A, Fukuura M, Nakato R, Kishimoto A, Takenaka T, Ogawa S, et al. Telomere-binding protein Taz1 controls global replication timing through its localization near late replication origins in fission yeast. Genes Dev 2012;26(18):2050–62.

[76] Zhang J, Xu F, Hashimshony T, Keshet I, Cedar H. Establishment of transcriptional competence in early and late S phase. Nature 2002;420(6912):198–202.

[77] Alabert C, Barth TK, Reverón-Gómez N, Sidoli S, Schmidt A, Jensen ON, et al. Two distinct modes for propagation of histone PTMs across the cell cycle. Genes Dev 2015;29(6):585–90.

[78] Sarkies P, Reams C, Simpson LJ, Sale JE. Epigenetic instability due to defective replication of structured DNA. Mol Cell 2010;40(5):703–13.

[79] Sarkies P, Murat P, Phillips LG, Patel KJ, Balasubramanian S, Sale JE. FANCJ coordinates two pathways that maintain epigenetic stability at G-quadruplex DNA. Nucleic Acids Res 2012;40(4):1485–98.

[80] Schiavone D, Guilbaud G, Murat P, Papadopoulou C, Sarkies P, Prioleau M-N, et al. Determinants of G quadruplex-induced epigenetic instability in REV1-deficient cells. EMBO J 2014;33(21):2507–20.

CHAPTER 7

Regulation of Cellular Identity by Polycomb and Trithorax Proteins

M. Wassef*,**,†, R. Margueron*,**,†

*Curie Institute, Paris, France
**INSERM, Paris, France
†CNRS, Paris, France

7.1 INTRODUCTION

Soon following fertilization and the first cell divisions, cells become committed to distinct fates. The specification and maintenance of these multiple cell identities require complex gene regulatory networks. Understanding the logic and precise events underlying such networks is a long-standing issue. With a few exceptions, cells with different fates retain the same genetic material; therefore initiation and maintenance of a given state occur independently of genetic changes and instead involve changes "above" the DNA sequence. A gene expression switch that can be maintained in the absence of the initiating signal(s) is thus termed "epigenetic" [1]. Such a property, also referred to as transcriptional memory, is required in order to perpetuate gene expression states and is believed to underlie cell fate stability (Fig. 7.1A). Transcription factors act as specificity factors in orchestrating switches in gene expression, but, in the absence of relay mechanisms, their action per se is reversible and ceases as soon as they are no longer active or present. The problem of maintenance of cell identity thus requires understanding the nature of such relay mechanisms that "remember" the initial inducing or silencing event (Fig. 7.1B). A straightforward mechanism of epigenetic switch entails a positive feedback loop, whereby the product of a gene stimulates, directly or not, its own expression [2]. Many developmental transcription factors are endowed with this property. Since transcription factors are diffusible molecules, such a relay mechanism is *trans* epigenetic (i.e., it can act from one chromosome to another). Genetic screens for genes involved in correct segmentation in *Drosophila* identified Polycomb (PcG) and Trithorax (TrxG) group as another class of antagonistic proteins that are required for the proper maintenance of gene expression programs. In contrast to transcription factors, PcG and TrxG proteins are largely devoid of DNA binding specificity and act broadly on specialized

165

Chromatin Regulation and Dynamics. http://dx.doi.org/10.1016/B978-0-12-803395-1.00007-1

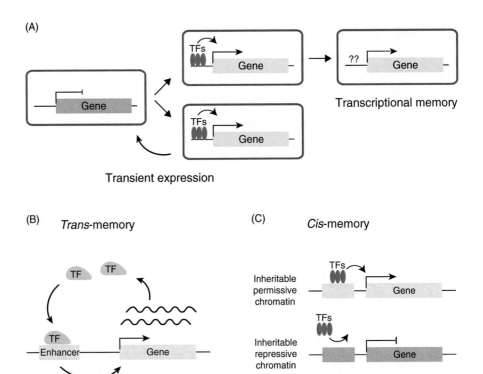

FIGURE 7.1 Schematic representation of inheritance of transcriptional states.
(A) Upon transcriptional stimulation, a gene becomes expressed. When the initiating transcriptions factors *(TFs)* are no longer present, expression of the target gene can be lost or maintained in subsequent stages. Maintenance in the absence of initiator TFs can be ensured in *trans* by a positive feedback loop (B) or in *cis* by modulation of chromatin structure into a state conducive to transcription (C).

sequences. It emerges that PcG and TrxG orchestrate a *cis* epigenetic switch whose logic has been complex to apprehend.

7.2   FUNCTIONS OF PcG AND TrxG PROTEINS DURING DEVELOPMENT

Genetic screens in *Drosophila* have provided an unprecedented opportunity to identify developmental regulators. Mutations in genes of the Hox family give rise to particular transformation termed "homeotic" (from the Greek word homoios, identical) which result in the conversion of the identity of one or several body segments into that of distinct segments. In flies, Hox genes are distributed in two gene clusters termed BX-C and ANT-C. Famous examples of homeotic mutations include the transformation of antennas into legs in *Antennapedia* mutants, which is caused by ectopic expression of the ANT-C Hox gene called *Antp* in head segments. In *Ultrabithorax* mutants, resulting from inactivation of the BX-C Hox gene called *Ubx*, the third thoracic segment containing the halteres is transformed into the identity of the second thoracic segment containing the wings. As a result, *Ultrabithorax* flies have two pairs of wings instead of one.

The function of Hox genes in defining anteroposterior identity is conserved in vertebrates and requires a precise control of the timing and territories in which their products are expressed (reviewed by Ref. [3]). This entails an elaborate interplay of upstream transcription factors that bind to regulatory sequences (e.g., enhancers) within Hox gene clusters to define the initial pattern of expression. However, while expression of Hox genes persists throughout embryogenesis and into adulthood, these early-acting transcription factors cease to be expressed much earlier, often during embryogenesis. Maintenance often involves positive and negative cross-regulations between Hox genes or even direct autoregulation (i.e., a Hox gene that stimulates its own expression). Such mechanisms seem to play a major role in the maintenance of Hox expression patterns in the vertebrate hindbrain (reviewed by Ref. [4]). In addition, correct maintenance of Hox gene expression involves PcG and TrxG proteins.

Like Hox genes, PcG and TrxG genes were identified in screens for homeotic mutants but they mapped outside of Hox clusters. The *Polycomb* (*Pc*) mutation was reported by Pam Lewis in 1947 [5]. Heterozygous male *Pc* flies present specialized bristles called sex combs on the second and third pair of legs in addition to the first pair, hence the name of the mutation. In addition, heterozygous *Pc* flies present partial antenna to leg transformations reminiscent of the *Antennapedia* mutant. Homozygous *Pc* embryos show a dramatic homeotic transformation of all thoracic and abdominal segments into the posterior-most segment. This phenotype is opposite to the phenotype resulting from complete deletion of BX-C, requires the presence of the BX-C cluster and can be exacerbated by extra copies of BX-C. This led E.B. Lewis to propose that *Pc* encodes a negative transregulator of BX-C [6]. This hypothesis was later validated through analysis of Hox gene expression in homozygous *Pc* mutants, which revealed ectopic expression of Hox transcripts anterior to their normal domain of expression [7,8]. Likewise, in heterozygous *Pc* flies, extra sex combs and antenna to leg transformations are respectively accompanied by the ectopic expression of the *Sex combs reduced* (*scr*) gene in the second and third leg imaginal discs and of the *Antp* gene in eye-antennal discs. Genetic screens yielded other mutants presenting phenotypic similarities with *Pc*. The latest screen reported about twenty genes that are classified as PcG [9]. Detailed analysis of PcG mutants revealed that misexpression of Hox genes is a late event in embryogenesis, the initial expression pattern being unaffected [10,11]. This indicated that contrary to classical repressors, such as Krüppel, which inhibit gene expression from the onset, Polycomb proteins act to maintain silencing once established.

Similarly, *Trithorax* (*trx*), the first TrxG gene, was found in a screen for drivers of lineage commitment [12]. Genetic analyses suggested that TrxG genes encode positive regulators of Hox gene expression [13,14]. Importantly, it was realized that the *Trithorax* mutation could suppress *Polycomb* phenotypes [13,15]. The majority of TrxG genes were subsequently identified in a screen for suppressors of *Pc* extra sex comb phenotype [16]. Thus, the antagonism with PcG proteins has been an essential criterion for the characterization of TrxG factors. Subsequent studies showed that TrxG proteins are required for the maintenance of Hox gene expression.

Mouse homologs of PcG and TrxG were later identified by sequence homology. Analyses of mouse PcG and TrxG mutants revealed a conserved role in the regulation of Hox gene expression. In particular, some of these mutants give rise to homeotic transformations of vertebrae that are caused by deregulated Hox gene expression in the embryonic mesoderm (reviewed by Ref. [17]).

Mapping experiments on polytene chromosomes of *Drosophila* salivary glands as well as genome-wide studies revealed that PcG and TrxG proteins are enriched at many genes in addition to Hox genes. However, contrasting with their ubiquitous expression and lack of discernable DNA-binding domain, the genomic localization of PcG and TrxG proteins was found to be cell type-specific. This raised the questions of how PcG and TrxG factors function and how cell-type specificity is achieved. A novel insight into the function of PcG proteins came from the sequence of the *Pc* gene, which uncovered a domain of homology with the heterochromatin-associated protein HP1 [18]. HP1 had been identified as a structural component of heterochromatin. The protein is involved in the heritable transmission of heterochromatic states underlying the phenomenon of position-effect variegation. The homology with HP1 suggested that Pc and, by extension, PcG proteins could be acting in a similar fashion in controlling heritable states of Hox gene expression. This hypothesis was supported by the observation of chromatin alterations in cells lacking the Polycomb protein Enhancer of zeste [19]. As we will see in the following sections, the regulation of chromatin structure by TrxG and PcG proteins is indeed pivotal for the inheritance of transcriptional states.

7.3   MOLECULAR FUNCTIONS OF PcG AND TrxG PROTEINS

Polycomb and Trithorax proteins act mostly within multiprotein complexes that are endowed with chromatin modifying activities. While Polycomb proteins seem to be acting in a concerted fashion in maintaining transcriptional silencing, Trithorax proteins are more diverse in their functions and comprise factors involved in counteracting PcGs as well as factors involved in promoting transcription.

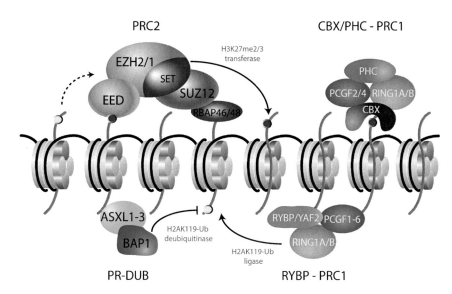

FIGURE 7.2 Schematic representation of the Polycomb machinery illustrating the interplay between the different complexes.

7.3.1 Polycomb Proteins

Biochemical analyses of Polycomb proteins indicate that they mainly belong to two multiprotein complexes, PRC1 and PRC2 (Fig. 7.2). Two additional complexes were characterized in *Drosophila*. The first, Pho-repressive complex (Pho-RC), contains Pho and SFMBT. This complex has no enzymatic activity but it recognizes specific histone modifications and it is involved in PRC1 and PRC2 recruitment [20,21]. Yet, whether such a complex exists in mammals remains an open question. YY1, the closest mammalian homolog of Pho, is required for embryonic development [22] but it has not been found to colocalize with Polycomb target regions as defined by PRC1 or PRC2 enrichment [23]. The second, Polycomb Repressive-Deubiquitinase (PR-DUB), is composed of Calypso and Asx. It was shown to deubiquitinate histone H2A [24]. While this complex is also found in mammals, its contribution to the Polycomb machinery remains unclear. Thus, most of the studies on Polycomb proteins have focused on PRC1 and PRC2.

7.3.1.1 PRC2 Complex

PRC2 is solely responsible for catalyzing the di- and trimethylation of histone H3 on its lysine 27, H3K27me2/3 [25–28]. The case of H3K27me1 is less clear. While PRC2 is responsible for this mark in *Drosophila*, other enzymes contribute to its deposition in plants and in mammals. H3K27me2 is widely distributed whereas H3K27me3 is mainly enriched at silent genes. In mammals, both PRC2 and H3K27me3 peak at the promoters of many silent genes. In *Drosophila*, *cis*-regulatory sequences called Polycomb Response Elements (PREs) that are necessary for PcG-mediated silencing are also clearly enriched for PRC2 components and H3K27me3. Several cofactors associate to the core PRC2 complex to modulate its recruitment and activity (reviewed by Ref. [29]).

For a long time, the question of whether H3K27me3 is a central effector for PRC2 function has remained unanswered. Hence, a number of genes that have the histone mark in their promoter region are nonetheless highly expressed [30], suggesting that it is not sufficient per se to enforce transcriptional silencing. The identification of several nonhistone substrates for a number of chromatin modifiers, including PRC2 [31,32], further suggests that chromatin might not always be the key substrate. While the role of a given histone modification can be assessed in yeast by introducing a point mutation in the DNA encoding the corresponding amino acid, hence preventing its modification, such an approach is difficult to implement in metazoans due to the multiple gene copies encoding each histone protein. This limitation has been recently alleviated in *Drosophila*. Deletions of the whole canonical histone gene cluster were generated that can be rescued by tandem copies of histone genes either wild-type or carrying a mutation in a histone residue [33]. Using this strategy, mutation of H3K27 preventing H3K27 methylation was shown to phenocopy loss of PRC2, demonstrating that H3K27 methylation is central to PRC2 function [34]. Intriguingly, H3K27me3 per se seems to neither alter the physical properties of chromatin nor counteract transcription in vitro (unpublished data), suggesting that the mark acts by promoting binding of proteins that are essential for the maintenance of transcriptional silencing.

7.3.1.2 *PRC1 Complex*

The case of PRC1 is more intricate since different PRC1 subcomplexes exist in *Drosophila* and in mammals, and since many subunits have undergone extensive diversification during evolution. Nonetheless, the E3 ubiquitin protein ligase Ring1 is always at the center of this family of complexes. Ring1 was shown to catalyze monoubiquitylation of histone H2A at lysine 119 (H2AK119ub1, [35]), a mark that overlaps with H3K27me3. In addition to this enzymatic activity, some of the PRC1 subcomplexes can modulate chromatin structure by promoting chromatin compaction and by inhibiting the nucleosome remodeling activity of SWI/SNF-related complexes [36,37].

Several studies have shown that the PRC1 component Pc could bind specifically to H3K27me3, supporting a model whereby PRC1 recruitment is downstream of PRC2 [26,38–40]. However, in mammals, this model was challenged by genome-wide analysis of PRC1 and PRC2 occupancy as well as biochemical analysis of PRC1 composition. Indeed, some PRC1 subcomplexes are devoid of Pc homologs (more precisely CBX4/6/7/8). These variant PRC1 complexes instead contain the RYBP protein or its paralog YAF2 at stoichiometric levels [41,42]. Genome-wide analyses in ES cells indicate that RYBP recruitment is unaffected by loss of H3K27me3 [41]. Yet, RYBP-containing PRC1 subcomplexes show a good correlation with H3K27me3, indicating the presence of convergent mechanisms to recruit PRC2 and RYBP-containing PRC1 subcomplexes at PcG target genes. Strikingly, while loss of H3K27me3 and thus of CBX-containing PRC1 subcomplex from chromatin has little impact on global and local levels of H2AK119ub, depletion of RYBP leads to a strong reduction of this mark. These data suggest that RYBP-containing PRC1 subcomplexes are the major H2AK119 ubiquitin ligases. Accordingly, RYBP–PRC1 complexes seem to be more efficient for H2AK119 ubiquitylation both in vitro and in vivo [42,43]. In *Drosophila*, dRAF, a variant PRC1 complex devoid of PC and PH (homologs of CBX and PHC proteins respectively) has also been characterized [44]. This complex does not contain the fly homolog of RYBP but, like RYBP-containing PRC1, dRAF is more active for H2AK119 ubiquitylation in vitro [44]. Recent studies in *Drosophila* and mammals have shown that H2AK119ub1 can also stabilize PRC2 binding to chromatin [43,45–47]. Thus, while PRC2 and RYBP-containing complexes can be recruited by independent mechanisms, they each promote or stabilize recruitment of the other complex. This could explain the high degree of overlap between PRC2 and PRC1 complexes.

In contrast to H2AK119ub ligase activity, the ability to promote chromatin compaction seems to be shared by the different PRC1 subcomplexes [42]. Chromatin compaction is independent of H2A ubiquitination since PRC1 compacts chromatin assembled from tailless histones in vitro [36]. This has led multiple groups to design experiments to evaluate the respective contribution of enzymatic activity versus chromatin compacting activity to PRC1 function. In *Drosophila*, the ubiquitylase activity of Sce, the ortholog of Ring1A/B is dispensable for Polycomb silencing and flies carrying an H2A-mutant that cannot be ubiquitylated by PRC1 do not show classical Polycomb phenotypes or loss of Polycomb target gene silencing [47]. Thus, in *Drosophila*, PRC1-mediated ubiquitylation of H2A is largely dispensable for PcG-mediated transcriptional repression. In mouse ES cells, Ring1A and Ring1B are needed for compaction and proper silencing of Hox loci [48,49]. In cells lacking both Ring1A and Ring1B, rescue experiments with ubiquitin ligase-deficient Ring1B showed that H2AK119ub1, although not required for compaction of Hox loci, is required for optimal silencing of Polycomb target genes [49]. Hence, H2AK119ub1 might have acquired additional importance in mammals. In this respect, it is worth mentioning that while a form of PRC2 catalyzing H3K27me3 is found in most eukaryotic organisms (including protists, filamentous fungi, algae and, some yeast strains [50,51]), homologs of PRC1 have not been found in all cases. Consistently, PRC1 appears to be more divergent between species than PRC2 [52].

7.3.2 Trithorax Proteins

As indicated previously, TrxG genes were identified both in screens for homeotic mutants and in a screen for suppressors of the *Polycomb* phenotype (reviewed by [53]). Yet, homologs of most TrxG genes are found in organisms, such as the

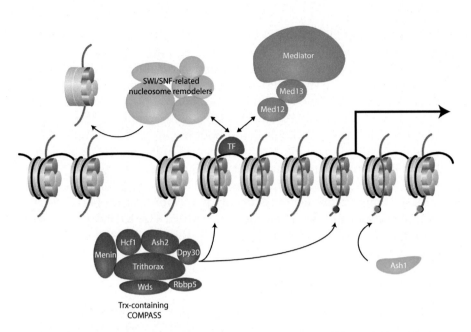

FIGURE 7.3 Schematic representation of the various chromatin-modifying activities of TrxG proteins and/or of their associated complexes.

budding yeast *Saccharomyces cerevisiae* that do not encode any Polycomb gene suggesting that antagonizing PcG is not the sole function of TrxG proteins. Since TrxG proteins are found in distinct complexes with proteins that are not themselves classified as TrxG, we will discuss the enzymatic activities and biological processes they are associated to (Fig. 7.3).

7.3.2.1   Mediator Proteins Kohtalo and Skuld

Kohtalo (*Kto*) and *Skuld* (*Skd*) genes were found in a screen for suppressors of the *Pc* phenotype in *Drosophila* [16]. They encode the Med12 and Med13 subunits of the Mediator complex respectively. The Mediator complex, conserved in yeast, *Drosophila* and mammals, is composed of more than 20 subunits and mediates interaction between gene-specific transcription factors and Pol II, promoting the formation of a preinitiation complex (reviewed by [54]). Mediator has also been implicated in downstream events, such as releasing Pol II pausing and promoting transcription elongation [55]. Med12 and Med13 are part of a submodule of Mediator called the CDK8 module also comprising cyclin-dependent kinase 8 (CDK8) and cyclin C (CCNC). The precise function of the CDK8 module is still under investigation and it seems to be able to both stimulate and repress transcription. While it colocalizes with the core Mediator complex genome-wide [56], biochemical studies have shown that association of this module with Mediator prevents binding of Pol II. It has been suggested that these seemingly contradictory observations might reflect a role for the CDK8 module in releasing paused Pol II and stimulating elongation while preventing untimely transcriptional reinitiation [57].

Considering the general role of mediator in transcription, it is surprising that specific deletion of two of its subunits is associated to a TrxG phenotype. The fact that different transcription factors interact with specific subunits of Mediator provides a possible explanation; deletion of these subunits abolishes the ability of the interacting transcription factor(s) to stimulate transcription without necessarily affecting the function of Mediator on other genes. Indeed, both Med12 and Med13 have been shown to be required for transcriptional activation of Wnt signaling targets in human cells [58] and *Drosophila* [59] through direct interactions with DNA-binding cofactors. While the aforementioned data support a role for Med12 and Med13 in participating in the maintenance of Hox gene expression, *Kto* and *Skd* were also identified in a distinct screen as being required for silencing of the Hox gene *Ubx* [9]. In conclusion, the role of Med12/Med13 in maintaining transcriptional states remains unclear and prompts further investigation.

7.3.2.2 SWI/SNF Chromatin Remodelers

Four of the TrxG genes identified in *Drosophila*, *Brm*, *Mor*, *Osa*, and *Kis*, encode ATP-dependent nucleosome remodeling enzymes. Nucleosome remodeling enzymes play a key role in handling nucleosomes and are implicated in virtually all processes that involve or deal with chromatin: chromatin assembly, replication, DNA repair, and transcription (reviewed by Ref. [60]). By using the energy of ATP hydrolysis, nucleosome-remodeling enzymes can evict or slide histones along DNA, or promote histone exchange. Several nucleosome-remodeling enzymes are recruited to regulatory sequences by transcription factors to facilitate DNA binding of other transcription factors and of the transcription machinery. Subsequently, nucleosome-remodeling enzymes are recruited by the transcription machinery to assist in histone eviction during elongation.

In *Drosophila*, heterozygous loss of the *Brahma (Brm)* gene, which encodes a nucleosome remodeler, suppresses all of the *Pc* phenotypes [61], indicating a fundamental role in antagonizing PcG silencing. For instance, ectopic expression of Scr in second and third thoracic leg disks consequent to *Pc* mutation is abrogated upon decreased Brm expression. The Brm protein is ubiquitously expressed and colocalizes with sites of active transcription on polytene chromosomes. Except for a few loci, loss of Brm protein results in a reduction of both initiating and elongating forms of Pol II on polytene chromosomes [62], suggesting that Brm plays a global role in promoting transcription. Brm belongs to the BAP (Brahma associated proteins complex) and PBAP (polybromo-containing BAP complex) nucleosome remodeling complexes. The Mor Trithorax protein is also a shared component of BAP and PBAP, while Osa is a specific subunit of the BAP complex. BAP and PBAP are the homologs of yeast

SWI/SNF and RSC nucleosome remodeling complexes respectively. SWI/SNF was characterized in *S. cerevisiae* for its role in the transcriptional initiation of inducible genes (reviewed by Ref. [63]). It is recruited by several transcriptional activators, such as Gal4 and facilitates removal of promoter-bound nucleosomes [64]. The RSC complex, which shares several subunits with SWI/SNF, is essential for cell cycle progression. Although SWI/SNF homologs are well conserved in *Drosophila* and mammals, they seem to have developed a high degree of modularity linked to the duplication of core proteins and to the incorporation of several novel subunits. In *Drosophila* and mammals, several subunits are expressed in a cell type–specific manner, and specific subunit composition is crucial for correct cell fate specification (reviewed by Ref. [65]). As in the case of Mediator subunits, cell type–specific subunits of SWI/SNF-related complexes are believed to mediate specific interactions with transcription factors [66]. Such an interaction has been reported between the BAP complex and the Zeste protein, a transcription factor that acts at the level of several Hox genes [67,68].

The function of Kismet (Kis) has been less extensively studied due to its large size that hinders biochemical studies. In addition to its role in maintaining transcription of Hox genes, Kis is also required during early embryogenesis for proper segmental expression of the *engrailed* (*en*) gene [69]. Studies on *Drosophila* polytene chromosomes have shown that Kis extensively overlaps with Brm at active genes [70]. However, in contrast to Brm, loss of Kis has little impact on transcription initiation but rather seems to specifically impair transcription elongation [71]. In vitro, CHD7, the human homolog of Kis and SWI/SNF have different properties; while SWI/SNF complexes are able to act on tightly packed nucleosomes, CHD7 can only function on loosely spaced nucleosomes [72]. Finally, loss of Kis also leads to a decrease of two other TrxG proteins, Ash1 and Trx, from polytene chromosomes [73], which as we will see later, play a key role in antagonizing PcG proteins. Thus, Kis seems to have a dual role in stimulating transcription and counteracting PcG proteins.

7.3.2.3 *Trx and Ash1 Histone Methyltransferases*

Trx is homologous to yeast Set1, a H3K4 methyltransferase belonging to the COMPASS complex (see Ref. [74] for an extended review). COMPASS catalyzes mono-, di-, and tri-methylation of H3K4. H3K4me3 localizes at the 5′ end of transcribed genes in yeast, fly, and mammals. Deposition of the mark is primarily a consequence of active transcription. Studies in yeast and mammalian systems have indeed suggested that COMPASS is recruited to transcribed genes through the polymerase-associated factor 1 (PAF1). In mammals and *Drosophila*, the presence of a CXXC DNA-binding domain in several COMPASS proteins is likely to provide an alternative recruitment pathway. H3K4me3 can in turn stimulate transcription in vitro [75] presumably by stabilizing the binding of factors, such as the chromatin remodeler CHD1 [76] or the general transcription factor TFIID [77,78]. In yeast, removal of Set1 has very little impact on steady state transcription and rather seems to affect the dynamics of gene expression in response to stimulus [79]. In flies, mutation of H3K4 has no major effect on the transcription of developmentally regulated genes [80], suggesting that loss of H3K4 methylation can be counterbalanced by other redundant processes.

Two other paralogs of Trx exist in *Drosophila*, Trr (Trithorax-related) and dSet1, the latter being most closely related to yeast Set1. Trx, Trr, and dSet1 each form a distinct COMPASS complex with shared core components as well as specific subunits. Another TrxG gene, Ash2, encodes a common subunit of *Drosophila* and mammalian COMPASS complexes. In mammals, each of the three Set1 homologs underwent another round of gene duplications, leading to a total of six COMPASS complexes. Initial studies into the roles of dSet1, Trx, and Trr in *Drosophila* reported that loss of dSet1 but not of the other two enzymes leads to a loss of bulk H3K4me3 in vivo [81–83]. Similar observations were made in mammalian cells, where the Set1 complexes SET1A and SET1B are responsible for the majority of H3K4me3 [84]. Genome-wide analyses revealed that the mammalian homologs of Trx, Mll1, and Mll2 are important for the deposition of H3K4me3 at a limited number of loci enriched for key developmental genes, including Hox genes [85–87]. Consistently, genes that require Mll2 for deposition of H3K4me3 are not bound by SET1A/B complexes in mES cells [87]. Of note these Mll2-specific targets are mostly silent and in a "bivalent" state, that is, also enriched for H3K27me3 [87]. Intriguingly, loss of Mll2 and hence of bivalency, does not seem to impact the induction of genes by retinoic acid [86,87]. Thus, Set1-dependant deposition of H3K4me3 at active genes can be viewed as an ancestral function that can facilitate transcription. Gene duplication in metazoans has led to the emergence of more specialized enzymes. Of these, Trx and its mammalian counterparts Mll1/2, play specific roles at the level of genes encoding developmental regulators.

The TrxG protein Ash1 is endowed with histone methyltransferase activity toward H3K36. Various enzymes in addition to Ash1 can catalyze H3K36 methylation. The mark, mostly present in transcribed regions, has been proposed to play multiple roles in transcription initiation, elongation, and splicing (reviewed by Ref. [88]). The Ash1 enzyme was shown to specifically di-methylate H3K36 in vitro and in vivo [73,89,90]. Loss of H3K36me2 resulting from the absence of Ash1 however, does not seem to globally impact transcription but instead leads to a substantial increase of H3K27me3 [73]. Antagonizing Polycomb silencing might in fact be the main function of Trx and Ash1 proteins. Hence, in *Drosophila*, loss of Hox gene expression consequent to the absence of the two methyltransferases is restored in PcG mutants [91]. In vitro studies have provided mechanistic insights into how Trx/Mll and Ash1 might antagonize PcG proteins. Both H3K4me3 and H3K36me2/3 inhibit PRC2 activity on the same histone tail [90,92,93]. Interestingly, these marks do not inhibit PRC2 on the other H3 tail of the nucleosome [93]. This could account for the cooccurrence of H3K27me3 and H3K4me-H3K36me2/3 at the level of "bivalent" promoters [93,94].

In conclusion, TrxG proteins can be subdivided into three functional groups; a first group comprising proteins of the Mediator complex which mediate interaction between transcription factors and the transcription machinery–containing RNA Pol II; a second group consisting of enzymes that can "remodel", that is, slide or remove nucleosomes, thus facilitating binding of transcription factors and of Pol II. These two groups of TrxG proteins are required to maintain the active state. A third group of proteins is specifically involved in antagonizing PcG proteins through covalent methylation of histone tails.

7.4   MECHANISMS OF PcG AND TrxG RECRUITMENT

The mechanisms of PcG/TrxG recruitment to their target genes are complex and appear to have diverged during evolution. An exception, which we alluded to in the previous section, concerns the interaction of TrxG proteins with transcription factors and/or the transcription machinery. Such interactions are found in both *Drosophila* and mammals (reviewed by Ref. [95]). We will here focus on the specificities of TrxG and PcG recruitment.

In *Drosophila*, specific sequences called PRE or TRE for PcG or TrxG Response Element respectively were shown to mediate the loading of PcG and TrxG factors. Functionally, PREs mediate PcG-dependent silencing while TREs participate in the maintenance of the active state. PREs and TREs often overlap, yet several studies have shown that in some instances, PRE function can be separated from TRE function [68,96]. At PRE/TREs, binding of PcG and of Trx is often concomitant and independent of the transcriptional status of the cognate gene [97–100]. PRE/TREs share many properties of enhancers [101] and a number of DNA-binding proteins involved in PcG recruitment are also important for TrxG recruitment and maintenance of transcription (reviewed by Ref. [102]). Hence, transcription factors, such as Zeste, Pipsqueak, or GAF are significantly enriched at PRE/TREs and may act both in transcriptional activation and repression. Focusing on PcG recruitment, the current model attributes an important role to the zinc finger Pho protein, which is the only known PcG protein capable of specific DNA binding. Yet, although PhoRC is required for optimal recruitment of PRC1 and PRC2 at some PREs, it cannot on its own explain the genome-wide distribution of PcG proteins. It is therefore proposed that specific combinations of DNA-binding proteins are required to mediate PcG recruitment [103].

The rules governing PcG/TrxG recruitment in mammals are quite different from *Drosophila*. Dedicated PRE-type *cis*-regulatory sequences seem to be an exception rather that the rule [104,105] and genome-wide studies have failed to identify consensus sequences through which PcG/TrxG proteins can be recruited. A few DNA-binding transcription factors, such as YY1, the mammalian homolog of Pho, Runx1, and REST, as well as several noncoding RNAs have been suggested to recruit PcGs but these can only account for a limited number of PcG-binding sites in the genome. The role of long noncoding RNAs (lncRNAs) in recruiting PRC2 remains debated [106]. Hence, it is well established that during X inactivation in mouse, PRC2 is recruited at the future inactive X as a consequence of X chromosome coating by the lncRNA Xist (reviewed by Ref. [107]). However, the precise underlying molecular mechanisms are still unclear. PRC2 is known to display high affinity for RNA. Yet, whether interactions between PRC2 and lncRNAs are specific and how lncRNAs could drive PRC2 toward specific loci are still open questions. Genome-wide analysis of PcG binding in mES cells revealed that CpG islands (CGIs) are important determinants of PcG recruitment [108]. CGIs are stretches of CpG-rich sequences that are found near the promoters of over 50% of vertebrate genes. Although most CpG dinucleotides in the genome are targets of DNA methyltransferases (DNMTs), CGIs are protected from DNA methylation. CGIs can be sufficient for PcG recruitment [23,109]. Furthermore, sequence motif analysis of PcG bound and unbound CGIs in mES cells revealed that PcG-bound CGIs are enriched for DNA motifs corresponding either to transcriptional activators that are not expressed or to transcriptional repressors that are expressed in ES cells [108], suggesting that absence of transcription rather than presence of specific DNA motifs is key to PcG recruitment. CGIs are bound by zinc finger CXXC DNA-binding domains. The CXXC-containing protein KDM2B, a subunit of a RYBP-containing PRC1 complex has been proposed to participate in the recruitment of PRC1 at CGIs [43,110,111]. However, a recent study does not support a major role for KDM2B in PRC1 recruitment and instead reveals an unexpected function of the protein in preventing DNA methylation at silent, PcG occupied CGIs [112]. Since DNA methylation antagonizes PRC2 recruitment [109,113,114], KDM2B binding is expected to indirectly favor PRC2 recruitment. In addition, KDM2B is a demethylase targeting H3K36me1/2 [115,116], the removal of this mark can thus help create a chromatin environment permissive to PRC2 activity.

Similarly, presence of a CXXC domain in both Mll1 and Mll2 might assist in the recruitment of the enzymes and subsequent deposition of H3K4me3 at bivalent CGIs [87]. However, this hypothesis still requires experimental testing. Hence, in mammals, CGI appears to serve as bait by default for PcG factors. Several parameters can then influence PcG binding, such as DNA methylation or ongoing transcription. This has led to the proposal that PcG recruitment is rather responsive than instructive [106]. While probably true in many instances, several independent processes control PcG recruitment and instructive recruitment is also likely to happen in cases, such as chromosome X inactivation.

7.5 INTERPLAY OF PcGs/TrxGs IN THE PROPAGATION OF TRANSCRIPTIONAL STATES

Studies in *Drosophila* have shown that the expression of Hox genes often follows two phases. In the embryo, initial Hox gene expression pattern is driven by early enhancers that are bound by transcription factors encoded by segmentation genes. Subsequently, expression of these transcription factors ceases and expression of Hox genes is maintained through the action of late-acting enhancers. These early and late enhancers sometimes originate from the same sequence. TREs, that often overlap with PREs can also sustain the active state and as such can be viewed as late-acting enhancers. Yet, two distinctive features characterize late enhancers. First, unlike early enhancers, which are active in specific tissues, late enhancers are, on their own, broadly active (reviewed by Refs. [117,118]). Their action therefore needs to be restricted. Second, contrary to early enhancers, late enhancers cannot overcome PRE-imposed silencing. They can only be active, and thus maintain transcription, in cells where PRE silencing has been overcome by early enhancers [119,120]. Therefore, it appears that one function of PREs is to relay information from early to late enhancers at the right place. This raises the question of how the memory is encoded.

The fact that PcG and TrxG proteins are chromatin modifiers provides an attractive hypothesis: the marks apposed by the antagonistic complexes would constitute chromatin-based "epigenetic signals" that once set, can be self-sustained throughout cell divisions in the absence of the initiating signal. In favor of this model, H3K27me3 is one of the most metabolically stable marks as assessed by isotopic labeling, with a half-life exceeding several cell cycles [121]. In *Caenorhabditis elegans*, H3K27me3 inherited from the gametes can be transmitted to daughter chromatids for several cell divisions in the absence of embryonic PRC2 [122]. These observations, among others [123,124], are in support of conservative segregation of the mark during S-phase whereby preexisting H3K27me3 is passed on to daughter cells in a fairly conservative fashion. Nevertheless, conservative segregation during chromatin replication is required but not sufficient to qualify a chromatin mark as epigenetic; it must be followed by the faithful replenishment of the mark in the absence of the initiating signal that led to its deposition at a given genomic location. Interestingly, the binding of the PRC2 subunit EED to H3K27me3 stimulates PRC2 activity and is required to sustain high levels of H3K27me3 [125]. Such a positive feedback loop could thus represent a mechanism to propagate preexisting methylation patterns. Whether the mark can indeed be propagated independently of the initiating signal remains unclear. While a study reported that H3K27me3 enrichment, imposed by transient recruitment of PRC2 in a defined locus, is maintained when the complex is no longer recruited [126], another study reported that the mark is gradually lost [127]. Hence, even in the case of H3K27me3, additional factors and/or a specific chromatin environment might be required for faithful propagation. However, self-reinforcing properties and stability of the repressive histone mark are likely to be important properties that allow genomic spreading of PcG-mediated silencing and that safeguard the stability of the repressed state, restricting the activity of late enhancers. Since active marks are less stable than repressive marks, they are less likely to explain memorization of the active state.

The next question is therefore how the initial activation is propagated. During this process three steps need to be considered: (1) how initial signals overcome PcG-mediated repression, (2) how these signals eventually potentiate late enhancers, and (3) how late enhancers counteract reestablishment of PcG-mediated silencing. We should first consider that chromatin acts as a barrier for transcription [128]. In particular, histones have been shown to limit binding of transcription factors and loading of the transcription machinery. Thus, chromatin constitutes a molecular threshold for gene activation [129]. By stimulating local compaction and through other chromatin regulating activities, Polycomb proteins might tighten chromatin-mediated silencing, further increasing the threshold for gene activation. This action on chromatin might block access of transcription factors to DNA. However, a strong exogenous activator, such as the yeast Gal4 transcription factor is sufficient to overcome PcG-mediated silencing [130], suggesting that strong activating inputs from embryonic enhancers in *Drosophila* and mammals [109,131] might be sufficient to overcome PcG silencing. Recruitment of chromatin modifying enzymes by transcription factors at enhancers and by the transcription machinery induce dramatic changes at the level of chromatin including histone eviction, histone exchange, and histone modifications that are likely to result in the conversion of PcG-mediated repressive chromatin into a transcriptionally permissive state. Reducing chromatin-based threshold could in turn pave the way to the second step, either enabling binding of transcription factors to late enhancers, allowing late enhancers to sustain recruitment of the transcription machinery at promoters and/or sustaining productive engagement of Pol II. The third step entails understanding how late enhancers, while not able to overcome PcG-mediated silencing, can nevertheless antagonize its reestablishment. The answer to this question might lie in the kinetics of chromatin changes. New histone incorporation during replication provides an opportunity for PcG proteins to reestablish a silent state and thus needs to be counteracted. In this respect, an important feature of H3K27me3 dynamics is that, although very stable, its replenishment following deposition of naive histones in S-phase is slow and in fact extends throughout the cell cycle up to the next S-phase [132–135]. This could provide an opportunity for late enhancers to maintain transcription although they are not initially able to overcome an established PcG-formed chromatin. We propose that the slow kinetics of H3K27me3 deposition after replication gives a competitive advantage for chromatin marks associated with transcription, such as H3K4me3, H3K36me3, and H3K27ac to inhibit deposition of H3K27me3 on new histones [90,92,93]. Mutations in TrxG proteins would tilt the balance in favor of H3K27me3 reestablishment, leading to a switch back to the silent state. Antagonism between nucleosome remodeling activities associated to transcription and PRC1-mediated compaction are also likely to participate in antagonizing reestablishment of PcG silencing as evidenced by the fact that, in vitro, PRC1 can only compact chromatin if added to chromatin prior to SWI/SNF [37].

Nevertheless, in contrast to mammals, *Drosophila* PREs and therefore recruitment of PcG proteins often occur outside the promoter and transcription unit. An important question is therefore how changes in chromatin composition resulting from transcriptional activation can efficiently counteract reestablishment of PcG-mediated silencing. Interestingly, several studies in *Drosophila* have shown that PREs are transcribed in a fashion closely matching transcription of the cognate gene [136–139]. In fact, forced transcription through PRE/TREs has been proposed to be sufficient to irreversibly switch a reporter gene to the active state [138,139], although recent reports suggest that transcription through a minimal PRE might not always be required or even sufficient to antagonize PcG recruitment and activity [140,141]. Alternatively, RNAs originating from TRE/PREs might simply stem from their enhancer-like properties. Antagonizing PcG-mediated silencing at target gene's promoter rather than at the PRE itself might in fact be the key parameter, as exemplified by the case of the *Ubx* gene [97]. Trx and several PcG proteins and H3K27me3 remain constitutively present at and around the PREs of the *Ubx* locus regardless of *Ubx* expression. In contrast, H3K27me3 abundance around the transcription start site of *Ubx* is anticorrelated with the transcriptional status of the gene.

An interesting property of this model is that PcG/TrxG action is by essence reversible. It is the dynamics of the system and the interplay between early acting signals, PcG/TrxG and late signals that orchestrate a switch endowed with "memory." In this respect, the activity of late enhancers is as important as PREs themselves in maintaining expression. Thus, the kinetics of H3K4/H3K36 methylation and of histone turnover imposed by late enhancers can determine whether the switch is irreversible (fast dynamics) or reversible (slow dynamics). If the signal provided by late enhancers is not strong enough or decreases at some point, H3K27me3 is expected to outcompete H3K4/H3K36 methylation and histone turnover and to switch the gene back to its repressed state. In support for this model, experiments in *Drosophila* imaginal wing disks show that strong ectopic Hox gene expression following long-term loss of PcG proteins cannot be rescued by resupplying PcG. However, Hox gene expression can be reversed to the silent state at an early time-point or in cells in which Hox genes are weakly expressed [142]. Finally, for a number of genes that are either transiently expressed or that can maintain their expression by autoregulation, rather than orchestrating a maintenance system, PcG proteins might simply act to impose a tighter threshold for initial activation and subsequently lock silencing upon cessation of transcription. In support for such a role, a study in *Drosophila* cells suggests that loss of PRC2 function leads to uncontrolled transcription, implicating H3K27me2 in transcriptional silencing of intergenic regions [143]. Similarly, using single cell analysis, we have recently observed that loss of Ezh2 in mouse cells leads to sporadic transcriptional events at a number of Polycomb target genes [144]. Hence, transcriptional memory following initial activation might represent a particular refinement of the system enabled by the presence of signals that counteract PcG reestablishment.

7.6 CONCLUSIONS

Gene regulatory networks, involving complex interactions between transcription factors and *cis*-regulatory modules (enhancers) have been shown to control most developmental processes and to specify cell identities. An additional layer of regulation in this network relies on the modulation of chromatin structure. Yet, the main actors in the modulation of chromatin structure are ubiquitously expressed and have no apparent DNA-binding specificity. This raises the question of whether chromatin modifiers are mostly effectors through which transcription factors modulate gene transcription or whether they play a more instructive role in the control of developmental processes. Focusing on the PcG and TrxG proteins, we highlighted here how the dynamic interplay between antagonizing chromatin modifiers can provide a way to encode memory of gene activation. Such a chromatin-based memory needs to be robust to ensure faithful maintenance of gene expression throughout lifetime but also flexible to allow the transcriptional status to be reset in response to a changing environment. However, many aspects concerning the molecular mechanisms underlying this epigenetic memory remain speculative. While it is clear that transcription factors can modulate chromatin structure (directly or indirectly) at their target genes, it is difficult to evaluate how important chromatin structure is in the control of transcription factors' function. Understanding the respective contribution of transcription factors and chromatin regulation in the control of cell identity is of particular importance for pathologies. Indeed despite frequent chromatin alterations in cancer, their consequences on tumor development are often still elusive.

Glossary

Epigenetic Characterizes a change that is perpetuated in the absence of the initiating signal without modification of the DNA sequence.
Homeotic transformation A change of the identity of a body part into that of another body part.

Imaginal discs Larval entities containing undifferentiated cells that give rise to adult structures.

Polytene chromosomes Chromosomes of *Drosophila* salivary glands that have undergone multiple rounds of endoreplications.

Abbreviations

BAP	Brahma associated proteins complex
Brm	Brahma
CGI	CpG island
Kis	Kismet
Kto	Kohtalo
PBAP	polybromo-containing BAP complex
Pc	Polycomb
PcG	Polycomb-group
PRE	Polycomb Response Element
Skd	Skuld
TRE	Trithorax Response Element
TrxG	Trithorax-group
Ubx	Ultrabithorax

References

[1] Ptashne M. On the use of the word 'epigenetic'. Curr Biol 2007;17(7):R233–6.

[2] Perrimon N, Pitsouli C, Shilo BZ. Signaling mechanisms controlling cell fate and embryonic patterning. Cold Spring Harb Perspect Biol 2012;4(8):a005975.

[3] Mallo M, Alonso CR. The regulation of Hox gene expression during animal development. Development 2013;140(19):3951–63.

[4] Tumpel S, Wiedemann LM, Krumlauf R. Hox genes and segmentation of the vertebrate hindbrain. Curr Top Dev Biol 2009;88:103–37.

[5] Lewis P. Pc: Polycomb. Drosophila Inf Serv 1949;21:69.

[6] Lewis EB. A gene complex controlling segmentation in *Drosophila*. Nature 1978;276(5688): 565–70.

[7] Beachy PA, Helfand SL, Hogness DS. Segmental distribution of bithorax complex proteins during *Drosophila* development. Nature 1985;313(6003):545–51.

[8] Wedeen C, Harding K, Levine M. Spatial regulation of Antennapedia and bithorax gene expression by the Polycomb locus in *Drosophila*. Cell 1986;44(5):739–48.

[9] Gaytan de Ayala Alonso A, Gutierrez L, Fritsch C, Papp B, Beuchle D, Muller J. A genetic screen identifies novel polycomb group genes in *Drosophila*. Genetics 2007;176(4):2099–108.

[10] Struhl G, Akam M. Altered distributions of Ultrabithorax transcripts in extra sex combs mutant embryos of *Drosophila*. EMBO J 1985;4(12):3259–64.

[11] Kuziora MA, McGinnis W. Different transcripts of the *Drosophila* Abd-B gene correlate with distinct genetic sub-functions. EMBO J 1988;7(10):3233–44.

[12] Ingham PW, Whittle R. Trithorax: a new homoeotic mutation of *Drosophila melanogaster* causing transformations of abdominal and thoracic imaginal segments. MGG 1980;179(3):607–14.

[13] Ingham PW. Differential expression of bithorax complex genes in the absence of the extra sex combs and trithorax genes. Nature 1983;306(5943):591–3.

[14] Shearn A. The ash-1, ash-2 and trithorax genes of *Drosophila melanogaster* are functionally related. Genetics 1989;121(3):517–25.

[15] Capdevila MP, García-Bellido A. Genes involved in the activation of the bithorax complex of *Drosophila*. Wilhelm Rouxs Arch Dev Biol 1981;190(6):339–50.

[16] Kennison JA, Tamkun JW. Dosage-dependent modifiers of polycomb and antennapedia mutations in *Drosophila*. Proc Natl Acad Sci USA 1988;85(21):8136–40.

[17] Gould A. Functions of mammalian Polycomb group and trithorax group related genes. Curr Opin Genet Dev 1997;7(4):488–94.

[18] Paro R, Hogness DS. The Polycomb protein shares a homologous domain with a heterochromatin-associated protein of *Drosophila*. Proc Natl Acad Sci USA 1991;88(1):263–7.

[19] Rastelli L, Chan CS, Pirrotta V. Related chromosome binding sites for zeste, suppressors of zeste and Polycomb group proteins in *Drosophila* and their dependence on Enhancer of zeste function. EMBO J 1993;12(4):1513–22.

[20] Wang L, Brown JL, Cao R, Zhang Y, Kassis JA, Jones RS. Hierarchical recruitment of polycomb group silencing complexes. Mol Cell 2004;14(5):637–46.

[21] Klymenko T, Papp B, Fischle W, Kocher T, Schelder M, Fritsch C, et al. A Polycomb group protein complex with sequence-specific DNA-binding and selective methyl-lysine-binding activities. Genes Dev 2006;20(9):1110–22.

[22] Affar el B, Gay F, Shi Y, Liu H, Huarte M, Wu S, et al. Essential dosage-dependent functions of the transcription factor yin yang 1 in late embryonic development and cell cycle progression. Mol Cell Biol 2006;26(9):3565–81.

[23] Mendenhall EM, Koche RP, Truong T, Zhou VW, Issac B, Chi AS, et al. GC-rich sequence elements recruit PRC2 in mammalian ES cells. PLoS Genet 2010;6(12):e1001244.

[24] Scheuermann JC, de Ayala Alonso AG, Oktaba K, Ly-Hartig N, McGinty RK, Fraterman S, et al. Histone H2A deubiquitinase activity of the Polycomb repressive complex PR-DUB. Nature 2010;465(7295):243–7.

[25] Cao D, Wang W, Duan X. Grand canonical Monte Carlo simulation for determination of optimum parameters for adsorption of supercritical methane in pillared layered pores. J Colloid Interface Sci 2002;254(1):1–7.

[26] Czermin B, Melfi R, McCabe D, Seitz V, Imhof A, Pirrotta V. *Drosophila* enhancer of Zeste/ESC complexes have a histone H3 methyltransferase activity that marks chromosomal Polycomb sites. Cell 2002;111(2):185–96.

[27] Muller J, Hart CM, Francis NJ, Vargas ML, Sengupta A, Wild B, et al. Histone methyltransferase activity of a *Drosophila* Polycomb group repressor complex. Cell 2002;111(2):197–208.

[28] Kuzmichev A, Nishioka K, Erdjument-Bromage H, Tempst P, Reinberg D. Histone methyltransferase activity associated with a human multiprotein complex containing the Enhancer of Zeste protein. Genes Dev 2002;16(22):2893–905.

[29] Margueron R, Reinberg D. The Polycomb complex PRC2 and its mark in life. Nature 2011;469(7330):343–9.

[30] Young MD, Willson TA, Wakefield MJ, Trounson E, Hilton DJ, Blewitt ME, et al. ChIP-seq analysis reveals distinct H3K27me3 profiles that correlate with transcriptional activity. Nucleic Acids Res 2011;39(17):7415–27.

[31] He A, Shen X, Ma Q, Cao J, von Gise A, Zhou P, et al. PRC2 directly methylates GATA4 and represses its transcriptional activity. Genes Dev 2012;26(1):37–42.

[32] Sanulli S, Justin N, Teissandier A, Ancelin K, Portoso M, Caron M, et al. Jarid2 methylation via the PRC2 complex regulates H3K27me3 deposition during cell differentiation. Mol Cell 2015;57(5):769–83.

[33] Gunesdogan U, Jackle H, Herzig A. A genetic system to assess in vivo the functions of histones and histone modifications in higher eukaryotes. EMBO Rep 2010;11(10):772–6.

[34] Pengelly AR, Copur O, Jackle H, Herzig A, Muller J. A histone mutant reproduces the phenotype caused by loss of histone-modifying factor Polycomb. Science 2013;339(6120):698–9.

[35] Wang H, Wang L, Erdjument-Bromage H, Vidal M, Tempst P, Jones RS, et al. Role of histone H2A ubiquitination in Polycomb silencing. Nature 2004;431(7010):873–8.

[36] Francis NJ, Kingston RE, Woodcock CL. Chromatin compaction by a polycomb group protein complex. Science 2004;306(5701):1574–7.

[37] Shao Z, Raible F, Mollaaghababa R, Guyon JR, Wu CT, Bender W, et al. Stabilization of chromatin structure by PRC1, a Polycomb complex. Cell 1999;98(1):37–46.

[38] Cao R, Wang L, Wang H, Xia L, Erdjument-Bromage H, Tempst P, et al. Role of histone H3 lysine 27 methylation in Polycomb-group silencing. Science 2002;298(5595):1039–43.

[39] Fischle W, Wang Y, Jacobs SA, Kim Y, Allis CD, Khorasanizadeh S. Molecular basis for the discrimination of repressive methyl-lysine marks in histone H3 by Polycomb and HP1 chromodomains. Genes Dev 2003;17(15):1870–81.

[40] Min J, Zhang Y, Xu RM. Structural basis for specific binding of Polycomb chromodomain to histone H3 methylated at Lys 27. Genes Dev 2003;17(15):1823–8.

[41] Tavares L, Dimitrova E, Oxley D, Webster J, Poot R, Demmers J, et al. RYBP-PRC1 complexes mediate H2A ubiquitylation at polycomb target sites independently of PRC2 and H3K27me3. Cell 2012;148(4):664–78.

[42] Gao Z, Zhang J, Bonasio R, Strino F, Sawai A, Parisi F, et al. PCGF homologs, CBX proteins, and RYBP define functionally distinct PRC1 family complexes. Mol Cell 2012;45(3):344–56.

[43] Blackledge NP, Farcas AM, Kondo T, King HW, McGouran JF, Hanssen LL, et al. Variant PRC1 complex-dependent H2A ubiquitylation drives PRC2 recruitment and polycomb domain formation. Cell 2014;157(6):1445–59.

[44] Lagarou A, Mohd-Sarip A, Moshkin YM, Chalkley GE, Bezstarosti K, Demmers JA, et al. dKDM2 couples histone H2A ubiquitylation to histone H3 demethylation during Polycomb group silencing. Genes Dev 2008;22(20):2799–810.

[45] Cooper S, Dienstbier M, Hassan R, Schermelleh L, Sharif J, Blackledge NP, et al. Targeting Polycomb to pericentric heterochromatin in embryonic stem cells reveals a role for H2AK119u1 in PRC2 recruitment. Cell Rep 2014;7(5):1456–70.

[46] Kalb R, Latwiel S, Baymaz HI, Jansen PW, Muller CW, Vermeulen M, et al. Histone H2A monoubiquitination promotes histone H3 methylation in Polycomb repression. Nat Struct Mol Biol 2014;21(6):569–71.

[47] Pengelly AR, Kalb R, Finkl K, Muller J. Transcriptional repression by PRC1 in the absence of H2A monoubiquitylation. Genes Dev 2015;29(14):1487–92.

[48] Eskeland R, Leeb M, Grimes GR, Kress C, Boyle S, Sproul D, et al. Ring1B compacts chromatin structure and represses gene expression independent of histone ubiquitination. Mol Cell 2010;38(3):452–64.

[49] Endoh M, Endo TA, Endoh T, Isono K, Sharif J, Ohara O, et al. Histone H2A mono-ubiquitination is a crucial step to mediate PRC1-dependent repression of developmental genes to maintain ES cell identity. PLoS Genet 2012;8(7):e1002774.

[50] Shaver S, Casas-Mollano JA, Cerny RL, Cerutti H. Origin of the polycomb repressive complex 2 and gene silencing by an E(z) homolog in the unicellular alga *Chlamydomonas*. Epigenetics 2010;5(4):301–12.

[51] Dumesic PA, Homer CM, Moresco JJ, Pack LR, Shanle EK, Coyle SM, et al. Product binding enforces the genomic specificity of a yeast polycomb repressive complex. Cell 2015;160(1–2):204–18.

[52] Schuettengruber B, Chourrout D, Vervoort M, Leblanc B, Cavalli G. Genome regulation by polycomb and trithorax proteins. Cell 2007;128(4):735–45.

[53] Kingston RE, Tamkun JW. Transcriptional regulation by trithorax-group proteins. Cold Spring Harb Perspect Biol 2014;6(10):a019349.

[54] Malik S, Roeder RG. The metazoan Mediator co-activator complex as an integrative hub for transcriptional regulation. Nat Rev Genet 2010;11(11):761–72.

[55] Conaway RC, Conaway JW. The Mediator complex and transcription elongation. Biochim Biophys Acta 2013;1829(1):69–75.

[56] Andrau JC, van de Pasch L, Lijnzaad P, Bijma T, Koerkamp MG, van de Peppel J, et al. Genome-wide location of the coactivator mediator: binding without activation and transient Cdk8 interaction on DNA. Mol Cell 2006;22(2):179–92.

[57] Poss ZC, Ebmeier CC, Taatjes DJ. The Mediator complex and transcription regulation. Crit Rev Biochem Mol Biol 2013;48(6):575–608.

[58] Kim S, Xu X, Hecht A, Boyer TG. Mediator is a transducer of Wnt/beta-catenin signaling. J Biol Chem 2006;281(20):14066–75.

[59] Carrera I, Janody F, Leeds N, Duveau F, Treisman JE. Pygopus activates Wingless target gene transcription through the mediator complex subunits Med12 and Med13. Proc Natl Acad Sci USA 2008;105(18):6644–9.

[60] Becker PB, Workman JL. Nucleosome remodeling and epigenetics. Cold Spring Harb Perspect Biol 2013;5(9):a017905.

[61] Tamkun JW, Deuring R, Scott MP, Kissinger M, Pattatucci AM, Kaufman TC, et al. Brahma: a regulator of *Drosophila* homeotic genes structurally related to the yeast transcriptional activator SNF2/SWI2. Cell 1992;68(3):561–72.

[62] Armstrong JA, Papoulas O, Daubresse G, Sperling AS, Lis JT, Scott MP, et al. The *Drosophila* BRM complex facilitates global transcription by RNA polymerase II. EMBO J 2002;21(19):5245–54.

[63] Fry CJ, Peterson CL. Chromatin remodeling enzymes: who's on first? Curr Biol 2001;11(5):R185–97.

[64] Bryant GO, Prabhu V, Floer M, Wang X, Spagna D, Schreiber D, et al. Activator control of nucleosome occupancy in activation and repression of transcription. PLoS Biol 2008;6(12):2928–39.

[65] Hargreaves DC, Crabtree GR. ATP-dependent chromatin remodeling: genetics, genomics and mechanisms. Cell Res 2011;21(3):396–420.

[66] Kadam S, McAlpine GS, Phelan ML, Kingston RE, Jones KA, Emerson BM. Functional selectivity of recombinant mammalian SWI/SNF subunits. Genes Dev 2000;14(19):2441–51.

[67] Kal AJ, Mahmoudi T, Zak NB, Verrijzer CP. The *Drosophila* brahma complex is an essential coactivator for the trithorax group protein zeste. Genes Dev 2000;14(9):1058–71.

[68] Dejardin J, Cavalli G. Chromatin inheritance upon Zeste-mediated Brahma recruitment at a minimal cellular memory module. EMBO J 2004;23(4):857–68.

[69] Daubresse G, Deuring R, Moore L, Papoulas O, Zakrajsek I, Waldrip WR, et al. The *Drosophila* kismet gene is related to chromatin-remodeling factors and is required for both segmentation and segment identity. Development 1999;126(6):1175–87.

[70] Srinivasan S, Armstrong JA, Deuring R, Dahlsveen IK, McNeill H, Tamkun JW. The *Drosophila* trithorax group protein Kismet facilitates an early step in transcriptional elongation by RNA Polymerase II. Development 2005;132(7):1623–35.

[71] Srinivasan S, Dorighi KM, Tamkun JW. *Drosophila* Kismet regulates histone H3 lysine 27 methylation and early elongation by RNA polymerase II. PLoS Genet 2008;4(10):e1000217.

[72] Bouazoune K, Kingston RE. Chromatin remodeling by the CHD7 protein is impaired by mutations that cause human developmental disorders. Proc Natl Acad Sci USA 2012;109(47):19238–43.

[73] Dorighi KM, Tamkun JW. The trithorax group proteins Kismet and ASH1 promote H3K36 dimethylation to counteract Polycomb group repression in *Drosophila*. Development 2013;140(20):4182–92.

[74] Shilatifard A. The COMPASS family of histone H3K4 methylases: mechanisms of regulation in development and disease pathogenesis. Annu Rev Biochem 2012;81:65–95.

[75] Jiang H, Lu X, Shimada M, Dou Y, Tang Z, Roeder RG. Regulation of transcription by the MLL2 complex and MLL complex-associated AKAP95. Nat Struct Mol Biol 2013;20(10):1156–63.

[76] Sims RJ III, Chen CF, Santos-Rosa H, Kouzarides T, Patel SS, Reinberg D. Human but not yeast CHD1 binds directly and selectively to histone H3 methylated at lysine 4 via its tandem chromodomains. J Biol Chem 2005;280(51):41789–92.

[77] Vermeulen M, Eberl HC, Matarese F, Marks H, Denissov S, Butter F, et al. Quantitative interaction proteomics and genome-wide profiling of epigenetic histone marks and their readers. Cell 2010;142(6):967–80.

[78] Lauberth SM, Nakayama T, Wu X, Ferris AL, Tang Z, Hughes SH, et al. H3K4me3 interactions with TAF3 regulate preinitiation complex assembly and selective gene activation. Cell 2013;152(5):1021–36.

[79] Weiner A, Chen HV, Liu CL, Rahat A, Klien A, Soares L, et al. Systematic dissection of roles for chromatin regulators in a yeast stress response. PLoS Biol 2012;10(7):e1001369.

[80] Hodl M, Basler K. Transcription in the absence of histone H3.2 and H3K4 methylation. Curr Biol 2012;22(23):2253–7.

[81] Mohan M, Herz HM, Smith ER, Zhang Y, Jackson J, Washburn MP, et al. The COMPASS family of H3K4 methylases in *Drosophila*. Mol Cell Biol 2011;31(21):4310–8.

[82] Ardehali MB, Mei A, Zobeck KL, Caron M, Lis JT, Kusch T. *Drosophila* Set1 is the major histone H3 lysine 4 trimethyltransferase with role in transcription. EMBO J 2011;30(14):2817–28.

[83] Hallson G, Hollebakken RE, Li T, Syrzycka M, Kim I, Cotsworth S, et al. dSet1 is the main H3K4 di- and tri-methyltransferase throughout *Drosophila* development. Genetics 2012;190(1):91–100.

[84] Wu M, Wang PF, Lee JS, Martin-Brown S, Florens L, Washburn M, et al. Molecular regulation of H3K4 trimethylation by Wdr82, a component of human Set1/COMPASS. Mol Cell Biol 2008;28(24):7337–44.

[85] Wang P, Lin C, Smith ER, Guo H, Sanderson BW, Wu M, et al. Global analysis of H3K4 methylation defines MLL family member targets and points to a role for MLL1-mediated H3K4 methylation in the regulation of transcriptional initiation by RNA polymerase II. Mol Cell Biol 2009;29(22):6074–85.

[86] Hu D, Garruss AS, Gao X, Morgan MA, Cook M, Smith ER, et al. The Mll2 branch of the COMPASS family regulates bivalent promoters in mouse embryonic stem cells. Nat Struct Mol Biol 2013;20(9):1093–7.

[87] Denissov S, Hofemeister H, Marks H, Kranz A, Ciotta G, Singh S, et al. Mll2 is required for H3K4 trimethylation on bivalent promoters in embryonic stem cells, whereas Mll1 is redundant. Development 2014;141(3):526–37.

[88] Wagner EJ, Carpenter PB. Understanding the language of Lys36 methylation at histone H3. Nat Rev Mol Cell Biol 2012;13(2):115–26.

[89] Tanaka Y, Katagiri Z, Kawahashi K, Kioussis D, Kitajima S. Trithorax-group protein ASH1 methylates histone H3 lysine 36. Gene 2007;397(1–2):161–8.

[90] Yuan W, Xu M, Huang C, Liu N, Chen S, Zhu B. H3K36 methylation antagonizes PRC2-mediated H3K27 methylation. J Biol Chem 2011;286(10):7983–9.

[91] Klymenko T, Muller J. The histone methyltransferases Trithorax and Ash1 prevent transcriptional silencing by Polycomb group proteins. EMBO Rep 2004;5(4):373–7.

[92] Schmitges FW, Prusty AB, Faty M, Stutzer A, Lingaraju GM, Aiwazian J, et al. Histone methylation by PRC2 is inhibited by active chromatin marks. Mol Cell 2011;42(3):330–41.

[93] Voigt P, LeRoy G, Drury WJ III, Zee BM, Son J, Beck DB, et al. Asymmetrically modified nucleosomes. Cell 2012;151(1):181–93.

[94] Bernstein BE, Mikkelsen TS, Xie X, Kamal M, Huebert DJ, Cuff J, et al. A bivalent chromatin structure marks key developmental genes in embryonic stem cells. Cell 2006;125(2):315–26.

[95] Smith E, Shilatifard A. The chromatin signaling pathway: diverse mechanisms of recruitment of histone-modifying enzymes and varied biological outcomes. Mol Cell 2010;40(5):689–701.

[96] Tillib S, Petruk S, Sedkov Y, Kuzin A, Fujioka M, Goto T, et al. Trithorax- and Polycomb-group response elements within an Ultrabithorax transcription maintenance unit consist of closely situated but separable sequences. Mol Cell Biol 1999;19(7):5189–202.

[97] Papp B, Muller J. Histone trimethylation and the maintenance of transcriptional ON and OFF states by trxG and PcG proteins. Genes Dev 2006;20(15):2041–54.

[98] Beisel C, Buness A, Roustan-Espinosa IM, Koch B, Schmitt S, Haas SA, et al. Comparing active and repressed expression states of genes controlled by the Polycomb/Trithorax group proteins. Proc Natl Acad Sci USA 2007;104(42):16615–20.

[99] Schuettengruber B, Ganapathi M, Leblanc B, Portoso M, Jaschek R, Tolhuis B, et al. Functional anatomy of polycomb and trithorax chromatin landscapes in *Drosophila* embryos. PLoS Biol 2009;7(1):e13.

[100] Schwartz YB, Kahn TG, Stenberg P, Ohno K, Bourgon R, Pirrotta V. Alternative epigenetic chromatin states of polycomb target genes. PLoS Genet 2010;6(1):e1000805.

[101] Tie F, Banerjee R, Saiakhova AR, Howard B, Monteith KE, Scacheri PC, et al. Trithorax monomethylates histone H3K4 and interacts directly with CBP to promote H3K27 acetylation and antagonize Polycomb silencing. Development 2014;141(5):1129–39.

[102] Kassis JA, Brown JL. Polycomb group response elements in *Drosophila* and vertebrates. Adv Genet 2013;81:83–118.

[103] Ringrose L, Paro R. Polycomb/Trithorax response elements and epigenetic memory of cell identity. Development 2007;134(2):223–32.

[104] Sing A, Pannell D, Karaiskakis A, Sturgeon K, Djabali M, Ellis J, et al. A vertebrate Polycomb response element governs segmentation of the posterior hindbrain. Cell 2009;138(5):885–97.

[105] Woo CJ, Kharchenko PV, Daheron L, Park PJ, Kingston RE. A region of the human HOXD cluster that confers polycomb-group responsiveness. Cell 2010;140(1):99–110.

[106] Klose RJ, Cooper S, Farcas AM, Blackledge NP, Brockdorff N. Chromatin sampling—an emerging perspective on targeting polycomb repressor proteins. PLoS Genet 2013;9(8):e1003717.

[107] Gendrel AV, Heard E. Noncoding RNAs and epigenetic mechanisms during X-chromosome inactivation. Annu Rev Cell Dev Biol 2014;30:561–80.

[108] Ku M, Koche RP, Rheinbay E, Mendenhall EM, Endoh M, Mikkelsen TS, et al. Genomewide analysis of PRC1 and PRC2 occupancy identifies two classes of bivalent domains. PLoS Genet 2008;4(10):e1000242.

[109] Jermann P, Hoerner L, Burger L, Schubeler D. Short sequences can efficiently recruit histone H3 lysine 27 trimethylation in the absence of enhancer activity and DNA methylation. Proc Natl Acad Sci USA 2014;111(33):E3415–21.

[110] Farcas AM, Blackledge NP, Sudbery I, Long HK, McGouran JF, Rose NR, et al. KDM2B links the Polycomb Repressive Complex 1 (PRC1) to recognition of CpG islands. Elife 2012;1:e00205.

[111] Wu X, Johansen JV, Helin K. Fbxl10/Kdm2b recruits polycomb repressive complex 1 to CpG islands and regulates H2A ubiquitylation. Mol Cell 2013;49(6):1134–46.

[112] Boulard M, Edwards JR, Bestor TH. FBXL10 protects Polycomb-bound genes from hypermethylation. Nat Genet 2015;47(5):479–85.

[113] Bartke T, Vermeulen M, Xhemalce B, Robson SC, Mann M, Kouzarides T. Nucleosome-interacting proteins regulated by DNA and histone methylation. Cell 2010;143(3):470–84.

[114] Hagarman JA, Motley MP, Kristjansdottir K, Soloway PD. Coordinate regulation of DNA methylation and H3K27me3 in mouse embryonic stem cells. PloS One 2013;8(1):e53880.

[115] He J, Kallin EM, Tsukada Y, Zhang Y. The H3K36 demethylase Jhdm1b/Kdm2b regulates cell proliferation and senescence through p15(Ink4b). Nat Struct Mol Biol 2008;15(11):1169–75.

[116] Blackledge NP, Zhou JC, Tolstorukov MY, Farcas AM, Park PJ, Klose RJ. CpG islands recruit a histone H3 lysine 36 demethylase. Mol Cell 2010;38(2):179–90.

[117] Maeda RK, Karch F. The ABC of the BX-C: the bithorax complex explained. Development 2006;133(8):1413–22.

[118] Steffen PA, Ringrose L. What are memories made of? How Polycomb and Trithorax proteins mediate epigenetic memory. Nat Rev Mol Cell Biol 2014;15(5):340–56.

[119] Chan CS, Rastelli L, Pirrotta V. A Polycomb response element in the Ubx gene that determines an epigenetically inherited state of repression. EMBO J 1994;13(11):2553–64.

[120] Poux S, Kostic C, Pirrotta V. Hunchback-independent silencing of late Ubx enhancers by a Polycomb Group Response Element. EMBO J 1996;15(17):4713–22.

[121] Zee BM, Levin RS, Xu B, LeRoy G, Wingreen NS, Garcia BA. In vivo residue-specific histone methylation dynamics. J Biol Chem 2010;285(5):3341–50.

[122] Gaydos LJ, Wang W, Strome S. Gene repression H3K27me and PRC2 transmit a memory of repression across generations and during development. Science 2014;345(6203):1515–8.

[123] Annunziato AT. Split decision: what happens to nucleosomes during DNA replication? J Biol Chem 2005;280(13):12065–8.

[124] Radman-Livaja M, Verzijlbergen KF, Weiner A, van Welsem T, Friedman N, Rando OJ, et al. Patterns and mechanisms of ancestral histone protein inheritance in budding yeast. PLoS Biol 2011;9(6):e1001075.

[125] Margueron R, Justin N, Ohno K, Sharpe ML, Son J, Drury WJ III, et al. Role of the polycomb protein EED in the propagation of repressive histone marks. Nature 2009;461(7265):762–7.

[126] Hansen KH, Bracken AP, Pasini D, Dietrich N, Gehani SS, Monrad A, et al. A model for transmission of the H3K27me3 epigenetic mark. Nat Cell Biol 2008;10(11):1291–300.

[127] Sarma K, Margueron R, Ivanov A, Pirrotta V, Reinberg D. Ezh2 requires PHF1 to efficiently catalyze H3 lysine 27 trimethylation in vivo. Mol Cell Biol 2008;28(8):2718–31.

[128] Struhl K. Fundamentally different logic of gene regulation in eukaryotes and prokaryotes. Cell 1999;98(1):1–4.

[129] Mirny LA. Nucleosome-mediated cooperativity between transcription factors. Proc Natl Acad Sci USA 2010;107(52):22534–9.

[130] Cavalli G, Paro R. The *Drosophila* Fab-7 chromosomal element conveys epigenetic inheritance during mitosis and meiosis. Cell 1998;93(4):505–18.

[131] Taberlay PC, Kelly TK, Liu CC, You JS, De Carvalho DD, Miranda TB, et al. Polycomb-repressed genes have permissive enhancers that initiate reprogramming. Cell 2011;147(6):1283–94.

[132] Aoto T, Saitoh N, Sakamoto Y, Watanabe S, Nakao M. Polycomb group protein-associated chromatin is reproduced in post-mitotic G1 phase and is required for S phase progression. J Biol Chem 2008;283(27):18905–15.

[133] Xu M, Wang W, Chen S, Zhu B. A model for mitotic inheritance of histone lysine methylation. EMBO Rep 2012;13(1):60–7.

[134] Zee BM, Britton LM, Wolle D, Haberman DM, Garcia BA. Origins and formation of histone methylation across the human cell cycle. Mol Cell Biol 2012;32(13):2503–14.

[135] Alabert C, Barth TK, Reveron-Gomez N, Sidoli S, Schmidt A, Jensen ON, et al. Two distinct modes for propagation of histone PTMs across the cell cycle. Genes Dev 2015;29(6):585–90.

[136] Rank G, Prestel M, Paro R. Transcription through intergenic chromosomal memory elements of the *Drosophila* bithorax complex correlates with an epigenetic switch. Mol Cell Biol 2002;22(22):8026–34.

[137] Bender W, Fitzgerald DP. Transcription activates repressed domains in the *Drosophila* bithorax complex. Development 2002;129(21):4923–30.

[138] Schmitt S, Prestel M, Paro R. Intergenic transcription through a polycomb group response element counteracts silencing. Genes Dev 2005;19(6):697–708.

[139] Herzog VA, Lempradl A, Trupke J, Okulski H, Altmutter C, Ruge F, et al. A strand-specific switch in noncoding transcription switches the function of a Polycomb/Trithorax response element. Nat Genet 2014;46(9):973–81.

[140] Pease B, Borges AC, Bender W. Noncoding RNAs of the Ultrabithorax domain of the *Drosophila* bithorax complex. Genetics 2013;195(4):1253–64.

[141] Erokhin M, Elizar'ev P, Parshikov A, Schedl P, Georgiev P, Chetverina D. Transcriptional read-through is not sufficient to induce an epigenetic switch in the silencing activity of Polycomb response elements. Proc Natl Acad Sci USA 2015;112(48):14930–5.

[142] Beuchle D, Struhl G, Muller J. Polycomb group proteins and heritable silencing of *Drosophila* Hox genes. Development 2001;128(6):993–1004.

[143] Lee HG, Kahn TG, Simcox A, Schwartz YB, Pirrotta V. Genome-wide activities of Polycomb complexes control pervasive transcription. Genome Res 2015;25(8):1170–81.

[144] Wassef M, Rodilla V, Teissandier A, Zeitouni B, Gruel N, Sadacca B, et al. Impaired PRC2 activity promotes transcriptional instability and favors breast tumorigenesis. Genes Dev 2015;29(24):2547–62.

CHAPTER 8

Interplay Between Chromatin and Splicing

A. Fiszbein, M.A. Godoy Herz, L.I. Gomez Acuña, A.R. Kornblihtt

Institute of Physiology, Molecular Biology and Neurosciences (IFIBYNE-CONICET) and Department of Physiology, Molecular and Cell, Faculty of Natural Sciences, University of Buenos Aires, Ciudad Universitaria, Buenos Aires, Argentina

8.1 INTRODUCTION

Alternative splicing (AS), is known as one of the fundamental mechanism involved in the increment of the complexity of multicellular organism, as it is a major contributor to protein diversity. Alternative splicing explains how different proteins can be synthesized from a single gene and is thought to be an important mechanism by which the expansion of coding capacities was achieved in higher eukaryotes during evolution. As eukaryotic genes are organized in exons and introns, splicing is a modification of the nascent mRNA in which introns are removed and exons are joined. Alternative splicing involves another level of regulation by which particular exons or introns may be included or excluded from the processed mature messenger RNA (mRNA).

Estimated to affect nearly 95% of human genes, alternative splicing is more the rule than the exception [1]. Indeed, it is known that more than 60% of human alternative splicing events are regulated between tissues [2] and it is not difficult to imagine that both splicing and alternative splicing must be specifically modulated in all developmental stages. Even more, many genetic disorders are caused by mutations that alter the function of splicing-regulatory elements. Variations in the abundance and activity of splicing regulatory proteins can lead to misregulation and have been often associated with some kind of cancer and disease [3]. Thus, the understanding of alternative splicing at the molecular level is not only important for the comprehension of gene expression, but it is also of medical relevance [4–6].

Alternative splicing is often the central target of different signalling pathways [1]. Its regulation not only depends on the interaction of splicing factors, transcription factors, and regulatory elements but it is also controlled by

CONTENTS

Chromatin Regulation and Dynamics. http://dx.doi.org/10.1016/B978-0-12-803395-1.00008-3

various extracellular and intracellular signals. Moreover, like other pre-mRNA processing reactions, splicing is also coupled to changes in its template. Since it is not the same to walk on a flat or a rocky road, transcription can lead to different splicing choices depending on the chromatin structure of the template gene. Therefore, regulation of alternative splicing involves chromatin structure states, changes in factor accessibility, posttranslational histone modifications, and RNA polymerase II (RNAPII) elongation rate [7,8].

8.2   INTERPLAY BETWEEN SPLICING AND TRANSCRIPTION

8.2.1   Alternative Splicing

In order to understand the coupling between splicing and transcription it is important to analyze the molecular basis of alternative splicing and its different levels of regulation. Introns are noncoding interspersed sequences that are normally not included in the mature mRNA. Oppositely, exons refer to the sequences that remain in the mature mRNA once the introns have been removed. Most times, exons are included in the mature mRNA and introns excluded, but alternative splicing sets the exception. In human cells, generally, 90% of the pre-mRNA is removed as introns, while 10% represents exonic sequences.

The enzyme responsible for transcription of mRNAs is RNAPII. Transcription starts at the +1 pointed by the first nucleotide to be synthesized and finishes downstream the cleavage and polyadenylation site. The unspliced mRNA precursor is recognized by a multiribonucleoprotein complex, the spliceosome, whose components have specific binding sites in each of the introns to be removed. The spliceosome is a megaparticle that includes five small nuclear ribonucleoproteins (U1, U2, U4, U5, and U6 snRNPs) and a set of auxiliary proteins. It is a highly dynamic structure; the variation of its conformation and composition allows accuracy and flexibility at the same time. The spliceosome structure has been recently resolved [9,10].

Conserved short sequences at the 5′ splice site (ss), 3′ss, and branch site in the recently synthesized mRNA contributes to spliceosome recognition and intron/exon definition. The branch site is followed by another recognition site known as a polypyrimidine tract. In the first step U1 snRNP is recruited to the 5′ss while auxiliary factors, such as SF1 and U2AF interact with the branch site and polypyrimidine tract, respectively. Then, U2 snRNP recognizes the branch site and subsequently, U4/U6 with U5 snRNPs are recruited (Fig. 8.1). Major rearrangements in RNA–RNA and RNA–protein

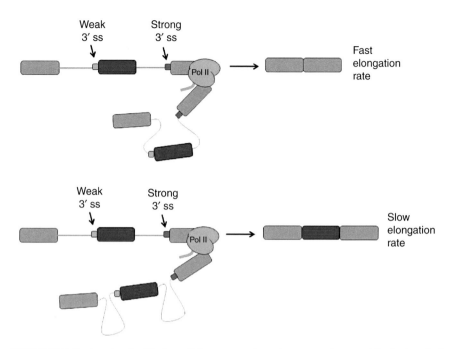

FIGURE 8.1 Regions involved in the splicing process in metazoans and some of the factors that recognize these regions.

The diagram shows two exons (gray boxes) flanking an intron with 3'and 5' splice sites and consensus sequences of the branch site. Different factors involved in the recognition of these sequences are shown sequentially as various complexes are formed in the process of splicing.

interactions, leading to the destabilization of the U1 and U4 snRNPs, give rise to the activated spliceosome [6]. The activated spliceosome then removes pre-mRNA introns by two consecutive transesterification reactions and the dissociated spliceosome releases the snRNPs, which can participate in additional rounds of splicing.

8.2.2 Connections Between Transcription and Splicing

Abundance of splicing factors, as well as their posttranslational modifications regulates constitutive and alternative splicing choices. Moreover, during the last decades it became clear that functional and physical interactions between transcription and splicing machineries constitute a new level of regulation. Evidence demonstrating a coupling between transcription and splicing suggest that promoters, transcription factors, coactivators, and inhibitors have a fundamental impact in alternative splicing decisions.

One of the first evidences for coupling between transcription and alternative splicing was the demonstration that different RNAPII promoters regulating the same transcriptional unit elicited different proportions of two splicing isoforms [11]. Moreover, in a similar way, sensitivity to splicing regulatory factors has been shown to depend on the promoter structure [12].

The first evidence of cooccurrence of splicing and transcription took place more than 20 years ago when electron microscopy observation of transcriptional units prepared from *Drosophila* embryos revealed the presence or RNA loops in the nascent transcript [13]. Since then, strong evidence reinforced the notion of cotranscriptional splicing. A much more recent work provides evidence of cotranscriptionality by high troughout transcriptome sequencing [14]. For long introns, the authors found higher levels of RNA present in the 5′ end of introns. The accumulation of reads formed a 5′–3′ decreasing slope along individual introns, rather than across the entire transcript. As this observation can be explained by the formation of nascent transcripts representing different stages of maturation across individual introns, the authors suggested a model where splicing takes place cotranscriptionally.

Studies in the past decade centered the attention on connections between transcription and splicing. Since recent publications revisited and demonstrated the fact that splicing could be a cotranscriptional process, Neugebauer and coworkers examined how frequent cotranscriptional splicing is among different genes and introns [15]. The first genome wide analysis on cotranscriptional splicing revealed that most introns are removed cotranscriptionally from intron-containing genes [16]. Moreover, the authors discovered that RNAP II pauses within terminal exons to yield highly efficient cotranscriptional splicing [16]. On the contrary, in a posterior work Rosbash and coworkers have shown that cotranscriptional splicing is deficient on short second exon genes in yeast [17]. The authors estimate that ≥90% of endogenous yeast splicing is posttranscriptional, consistent with an analysis of posttranscriptional snRNP-associated pre-mRNA. The question is thus not yet resolved, as contradictory evidence has been obtained by different approaches and in different species.

Nevertheless, the prevalence of splicing catalysis (regarding its cotranscriptional or posttranscriptional status) depends on the organism. In mammals the evidence suggests convincingly that splicing is mainly cotranscriptional [18–20]. Using a quantitative RT-PCR assay, Black and coworkers demonstrated that the majority of introns are already excised from the human c-Src and fibronectin pre-mRNAs that are still in the process of synthesis [19]. The same was seen for introns flanking alternative exons or constitutive exons, but the first case shows differences in excision efficiency between cell lines with different regulatory conditions [19]. In a similar scenario, Wieslander and coworkers show that in flies, cotranscriptionality could depend on the location of the intron in the gene. While intron 3 of Balbiani ring 1 gene is excised simultaneously with transcription, intron 4 is excised posttranscriptionally in most of the cases [18]. In general, however, pre-mRNA splicing has been reported to occur co-transcriptionally on the nascent transcripts and splice site selection may generally precede polyadenylation [13].

Altogether, the results demonstrate that despite transcription and splicing are carried out by two distinct macromolecular machines capable of functioning independently of one another, spliceosome assembly is a cotranscriptional process, in vivo and splicing may occur during transcription but also after transcription is completed. It is now believed that some introns could be processed cotranscriptionally while others could be spliced posttranscriptionally and it is still a question whether the same intron is always processed in the same way.

The cotranscriptionality does not necessarily mean that the first intron exposed to the spliceosome will be removed on the first place. Moreover, studying the relative order of removal of the introns flanking an alternative cassette exon, it has been found that there is a preferential removal of the intron downstream from the cassette exon before the upstream intron is removed [20]. The commitment to be spliced could thus be completed in the exact order the introns are transcribed and marked by spliceosome assembly, serving as a guarantee that the intron will be spliced. The specific order this process might, however, be carried out independently from the order of transcription.

8.2.3   Coupling Alternative Splicing to Transcription: Models

In order to explain how coupling between transcription and splicing works, two nonmutually exclusive models have been proposed for metazoans. The first model—the recruitment model—centers the attention on the factors recruited to the transcription machinery, while the second one—the kinetic model—focuses on the opportunity that the splicing factors have to recognize splice sites.

The recruitment model involves the association of splicing factors to transcription sites by the transcription machinery. The carboxy-terminal domain (CTD) of the catalytic subunit of RNAPII plays a pivotal role in the process. This highly repetitive domain is subject to multiple regulatory posttranslational modifications affecting transcriptional properties of RNAPII and its protein interactions. The mammalian CTD comprises 52 tandemly repeated heptapeptides with a consensus sequence enriched in the aminoacid serine. In one of the founding works in 1995, Gerber and coworkers discovered that the CTD is necessary for enhancer-driven transcription [21]. Moreover, some years later, McCracken et al. [22,23] provided strong evidence about a fundamental function of the CTD not only in transcriptional activation, but also in capping, splicing, and 3'-end processing. These results promoted the idea that the CTD couples mRNA processing to transcription supporting the concept that both processes are controlled simultaneously and can influence each other by being regulated in a coordinated manner. Many studies have characterized the functional relevance of the CTD in transcription regulation and most centralized the evidence in the posttranslational modifications of CTD residues [21,22,24–28]. It is known that phosphorylation on serine 5 plays a key role in the recruitment of capping enzymes, while serine 2 has a fundamental role in 3' mRNA processing [29]. Furthermore, phosphorylation of serine 7 and threonine 4 also has a role in 3' processing of small nuclear RNAs [30] and histone mRNAs [31]. Depending on the mentioned phosphorylation, pattern of the CTD, it is known that splicing factors are recruited to the sites of transcription [24,32,33]. However, the mechanism is not well elucidated since the interactions could be direct or mediated by the nascent mRNA [34]. Most importantly, it has been demonstrated that transcription by an RNAPII lacking its CTD results in inefficient splicing [23].

More specifically, there is a lot of evidence supporting a recruitment model in which splicing factors bind the transcription machinery affecting alternative splicing choices. For example, it was found that the CTD is required for the inhibitory action of the serine/arginine-rich (SR) protein SRSF3 on the inclusion of a fibronectin cassette exon in the mature mRNA, as a mutant polymerase with a truncated CTD abolishes cassette exon upregulation upon SRSF3 knockdown [35]. Another example of the recruitment mechanism, in this case in a CTD-independent manner, is illustrated by the observation that the termogenic activator PGC-1 favours the exclusion of an alternatively spliced exon 25 of the fibronectin gene by binding to its promoter [36].

Finally, a more recent work shows how Mediator regulates alternative splicing decisions by recruiting mRNA processing factors, such as hnRNP L. Moreover, most alternative splicing events regulated by hnRNP L are also regulated by one of the mediator subunits called MED23. These results demonstrate a cross talk between Mediator and the splicing machinery [37].

In relation to the kinetic model of coupling between transcription and splicing, it is important to take into account that RNAPII elongates the nascent transcript at an average of 2–3 kb/min, but its speed undergoes changes along the genes. The kinetic coupling might thus provide an explanation for how the transcription rate could influence alternative splicing choices (Fig. 8.2).

The model in Fig. 8.2 presents the case in which the 3' splice site of the intron upstream of a skipped exon is weaker compared to the 3' splice site of the intron downstream. A fast RNAPII or without pauses would expose both 3' sites to the spliceosome components almost simultaneously, leading to a competition between them. This competition would favor the use of the stronger 3' splice site of the intron downstream the skipped exon. On the contrary,

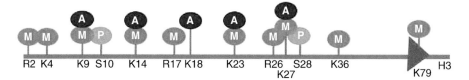

FIGURE 8.2 The kinetic model of functional coupling between transcription and alternative splicing.

The diagram shows the simplest model proposed to explain why the rate of elongation may favor the inclusion of an alternative exon with a weak site of splicing.

a slower RNAPII or pauses during transcription would allow more time for splicing factor assembly at the first transcribed weak 3′ splice site of the intron upstream the skipped exon, committing it to splicing in spite of its weakness.

However, the order of intron removal does not have to be the same as the order of transcription. As mentioned before, it has been shown for the fibronectin alternative exon 33 (also known as EDI) that there is a preferential removal of the intron downstream of the skipped exon before the upstream intron has been removed [20] and, thus, slow elongation rate of the polymerase favors commitment of intron removal rather than splicing during spliceosome assembly. Overall, the kinetic coupling explains how low elongation rates or transcriptional pausing regulate the inclusion of alternative exons, whereas rapid elongation rates or the absence of transcriptional pausing favor its exclusion (Fig. 8.2).

The first direct demonstration that transcription elongation affects alternative splicing in human cells was provided by the use of a mutant form of RNAPII with a point mutation that confers a lower elongation rate. The slow polymerase stimulated the inclusion of the fibronectin EDI exon, confirming the inverse correlation between elongation rate and inclusion of this alternative exon [38]. Moreover, it has been demonstrated that CCCTC-binding factor, more known as CTCF, can bind to DNA sequences and promote inclusion of weak upstream exons by mediating local RNAPII pausing. The accessibility of the DNA-binding protein to its target region is inhibited by DNA methylation, providing another mechanism by which a chromatin mark could change RNAPII processivity, and thus affect alternative splicing choices [39].

Furthermore, since the RNAPII machinery itself recruits many splicing regulatory proteins, the transcription elongation rate would affect the action of these *trans*-acting regulators as well. Instead of being mutually exclusive both models of coupling are therefore complementary to each other, and there is a cross talk between RNAPII elongation rate, the splicing machinery, and the recruitment of factors. The DBIRD complex provides an example highlighting the interface between the two models. This complex binds directly to RNAPII and the nascent transcript favoring the exclusion of alternative exons by enhancing elongation rate.

8.3   CHROMATIN STRUCTURE AND ITS IMPACT ON SPLICING

8.3.1   Alternative Splicing Regulation by Chromatin Structure

Changes in RNAPII elongation rates could be caused by changes in the transcription template. This template is composed of chromatin structural units called nucleosomes that consist of 147 pb of double-strand DNA wrapped around a histone octamer unit. The histones forming a nucleosome could undergo different posttranslational modifications that lead to modulation of the chromatin structure (Fig. 8.3). It is well established now that changes in chromatin structure caused by more compact or more relaxed nucleosome organization affect the transcription process. Besides, accumulated evidence

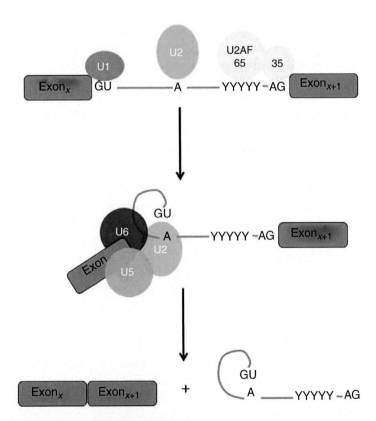

FIGURE 8.3 Covalent modifications in histone H3.
Different modifications that histone H3 may carry at each of its residues belonging to the amino terminal domain are shown. Each modification is indicated by the one-letter code, while M refers to a possible methylation, P refers to a possible phosphorylation and A refers to a possible acetylation.

indicates that a high number of nucleosomes are not randomly distributed, but instead are preferentially positioned in exons [40–43]. Moreover, nucleosome positioning seems to be stronger in constitutive exons compared to alternative exons [44,45]. Overall, these results suggest that nucleosomes may play a role in exon definition as they are positioned on exons with stronger definition and that this positioning could influence RNAPII elongation rate by creating internal pauses during transcription.

8.3.2   Modulation of RNAPII Elongation Rate by Chromatin

As previously described, the kinetic model postulates that the rate of RNAPII elongation affects alternative splicing choices. In most cases, a slow RNAPII elongation rate is linked with more efficient recognition of the weak splice sites, leading to a higher inclusion of alternative exons [38]. However, in some cases a slow RNAPII has been shown to promote exon skipping by promoting the recruitment of negative factors to splice sites [46]. In both cases, however, the elongation rate of RNAPII emerges as a key regulator of alternative splicing.

The kinetic model was first tested in vivo using a human mutant RNAPII carrying a single point mutation, which is equivalent to *Drosophila's* C4 mutation that was previously studied [47,48]. This mutation causes a slow elongation rate. Human cell lines cotransfected with a fibronectin EDI reporter minigene and the RNAPII mutant exhibit a higher inclusion of the alternative exon, confirming the complex relationship between elongation rates and splicing.

The chromatin template has been shown to affect RNAPII enlongation also in vivo. For example, alternative splicing can be modulated by targeting an siRNAs to an intronic sequence close to an alternative exon and subsequent heterochromatin formation [49]. Hence, an siRNA designed to target intron 33 of the fibronectin 1 gene affected not only the local chromatin structure, but also alternative splicing. In line with the hypothesis that the effects of the siRNA depended on chromatin alterations, treatments that prevent the deposition of repressive chromatin modifications counteracted the consequences of the siRNA on alternative splicing. For example, treatments with the histone deacetylase inhibitor trichostatin A, the DNA methyltransferase inhibitor 5-azadeoxycytidine, and the histone methyltransferase inhibitor BIX 01294 that inhibits the deposition of H3K9 dimethylation counteracted the effects of the siRNA on alternative splicing. Furthermore, chromatin immunoprecipitation assays measuring H3K9me2 and H3K27me3 levels confirmed that the intronic siRNA leads to modification of the chromatin context within the coding region of the FN1 gene. Interestingly, Argonaute proteins are also required in this particular case to couple chromatin modifications to alternative splicing [49,50]. These results thus confirm that an alternative chromatin template affects alternative splicing.

Physiological conditions can also change chromatin structure and thus have the potential to modulate alternative splicing: this is illustrated by the example of the neural cell adhesion molecule (NCAM) gene. Neuronal cell depolarization produces intragenic chromatin modifications on the NCAM gene and, at the same time, modulates alternative splicing [51]. Depolarization thus not only produces an increase in global histone acetylation in the N2a cell line, but also increases H3K9 acetylation in an internal region of the NCAM gene, surrounding the alternative exon 18, while promoting exon 18 skipping. These results are in agreement with the model of kinetic coupling between transcription and alternative splicing in a relevant physiological context. Hence, an external stimulus (depolarization) produces changes in specific chromatin marks inside the body of the NCAM gene. These epigenetic changes in turn modulate alternative splicing trough modulation of RNAPII elongation.

Another study shows how alternative splicing of the NCAM gene is sensitive to epigenetic changes in a neuronal differentiation model [52]. Differentiation of mouse N2a neural cells thus promotes not only the formation of repressed chromatin within the NCAM gene, but also the inclusion of exon E18. Treatments with drugs that promote the formation of open chromatin, such as TSA, decrease E18 inclusion in undifferentiated and differentiated N2a cells. E18 inclusion also decreases with 5-azacytidine in differentiated N2a cells. These results led to propose that the chromatin landscape of the NCAM gene changes upon neuronal differentiation with consequences on NCAM alternative splicing. Indeed, native chromatin immunoprecipitation assays measuring the presence of the repressive H3K9 dimethylation and H3K27 trimethylation in undifferentiated and differentiated cells along the NCAM gene body confirmed an increase in both repressive chromatin marks during neuronal differentiation [52]. The effect of repressive chromatin marks on alternative splicing can be explained by the inhibitory effect of chromatin on RNAPII elongation, which, in turn, might modulate alternative splicing of E18. To assess this hypothesis, an elongation assay was performed to analyze pre-mRNA levels after releasing cells from a transcriptional blockage. As expected, differentiated N2a cells exhibit a slower RNAPII elongation rate than undifferentiated cells. Differentiation induces a repressive chromatin state, and this correlates with inhibition of RNAPII elongation in the alternative exon region, which could regulate alternative splicing of the NCAM gene [52].

The kinetic model provides an attractive explanation for the mechanism by which RNAPII elongation affects alternative splicing choices. As mentioned earlier, a slow RNAPII is linked with higher inclusion of alternative exons in most cases. However, in some cases a slow RNAPII can promote exon skipping by promoting the recruitment of negative factors to splice sites [46]. Using the alternative exon 9 of the CFTR gene as a model, it was demonstrated that a slower elongation can cause exon skipping. Drugs that inhibit transcription elongation, such as DRB or CPT, promote a higher rate of exon skipping of the CFTR minigene. On the other hand, increasing the concentration of TSA (a drug that creates a more open chromatin complex) causes higher exon inclusion in this example. Furthermore, a slow RNAPII (the human mutant RNAPII which is resistant to alfa-amanitin and shows a slower elongation rate, as was previously described) also promotes exon skipping in the CFTR minigene. Hence, a slow RNAPII promotes, in this case, recruitment of negative factors, such as ETR-3, to splice sites. This negative factor displaces the constitutive splicing factor U2AF65 from the polypirimidine tract, as demonstrated by using RNA-IP techniques along the CFTR gene [46].

8.3.3   Regulation of Splicing Factor Recruitment by Chromatin

Chromatin modifications can not only affect RNAPII elongation, but also promote differential recruitment of splicing factors to modulate alternative splicing choices. This is demonstrated by studies of histone modifications along genes whose splicing depends on the polypyrimidine tract-binding protein (PTB) splicing factor [53]. PTB-dependent genes are thus enriched in H3K-36me3 and depleted in H3K4me3 at the alternatively spliced regions. Modulation of these marks produces changes on alternative splicing of the involved genes. However, the mechanism of action of these histone modifications in the regulation of alternative splicing does not seem to involve change in the rate of RNA Pol II elongation. Instead, chromatin has been shown to create a platform for the recruitment of the PTB splicing factor to the RNA. An adaptor protein, MRG15, participates in this complex by binding to H3K36me3. High levels of H3K36me3 promote binding of the adaptor protein MRG15, which, in turn, facilitates recruitment of the PTB splicing factor, and regulates alternative splicing. The chromatin mark (H3K36me3), the adaptor protein (MRG15) and the splicing factor (PTB) perform together as a chromatin-splicing adaptor complex [53,54]. This particular combination that constitutes the chromatin-splicing adaptor complex is probably one of many adaptor systems. Other combinations of histone modifying enzymes, chromatin binding protein, and splicing factors may, in turn, interact to form a chromatin-adaptor network [54].

A recent study shows how a long noncoding RNA (lncRNA) can recruit chromatin modifying enzymes to the FGFR2 gene, creating a chromatin environment that modulates alternative splicing of this gene [55]. The antisense FGFR2 lncRNA (asFGFR2) overlapping with exon IIIc of the FGFR2 gene has thus been shown to promote the recruitment of several Polycomb-group proteins and their binding partner, the histone demethylase KDM2a, to the FGFR2 gene. KDM2a presence in the locus reduces H3K36me2 and H3K36me3 levels, which in turn impairs the binding of the chromatin-splicing adaptor complex

MRG15-PTB, a splicing repressor, to the FGFR2 gene. As a consequence, repression of exon IIIb inclusion is inhibited. The role of KDM2a in the regulation of alternative splicing has been further supported by the observation that its downregulation reduces the inclusion of the alternative exon IIIb of the FGFR2 gene despite the presence of the asFGDR2. As the expression of asFGFR2 is cell type-specific, this example illustrates that a lncRNA can regulate cell type-specific patterns of alternative splicing by inducing the formation of a specific local chromatin status that modulates the recruitment of chromatin-splicing adaptor complexes and splicing factors to a specific locus [55].

8.3.4   The Other way Around: From Splicing to Chromatin

A novel perspective on the link between splicing and chromatin is represented by the idea, that not only chromatin structure influences alternative splicing choices, but also there is a reciprocal relationship in which splicing could regulate epigenetic mark deposition. One of the first examples focuses on how splicing factors regulate nucleosome organization [43]. The authors show that overexpression of SRSF1 and SRSF2 promotes nucleosome depletion, whereas overexpression of hnRNP A1 causes an increase in nucleosome occupancy. In addition, disruption of the binding sites for the Hu splicing regulator, a protein that binds to HDAC2 and inhibits its enzymatic activity, leads not only to upregulation of alternative exon inclusion but also to the downregulation of acetylation on the histones of surrounding nucleosomes [56].

Furthermore, it is known that intron-less genes show lower levels of H3K-36me3 and that this observation is independent of transcription [57]. The authors propose that splicing couples the recruitment of HYPB/Setd2 methyltransferase to the actively transcribing RNAPII and demonstrate that disruption of splicing cause a reduction in H3K36me3 levels. These results suggest that splicing could establish deposition of epigenetic marks and thus regulate chromatin structure.

8.4   NEW INSIGHTS ON EPIGENETIC REGULATION OF ALTERNATIVE SPLICING

8.4.1   Interplay Between Chromatin and Splicing in Plants

Although the interplay between chromatin and alternative splicing has not been studied as deeply in plants, as it has in mammalian cells, there is a growing evidence that supports that these processes are connected. Since transcription by RNAPII is affected by chromatin structure it is not extraordinary to expect that chromatin modifications impact on alternative splicing choices in plants as well. Here we focus on connections between chromatin modifications and alternative splicing that have been recently established in plants.

The splicing machinery can affect RNA-directed DNA methylation (RdDM) [58], a mechanism that involves epigenetic modifiers, which are guided to target loci by smalls RNAs in plants. In RdDM, methyltranferases are guided to their target loci by ribonucleoproteins and small RNAs are key to trigger DNA methylation [59]. This process plays a major role in maintaining genome integrity by silencing transposable elements.

There is strong evidence on the participation of splicing factors in the accumulation of siRNA and the formation of RdDM. For example, mutant plants for SR45 (ARGININE/SERINE-RICH45), a plant specific splicing factor, show altered levels of DNA methylation at endogenous RdDM targets [60]. This suggests that SR45 is needed for the establishment of RdDM.

This is not the only study showing the involvement of the splicing machinery in the establishment of this kind of DNA methylation: mutants of other components of the splicing machinery are defective in siRNA accumulation and DNA methylation [61]. Although the need for these splicing factors has been established, these mutations do not alter transcription or alternative splicing of known RdDM factors, suggesting that their contribution to RdDM is probably direct [58]. Although the mechanisms remain unclear, a connection between chromatin and splicing has been established.

We have previously described examples that show how posttranslational modifications of histones can affect chromatin structure, and therefore alter alternative splicing choices in mammalian cells. Different chromatin structures can affect alternative splicing by altering RNAPII elongation rate or by allowing recruitment of distinct splicing factors. This has been widely studied in animals. However, it is not clear if these processes apply to plants as well. For instance, the relationship between RNAPII elongation and alternative splicing still needs to be addressed in plants [58]. Nevertheless, this emerging field is starting to grow.

Recent studies have identified a plant histone methyltransferase that affects alternative splicing. *Arabidopsis* PRMT5 (protein arginine methyltransferase 5) is a methyltransferase that methylates histone residues, as well as small nuclear ribonucleoproteins [62]. PRMT5 is also able to methylate spliceosomal components, and thereby modulate different alternative splicing events, such as exon skipping, alternative donor, and alternative acceptor splicing. Interestingly, prmt5 mutants strongly affect alternative splicing of PRR9, which is one of the core components of the plant circadian clock. As a result, mutations in PRMT5 affect several circadian rythms in *Arabidopsis*. The *Drosopihila* PRMT5-homolog, DART5, also shows defects in certain alternative splicing-regulated transcripts that belong to the circadian clock. Thus, the identification of PRMT5 as an arginine methyltransferase—that modifies histones and spliceosomal components—not only supports the idea of a convergent evolutionary process but also provides a starting point for the understanding of chromatin modifications and alternative splicing regulation in plants [63].

8.5   BEYOND CHROMATIN STRUCTURE

8.5.1   Formation of Circular RNAs

There are many ways in which RNAs can be processed within the cell. The recently discovered circular RNAs, for example, emerged as fundamental regulators of microRNA action, and are themselves regulated by splicing and alternative splicing modulation. Circular RNAs are defined as RNA molecules without ends that form closed continuous strands. These very stable forms of RNA are considered noncoding RNAs and have been described to act as antagonists of microRNAs suggesting an important role in the regulation of genetic programs during cell differentiation.

A foundational manuscript in 1993 found circular RNA molecules containing exons in genomic order without introns and suggest that this mechanisms could present interesting clues about the regulation of splicing [64]. We now know that the vast majority of circular RNA production occurs at major spliceosome splice sites and that during splicing various mechanisms can lead to the generation of circular RNAs. Using bioinformatic analyses of large RNA sequencing datasets, Kelly and coworkers found that exon skipping was positively correlated with the formation of circular RNAs that contained the skipped exon [65]. Since splicing occurs mostly cotranscriptionally, RNA binding proteins and other factors involved in the formation of circular RNAs could slow down the splicing process to allow juxtaposition of both splice sites flanking an individual exon to upregulate circular RNA formation [65]. While new evidence about circular RNAs and their relationship with splicing is being generated at a rapid pace, there are still questions to answer. Importantly, the regulation of alternative splicing by chromatin could prove to be key to our understanding of the formation of circular RNAs.

8.5.2   *Trans*-splicing

Recently, intriguing noncanonical forms of RNA processing have been discovered in eukaryotes that might contribute to nongenetic, phenotypic heterogeneity within a cell population. While *cis*-splicing processes a single pre-mRNA molecule, the rare events of *trans*-splicing generate mature RNA transcripts from multiple pre-mRNAs by joining and ligating exons from these different molecules. Mongelard and coworkers found, for example, that two mutant alleles of the same gene, each located in one of the two homologous chromosomes, restore the wild-type function of the gene under some conditions [66]. The authors suggested that the two separate transcription units can be combined in a *trans*-splicing reaction to form the mature mRNA that, in turn, encodes the wild-type protein. Similarly to what happens with *cis*-splicing, it has been demostrated that *trans*-splicing can take place cotranscriptionally. In the processing of *Trypanosoma* tubuline, the pre-mRNA has been reported to undergo *trans*-splicing cotranscriptionally, preceding polyadenilation [67].

Not only *cis*-alternative splicing, but also RNA editing, non sense mediated decay, alternative *trans*-splicing and all other regulatory mechanisms of gene expression, contribute to the generation of protein diversity. Chimeric RNA species generated by alternative *trans*-splicing could potentially have important functional consequences on the diversification of the proteome acting, as novel sources of protein diversity. Apart from being generated by *trans*-splicing, chimeric mRNAs can also be produced by chromosomal translocations which are very frequent events in cells of the inmune system and in cancer cells. Moreover, evidence now suggests that chimeric proteins generated by *trans*-splicing mechanisms may also be expressed under physiological circumstances, and represent a novel source of diversity [68].

8.6 CONCLUDING REMARKS

It is clear now that splicing and alternative splicing are tightly regulated processes controlled by multiple factors and regulatory elements acting in concerted ways. As alternative splicing leads to multiple proteins encoded by a single gene, the process expands the coding capacity of the genomes and thus could stimulate the complexity of higher eukaryotes. From its discovery in 1977 [69,70] remarkable progress was made in the alternative splicing field and our understanding of its regulation leads us to consider it as a dynamic and flexible process.

Accumulating evidence highlights the role of chromatin structure as a key regulator of alternative splicing choices. We have discussed how intergenic chromatin marks could affect alternative splicing decisions by regulating RNAPII elongation rate and altering the recruitment of regulatory factors and *cis*-acting elements. Moreover, posttranslational histones modifications were shown to control splicing by directly binding to splicing factors and mediating their recruitment to the nascent mRNA being synthesized.

However, our understanding of the mechanism that regulate alternative splicing is far from complete, as illustrated by the existence of splicing patterns that cannot be explained by our current knowledge. The observation that nucleosome positioning and specific histone marks are enriched in exons over introns opens the possibility that chromatin marks are not only regulating alternative splicing decision, but also exon recognition and genome organization. To what extent this evidence supports a model in which histone marks have been relevant in exon recognition during evolution remains unexplored.

Most importantly, understanding how combinations of dynamic chromatin signatures determine splicing patterns will require the simultaneous analysis of histone modifications and splicing events in small cell populations, or even in single cells.

Abbreviations

AS	alternative splicing
RNPII	RNA polymerase II
mRNA	messenger RNA
pre-mRNA	precursor messenger RNA
snRNPs	small nuclear ribonucleoproteins
PCR	polymerase chain reaction
CTD	carboxy-terminal domain of the catalytic subunit of RNAPII
SR	serine/arginine-rich protein
EDI	fibronectin alternative exon 33
siRNAs	small interfering RNA
NCAM	neural cell adhesion molecule
H3K9	histone 3 lysine 9
H3K27	histone 3 lysine 27
H3K36me3	histone 3 lysine 36 trimethylation
lncRNA	long noncoding RNA

References

[1] Kornblihtt AR, Schor IE, Alló M, Dujardin G, Petrillo E, Muñoz MJ. Alternative splicing: a pivotal step between eukaryotic transcription and translation. Nat Rev Mol Cell Biol 2013;14:153–65.

[2] Wang ET, Sandberg R, Luo S, Khrebtukova I, Zhang L, Mayr C, Kingsmore SF, Schroth GP, Burge CB. Alternative isoform regulation in human tissue transcriptomes. Nature 2008;456:470–6.

[3] Srebrow A, Kornblihtt AR. The connection between splicing and cancer. J Cell Sci 2006;119:2635–41.

[4] Novoyatleva T, Tang Y, Rafalska I, Stamm S. Pre-mRNA missplicing as a cause of human disease. Prog Mol Subcell Biol 2006;44:27–46.

[5] Ward AJ, Cooper TA. The pathobiology of splicing. J Pathol 2010;220:152–63.

[6] Will CL, Lührmann R. Spliceosome structure and function. Cold Spring Harb Perspect Biol 2011;3:1–2.

[7] Naftelberg S, Schor IE, Ast G, Kornblihtt AR. Regulation of alternative splicing through coupling with transcription and chromatin structure. Annu Rev Biochem 2015;84:165–98.

[8] Gómez Acuña LI, Fiszbein A, Alló M, Schor IE, Kornblihtt AR. Connections between chromatin signatures and splicing. Wiley Interdiscip Rev RNA 2013;4:77–91.

[9] Yan C, Hang J, Wan R, Huang M, Wong CCSY. Structure of a yeast spliceosome at 3.6-angstrom resolution. Science 2015;80-:1–16.

[10] Hang J, Wan R, Yan C. Structural basis of pre-mRNA splicing. Science 2015;349:1191–8.

[11] Cramer P, Pesce CG, Baralle FE, Kornblihtt AR. Functional association between promoter structure and transcript alternative splicing. Proc Natl Acad Sci USA 1997;94:11456–60.

[12] Cramer P, Cáceres JF, Cazalla D, Kadener S, Muro AF, Baralle FE, Kornblihtt AR. Coupling of transcription with alternative splicing: RNA pol II promoters modulate SF2/ASF and 9G8 effects on an exonic splicing enhancer. Mol Cell 1999;4:251–8.

[13] Beyer AL, Osheim YN. Splice site selection, rate of splicing, and alternative splicing on nascent transcripts. Genes Dev 1988;2:754–65.

[14] Ameur A, Zaghlool A, Halvardson J, Wetterbom A, Gyllensten U, Cavelier L, Feuk L. Total RNA sequencing reveals nascent transcription and widespread co-transcriptional splicing in the human brain. Nat Struct Mol Biol 2011;18:1435–40.

[15] Brugiolo M, Herzel L, Neugebauer KM. Counting on co-transcriptional splicing. F1000Prime Rep 2013;5:9.

[16] Carrillo Oesterreich F, Preibisch S, Neugebauer KM. Global analysis of nascent RNA reveals transcriptional pausing in terminal exons. Mol Cell 2010;40:571–81.

[17] Tardiff DF, Lacadie SA, Rosbash M. A genome-wide analysis indicates that yeast pre-mRNA splicing is predominantly posttranscriptional. Mol Cell 2006;24:917–29.

[18] Baurén G, Wieslander L. Splicing of Balbiani ring 1 gene pre-mRNA occurs simultaneously with transcription. Cell 1994;76:183–92.

[19] Pandya-Jones A, Black DL. Co-transcriptional splicing of constitutive and alternative exons. RNA 2009;15:1896–908.

[20] De la Mata M, Lafaille C, Kornblihtt AR. First come, first served revisited: factors affecting the same alternative splicing event have different effects on the relative rates of intron removal. RNA 2010;16:904–12.

[21] Gerber HP, Hagmann M, Seipel K, Georgiev O, West MA, Litingtung Y, Schaffner W, Corden JL. RNA polymerase II C-terminal domain required for enhancer-driven transcription. Nature 1995;374:660–2.

[22] McCracken S, Fong N, Rosonina E, Yankulov K, Brothers G, Siderovski D, Hessel A, Foster S, Shuman S, Bentley DL. 5′-Capping enzymes are targeted to pre-mRNA by binding to the phosphorylated carboxy-terminal domain of RNA polymerase II. Genes Dev 1997;11:3306–18.

[23] McCracken S, Fong N, Yankulov K, Ballantyne S, Pan G, Greenblatt J, Patterson SD, Wickens M, Bentley DL. The C-terminal domain of RNA polymerase II couples mRNA processing to transcription. Nature 1997;385:357–61.

[24] Muñoz MJ, de la Mata M, Kornblihtt AR. The carboxy terminal domain of RNA polymerase II and alternative splicing. Trends Biochem Sci 2010;35:497–504.

[25] Buratowski S. Progression through the RNA polymerase II CTD cycle. Mol Cell 2009;36:541–6.

[26] Egloff S, Dienstbier M, Murphy S. Updating the RNA polymerase CTD code: adding gene-specific layers. Trends Genet 2012;28.333–41.

[27] Heidemann M, Hintermair C, Voß K, Eick D. Dynamic phosphorylation patterns of RNA polymerase II CTD during transcription. Biochim Biophys Acta 2013;1829:55–62.

[28] Hsin J-P, Manley JL. The RNA polymerase II CTD coordinates transcription and RNA processing. Genes Dev 2012;26:2119–37.

[29] Kim M, Krogan NJ, Vasiljeva L, Rando OJ, Nedea E, Greenblatt JF, Buratowski S. The yeast Rat1 exonuclease promotes transcription termination by RNA polymerase II. Nature 2004;432:517–22.

[30] Egloff S, O'Reilly D, Chapman RD, Taylor A, Tanzhaus K, Pitts L, Eick D, Murphy S. Serine-7 of the RNA polymerase II CTD is specifically required for snRNA gene expression. Science 2007;318:1777–9.

[31] Hsin J-P, Sheth A, Manley JL. RNAP II CTD phosphorylated on threonine-4 is required for histone mRNA 3′ end processing. Science 2011;334:683–6.

[32] Moore MJ, Proudfoot NJ. Pre-mRNA processing reaches back to transcription and ahead to translation. Cell 2009;136:688–700.

[33] Perales R, Bentley D. Cotranscriptionality: the transcription elongation complex as a nexus for nuclear transactions. Mol Cell 2009;36:178–91.

[34] Das R, Yu J, Zhang Z, Gygi MP, Krainer AR, Gygi SP, Reed R. SR proteins function in coupling RNAP II transcription to pre-mRNA splicing. Mol Cell 2007;26:867–81.

[35] De la Mata M, Kornblihtt AR. RNA polymerase II C-terminal domain mediates regulation of alternative splicing by SRp20. Nat Struct Mol Biol 2006;13:973–80.

[36] Monsalve M, Wu Z, Adelmant G, Puigserver P, Fan M, Spiegelman BM. Direct coupling of transcription and mRNA processing through the thermogenic coactivator PGC-1. Mol Cell 2000;6:307–16.

[37] Huang Y, Li W, Yao X, Lin QJ, Yin JW, Liang Y, Heiner M, Tian B, Hui J, Wang G. Mediator complex regulates alternative mRNA processing via the MED23 subunit. Mol Cell 2012;45:459–69.

[38] De La Mata M, Alonso CR, Kadener S, Fededa JP, Blaustein M, Pelisch F, Cramer P, Bentley D, Kornblihtt AR. A slow RNA polymerase II affects alternative splicing in vivo. Mol Cell 2003;12:525–32.

[39] Shukla S, Kavak E, Gregory M, Imashimizu M, Shutinoski B, Kashlev M, Oberdoerffer P, Sandberg R, Oberdoerffer S. CTCF-promoted RNA polymerase II pausing links DNA methylation to splicing. Nature 2011;479:74–9.

[40] Amit M, Donyo M, Hollander D, Goren A, Kim E, Gelfman S, Lev-Maor G, Burstein D, Schwartz S, Postolsky B, et al. Differential GC content between exons and introns establishes distinct strategies of splice-site recognition. Cell Rep 2012;1:543–56.

[41] Tilgner H, Knowles DG, Johnson R, Davis CA, Chakrabortty S, Djebali S, Curado J, Snyder M, Gingeras TR, Guigó R. Deep sequencing of subcellular RNA fractions shows splicing to be predominantly co-transcriptional in the human genome but inefficient for lncRNAs. Genome Res 2012;22:1616–25.

[42] Spies N, Nielsen CB, Padgett RA, Burge CB. Biased chromatin signatures around polyadenylation sites and exons. Mol Cell 2009;36:245–54.

[43] Keren-Shaul H, Lev-Maor G, Ast G. Pre-mRNA splicing is a determinant of nucleosome organization. PLoS One 2013;8.

[44] Schwartz S, Meshorer E, Ast G. Chromatin organization marks exon–intron structure. Nat Struct Mol Biol 2009;16:990–5.

[45] Huang H, Yu S, Liu H, Sun X. Nucleosome organization in sequences of alternative events in human genome. BioSystems 2012;109:214–9.

[46] Dujardin G, Lafaille C, de la Mata M, Marasco LE, Muñoz MJ, Le Jossic-Corcos C, Corcos L, Kornblihtt AR. How slow RNA polymerase II elongation favors alternative exon skipping. Mol Cell 2014;54:683–90.

[47] Coulter DE, Greenleaf AL. A mutation in the largest subunit of RNA polymerase II alters RNA chain elongation in vitro. J Biol Chem 1985;260:13190–8.

[48] Chen Y, Chafin D, Price DH, Greenleaf AL. Drosophila RNA polymerase II mutants that affect transcription elongation. J Biol Chem 1996;271:5993–9.

[49] Alló M, Buggiano V, Fededa JP, Petrillo E, Schor I, de la Mata M, Agirre E, Plass M, Eyras E, Elela SA, et al. Control of alternative splicing through siRNA-mediated transcriptional gene silencing. Nat Struct Mol Biol 2009;16:717–24.

[50] Ameyar-Zazoua M, Rachez C, Souidi M, Robin P, Fritsch L, Young R, Morozova N, Fenouil R, Descostes N, Andrau J-C, et al. Argonaute proteins couple chromatin silencing to alternative splicing. Nat Struct Mol Biol 2012;19:998–1004.

[51] Schor IE, Rascovan N, Pelisch F, Alló M, Kornblihtt AR. Neuronal cell depolarization induces intragenic chromatin modifications affecting NCAM alternative splicing. Proc Natl Acad Sci USA 2009;106:4325–30.

[52] Schor IE, Fiszbein A, Petrillo E, Kornblihtt AR. Intragenic epigenetic changes modulate NCAM alternative splicing in neuronal differentiation. EMBO J 2013;32:2264–74.

[53] Luco RF, Pan Q, Tominaga K, Blencowe BJ, Pereira-Smith OM, Misteli T. Regulation of alternative splicing by histone modifications. Science 2010;327:996–1000.

[54] Luco RF, Allo M, Schor IE, Kornblihtt AR, Misteli T. Epigenetics in alternative pre-mRNA splicing. Cell 2011;144:16–26.

[55] Gonzalez I, Munita R, Agirre E, Dittmer Ta, Gysling K, Misteli T, Luco RF. A lncRNA regulates alternative splicing via establishment of a splicing-specific chromatin signature. Nat Struct Mol Biol 2015;5:370–6.

[56] Zhou H-L, Hinman MN, Barron VA, Geng C, Zhou G, Luo G, Siegel RE, Lou H. Hu proteins regulate alternative splicing by inducing localized histone hyperacetylation in an RNA-dependent manner. Proc Natl Acad Sci USA 2011;108:E627–35.

[57] De Almeida SF, Grosso AR, Koch F, Fenouil R, Carvalho S, Andrade J, Levezinho H, Gut M, Eick D, Gut I, et al. Splicing enhances recruitment of methyltransferase HYPB/Setd2 and methylation of histone H3 Lys36. Nat Struct Mol Biol 2011;18:977–83.

[58] Mathieu O, Bouché N. Interplay between chromatin and RNA processing. Curr Opin Plant Biol 2014;18:60–5.

[59] Bond DM, Baulcombe DC. Epigenetic transitions leading to heritable, RNA-mediated de novo silencing in Arabidopsis thaliana. Proc Natl Acad Sci USA 2015;112:917–22.

[60] Ausin I, Greenberg MVC, Li CF, Jacobsen SE. The splicing factor SR45 affects the RNA-directed DNA methylation pathway in Arabidopsis. Epigenetics 2012;7:29–33.

[61] Zhang C-J, Zhou J-X, Liu J, Ma Z-Y, Zhang S-W, Dou K, Huang H-W, Cai T, Liu R, Zhu J-K, et al. The splicing machinery promotes RNA-directed DNA methylation and transcriptional silencing in *Arabidopsis*. EMBO J 2013;32:1128–40.

[62] Henriques R, Mas P. Chromatin remodeling and alternative splicing: pre- and post-transcriptional regulation of the Arabidopsis circadian clock. Semin Cell Dev Biol 2013;24:399–406.

[63] Sanchez SE, Petrillo E, Beckwith EJ, Zhang X, Rugnone ML, Hernando CE, Cuevas JC, Godoy Herz MA, Depetris-Chauvin A, Simpson CG, et al. A methyl transferase links the circadian clock to the regulation of alternative splicing. Nature 2010;468:112–6.

[64] Cocquerelle C, Mascrez B, Hétuin D, Bailleul B. Mis-splicing yields circular RNA molecules. FASEB J 1993;7:155–60.

[65] Wilusz J. Circular RNA and splicing: skip happens. J Mol Biol 2015;427:2411–3.

[66] Mongelard F, Labrador M, Baxter EM, Gerasimova TI, Corces VG. Trans-splicing as a novel mechanism to explain interallelic complementation in *Drosophila*. Genetics 2002;160:1481–7.

[67] Ullu E, Matthews KR, Tschudi C. Temporal order of RNA-processing reactions in trypanosomes: rapid trans splicing precedes polyadenylation of newly synthesized tubulin transcripts. Mol Cell Biol 1993;13(1):720–5.

[68] Casado-Vela J, L acal JC, Elortza F. Protein chimerism: novel source of protein diversity in humans adds complexity to bottom-up proteomics. Proteomics 2013;13:5–11.

[69] Berget SM, Sharp Pa. A spliced sequence at the 5′-terminus of adenovirus late mRNA. Brookhaven Symp Biol 1977;74:332–44.

[70] Chow LT, Gelinas RE, Broker TR, Roberts RJ. An amazing sequence arrangement at the 5′ ends of adenovirus 2 messenger RNA. Cell 1977;2:1–8.

CHAPTER 9

Crosstalk Between Non-Coding RNAs and the Epigenome in Development

M. Berdasco*, M. Esteller*,**,†

*Cancer Epigenetics Group, Cancer Epigenetics and Biology Program (PEBC), Bellvitge Biomedical Research Institute (IDIBELL), Barcelona, Catalonia, Spain
**Department of Physiological Sciences II, School of Medicine, University of Barcelona, Barcelona, Catalonia, Spain
†Catalan Institution for Research and Advanced Studies (ICREA), Barcelona, Catalonia, Spain

9.1 INTRODUCTION

9.1.1 The ncRNA Landscape

One of the major discovery of the Encyclopedia of DNA Elements project was that at least 90% of the human genome is actively transcribed [1]. A large fraction of the genome is thus constituted by transcribed noncoding elements that might have critical roles in cellular functions [2] and include, for example, gene regulatory regions,or noncoding RNAs (ncRNAs). Although initially considered nonfunctional elements (*the dark matter*), recent evidence suggests that ncRNAs—defined as RNA sequences that do not code for any protein—play major roles in cellular biology and development, and their deregulation is associated with human diseases [3,4]. In recent years, "the RNA world" has received increased attention from the scientific community, which has changed our view of the central dogma of molecular biology. RNA is thus not solely an intermediate between the DNA sequence and proteins, but has crucial roles in development and cell differentiation.

It is, however, not a new concept that ncRNAs can have important cellular functions. The so-called "structural ncRNAs" have thus been extensively studied in the last 50 years and include transfer RNAs associated with aminoacid transport, ribosomal RNAs involved in protein synthesis, small nuclear RNAs with a role in RNA splicing or small nucleolar RNAs linked to RNA modifications. Although these structural ncRNAs are ubiquitously expressed in all cell types and play essential roles in development and cell differentiation, they will not be discussed further in this chapter, which instead focuses on the classification

CONTENTS

Chromatin Regulation and Dynamics. http://dx.doi.org/10.1016/B978-0-12-803395-1.00009-5

and functions of the "regulatory ncRNAs" in diverse transcriptional and post-transcriptional processes.

9.1.2   Classification of Regulatory Noncoding RNAs

In contrast to structural ncRNAs, "regulatory" ncRNAs have a well-defined expression pattern that depends on the developmental stage and cell type. Whereas most of the regulatory ncRNAs have yet to be functionally characterized, they have been classified based on their size with an arbitrary cut-off of 200 nucleotides (nt), distinguishing between short ncRNAs (<200 nt) and long noncoding RNAs (lncRNAs) (> 200 nt). Most of the studies in the last decade have been focused on the shortest of all ncRNAs: microRNAs (miRNA). MiRNAs are small ncRNAs approximately 22 nucleotides in length, which are processed from precursor RNAs by the RNAse III enzymes DROSHA and DICER, and function as endogenous posttranscriptional silencers of their target genes [5]. MiRNAs are expressed in a tissue-dependent manner and are important regulators of several cellular processes, including proliferation, differentiation, apoptosis, and development [6]. The deregulation of miRNA expression is often linked to human developmental defects and diseases through imperfect pairing with their respective target mRNAs, which can have deleterious consequences on miRNA functions, such as translational repression, regulation of mRNA decay or cleavage [7]. Due to their impact on genetic engineering tools, special attention has been paid in the last decade to the small-interfering RNAs (siRNAs). SiRNAs are 20–24 nt in length, and they are also generated by Dicer processing from long double-stranded RNAs and small hairpin RNAs [3]. Although siRNAs can be involved in different regulatory functions, their most significant role is mediated by the RNA interference (RNAi) pathway. SiRNAs thus interfere with the expression of specific genes with complementary nucleotide sequences mainly by promoting mRNA cleavage via the RNA-induced silencing complex (RISC), Argonaute-2 (AGO-2) and other auxiliary RISC proteins. This property has been employed as a tool for achieving transient gene knockdown in in vitro cell cultures. Other notable small ncRNAs include the Piwi-interacting RNAs, which are ncRNAs of 24–30 nt in length and, in contrast to miRNA and siRNA, are generated in a Dicer-independent manner. They received their name because of their ability to form a complex with the PIWI family of proteins [3], and act as suppressors of transposable element expression and also prevent transposon jumping [3].

Apart from these small ncRNAs, widespread application of high-throughput RNA sequencing, together with computational tools, has identified a large number of lncRNAs. Based on the NONCODE database that represents the most complete collection and annotation of noncoding RNAs, the number of identified lncRNAs in the human genome reached 95,135 [8]. The length of lncRNAs varies between 200 bp and >100 kb. Although lncRNAs display cell-type specific expression pattern, in contrast to small ncRNAs they are less evolutionary conserved than protein-coding genes [9]. The lower level of sequence conservation in lncRNAs could reflect that their functions depend to a higher degree on the three-dimensional (3D) structure of the RNA rather than its sequence. LncRNAs are important regulators of multiple cellular processes, including development, cell fate, inflammation, stress response, chromatin conformation or imprinting, among others [10–12].

The most well-known classification of lncRNAs is based on their genomic location with respect to protein-coding genes [11] and includes the following six categories: (1) intergenic lncRNAs (lincRNAs) that are located between two protein-coding genes and constitute the majority of the known lncRNAs; (2) intronic lncRNAs that are located within introns of protein-coding genes; (3) bidirectional promoter lncRNAs defined as lncRNAs transcribed within 1 kb of protein-coding gene promoters, in the opposite direction from the protein-coding transcript; (4) enhancer lncRNAs (elncRNAs) that are generally <2 kb long and are transcribed from enhancer elements; (5) sense lncRNAs that intersect protein-coding exons or introns that have the same orientation as the lncRNA transcript; and (6) natural antisense transcripts (NATs) that contain sequences complementary to other endogenous RNAs.

A novel classification of lncRNAs based on their function has recently been proposed [12], which subdivides lncRNAs into signals, decoys, scaffolds, guides and enhancer lncRNAs (Fig. 9.1). The transcription of lncRNAs within the group termed "signals" is fine-tuned by environmental signals and displays spatiotemporal and cell type-specific regulation. This feature is a cue for integrating developmental signals and stimulus-specific cellular information during cell differentiation. We also know that specific lncRNAs act as "decoys", because they are able to sequester a range of RNA-dependent effectors and chromatin proteins (transcription factors, chromatin modifiers, splicing factors, etc.), thereby inhibiting their functions. LncRNAs could also serve as "scaffolds" for assembling two or more proteins into discrete complexes. Moreover, it has been described that lncRNAs can serve as "guides" for target recognition, and are able to recruit chromatin modifiers and other proteins to specific genomic loci. In the next sections we will summarize examples of lncRNAs acting as signals, decoys, scaffolds, and guides.

9.2 FUNCTIONAL ROLES OF lncRNAs

Although we still have limited information about their function, our current knowledge suggests that lncRNAs play regulatory roles in diverse biological processes [13]. Since many lncRNAs are transcribed from segments of

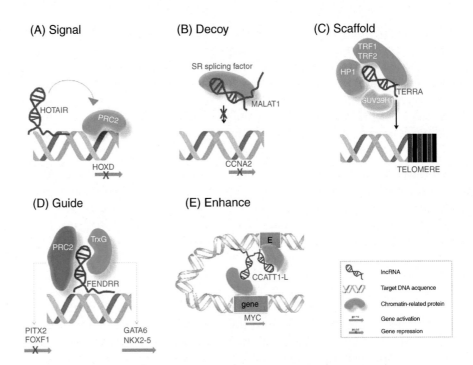

FIGURE 9.1 Classification of long noncoding RNAs based on their mechanism of action.
(A) Signal lncRNAs regulate gene expression in response to specific stimuli and enable the coordination of gene transcription in space and time. *HOTAIR* contributes to the regulation of the temporally and spatially restricted pattern of *Hox* gene expression, and represses, for example, *HOXD* transcription by recruiting PRC2 and other chromatin modifiers to the *HOXD* promoter in order to increase the level of trimethylation at lysine 27 of histone H3 and induce the formation of repressed chromatin state during development; (B) decoy lncRNAs titrate away DNA-binding proteins, such as transcription factors, leading to repression of gene expression. *MALAT1* lncRNA, for example, traps SR splicing factors altering the alternative splicing of a set of endogenous pre-mRNAs; (C) scaffold lncRNAs serve as adaptors to facilitate the formation of complexes between two or more proteins. For example, *TERRA* associates with the shelterin components TRF1 and TRF2 and with members of the epigenetic machinery (HP1 and SUV39H1) to promote heterochromatin formation at chromosome ends; (D) guide lncRNAs form complexes with chromatin proteins and facilitate their localization to specific target genes. *FENDRR*, for example, forms a complex with PRC2 to repress the expression of the *PITX2* or *FOXF1* genes. It could also bind to TrxG proteins to activate the expression of the *GATA6* or *NKX2-5* genes; and (E) enhancer lncRNAs facilitate the interactions between enhancers and promoter regions of target genes through chromosomal looping. The lncRNA *CCATT1-L* is expressed from a super-enhancer upstream of the *MYC* locus and positively regulates *MYC* transcription.

protein-coding genes, the question is whether all expressed lncRNAs have a regulatory role, or a subset of them is just transcriptional noise or by-product of mRNA processing [14]. As the biogenesis, processing and function of small ncRNAs have been extensively reviewed, in this chapter we will focus on the roles of lncRNAs.

9.2.1 The Role of lncRNAs in Transcriptional Regulation

Whereas miRNAS mostly downregulate gene expression by suppressing translation and inducing mRNA degradation, the control of gene regulation by lncRNAs could be exerted by diverse molecular mechanisms. First, lncRNAs can directly silence or activate the transcription of a particular gene by regulating the activity of transcription factors and RNA polymerases. Second, lncRNAs can regulate gene transcription by serving as "guides" to direct chromatin proteins to specific genomic loci either in *cis*, within the same genomic locus where the lncRNA gene is located, or in *trans*, affecting gene expression at a different locus or even on a different chromosome. This second mechanism will be discussed in Section 9.2.1.2 that describes the epigenetic control exerted by lncRNAS.

9.2.1.1 *Transcriptional Control Mediated by Interactions Between lncRNAs and the Transcriptional Machinery*

Some lncRNAs can directly bind to specific transcription factors and regulate the expression of their target genes by *cis* mechanisms. An example of lncRNA-mediated transcriptional activation involves *HSR1*, a lncRNA that can form a complex with the heat shock factor heat shock transcription factor 1 (HSF1) and the translation elongation factor eEF1A to induce the expression of heat-shock proteins and other cytoprotective proteins under specific environmental signals [15]. Another example of a ncRNA that upregulates gene expression via interacting with a transcription factor is represented by *Evf-2*, which is transcribed from an ultraconserved intergenic region located between the *Dlx5* and 6 genes, and functions as a transcriptional coactivator of DLX-2 protein to increase the activity of the *Dlx-5/6* enhancer element [16]. Proteins of the transcription machinery (polymerases, helicases, or promoter binding proteins) also form complexes with lncRNAs to regulate expression. The RNA helicases p68/p72 and the noncoding RNA Steroid receptor RNA Activator (*SRA*) act as co-activators of MYOD, a master regulator of skeletal muscle differentiation, and form a complex with MYOD to regulate the expression of its target genes [17]. The list of lncRNA activators of gene transcription has been recently expanded by an analysis of the GENCODE annotation in a panel of human cell lines [18]. Surprisingly, a set of lncRNAs was found to possess enhancer-like functions in human cell lines. Transcription of these eRNAs increases the expression of genes located in their vicinity, including the expression of the master regulator of hematopoiesis, *SCL* (also called *TAL1*), *SNAI1* and *SNAI2* [18]. Similarly, the lncRNA *CCATT1-L* is expressed from a super-enhancer upstream of the *MYC* locus and positively regulates *MYC* transcription [19].

LncRNAs can also act as negative regulators of transcription and silence gene expression by direct binding to corepressors. This mechanism of lncRNA action contributes to the selective regulation of alternative promoters within the same gene. For example, the major promoter of the dihydrofolate reductase

(*DHFR*) gene is repressed in quiescent cells by a lncRNA transcribed upstream of the minor promoter [20]. The lncRNA thus forms a stable complex with the general transcription factor IIB (TFIIB), and induces the dissociation of the preinitiation complex from the major promoter. Furthermore, certain lncRNAs can alter global gene expression by *trans* mechanisms. For example, Alu RNAs, which are transcribed from short interspersed elements (SINEs), bind to RNA polymerase II complexes at specific promoters and repress transcription upon heat-shock in human cells [21]. Although the molecular mechanism by which these ncRNAs affect transcription is not fully understood, the abundance of ultraconserved regions and Alu elements in the human genome suggests that this mechanism might be a more common strategy for gene regulation, than previously anticipated.

9.2.1.2   Epigenetic Control of Transcription Mediated by lncRNA

As mentioned before, lncRNAs can also associate with complexes that modify chromatin. Indeed, chromatin modifiers are among the most frequent protein partners of lncRNAs that are identified until now. Hence, 30% of the lincRNAs that are expressed in mouse embryonic stem (ES) cells are associated with at least 1 of 12 different protein complexes that act as chromatin readers, writers or erasers [22]. Most of the lncRNAs that have been found to associate with chromatin modifying complexes influence the function and localisation of the Polycomb complex PRC2 [23,24]. Generally, Polycomb complexes are responsible for maintaining a repressed state, while Tritorax group (TrxG) of proteins act in active transcriptional contexts (see also Chapter 7) [25]. While the PRC2 subunit Enhancer of zeste 2 (EZH2) catalyzes the mono, di- and tri-methylation of histone H3 at lysine 27 (H3K27me1, H3K27me2 or H3K27me3), H3K27me3 acts as a signal for PRC1 binding. The PRC1 complex can further induce the monoubiquitylation of histone H2A subunits, resulting in chromatin compaction and gene repression of specific loci. It has been estimated that more than 20% of human lncRNAs can associate with PRC2 [23], although lncRNAs with binding affinity to TrxG proteins were also described [26]. One of the best known example of lncRNAs that interacts with PCR2 in *trans* is represented by *HOTAIR*, a 2.2 kb ncRNA transcribed from the *HOXC* ncRNA cluster that harbors more than 230 lncRNAs. *HOTAIR* induces heterochromatin formation and represses transcription by recruiting PCR2 to increase the level of H3K27me3 marks in the *HOXD* locus [27]. Interestingly, the 5' domain of *HOTAIR* binds PRC2, whereas a 3' domain of *HOTAIR* binds the LSD1/CoREST complex that also contributes to *HOXD* repression by demethylation of H3K4me2 [28]. More examples of lncRNAs that mediate transcriptional regulation include the lncRNA *ANRIL* (or *CDKN2B-AS1*) that recruits PRC2 leading to the silencing of the tumor suppressor locus *INK4b-ARF-INK4a* [29]; the NAT *APOA1-AS* that acts as a negative transcriptional regulator of the apolipoprotein A1 (*APOA1*) both in vitro and in vivo [30]; the NAT *SCAANT1* that inhibits ataxin-7 expression by interacting with CTCF [31] or the repression of *BDNF*, *GDNF*, and *EPHB2* by their corresponding natural antisense (NAT) transcripts [32]. One elegant example of lncRNA complexity reflecting their potential dual role in gene transcription (i.e., either activation or repression), is represented by the lateral mesoderm-specific lncRNA *FENDRR*. It is essential for adequate heart and body wall development in the mouse and not only represses the expression of *FOXF1* and *PITX2* genes via PRC2 binding, but can also interact with TrxG/MLL complexes to maintain gene expression [33].

Gene repression by lncRNAs also involves epigenetic complexes other than the PRC2 complex. The RNA-binding protein, translocated in liposarcoma (TLS) specifically binds to and inhibits the acetyltransferase activity of the CREB-binding protein and p300 at the cyclin D1 (*CCND1*) gene in human cell lines during induced DNA damage [34]. Recruitment of TLS to the *CCND1* promoter is directed by single-stranded ncRNA transcripts that are induced by DNA damage and are tethered to the 5′ regulatory regions of the *CCND1* gene, resulting in gene repression [34].

There are fewer examples that illustrate the role of lncRNAs in the recruitment of chromatin modifiers associated with gene activation to specific loci. *HOTTIP*, a lincRNA transcribed from the 5′ end of the homeobox *HOXA* locus, has been described to activate the expression of several *HOXA* genes in vivo [35] by recruiting the chromatin modifiers WDR5/MLL. *HOTTIP* ncRNA thus binds to the adaptor protein WDR5 directly and targets the WDR5/MLL complexes to *HOXA*, thereby increasing the level of histone H3 lysine 4 trimethylation (H3K4me3) at the *HOXA* promoter [35]. Another recent evidence of lncRNA involvement in gene activation is represented by the regulation of gene expression at the *INK4* locus by the very long intergenic ncRNA (vlincRNA) *VAD* in senescent cells [36]. *VAD* thus binds to and inhibits the incorporation of the repressive histone variant H2A.Z at the *INK4* locus, leading to increased *INK4* expression and senescence-associated cell cycle arrest.

9.2.2 Posttranscriptional Regulation by lncRNAs

Accumulating evidence supports the role of small-ncRNAs (i.e., miRNAs) in posttranscriptional regulation of gene expression [37]. In addition, we begin to elucidate the role of lncRNAs in the posttranscriptional processing of mRNAs, such as mRNA splicing, maturation and decay, mRNA transport, mRNA stability and degradation.

9.2.2.1 *mRNA Splicing*

Immediately after transcription, nascent pre-mRNAs can be spliced and processed into one or many different isoforms contributing to the diversification of gene functions (see also Chapter 8). The best studied lncRNA associated with pre-mRNA splicing is the metastasis associated lung carcinoma transcript

1 (*MALAT1*). It is preferentially found in the nucleus and controls alternative splicing by regulating the cellular levels of the phosphorylated forms of serine/arginine (SR) splicing factors and their localisation in nuclear speckles [38]. In line with these observations, depletion of *MALAT1* or overexpression of an SR protein changes the alternative splicing of a similar set of endogenous pre-mRNAs [38]. Moreover, many lncRNAs belonging to the NAT category are more directly involved in alternative splicing in human cells. They can thus form RNA duplexes to mask *cis*-regulatory elements in the pre-mRNA of overlapping genes, thereby regulating alternative splicing. Other NATs involved in splicing include the tyrosine kinase containing immunoglobulin and epidermal growth factor homology domain-1 antisense RNA (*TIE-1AS*) [39], the zinc-finger E-box binding homeobox 2 antisense RNA (*ZEB2-AS*) associated to E-cadherin dowregulation during epithelial-mesenchymal transition [40], or the beta-amyloid precursor protein-cleaving enzyme 1 antisense RNA (*BACE1-AS*) directly implicated in the increased abundance of amyloid-beta 1-42 protein in Alzheimer's disease [41].

9.2.2.2 mRNA Editing

NATs also have the potential to guide mRNA editing, a molecular process through which cells can introduce changes to specific nucleotide sequences within an mRNA molecule (e.g., 5'-capping and 3'-polyadenylation). A-to-I editing is a process that involves the conversion of adenosine to inosine in double-stranded RNA resulting in changes in the RNA structure, mRNA splicing or targeting by miRNAs, among others [42]. It is regulated by the proteins paraspeckle component 1 (PSPC1) and non-POU domain-containing octamer-binding (NONO). Moreover, mRNA editing is facilitated by the lncRNA nuclear paraspeckle assembly transcript 1 (NEAT1) that binds to PSPC1 and NONO, and colocalizes with paraspeckeles, which are nuclear sub-compartments that retain mRNAs in the nucleus, for example, for splicing and editing [43].

9.2.2.3 mRNA Stability and mRNA Degradation

The level of gene expression can also be regulated by the balance between mRNA synthesis, stabilization and mRNA degradation or decay [43a], a mechanism responsible for the rapid degradation of mRNAs containing premature termination codons in order to prevent the synthesis of truncated proteins. It is well established that AU-rich elements found in the 3'UTRs of many unstable mRNAs act as *cis*-acting elements regulating mRNA degradation [44]. Specific NATs can thus bind to these AU-elements, thereby influencing the rate of mRNA degradation. For example, the antisense *BCL2/IGH* transcript, which encompasses the *IGH* locus, the t(14;18) fusion site and the translocated *BCL2* mRNA, has been suggested to mask the AU-rich motif present in the 3'UTR of the *BCL2* mRNA to stabilize *BCL2* mRNA levels in follicular lymphoma [45]. Although many other examples of lncRNAs acting as mRNAs "stabilizers" have been described [46,47], we still lack an understanding of their mechanism of action. The inducible nitric oxide synthase (iNOS), an important mediator of nitric oxide generation during inflammation, has an antisense strand corresponding to the 3'UTR of the *iNos* gene. This *iNos-As* transcript interacts with *iNos* mRNA and facilitates its stability *via* binding to the AU-rich element-binding HuR protein in the cytoplasm of rat hepatocytes after interleukin-1 induction [46]. In a similar manner, the beta-secretase-1 (*BACE1*) mRNA, a crucial enzyme upregulated in Alzheimer's disease, is under the control of its lncRNA transcript (*BACE1-AS*) [47].

In contrast to the above mentioned examples, lncRNAs can also potentiate mRNA degradation, or mediate the effects of NMD inhibition on cellular functions. For example, the lncRNA growth arrest specific transcript 5 (*GAS5*) has been shown to induce growth arrest upon the inhibition of NMD in several human lymphocyte cell lines [48,48a]. Moreover, translationally active mRNAs, whose 3'-UTRs bind to STAU1, a protein that binds to double-stranded RNA, can be degraded by STAU1-mediated decay or SMD. STAU1-binding sites can be formed by imperfect base-pairing between an Alu element in the 3' UTR of an SMD target and another Alu element in a lncRNA mediating mRNA decay [49].

9.2.2.4 mRNA Translation

LncRNAs are mostly localized in the nucleus, where they regulate gene expression. In addition, a minority (~15%) of the known lncRNAs are present in the cytoplasm where they can regulate translation, the process by which mRNA is decoded by a ribosome to produce a specific amino acid chain. For example, an antisense RNA inhibits the expression of the transcription factor PU.1 by modulating ribosomal entry or by aborting translation in the first steps of initiation and elongation [50]. The translation process may also be stopped by displacing mRNAs from translation-associated proteins. For example, the lncRNA urothelial carcinoma-associated 1 (*UCA1*) forms a complex with heterogeneous nuclear ribonucleoprotein I (hnRNP I), which increases the stability of the *UCA1* mRNA [51]. As hnRNP I also facilitates the translation of p27 (Kip1) *via* an interaction with the 5'-untranslated region (5'-UTR) of p27 mRNAs, the binding of *UCA1* to hnRNP I decreases p27 protein levels by competitive inhibition. Certain antisense lncRNAs can also upregulate the translation of their sense mRNA, thereby enabling a rapid increase in protein levels in response to environmental signals without the need for *de novo* RNA synthesis. This mechanism has been demonstrated to be responsible for the upregulation of the translation of the ubiquitin carboxy-terminal hydrolase L1 (*Uchl1*) mRNA by its antisense lncRNA upon stress induction [52].

9.2.2.5 Subcellular Localization of Proteins and Protein Stability

LncRNAs might modulate protein functions also by regulating the subcellular localization of the proteins in question. This is exemplified by the *NRON* ncRNA, which functions as a repressor of the nuclear factor of activated T cells (NFAT). *NRON*, however, does not directly influence the transcriptional activation of the *NFAT* gene, but instead it specifically inhibits the nuclear accumulation of NFAT protein via interactions, for example, with factors involved in nucleo-cytoplasmic transport [53]. Furthermore, lncRNAs might also affect protein degradation. Thus, the lncRNA *HOTAIR* facilitates the degradation of ataxin-1 and snurportin-1 proteins by targeting them to members of the ubiquitination pathway, such as DZIP3 and MEX3B, respectively, for proteasomal degradation [54].

9.2.3 LncRNAs Control the Nuclear Architecture

Although only a small number of lncRNAs have been well characterized, it is evident that they also participate in the control of the nuclear architecture and higher-order chromosome structures, including X-inactivation (Xi) in females, genomic imprinting, telomere stability, formation of nuclear subcompartments, or generation of long-range DNA looping (Table 9.1). Some examples are described in the following sections.

9.2.3.1 X-Chromosome Inactivation

Mammalian females must inactivate one of their two X chromosomes to achieve dosage compensation. The Xi mechanism provided one of the first examples of a lncRNA that is directly involved in establishing repressive chromatin domains (see also Chapter 14). A noncoding RNA, *Xist*, is known to spread in *cis* along the regions of the X chromosome destined to be silenced (more than 1000 genes must be silenced across multiple cell divisions) in a mono-allelic manner [55]. *Xist* is required, however, mainly for the initiation step and not for the maintenance of X-inactivation. The transcription of *Xist* and its binding to the inactive-X is promoted by the transcription factor Yin Yang 1 (YY1) [56,57,67a]. Furthermore, *Xist* expression is also enhanced by a set of lncRNAs: *RepA*, *Ftx* and *Jpx* [56,57]. *Jpx* is transcriptionally induced concomitantly with X inactivation, and its heterozygous deletion in an ES cell model led to high levels of cell death that was correlated with a failure to express *Xist* in female cells [56]. *Jpx* thus induces *Xist* transcription by the sequestration of its transcriptional repressor CCCTC-binding factor (CTCF) [68]. Once *Xist* has coated the inactive-X, it induces the formation of repressive heterochromatin by a mechanism dependent on the PRC2 complex and histone methylation. Looping the loop, *Xist* expression is also regulated by another lncRNA, *Tsix*, which is transcribed from the *Xist* locus in antisense direction. *Tsix* acts in *cis* and counteracts *Xist* expression by inducing repressive histone marks at the *Xist* promoter [69].

9.2.3.2 Telomere Stability

Telomeres shorten after each mitotic cell division in adult somatic cells leading to cell senescence and apoptosis (see also Chapter 13). However, telomeres can

Table 9.1 Selected Long Noncoding RNAs (lncRNAs) that can Shape 3D Nuclear Architecture at Different Levels of Nuclear Organization

lncRNA	Level of Nuclear Organization	Functional Effect	References
Xist	Intrachromosomal regulatory role	X-inactivation in females	[55–57]
TERRA	Intrachromosomal regulatory role	Telomere stability	[58–60]
NEAT1	Nuclear bodies	Paraspeckle structural integrity	[61]
MALAT1	Nuclear bodies	Splicing speckles	[38]
Kncq1ot1/Lit1	Perinucleolar region	Transcriptional silencing of imprinted genes	[62]
Air	Nuclear compartments	Transcriptional silencing of distal promoter	[63]
SatIII (flies)	Nuclear stress bodies	Centromere identity in response to stress	[64,65]
PRNCR1	Enhancer–promoter interactions	Transcriptional regulation	[66]
PCGEM1	Enhancer–promoter interactions	Transcriptional regulation	[66]
CCATT1-L	Enhancer–promoter interactions	Transcriptional regulation	[19]
FIRRE	Interchromosomal interactions	Transcriptional regulation	[67]

(A) Some lncRNAs regulate chromatin states within individual chromosomes (i.e., intrachromosomally) to assure, for example, heterochromatin formation including telomere stability (TERRA) or X-inactivation in females (Xist); (B) lncRNA expression has been also associated with the integrity of nuclear bodies, such as paraspeckles (NEAT1), splicing speckles (MALAT1), or nuclear stress bodies (SatIII); (C) lncRNAs could also guide target sequences to specific nuclear regions to control gene expression (Kncq1ot1/Lit1 or Air); (D) lncRNAs can mediate chromatin looping to promote interactions between distal enhancers and their promoters; and (E) lncRNAs can also facilitate interactions between sequences located in different chromosomes (Firre).

be replenished by the telomerase reverse transcriptase enzyme, an enzyme that is very active in stem cells and also in tumor cells. The shelterin complex of proteins is known to regulate telomerase activity by binding to telomeres and inducing the formation of a t-loop, a cap structure of 300 bp single-stranded DNA that stabilizes the telomere, preventing the telomere ends from being recognized as potential break points by the DNA repair machinery [70]. Six proteins form the shelterin complex in humans: TRF1, TRF2, POT1, TIN2, TPP1, and RAP1 [71]. Telomere repeat-containing RNAs (*TERRAs*) are a group of lncRNAs transcribed at the telomeres that actively participate in the regulation of telomere homeostasis and telomere function [58] through different mechanisms, including the stabilization of chromosome ends or the prevention of senescence induced by DNA damage response. *TERRA* transcripts are actively recruited to chromosome ends through interactions with stable components of the telomeric structure [59]. First, *TERRA* associates with the shelterin components TRF1 and TRF2 [59], but also with members of the epigenetic machinery including the heterochromatin protein 1 (HP1), the lysine 9-histone 3 methyltransferase SUV39H1 or the component of the NuA2 histone acetyltransferase complex, MORF4L2 [72,73]. Recent findings thus suggest that *TERRA* could act as a scaffold molecule promoting the binding of proteins associated with heterochromatin formation to the chromosome ends. Second, *TERRA* transcripts can base-pair with their template DNA strand, forming RNA:DNA hybrid structures known as R-loops at telomere regions [60]. Interestingly, accumulation of telomeric R-loops promotes also the alternative lengthening of telomeres (ALT) pathway by facilitating homologous recombination between telomeric repeats of different chromosomes, and delays senescence in cells lacking telomerase [60]. Furthermore, tight regulation of the levels of *TERRA* during the cell cycle is involved in telomere protection and in the suppression of the DNA damage response at telomeres [74]. During telomeric DNA replication, exposed ssDNA is thus bound by the ssDNA-binding protein RPA, which is essential for DNA replication and activation of the ATR checkpoint. RPA can be displaced by another ssDNA binding protein, hnRNAP1, which can be sequestered by *TERRA* [75]. *TERRA* expression changes during the cell cycle: it is expressed at low levels in the late S phase while its expression increases in early G1 phase. As *TERRA* binds hnRNPA1, decreased *TERRA* during S phase facilitates hnRNPA1 binding to telomeric DNA and RPA displacement [75]. Release of RPA from telomeres prevents erroneous activation of ATR during cell cycle progression, enhancing telomere protection. Finally, it has been postulated that *TERRA* is also involved in the direct regulation of telomerase activity. In vitro assays showed that *TERRA* inhibits telomerase activity [76]; however, its regulatory role in vivo is still unclear.

9.2.3.3 *Formation of Nuclear Subcompartments*

In the eukaryotic nucleus, nuclear functions tend to be compartmentalized into numerous subnuclear bodies including nucleoli, splicing speckles, paraspeckles, Cajal bodies or promyelocytic leukaemia bodies. However, these subnuclear bodies are not surrounded by a membrane, enabling their components to more freely exchange with the content of the surrounding nucleoplasm [76a]. Although relatively little is known about the underlying mechanism of nuclear compartmentalization, lncRNAs are involved in the building up and function of several nuclear bodies. Prominent examples include the paraspeckles, which are nuclear foci that depend on the lncRNA *NEAT1* for their structural integrity. *NEAT1* is thus a scaffold lncRNA that serves as a platform for the assembly of large macromolecular complexes in paraspeckles [61]. Another example is represented by the lncRNA *MALAT1*, which is specifically localized to splicing speckles [38]. Although it has been demonstrated that *NEAT1* is required for paraspeckle integrity, *MALAT1* is not essential for the structure of splicing compartments, suggesting a different mechanisms of action.

Finally, nuclear stress bodies (NSBs) are transient nuclear subcompartments formed in response to heat shock [64]. Their formation is initiated by a direct interaction between the heat shock transcription factor 1 (HSF1) and pericentric tandem repeats of satellite III DNA sequences [64]. In response to stress, NSBs change their epigenetic status from heterochromatin to euchromatin and allow the transcription of noncoding satellite III transcripts that remain localized in this specific nuclear region [64]. Interestingly, it has recently been demonstrated in flies that SatIII ncRNAs also localize to the centromeric regions of all chromosomes. Specifically, SAT III RNA binds to the kinetochore component CENP-C regulating centromere identity [65].

9.2.3.4 *Targeting of Genomic Regions to Nuclear Subcompartments*

Apart from their roles in the formation and maintenance of nuclear subcompartments, lncRNAs have also been shown to regulate the sub-nuclear localisation of their target regions or the regions which they are transcribed from. For example, the imprinted *Kcnq1ot1/Lit1* ncRNA mediates transcriptional silencing in *cis* by targeting its locus to the repressive environment of the perinucleolar region [62]. Indeed, the silencing domain in the *Kcnq1ot1* transcript, rather than the DNA sequence, is crucial for targeting episomes to the vicinity of the nucleolus, a process that is dependent on the cell-cycle phase [62]. Similarly, the imprinted *Air* ncRNA silences the transcription of the paternal *Slc22a3* gene locus through a *cis* interaction between the ncRNA and chromatin modifiers (G9a H3K9 methyltransferase) in *Air* nuclear compartments [63].

Interestingly, the lncRNA *Xist* also acts in part by guiding the inactive X chromosome to subnuclear compartments specialised in replicating and maintaining repressed chromatin states [77]. Prior to the onset of the Xi process, the two female X chromosomes assume random positions relative to each other in the nucleus. During the initial phase of X inactivation, homologous X chromosomes pair transiently, which is followed by symmetry-breaking and the establishment of monoallelic *Xist* expression [78]. In an elegant assay, Zhang and coworkers [77] demonstrated that once the X-X pair dissociates, the Xa and Xi move to distinct subnuclear positions and acquire opposite epigenetic states. The Xi thus migrates to the perinucleolar space specialized in replicating condensed chromatin during S phase in a *Xist*-dependent manner, followed by its localisation to the repressive environment of the nuclear periphery. This model thus suggests that lncRNA-mediated nuclear localisation is highly specific, depends on the cell cycle phase and promotes the copying and maintenance of the epigenetic marks after each cell division.

9.2.3.5   Short- and Long-Range Chromosomal Looping and Inter-chromosomal Interactions

The 3D structure of the nucleus influences and is influenced by gene transcription [79]. It is well established that promoters and enhancers that are located on the same or even on different chromosomes can form complex chromatin fibre interactomes by means of chromatin loops and bridges, respectively. Such chromatin loops contribute to the formation of 3D nuclear compartments known as topologically associating domains that are demarcated by a high density of chromatin architectural proteins, such as CTCF, and provide a 3D framework where regulatory elements and genes cooperate to coordinate gene expression and increase the efficiency of transcription [66]. Although the mechanisms by which these chromatin loops are established and maintained, and their effect on gene transcription are not fully understood, an involvement of lncRNAs cannot be ruled out. Hence, enhancer RNAs (eRNA; also known as activating RNAs) are a class of ncRNAs transcribed from active enhancers [66] that could influence not only the transcription of nearby and distant coding genes, but also 3D genome folding. Regarding how this control might be exerted, it has been postulated that eRNAs interact with the Mediator complex, which is a protein complex that connects transcription factors at enhancers to the basic transcriptional machinery located at target promoters. Moreover, the Mediator complex is functionally linked with both DNA bending and phosphorylation of H3S10 that favors enhancer-promoter loop formation resulting in active transcription of eRNA target genes [80]. Enhancer-promoter contacts can be further stabilized by the cohesin complex. In murine ES cells and mouse embryonic fibroblasts, it was thus observed that the cohesin-loading factor NIPBL is associated with Mediator–cohesin complexes at specific promoters [81]. Mediator and cohesin cooccupy different promoters in different cell types, thus generating cell-type-specific DNA loops regulating the gene expression program of each cell [81]. In summary, the combination of CTCF, Mediator complex, eRNAs and cohesin appears to determine chromatin interactions at different length scales. Whereas the CTCF/cohesin pair anchors long-range constitutive interactions that might form the topological basis for invariant subdomains; the mediator/cohesin complex bridges short-range enhancer–promoter interactions within and between larger subdomains [82].

It will be important to explore further the potential role of lncRNAs in the modulation of nuclear architecture within and between chromosomes. A seminal example of long-range lncRNA action is represented by the lncRNA *CCAT1-L*, which is transcribed specifically in human colorectal cancers from a locus upstream of the *MYC* gene, and controls *MYC* transcription by a mechanism involving long-range chromatin looping together with CTCF [19]. Finally, an example of *trans*-acting lncRNA is represented by *Firre* that is localized in an

~5-Mb domain on the X chromosome, but interacts with the heterogeneous nuclear ribonucleoprotein U through a 156-bp repeating sequence to coordinate the expression of at least three gene loci located on chromosomes 2, 9, 15, and 17, which overlap known genes including *Slc25a12*, *Ypel4*, *Eef1a1*, *Atf4*, and *Ppp1r10* [67].

9.3 REGULATORY MECHANISMS OF lncRNA EXPRESSION

LncRNA expression is tightly regulated in a spatiotemporal manner, which is reflected in their differentiation stage- and cell type-specific expression pattern during development [83]. Despite of their importance, the mechanisms that govern the spatiotemporal control of lncRNA expression are largely unknown. There is, however, evidence demonstrating that the expression of lncRNAs is orchestrated by regulatory mechanisms similar to those acting at coding regions including transcription factor binding, RNA splicing, RNA stabilization, or epigenetic regulation [84–86].

A seminal example of differentiation stage- and cell type-specific expression of lncRNAs is represented by the highly dynamic expression pattern of lincRNAs during the development and differentiation of T cells [87]. Specific transcription factors, such as T-BET, GATA-3, STAT4, and STAT6 thus bind to lincRNA genes to guide their expression during the different stages of T cell differentiation, from the early T cell progenitors to the terminally differentiated helper T cell subsets [87]. Furthermore, these lincRNAs are located in genomic regions that are also enriched for genes that encode proteins with immunoregulatory functions, suggesting a possible coevolution of protein coding and lincRNA genes [87]. LincRNAs are also regulated by pluripotency factors and have been implicated in the regulation of the pluripotent state. Pharmacologic downregulation of *Oct4* and *Nanog* RNA levels in mouse ES cells thus increased the expression of the lncRNAs *AK028326* and *AK141205* [88]. Furthermore, OCT4 was found to be necessary for the transcription of several lncRNAs that modulate the pluripotent state in ESCs and induced pluripotent stem (iPS) cells [89]. Finally, the promoter region of the *lincRNA-p21* contains canonical binding sites for the tumor suppressor protein p53 that activates *lincRNA-p21* expression, which is essential for p53-dependent apoptosis [90].

Apart from transcription factor-mediated control, lncRNAs are also regulated by epigenetic mechanisms, including CpG methylation and histone marks. LncRNAs transcribed from ultraconserved regions of the genome are thus subjected to transcriptional silencing by CpG hypermethylation [91]. Specifically, the lncRNAs *Uc.160 +*, *Uc283 + A*, and *Uc.346+* were found to undergo specific

CpG island hypermethylation-associated silencing in cancer cells compared with normal tissues [91]. Regarding histone marks, a study based on custom-designed microarrays showed that the expression profile of lncRNAs in ES cells, lineage-restricted neuronal progenitor cells, and terminally differentiated fibroblasts was under the control EZH2-mediated H3K27 methylation [92]. Finally, in recent years many studies have shown that lncRNAs are under the regulation of miRNAs [93]. The first study to show that a noncoding antisense transcript can be a direct miRNA target uncovered the role of miR-671 in the regulation of the circular antisense transcript of the Cerebellar Degeneration-Related protein 1 (*CDR1*) locus in an Ago2-slicer-dependent manner [94]. Importantly, the list of miRNAs controlling lncRNAs and their deregulation in cancer is increasing to include the following examples: miR125 that controls *HOTTIP* expression in hepatocelular carcinoma [95]; miR-101 and miR-217 that target *MALAT1* in esophageal carcinoma [96]; miR-211 that controls the lncRNa *LOC285194* expression in colorectal cancer [97] or the miR-1-mediated regulation of *UCA1* in bladder cancer [98].

9.4   THERAPEUTIC APPLICATIONS OF lncRNAs

As lncRNAs are important players in human pathologies (reviewed in [3,11,99,100]), their potential as disease biomarker is widely explored. For example, lncRNAs can be present in exosomes and detected in minimally invasive or noninvasive biological samples, such as blood or urine. Exosomes are extracellular vesicles containing proteins, lipids, and RNA derived from their donor cell and can be taken up by other cells. The intercellular transfer of exosome contents thus provides a mechanism by which cells can communicate with each other in their local microenvironment. The presence of lncRNAs within exosomes is very interesting, because it provides a potential explanation for how exosomes can modulate gene expression in the recipient cells. Although initially thought to be unstable molecules, a half-life study examining approximately 800 lncRNAs demonstrated that they are more stable than expected [101]. To date, a few studies have analyzed altered patterns of circulating lncRNAs in patients. For example, the lncRNA *LIPCAR* is increased in the plasma of patients with chronic heart failure and strongly associates with mortality rates [102]. Inflammatory conditions, such as acute kidney injury (AKI) also leads to the release of lncRNAs. Moreover, alteration of circulating concentrations of the NAT *TapSAKI* in blood from patients with AKI was documented to be a predictor of mortality in this patient cohort [103]. The deregulated expression of circulating lncRNAs has also been explored in the field of oncology, and has proved to be useful diagnostic and prognostic marker. For example, the level of *DD3*^{PCA3} lncRNA in the urine of patients with varying stages of prostate cancer can be used as a predictor of malignancy that complements the well-established PSA test and facilitates patient stratification

[104]. Furthermore, in the serum of patients with cervical cancer, circulating *HOTAIR* was commonly upregulated, and its expression was associated with advanced tumor stages, lymphatic vascular space invasion, and lymphatic node metastasis [105]. Circulating lncRNAs are also found in the medium of cultured cells and serve as interesting models for studies on drug resistance [106]. For example, *LINC-ROR* is among the most significantly upregulated lncRNAs in hepatocelular carcinoma, and it is released into the culture medium of HepG2 cells upon treatmentt with TGFβ [106]. Most importantly, *LINC-ROR* expression was associated with cellular response to chemotherapy agents used against hepatocarcinoma including sorafenib, camptothecin, or doxorubicin. Knockdown models of *LINC-ROR* showed a significant reduction of cell viability and increased apoptosis, while incubation of naïve cells with extracellular exosome preparations containing *LINC-ROR* reduced cell death induced by chemotherapy [106].

Taking together the above considerations, it is clear that lncRNAs (especially circulating lncRNAs) are ideal candidates as diagnostic and prognostic biomarkers of human diseases. In addition, new uses of lncRNAs are being explored including their potential as therapeutic targets. Two features make them especially interesting candidates in this respect. First, they provide a solution for targeting "undruggable" molecules, such as those factors that need to be upregulated to achieve therapeutic effects. Second, lncRNA expression shows high specificity, because they are expressed at specific stages of development and in a cell type-specific manner. Furthermore, they usually target one gene or a small group of related genes. To date, the possibility of activating the expression of specific proteins by small molecule drugs remains unsuccessful. Specific lncRNAs, however, can be silenced using synthetic antisense oligonucleotides (ASO) resulting in the upregulation of their target genes. In this regard, a recent work described how deficits of the UBE3A protein in Angelman syndrome could be restored with lncRNA-based therapy [107]. Angelman syndrome is an imprinting disorder characterized by intellectual disability, developmental delay, behavioural uniqueness, speech impairment, seizures and ataxia, which is caused by maternal deficiency of the *UBE3A* gene product that encodes an E3 ubiquitin ligase. As the paternal *UBE3A* allele is normally silenced by a nuclear antisense lncRNA (*UBE3A-ATS*), a strategy that upregulates the intact paternal allele could ameliorate UBE3A deficiency. The authors demonstrated that *Ube3a-ATS* levels could be reduced with ASO in murin models of Angelman syndrome [107]. ASO treatment thus achieved specific reduction of *Ube3a-ATS* and sustained the expression of the paternal *Ube3a* allele in neurons in vitro and in vivo. Interestingly, partial restoration of UBE3A protein levels in the Angelman syndrome mouse model also ameliorated some of the cognitive deficits associated with the disease. Finally, due to the specific interactions between certain lncRNAs and the PRC2 complex, ASOs can also be used for blocking the interaction site between a lncRNA and PRC2 proteins to interfere with lncRNA-mediated gene repression [107].

Apart from ASOs, lncRNAs could also be targeted by siRNAs or viral vectors containing shRNAs [108]. The main limitation of these approaches is that they require a delivery vehicle (i.e., liposomes, nanoparticles, or viruses) to protect the siRNA or shRNA vectors from nuclease-mediated degradation in in vivo models. Notwithstanding these limitations, several studies have addressed this matter and currently several clinical trials based on RNAi technology are being conducted [109]. Finally, it must be considered that although lncRNAs represent attractive pharmacological and therapeutic targets, the effects on global gene expression of their targets are in most cases unknown and should also be carefully considered.

9.5 CONCLUSIONS

Next-generation technology allowed the identification of a plethora of novel lncRNAs expressed in different human cell types during development and also in pathological situations. In relation with lncRNA functions, we begin to elucidate their role in transcriptional regulation through interactions with transcription factors and chromatin remodeling complexes or other epigenetic factors. However, we still do not fully understand how these lncRNAs are recruited to specific genomic loci. While the role of lncRNAs in 3D nuclear organization is very attractive, many questions remain unsolved. In spite of the large number of lncRNA sequences, we have a reduced list of specific lncRNAs involved in nuclear domain organization and we have even less knowledge about how they organize these nuclear subcompartments under variable cellular conditions.

Nevertheless, many important questions regarding lncRNA complexity remain unanswered. First, it must be highlighted that the same lncRNA can be involved in different but complementary biological processes under different cellular conditions. An example is represented by the dual role of the *CCND1* lncRNA in serving as a "signal" for TLS activation, or acting as "guide" for repressor complexes to form on chromatin. It is plausible that an individual lncRNA may display different functions depending on its spatial and temporal expression profile or under different external and internal stimuli. Moreover, many lncRNAs overlap regulatory elements that could be functionally relevant independently of the lncRNA itself. Such complexity strongly hinders the interpretation of the resulting phenotypes.

To solve the major questions in lncRNA biology, it will be necessary to develop experimental systems including inducible lncRNA expression and experimentally alter specific lncRNA domains to explore lncRNA function. It will be also important to move from in vitro studies, which are performed in homogeneous cell populations, to in vivo experimental systems that will lead to a better understanding of why lncRNAs are expressed in a tight spatial and temporal manner. Thus, we have now numerous possibilities to dissect the "dark matter" of the genome and understand their mechanisms of action.

Abbreviations

AGO-2	Argonaute-2
AIR	Antisense Igf2r RNA
ANRIL	Antisense noncoding RNA in the INK4 locus
APOA1	Apolipoprotein A1
ASO	Antisense oligonucleotides
BACE1	Beta-Secretase-1
BDNF	Brain-derived neurotrophic factor
CCND1	Cyclin D1
CTCF	CCCTC-binding factor
DHFR	Dihydrofolate reductase
Element Kcnq1ot1/Lit1	KCNQ1 opposite strand/antisense transcript 1 (Non-Protein Coding)
elncRNAs	Enhancer lncRNAs
EPHB2	EPH receptor B2
EZH2	Enhancer of zeste 2
FENDRR	FOXF1 adjacent noncoding developmental regulatory RNA
FIRRE	Functional intergenic repeating RNA
GAS5	Growth arrest specific transcript 5
GDNF	Glial cell derived neurotrophic factor
HOTAIR	HOX transcript antisense RNA
HOTTIP	HOXA distal transcript antisense RNA (nonprotein coding)
HP1	Heterochromatin protein 1
HSF1	Heat shock transcription factor 1
LINC-ROR	Long intergenic nonprotein coding RNA, regulator of reprogramming
lncRNA	Long noncoding
MALAT1	Metastasis associated lung adenocarcinoma transcript 1 (nonprotein coding)
MYOD	Myogenic differentiation 1
NAT	Natural antisense transcripts
NEAT1	Nuclear paraspeckle assembly transcript 1
NFAT	Nuclear factor of activated T cells
NONO	Non-POU domain containing, octamer-binding
PRC1, 2	Polycomb repressive complex 1, 2
PSPC1	Paraspeckle component 1
RISC	RNA-induced silencing complex
RNA miRNA	MicroRNA
RNAi	RNA interference
SINE	Short interspersed elements
SMD	STAU1-mediated decay
SRA	Steroid receptor RNA activator

TERRA	Telomere repeat-containing RNAs
UBE3A	Ubiquitin protein ligase E3A
UCA1	Urothelial carcinoma-associated 1
Uchl1	Ubiquitin carboxy-terminal hydrolase L1
YY1	Yin yang 1

References

[1] ENCODE Project Consortium. An integrated encyclopedia of DNA elements in the human genome. Nature 2012;489(7414):57–74.

[2] Djebali S, et al. Landscape of transcription in human cells. Nature 2012;489(7414):101–8.

[3] Esteller M. Non-coding RNAs in human disease. Nat Rev Genet 2011;12(12):861–74.

[4] Khorkova O, et al. Basic biology and therapeutic implications of lncRNA. Adv Drug Deliv Rev 2015;87:15–24.

[5] Ambros V. The functions of animal microRNAs. Nature 2004;431(7006):350–5.

[6] He L, Hannon GJ. MicroRNAs: small RNAs with a big role in gene regulation. Nat Rev Genet 2004;5(7):522–31.

[7] Melo SA, Esteller M. Dysregulation of microRNAs in cancer: playing with fire. FEBS Lett 2011;585(13):2087–99.

[8] Xie C, et al. NONCODEv4: exploring the world of long non-coding RNA genes. Nucleic Acids Res 2014;42(Database issue):D98–103.

[9] Pang KC, et al. Rapid evolution of noncoding RNAs: lack of conservation does not mean lack of function. Trends Genet 2006;22(1):1–5.

[10] Tsai MC, et al. Long noncoding RNA as modular scaffold of histone modification complexes. Science 2010;329(5992):689–93.

[11] Clark BS, Blackshaw S. Long non-coding RNA-dependent transcriptional regulation in neuronal development and disease. Front Genet 2014;5:164.

[12] Schmitz U, et al. The RNA world in the 21st century—a systems approach to finding non-coding keys to clinical questions. Brief Bioinform 2015;1–13.

[13] Qu Z, Adelson DL. Evolutionary conservation and functional roles of ncRNA. Front Genet 2012;3:205.

[14] Struhl K. Transcriptional noise and the fidelity of initiation by RNA polymerase II. Nat Struct Mol Biol 2007;14(2):103–5.

[15] Shamovsky I, et al. RNA-mediated response to heat shock in mammalian cells. Nature 2006;440(7083):556–60.

[16] Feng J, et al. The Evf-2 noncoding RNA is transcribed from the Dlx-5/6 ultraconserved region and functions as a Dlx-2 transcriptional coactivator. Genes Dev 2006;20(11):1470–84.

[17] Caretti G, et al. The RNA helicases p68/p72 and the noncoding RNA SRA are coregulators of MyoD and skeletal muscle differentiation. Dev Cell 2006;11(4):547–60.

[18] Ørom UA, et al. Long noncoding RNAs with enhancer-like function in human cells. Cell 2010;143(1):46–58.

[19] Xiang JF, et al. Human colorectal cancer-specific CCAT1-L lncRNA regulates long-range chromatin interactions at the MYC locus. Cell Res 2014;24(5):513–31.

[20] Martianov I, et al. Repression of the human dihydrofolate reductase gene by a non-coding interfering transcript. Nature 2007;445(7128):666–70.

[21] Mariner PD, et al. Human Alu RNA is a modular transacting repressor of mRNA transcription during heat shock. Mol Cell 2008;29(4):499–509.

[22] Guttman M, et al. lincRNAs act in the circuitry controlling pluripotency and differentiation. Nature 2011;477(7364):295–300.

[23] Khalil AM, et al. Many human large intergenic noncoding RNAs associate with chromatin-modifying complexes and affect gene expression. Proc Natl Acad Sci USA 2009;106(28): 11667–72.

[24] Zhao J, et al. Genome-wide identification of polycomb-associated RNAs by RIP-seq. Mol Cell 2010;40(6):939–53.

[25] Schuettengruber B, et al. Trithorax group proteins: switching genes on and keeping them active. Nat Rev Mol Cell Biol 2011;12(12):799–814.

[26] Wang KC, Chang HY. Molecular mechanisms of long noncoding RNAs. Mol Cell 2011;43(6):904–14.

[27] Rinn JL, et al. Functional demarcation of active and silent chromatin domains in human HOX loci by noncoding RNAs. Cell 2007;129(7):1311–23.

[28] Tsai MC, et al. Long noncoding RNA as modular scaffold of histone modification complexes. Science 2010;329(5992):689–93.

[29] Aguilo F, et al. Long noncoding RNA, polycomb, and the ghosts haunting INK4b-ARF-INK4a expression. Cancer Res 2011;71(16):5365–9.

[30] Halley P, et al. Regulation of the apolipoprotein gene cluster by a long noncoding RNA. Cell Rep 2014;6(1):222–30.

[31] Sopher BL, et al. CTCF regulates ataxin-7 expression through promotion of a convergently transcribed, antisense noncoding RNA. Neuron 2011;70(6):1071–84.

[32] Modarresi F, et al. Inhibition of natural antisense transcripts in vivo results in gene-specific transcriptional upregulation. Nat Biotechnol 2012;30(5):453–9.

[33] Grote P, et al. The tissue-specific lncRNA Fendrr is an essential regulator of heart and body wall development in the mouse. Dev Cell 2013;24(2):206–14.

[34] Wang X, et al. Induced ncRNAs allosterically modify RNA-binding proteins in *cis* to inhibit transcription. Nature 2008;454:126–30.

[35] Wang KC, et al. A long noncoding RNA maintains active chromatin to coordinate homeotic gene expression. Nature 2011;472(7341):120–4.

[36] Lazorthes S, et al. vlincRNA participates in senescence maintenance by relieving H2AZ-mediated repression at the INK4 locus. Nat Commun 2015;6:5971.

[37] Filipowicz W, et al. Mechanisms of post-transcriptional regulation by microRNAs: are the answers in sight? Nat Rev Genet 2008;9(2):102–14.

[38] Tripathi V, et al. The nuclear-retained noncoding RNA MALAT1 regulates alternative splicing by modulating SR splicing factor phosphorylation. Mol Cell 2010;39(6):925–38.

[39] Li K, et al. A noncoding antisense RNA in tie-1 locus regulates tie-1 function in vivo. Blood 2010;115(1):133–9.

[40] Beltran M, et al. A natural antisense transcript regulates Zeb2/Sip1 gene expression during Snail1-induced epithelial-mesenchymal transition. Genes Dev 2008;22(6):756–69.

[41] Faghihi MA, et al. Expression of a noncoding RNA is elevated in Alzheimer's disease and drives rapid feed-forward regulation of beta-secretase. Nat Med 2008;14(7):723–30.

[42] Hundley HA, Bass BL. ADAR editing in double-stranded UTRs and other noncoding RNA sequences. Trends Biochem Sci 2010;35(7):377–83.

[43] Clemson CM, et al. An architectural role for a nuclear noncoding RNA: NEAT1 RNA is essential for the structure of paraspeckles. Mol Cell 2009;33(6):717–26.

[43a] Lykke-Andersen S, Jensen TH. Nonsense-mediated mRNA decay: an intricate machinery that shapes transcriptomes. Nat Rev Mol Cell Biol 2015;16(11):665–77.

[44] Bevilacqua A, et al. Post-transcriptional regulation of gene expression by degradation of messenger RNAs. J Cell Physiol 2003;195(3):356–72.

[45] Capaccioli S, et al. A bcl-2/IgH antisense transcript deregulates bcl-2 gene expression in human follicular lymphoma t(14;18) cell lines. Oncogene 1996;13(1):105–15.

[46] Matsui K, et al. Natural antisense transcript stabilizes inducible nitric oxide synthase messenger RNA in rat hepatocytes. Hepatology 2008;47(2):686–97.

[47] Faghihi MA, et al. Expression of a noncoding RNA is elevated in Alzheimer's disease and drives rapid feed-forward regulation of beta-secretase. Nat Med 2008;14(7):723–30.

[48] Mourtada-Maarabouni M, Williams GT. Growth arrest on inhibition of nonsense-mediated decay is mediated by noncoding RNA GAS5. Biomed Res Int 2013;358015.

[48a] Popp MW, Maquat LE. Leveraging Rules of Nonsense-Mediated mRNA Decay for Genome Engineering and Personalized Medicine. Cell 2016;165(6):1319–22.

[49] Gong C, Maquat LE. lncRNAs transactivate STAU1-mediated mRNA decay by duplexing with 3′ UTRs via Alu elements. Nature 2011;470(7333):284–8.

[50] Ebralidze AK, et al. PU.1 expression is modulated by the balance of functional sense and antisense RNAs regulated by a shared *cis*-regulatory element. Genes Dev 2008;22(15):2085–92.

[51] Huang J, et al. Long non-coding RNA UCA1 promotes breast tumor growth by suppression of p27 (Kip1). Cell Death Dis 2014;5:e1008.

[52] Carrieri C, et al. Long non-coding antisense RNA controls Uchl1 translation through an embedded SINEB2 repeat. Nature 2012;491(7424):454–7.

[53] Willingham AT, et al. A strategy for probing the function of noncoding RNAs finds a repressor of NFAT. Science 2005;309(5740):1570–3.

[54] Yoon JH, et al. Scaffold function of long non-coding RNA HOTAIR in protein ubiquitination. Nat Commun 2013;4:2939.

[55] Froberg JE, et al. Guided by RNAs: X-inactivation as a model for lncRNA function. J Mol Biol 2013;425(19):3698–706.

[56] Tian D, et al. The long noncoding RNA, Jpx, is a molecular switch for X chromosome inactivation. Cell 2010;143(3):390–403.

[57] Jeon Y, Lee JT. YY1 tethers Xist RNA to the inactive X nucleation center. Cell 2011;146(1):119–33.

[58] Cusanelli E, Chartrand P. Telomeric repeat-containing RNA TERRA: a noncoding RNA connecting telomere biology to genome integrity. Front Genet 2015;6:143.

[59] Deng Z, et al. TERRA RNA binding to TRF2 facilitates heterochromatin formation and ORC recruitment at telomeres. Mol Cell 2009;35:403–13.

[60] Yu T, et al. Telomeric transcripts stimulate telomere recombination to suppress senescence in cells lacking telomerase. Proc Natl Acad Sci USA 2014;111:3377–82.

[61] Hirose, et al. NEAT1 long noncoding RNA regulates transcription via protein sequestration within subnuclear bodies. Mol Biol Cell 2014;25(1):169–83.

[62] Mohammad F, et al. Kcnq1ot1/Lit1 noncoding RNA mediates transcriptional silencing by targeting to the perinucleolar region. Mol Cell Biol 2008;28(11):3713–28.

[63] Nagano T, et al. The Air noncoding RNA epigenetically silences transcription by targeting G9a to chromatin. Science 2008;322(5908):1717–20.

[64] Valgardsdottir R, et al. Structural and functional characterization of noncoding repetitive RNAs transcribed in stressed human cells. Mol Biol Cell 2005;16(6):2597–604.

[65] Rošić S, et al. Repetitive centromeric satellite RNA is essential for kinetochore formation and cell division. J Cell Biol 2014;207(3):335–49.

[66] Shibayama Y, et al. lncRNA and gene looping: what's the connection? Transcription 2014;5(3):e28658.

[67] Hacisuleyman E, et al. Topological organization of multichromosomal regions by the long intergenic noncoding RNA Firre. Nat Struct Mol Biol 2014;21(2):198–206.

[67a] Makhlouf M, Ouimette JF, Oldfield A, Navarro P, Neuillet D, Rougeulle C. A prominent and conserved role for YY1 in Xist transcriptional activation. Nat Commun 2014;5:4878.

[68] Sun S, et al. Jpx RNA activates Xist by evicting CTCF. Cell 2013;153(7):1537–51.

[69] Zhao J, et al. Polycomb proteins targeted by a short repeat RNA to the mouse X chromosome. Science 2008;322(5902):750–6.

[70] de Lange T. Shelterin: the protein complex that shapes and safeguards human telomeres. Genes Dev 2005;19:2100–10.

[71] Ye J, et al. Transcriptional outcome of telomere signalling. Nat Rev Genet 2014;15(7): 491–503.

[72] Arnoult N, et al. Telomere length regulates TERRA levels through increased trimethylation of telomeric H3K9 and HP1alpha. Nat Struct Mol Biol 2012;19:948–56.

[73] Porro A, et al. Functional characterization of the TERRA transcriptome at damaged telomeres. Nat Commun 2014;5:5379.

[74] Pennarun G, et al. ATR contributes to telomere maintenance in human cells. Nucleic Acids Res 2010;38(9):2955–63.

[75] Flynn RL, et al. TERRA and hnRNPA1 orchestrate an RPA-to-POT1 switch on telomeric single-stranded DNA. Nature 2011;471:532–6.

[76] Redon S, et al. The non-coding RNA TERRA is a natural ligand and direct inhibitor of human telomerase. Nucleic Acids Res 2010;38(17):5797–806.

[76a] Mao YS, Zhang B, Spector DL. Biogenesis and function of nuclear bodies. Trends Genet 2011;27(8):295–306.

[77] Zhang LF, et al. Perinucleolar targeting of the inactive X during S phase: evidence for a role in the maintenance of silencing. Cell 2007;129(4):693–706.

[78] Xu N, et al. Transient homologous chromosome pairing marks the onset of X inactivation. Science 2006;311(5764):1149–52.

[79] Fanucchi S, et al. Chromosomal contact permits transcription between coregulated genes. Cell 2013;155(3):606–20.

[80] Lai F, et al. Activating RNAs associate with Mediator to enhance chromatin architecture and transcription. Nature 2013;494(7438):497–501.

[81] Kagey MH, et al. Mediator and cohesin connect gene expression and chromatin architecture. Nature 2010;467(7314):430–5.

[82] Phillips-Cremins JE, et al. Architectural protein subclasses shape 3D organization of genomes during lineage commitment. Cell 2013;153(6):1281–95.

[83] Dinger ME, et al. Long noncoding RNAs in mouse embryonic stem cell pluripotency and differentiation. Genome Res 2008;18(9):1433–45.

[84] Jiang Q, et al. TF2LncRNA: identifying common transcription factors for a list of lncRNA genes from ChIP-Seq data. Biomed Res Int 2014;2014:317642.

[85] Kang MJ, et al. HuD regulates coding and noncoding RNA to induce APP→Aβ processing. Cell Rep 2014;7(5):1401–9.

[86] Guil S, Esteller M. RNA–RNA interactions in gene regulation: the coding and noncoding players. Trends Biochem Sci 2015;40(5):248–56.

[87] Hu, et al. Expression and regulation of intergenic long noncoding RNAs during T cell development and differentiation. Nat Immunol 2013;14(11):1190–8.

[88] Sheik MJ, et al. Conserved long noncoding RNAs transcriptionally regulated by Oct4 and Nanog modulate pluripotency in mouse embryonic stem cells. RNA 2010;16(2):324–37.

[89] Loewer S, et al. Large intergenic non-coding RNA-RoR modulates reprogramming of human induced pluripotent stem cells. Nat Genet 2010;42(12):1113–7.

[90] Huarte M, et al. A large intergenic noncoding RNA induced by p53 mediates global gene repression in the p53 response. Cell 2010;142(3):409–19.

[91] Lujambio A, et al. CpG island hypermethylation-associated silencing of non-coding RNAs transcribed from ultraconserved regions in human cancer. Oncogene 2010;29(48): 6390–401.

[92] Wu SC, et al. Role of H3K27 methylation in the regulation of lncRNA expression. Cell Res 2010;20(10):1109–16.

[93] Liz J, Esteller M. lncRNAs and microRNAs with a role in cancer development. Biochim Biophys Acta 2015;1859(1):169–76.

[94] Hansen TB, et al. miRNA-dependent gene silencing involving Ago2-mediated cleavage of a circular antisense RNA. EMBO J 2011;30(21):4414–22.

[95] Tsang FH, et al. Long non-coding RNA HOTTIP is frequently up-regulated in hepatocellular carcinoma and is targeted by tumour suppressive miR-125b. Liver Int 2015;35(5): 1597–606.

[96] Wang X, et al. Silencing of long noncoding RNA MALAT1 by miR-101 and miR-217 inhibits proliferation, migration, and invasion of esophageal squamous cell carcinoma cells. J Biol Chem 2015;290(7):3925–35.

[97] Liu Q, et al. LncRNA loc285194 is a p53-regulated tumor suppressor. Nucleic Acids Res 2013;41(9):4976–87.

[98] Wang T, et al. Hsa-miR-1 downregulates long non-coding RNA urothelial cancer associated 1 in bladder cancer. Tumour Biol 2014;35(10):10075–84.

[99] Li X, et al. lncRNAs: insights into their function and mechanics in underlying disorders. Mutat Res Rev Mutat Res 2014;762:1–21.

[100] Devaux Y, et al. Long noncoding RNAs in cardiac development and ageing. Nat Rev Cardiol 2015;12(7):415–25.

[101] Clark MB, et al. Genome-wide analysis of long noncoding RNA stability. Genome Res 2012;22(5):885–98.

[102] Kumarswamy R, et al. Circulating long noncoding RNA, LIPCAR, predicts survival in patients with heart failure. Circ Res 2014;114(10):1569–75.

[103] Lorenzen JM, et al. Circulating long noncoding RNATapSaki is a predictor of mortality in critically ill patients with acute kidney injury. Clin Chem 2015;61(1):191–201.

[104] Tinzl M, et al. DD3PCA3 RNA analysis in urine—a new perspective for detecting prostate cancer. Eur Urol 2004;46:182–6.

[105] Li J, et al. A high level of circulating HOTAIR is associated with progression and poor prognosis of cervical cancer. Tumour Biol 2015;36(3):1661–5.

[106] Takahashi K, et al. Extracellular vesicle-mediated transfer of long non-coding RNA ROR modulates chemosensitivity in human hepatocellular cancer. FEBS Open Bio 2014;4: 458–67.

[107] Meng L, et al. Towards a therapy for Angelman syndrome by targeting a long non-coding RNA. Nature 2015;518(7539):409–12.

[108] Fatemi RP, et al. De-repressing LncRNA-targeted genes to upregulate gene expression: focus on small molecule therapeutics. Mol Ther Nucleic Acids 2014;3:e196.

[109] Davidson BL, McCray PB Jr. Current prospects for RNA interference-based therapies. Nat Rev Genet 2011;12(5):329–40.

CHAPTER 10

Epigenetic Regulation of Nucleolar Functions

A-.K. Östlund Farrants

Department of Molecular Biosciences, The Wenner–Gren Institute,
Stockholm University, Stockholm, Sweden

10.1 INTRODUCTION

Ribosomal transcription is a fundamental process that is essential for protein synthesis. In particular, proliferating and metabolically active cells require high levels of ribosomal transcription. The transcription of ribosomal RNA (rRNA) genes constitutes the major part of all transcription in cells and is essential for both cell growth and maintenance of the cellular protein content. These genes exist in large tandem repeats and are transcribed by different polymerases than that responsible for the transcription of the protein-coding genes. Four rRNAs are incorporated into the small- and the large ribosomal proteins in the nucleus before being exported into the cytoplasm where translation occurs. The four rRNA species are transcribed by two different kinds of RNA polymerases (pol). The large ribosomal transcript, that is, the 47/45S rRNA in mammalian cells (35S rRNA in yeast) is produced by RNA pol I, with the auxiliary factors upstream binding factor (UBF), selectivity factor 1 (SL1) complex [equivalent to mouse transcription initiation factor IB (TIF1B)], and hRRN3 [equivalent to mouse transcription initiation factor IA (TIF1A)] that marks the initiation competent form of pol I. This transcript is subsequently processed into the 18S, 5.8S, and 28S rRNAs (25S rRNA in yeast), which are assembled with ribosomal proteins into the small- and the large ribosomal subunits (for reviews see Refs. [1–3]). The fourth rRNA, the 5S rRNA, is transcribed by RNA pol III, which employs its own specific transcription factors, such as TFIIIA (specific for 5S rRNA genes), TFIIIB, and TFIIIC. The 5S rRNA is incorporated into the large ribosomal subunit together with the ribosomal proteins RPL5 and RPL5 [4–6]. The variation in auxiliary factors used by the different RNA pols reflects differences in the promoter architecture of the genes transcribed by RNA pol I, RNA pol II, or RNA pol III. For example, the RNA pol III is able to use internal promoters within short genes. The RNA pol I and its associated factors are mainly concentrated in the nucleolus, forming a nuclear body

Chromatin Regulation and Dynamics. http://dx.doi.org/10.1016/B978-0-12-803395-1.00010-1

dedicated to the transcription of rRNA genes, processing and modifications of the rRNA, and assembly of the ribosomal subunits. Although this is the main function of the nucleolus, other RNPs, such as the signal recognition particle and the telomerase, also are assembled in the nucleolus and many nonribosomal proteins might be present (for review see Refs. [7–9]).

10.2 ASSEMBLY OF THE NUCLEOLUS

The RNA pol I genes are organized in tandem repeats in several loci located on different chromosomes, around which the nucleolus is formed. Human cells have approximately 200 copies of RNA pol I genes in the haploid genome, which are found on five acrocentric chromosomes [1,10,11]. *Saccharomyces cerevisiae* has around 200 copies of rRNA genes in one locus, and the nucleolus is formed around the transcribed copies of rRNA genes. The rRNA genes are present in a different chromatin configuration than the rest of the chromosomes, and this is reflected in the formation of a secondary constriction on the metaphase chromosome [12,13]; Miller spreads, that is, electron micrcopic images of transcription units in chromatin [14], suggest that these genes exist in a very open chromatin state [15,16]. The loci in mammalian cells also form constrictions, referred to as the nucleolar-organizing centers (NORs) (for review see Refs. [1,17,18]), which are responsible for the assembly of the nucleolar factors after each cell cycle. The determining event in the assembly of the nucleolus is the binding of the HMG protein UBF to the NORs [17,19,20]. Depletion of UBF thus abolishes the mitotic constrictions, the hallmark of competent chromatin state at NORs [20]. UBF binds to the promoter and along the transcribed region as a dimer, forming a structure that resembles a histone fold, and it remains at the rRNA gene promoter during mitosis [21,22]. This has led to the suggestion that one of the functions of UBF is to act as a mitotic bookmark of active rDNA throughout the cell cycle [23]. A yeast ortholog to UBF, Hmo1, has been shown to assemble on active rRNA genes also in yeast [24].

To obtain a complete nucleolus with all its factors, transcription must proceed and both the promoter- and the transcribed rRNA gene must be present. As the formation of the nucleolus requires transcription, proliferating cells with a high level of rRNA transcription have an increased number of nucleoli compared to quiescent cells (for review see Refs. [25,26]). This indicates that transcription or the transcript itself is important for the assembly process [17,20]. Flanking regions of the rRNA genes are also important for the proper assembly of the nucleolus [27]. Particularly important is the distal sequence called the distal junction (DJ), which is an inverted repeat at the telomeric side of the rDNA locus and is packed in chromatin containing active histone marks. The sequence encodes a transcript transcribed by RNA pol II, and has been suggested to anchor the rDNA to the perinucleolar space, most probably by a process involving the DJ RNA [27]. Recently, it was shown that inverted Alu repeats transcribed by RNA pol II in human cells (or B1 elements in mouse) contribute to nucleolar structure and function. The Alu sequence binds to nucleolar proteins, in particular to nucleolin, which localizes to the nucleolus and associates with nucleophosmin and UBF to support nucleolar structure [28]. Other chromosome structures also influence the formation and integrity of the nucleolus. Telomere disruption caused by DNA damage or by low levels of the telomere protein TRF2 affects nucleolar stability, possibly by disrupting TRF2, UBF, and cohesin interactions [29]. The integrity of the nucleolus is important for proper cellular functions. At the same time several stress signals affect the nucleolus, such as transcription disruptions, hypoxia, heat shock, and viral infections [30]. The cellular stress response converges on the RNA pol I transcription machinery and inhibits transcription and ribosomal biogenesis, possibly to save energy and rescue cell homeostasis in the short term. This also affects the structure of the nucleolus, which is disintegrated into parts. The integrity of the nucleolus and NORs is thus affected by cellular homeostasis, with UBF in the center of the assembling processes and the maintenance the nucleolus.

10.3 REGULATION OF THE RNA POL I MACHINERY

UBF belongs to the RNA pol I machinery, and associates to the upstream control element (UCE) sequence at the promoter of mammalian rRNA genes. The UCE sequence is situated just upstream of the core promoter, which also binds to SL1/TIF-1B [31] (Fig. 10.1). UBF and SL1 interact directly with subunits of RNA pol I or with the mediating factor RRN3/TIF-1A. Hence, the RPA43 subunit of RNA pol I interacts with SL1 via RRN3/TIF1A, and the PAF53 subunit binds to UBF. Moreover, PAF49 interacts with both UBF and SL1/TIF-1B (for reviews see Refs. [2,32–36]. UBF and SL1/TIF-1B are both required for enhanced RNA pol I transcription, where they promote the binding of factors and RNA pol I to rDNA promoters (reviewed in Refs. [36–38]). Moreover, they are involved in promoter clearance, that is, they promote the conversion of RNA pol I from the initiation competent form to the elongating form [39,40]. The interactions between these factors are regulated by posttranslational modifications, such as phosphorylation and acetylation (Fig. 10.1). Some of these modifications upregulate rRNA transcription during cell growth and proliferation, whereas others downregulate rRNA expression in response to starvation and stress. UBF is phosphorylated by kinases upon the activation of several signaling pathways, for example, by the PI3-kinase stimulated by insulin-like growth factor [41], by the mTOR kinase [42], by ERK in the MAP kinase pathway [43] and by CK2 [44]. All of these modifications promote transcription by stabilizing the interactions within the RNA polymerase assembly and with DNA. In addition, proliferative signals facilitate rRNA transcription by the phosphorylation of RRN3/TIF-IA by kinases, such as mTOR [45], ERK [46], and CK2 [47], leading

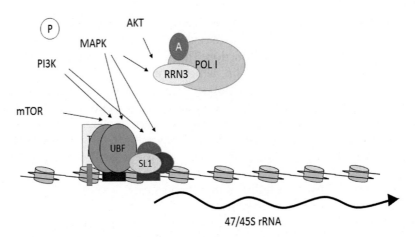

FIGURE 10.1 Signalling pathways regulate ribosomal transcription.
RNA pol I auxillary factors can be phosphorylated by kinases participating in several signaling pathways to enhance or reduce the interaction between the proteins and thereby regulate rRNA transcription. The signaling pathway members shown are responsible for activating phosphorylations. A dimer of upstream binding factor (UBF) binds the upstream control element (UCE), which is stabilized by selectivity factor 1 (SL1) on the promoter. The RRN3 interacts with RNA pol I and actin (A), and mediates the recruitment of RNA pol I to rDNA via binding to the UBF–SL1 complex to form the preinitiation complex.

to its dissociation from pol I after transcription initiation, which is a prerequisite for efficient elongation. On the other hand, phosphorylation of TIF-IA by JNK in response to stress signals disrupts its interaction with RNA pol I and leads to the translocation of TIF-IA from the nucleolus to the nucleoplasm, reducing rRNA transcription [48]. Acetylation of certain factors by histone acetyl transferases (HATs) also contributes to the assembly of the pol I machinery at the promoter. UBF is thus acetylated by p300 and this inhibits the binding of and subsequent repression by the retinoblastoma protein (Rb) [49,50]. TAFI-68 in SL1 is acetylated by PCAF, which enhances the binding of SL1/TIF-1B to DNA and increases transcription [51]. This is reversed via its deacetylation by the NAD+-dependent deacetylase ortholog to SIR2, SIRT1. Other modifications also occur during RNA pol I transcription, such as the ubiquitination of RNA pol I subunits in yeast [52]. The proteasome is recruited to the rRNA genes also in mammalian cells by RRN3/TIF-1A [53] to potentially degrade factors at RNA pol I promoters in an ubiquitin/proteasome-dependent manner [54]. Other nucleolar proteins also contribute to RNA pol I transcription. Nucleolin and nucleophosmin have thus been assigned major regulatory roles in transcription by binding to the promoters and the transcribed regions of RNA pol I genes and affecting transcription rate and processing (reviewed in Refs. [55–57]). It is now believed that these proteins have histone-chaperone activity and act by affecting the chromatin state of rRNA genes [58–60].

10.4 CHROMATIN AT RRNA GENES

10.4.1 Role of UBF

The activity or inactivity of genes within the rDNA loci correlates with an accessible or less accessible chromatin state, respectively [15]. The nature of the chromatin at the rDNA locus is slightly different from the chromatin at other locations in the genome [1,57]. The active rRNA genes are thus not organized in nucleosomes or have a low level of nucleosome occupancy. On the contrary, nucleosomes and silencing proteins are abundant on the silent rRNA genes [1,57]. The active chromatin state at rDNA loci in mammalian cells is marked by UBF. In addition to binding to the promoter where it marks active NORs, UBF binds throughout the transcribed region [12,61,62]. The effect of UBF on chromatin is elicited by its several HMG domains, which help UBF to introduce a bend in the DNA [43] and to compete with histone H1 for the binding to linker DNA [63,64]. Interestingly, UBF has recently been associated with highly transcribed RNA pol II genes, such as the clustered histone genes, suggesting that UBF introduces a chromatin state compatible with a high level of transcription [62,65]. The number of active rRNA genes changes during development, and in some differentiated cells it can be as few as 30%. The amount of UBF also declines during differentiation [64], suggesting that the level of UBF is involved in regulating the number of active rDNA genes (reviewed in Refs. [1,61]).

Other nuclear proteins have also been associated with the active chromatin configuration at rDNA. Nucleolin and nucleophosmin are thus found on the active gene copies, and silencing of these proteins induces a higher density of nucleosomes followed by reduced rRNA transcription [58,60,66]. Nucleolin is a histone H2A/H2B chaperone [59,67], which also associates with H1 [68] and the chromatin remodeling complexes SWI/SNF and ACF [67]. While nucleolin enhances the binding of UBF and opposes the silencing of rRNA genes [58], nucleophosmin requires its RNA-binding domain and UBF to be recruited to the active rRNA genes and facilitates transcription [60].

The yeast HMG protein, Hmo1, is also associated with active rRNA genes, and has been suggested to maintain an open chromatin configuration [69,70]. Although Hmo1 and UBF share conserved functions in Pol I transcription, UBF has also mammalian-specific roles that cannot be substituted by the ectopic expression of Hmo1 in mammalian cells lacking UBF. This might in part be explained by the fact that yeast has a closed mitosis, and therefore the active genes may not need a mitotic bookmarking [23]. Although the binding of HMG proteins, UBF and Hmo1, contribute to the formation of an accessible chromatin structure at active rRNA genes, histones have also been found to be present at these loci. Whether nucleosomes and histones contribute to active ribosomal transcription, similarly to their role in RNA pol II transcription, is still not fully elucidated. Nucleosomes are thus found outside of the transcribed region of active rRNA genes both in *S. cerevisiae* and mammalian cells. In mammalian cells, in vitro- and in vivo experiments show that nucleosomes are positioned at the rRNA gene promoters, and nucleosome positioning functions as a transcriptional switch [71–73]. Finally, inactive rRNA genes do not bind UBF, but are instead configured in heterochromatin that resembles chromatin at other silent regions of the genome, as discussed below.

10.4.2 Chromatin and Silencing of rRNA Genes: NuRD and NoRC

At least two different silent chromatin states have been identified at the rRNA genes [64,74,75]. The first category of chromatin states associated with lack of transcription is termed poised states, which are defined by hypoacetylated histones and unmethylated DNA at the rRNA promoters as common denominators. The second category of chromatin states is termed silent states, which is defined by DNA methylation (in mammalian cells), heterochromatic histone marks, and the presence of HP1α at the rRNA promoters (reviewed in Refs. [1,75]) (Fig. 10.2). One type of poised state is induced by the recruitment of the chromatin remodeling complex NuRD, which comprises CHD4, HDACs, TFIIH, and methyl CpG–binding proteins [76]. The binding of NuRD to the rRNA promoters leads to the hypoacetylation of histones H4 and H3, and to the positioning of the promoter-bound nucleosomes toward a repressed position while exhibiting bivalent histone marks including H3K4me3 and H3K27me3 [75]. SL1 and UBF are also present at such rRNA promoters, further supporting the notion that this state represents a poised state that can easily be activated (Fig. 10.2). This chromatin state is established upon serum starvation. The binding of NuRD thus increases when NIH3T3 cells are serum-starved, with a consequent increase in the fraction of nucleosomes that are in a repressed state. This is rapidly reversed when the cells are fed serum again [75]. A second type of pseudosilent poised state can be observed when UBF is knocked down with siRNA (Fig. 10.2). This state is defined by the binding of the canonical H1, presence of H3K9me3 with consequent binding of HP1α and the lack of DNA methylation [64,77,78] (Fig. 10.2). Similarly to the poised state, the pseudosilent state is also likely to be quickly reversed to an active state on transcriptional activation. Initial studies have suggested that the number of rRNA genes that are in active chromatin configuration does not increase on rRNA activation [79], suggesting that the regulation of ribosomal genes occurs mainly by phosphorylation and acetylation of the RNA pol I machinery. However, the finding that the poised state can be activated when cells are reintroduced to serum after serum starvation [75] uncovered that both more genes and a higher transcription output at each gene can contribute to increased rRNA transcription. This indicates that the poised- and pseudosilent poised states represent a pool of temporarily silent rRNA copies that can be activated on external stimuli.

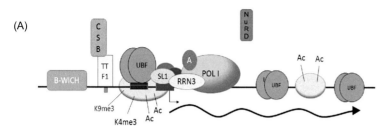

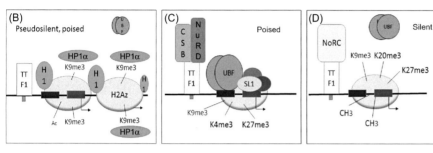

FIGURE 10.2 The RNA pol I genes adopt several different chromatin states.
(A) The active state is open with the nucleosomes in activating position, not covering the transcription start site. hRRN3/TIF-1B and RNA pol I are associated to the promoter. Cockayne syndrome protein B (CSB) and B-WICH are present at the promoter, CSB recruited by transcription terminator factor-1 (TTF1). At least three different transcriptionally inactive chromatin states have been described; (B) the pseudosilent poised; and (C) poised states are dynamic states characterised by a nucleosome in a repressive state, covering the transcription start site, presence of silent or bivalent histone marks and no DNA methylation. The poised state is established by the NuRD complex and the pseudosilent poised state is defined by canonical H1 and HP1α. (D) The stable silent state is defined by repressive histone marks and DNA methylation. This state is established by NoRC, recruited by TTF-I and pRNA (see Fig. 10.3). Interestingly, TTF-I can also be involved in the recruitment of CHD4 present in NuRD and CSB.

The most studied silent chromatin state at the rDNA is a heterochromatin configuration associated with DNA methylation, which is established by the chromatin remodeling complex NoRC, comprising TIP5/BAZ2A and the ISWI–ATPase SNF2h [80]. NoRC is thus responsible for positioning a nucleosome to a repressed position at the transcription start site in the rRNA promoter and for recruiting several silencing factors that deposit repressive histone modifications and DNA methylation (reviewed in Ref. [81]) (Fig. 10.2). This complex is recruited to the rRNA genes by the terminator protein transcription terminator factor-I (TTF-1) [82], H4K16Ac modifications [83,84], and the promoter RNA (pRNA) [85] (Fig. 10.2). TTF-I has several DNA-bindings sites at the end of the rRNA genes and can efficiently terminate RNA pol I transcription. It has also one binding site upstream of the core promoter, which is termed T_0 [86]. Interestingly, TTF-I proteins bound at the promoter site of the rRNA genes can bind not only to a number of silencing factors, but also to activating proteins, which enables TTF-I to act as a transcriptional switch. It has also been shown that TTF-I proteins binding at the T_0 site and at the end of the rRNA gene can interact with each other to form a loop to promote the efficiency transcriptional reinitiation [86,87]. The dynamic chromatin loop is thus formed between the 3′ and 5′ ends of the active rRNA genes, which likely serves to enhance the efficiency of the reloading of the RNA pol I machinery to the promoter after each round of transcription [81,86]. Loop formation has been observed also at silent rDNA copies with the base for the loops formed at the periphery of the nucleolus, which is postulated to be important for nucleolar architecture [25]. TTF-I-mediated repression is a multistep process. First, TTF-I binds TIP5 present in the NoRC complex, which leads to the remodeling of the positioned nucleosome at the promoter to its repressed state covering the SL1/TIF-1B site [71–73]. At a later stage, TIP5 binds the pRNA that stabilizes the NoRC complex at the promoter by forming a DNA–RNA hybrid with the T_0 site and the promoter [88]. In turn, the pRNA inhibits the chromatin remodeling activity of NoRC, and instead promotes the recruitment of silencing factors [73]. It has also been suggested that nucleolin maintains the active state by blocking the binding of TTF-I to its T_0 site [58]. TTF-I thus seems to play an important role in regulating the chromatin state at rRNA genes, and it has been suggested that the amount of TTF-I bound to the T_0 site determines the recruitment of different factors regulating rRNA transcription. The level of TTF-I in the nucleolus can be regulated by, for example, changes in its subcellular localization or the degradation rate in response to external stimuli, such as p19Arf and p53 activation [89,90]. Apart from TTF-I, other proteins can also stabilize the NoRC at rRNA genes, such as the BEND3 protein that binds to NoRC and stabilizes its association to TIP5/BAZ2A [91]. This interaction relies on sumoylation of BEND3. Sumoylation also induces the biding of phosphatidylinositol-4-phosphate 5-kinase (PIP5K) to H3K9me3- and HP1-marked rDNA, but the implication of this is not clear [92]. Proteins regulating rRNA transcription can be sumoylated also in yeast, which influences their binding to rDNA [93]. These results indicate that multiple posttranslational modifications of nucleolar proteins influence the function and regulation of chromatin states at the rDNA.

RNA moieties are important regulators of rRNA silencing (Fig. 10.3). An example is represented by the pRNA, which is an important regulatory component of the transcriptional silencing model presented by Ref. [73]. The pRNA is a noncoding RNA transcribed by RNA pol I from the spacer promoter, which is located 2000 bp upstream of the transcription start site of the rRNA genes in mice [94]. During the multistep process of pRNA production, the spacer RNA is degraded by the RNA exosome complex, which degrades RNAs mainly in RNA quality control [95]. The remaining 150–200-nucleotide long pRNA is complementary to the rRNA promoter site [85,96] and binds to the TAM domain in TIP5 [97]

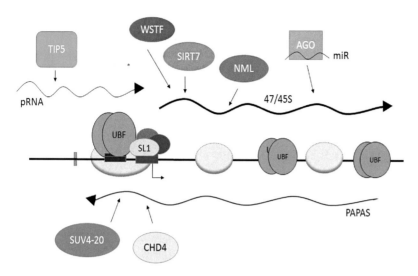

FIGURE 10.3 Regulatory proteins are recruited to the rDNA loci by RNAs.
RNAs originating from the rDNA recruit several regulatory proteins. For example, the silencing proteins CHD4 in NuRD and the histone methyltransferase Suv4-20h2 can be recruited by the antisense ncRNA, *PAPAS*. TIP5 in the silencing NoRC is recruited by the pRNA transcribed from the spacer RNA originating from the spacer promoter 2000 bp upstream of the transcription start site in mouse cells. The silencing by NoRC is associated with DNA methylation, H3K9me3 and H4K20me3, where the latter is added by the Suv4-20h2. The nucleomethylin *(NML)*, a component of the eNOSC complex is involved in the silencing of rRNA genes in response to energy depletion, and is kept inactive by binding to the 47/45S rRNA. AGO, toghether with miRs, downregulates the transcription of the 47/45S RNA. Activating proteins are also recruited to the rRNA genes; SIRT7 and William syndrome transcription factor (WSTF) associate with 47/45S rRNAs. In many cases, however, the RNAs act in collaboration with other recruiting factors and other proteins, such as TTF-I (see text for more details).

when TIP5 is deacetylated by SIRT1 [98]. Computer predictions have now identified pRNA in several species, including human cells [99], although the spacer promoter for the human transcript has not been identified. The NoRC complex loaded with pRNA in turn recruits the histone deacetylase HDAC1, the histone methyltransferase EZH2 [100], and DNA methyltransferases DNMT1 and DMNT3B [88,101,102], resulting in DNA methylation of the rRNA promoter and the deposition of repressive histones marks. The NoRC complex and pRNA are also involved in the stable propagation of heterochromatin at silent rRNA copies during S phase, because the production of pRNA is coordinated with the late replication timing of silent rRNA copies to promote heterochromatin formation in *trans* on the newly replicated silent genes [71,103]. In addition, NoRC and pRNA interacts with poly(ADP-ribose)-polymerase I (PARP1) at silent rRNA genes after replication, resulting in ribosylated histones, which also promotes the maintenance of heterochromatin [104].

Another example of noncoding RNAs (ncRNAs) involved in the transcriptional repression of rRNA genes is represented by the antisense RNA promoter and pre-rRNA antisense (*PAPAS*), which is transcribed by RNA pol II in antisense orientation to the 47/45S RNA [105]. *PAPAS* transcription has been observed in growth arrested and contact inhibited cells, where it silences gene expression by recruiting Suv4-20h2, the enzyme responsible for the deposition of the repressive H4K20me3 modification. The transcription of *PAPAS* thus correlates with high levels of H4K20me3 modifications present at the rRNA promoters and with low RNA pol I transcription rates [105,106]. Recently, PAPAS transcription has also been observed upon hypotonic stress and heat shock, where it contributes to the stable repression of rRNA genes via an H4K20me3-independent mechanism [107,108] (Fig. 10.3). Hypotonic stress thus induces the ubiquitinylation of Suv4-20h2 by NEDD4, which is followed by its degradation, enabling *PAPAS* to bind to and recruit the NuRD complex instead. Heat shock leads to the inhibition of RRN3/TIF-1A phosphorylation and to the subsequent downregulation of 45S rRNA transcription, which is paralleled by an increase in *PAPAS* transcription. Similarly to hypotonic stress, heat shock also induces the degradation of Suv4-20h2, allowing *PAPAS* to bind to the NuRD complex. It is important to note, however, that *PAPAS* is not upregulated during all kinds of stress responses and is not formed, for example, in response to transcriptional stress induced by actinomycin D [107,108].

Recently, it was shown that genes loaded with UBF can also exist in a more condensed state, as trimethylation of UBF by the methyltransferase ESET leads to the condensation of chromatin [109]. This suggests that transition from the transcriptionally active state into poised- or pseudosilent poised states may occur via several different mechanisms, which can involve the activity of the NuRD complex, the recruitment of H1 and HP1α, or the modifications of UBF (Fig. 10.2), which all induce a poised chromatin state that can be easily activated on external stimuli. It remains to be shown, however, whether genes silenced by NoRC and repressed by a combination of histone modifications and DNA methylation could also be activated on external stimuli at times when more ribosomes are required. Stable heterochromatin states can be altered in diseases, such as cancer, which is associated with the dysregulation of ribosomal transcription. Deregulated heterochromatin formation in cancer was thus proposed to cause the reactivation of silent rRNA genes [61,110,111]. Moreover, as heterochromatin plays important roles in the maintenance of genome stability at repetitive sequences, deregulation of heterochromatin formation in cancer at rRNA genes predisposes to DNA damage [106]. As described earlier, NoRC is involved in the maintenance of heterochromatin states at the silent rDNA loci. Furthermore, NoRC binds to and contributes to the silencing of also other repetitive sequences in the genome, such as telomeres, pericentromers, and centromeres [106,110]. The NoRC complex thus has a more general role in the maintenance of heterochromatin and genomic integrity by promoting proper chromosome segregation and counteracting recombination events at repetitive loci that might result in the loss of genetic material.

The silencing of rRNA genes in yeast is tightly linked to genomic stability. Silencing thus prevents loss of genes in extrachromosomal rDNA circles that emerge as a result of recombination between repetitive sequences. Most of the 200 rRNA genes in yeast are silenced by chromatin compaction, which is induced by the Sir2-containing RENT complex [112]. Sir2 is a NAD^+-dependent histone deacetylase that is important for the spreading of silent chromatin states, characterized by deacetylated histones. The Sir2 protein is targeted to the rDNA locus together with the NET1 protein, and binds to two sites in the intergenic spacer between rRNA genes [113]. Interestingly, sumoylation of the Sir2 protein regulates the distribution of Sir2 between different silenced regions. Hence, the sumoylated form of Sir2 does not bind to telomers, whereas its binding to NET1 is not affected [114]. Similarly to the silencing of other repetitive regions, Sir2-mediated silencing of the rDNA locus has been coupled to the aging process in yeast, and is considered to be important for the protection of gene copy number to prevent premature aging [115,116].

It is still not clear why mammalian cells have evolved to carry many copies of the RNA pol I–regulated genes. On one hand it may enable the cells to meet the need to synthesize high levels of proteins. This idea is supported by the reduced number of active rRNA genes present in differentiated cells that require less protein synthesis than proliferating cells. *S. cerevisiae* also has many rRNA gene copies, although it is unclear if every gene is ever used in every cell. According to a second theory, the high copy number of rRNA genes might be linked with the unique properties of the ribosomal loci, that is, that they are both repetitive and highly transcribed. The high level of transcriptional activity at the repetitive sequences makes the rDNA loci thus particularly vulnerable for DNA damage. Importantly, it was recently shown that active and inactive genes coexist at each rDNA locus [117], suggesting that alternation of silent- and active genes would reduce the susceptibility of the whole locus for DNA damage.

10.4.3 Silencing Complexes Functioning in Response to External Stimuli: eNoSC and RUNX2

External stimuli activate a number of signaling pathways that regulate ribosomal biogenesis by phosphorylating and acetylating the transcription machinery, and also by activating specific proteins or complexes that change the chromatin structure. Several pathways respond to changes in the level of nutrients, such as the mTOR pathway, and the PI3-kinase- and AMP-activated protein kinase (AMPK) pathways. AMPK responds to glucose deprivation and directly signals to the transcription machinery by phosphorylating RRN3/TIF-IA, thereby preventing RRN3/TIF-IA binding to SL1/TIF-1B [118]. A second pathway responding to glucose deprivation involves the NAD^+-dependent deacetylase, SIRT1, which senses the reduction in the energy levels and reduces chromatin accessibility [119]. SIRT1 is a component of the eNoSC complex, which also comprises nucleomethylin (NML) and histone methyl transferase Suv3-9H1. The eNoSC complex senses low ATP/ADP ratio and suppresses rRNA transcription. NML is thus bound to rRNA at high ATP levels, when rRNA transcription is active, which inhibits its interaction with SIRT1 [120] (Fig. 10.3). Reduction in rRNA levels at low energy conditions enables NML to recruit SIRT1 to the eNoSC complex, which leads to the formation of repressed chromatin at the rRNA promoters. The eNoSC complex thus deacetylates histone H3 and methylates histone H3 at lysine 9 (H3K9me2), converting an active chromatin state into heterochromatin [119]. The fact that SIRT1 is not recruited to the eNoSC complex until the rRNA level has been significantly reduced implies that this complex acts secondarily to the downregulation of rRNA transcription. The action of eNoSC is reversible, as NML is released from SIRT1 binding by nascent rRNA transcripts when glucose is reintroduced. In addition, 5S rRNA also binds to NML [120], suggesting that this mechanism could contribute to the coordination of transcription at the two types of rRNA genes. Further proteins that have been identified binding to NML include nicotinamide mononucleotide adenyltransferase 1 (NMNAT1), which promotes SIRT1 activity by increasing local NAD+ levels [121]. In addition to its role in histone deacetylation, SIRT1 is also involved in the repression of rRNA transcription during mitosis by deacetylating the TAF168 subunit of the pol I–specific transcription factor, SL1/TIF-1B [122]. This deacetylation together with cdk1/cyclin B–dependent phosphorylation of TAF1110 within SL1/TIF-1B impairs the binding of SL1/TIF-1B to the rRNA promoters and thereby inhibits pol I transcription.

The number of active rRNA genes is reduced during differentiation, a process that is likely regulated by multiple factors in addition to the NuRD- and the NoRC complexes. For example, silencing of rRNA genes during differentiation of osteoblasts requires the osteogenic transcription factor RUNX2 [123], which binds to the promoter of RNA pol I genes and recruits HDAC1. HDAC1 deacetylates UBF and histone H3 and H4, which results in repression of rRNA transcription [124]. Although the NuRD complex is required for the silencing of rRNA gene copies during the differentiation of C2C12 cells [75], it is not known if other cell type- specific factors are also involved. The differentiation process seems to affect the chromatin states at the rRNA genes in a complex manner that is either dependent or independent of the NuRD- and NoRC complexes. Differentiation induced downregulation of rRNA transcription has thus been linked with increased DNA methylation at the rRNA genes, which is regulated by the NoRC complex and is considered to result in stable heterochromatinization. Silencing by the NuRD complex would, on the other hand, represent a dynamic silent state that could be altered and switched to an active chromatin state, as demonstrated when NIH3T3 cells were stimulated with serum after starvation [75]. DNA methylation of the rRNA genes is prevented by the NuRD component CHD4, which inhibits the expression of TIP5, a member of the

NoRC complex that recruits DNA methyltransferases to rRNA genes [125]. Different RNA species may also contribute to the initiation of the silencing, such as the noncoding RNA *PAPAS* that recruits SUV4-20h2 to rRNA genes [105].

10.4.4 Chromatin and Actively Transcribed rRNA Genes: Histones and UBF

The actively transcribed rRNA genes are maintained in an open chromatin structure by UBF (or Hmo1 in yeast) (reviewed in Refs. [126,127]); whether histones are present is not fully elucidated. Several studies have indicated that in yeast the active rRNA genes are devoid of nucleosomes. Miller spreads performed on *S. cerevisiae* moreover suggest that RNA pol I is present only on the transcribed genes [128,129]. Furthermore, experiments using Mnase fused to histones have shown that no histones are present on the active rRNA genes in yeast [24,130]. This suggests that the chromatin state of the active genes in yeast is in a very open configuration, free of nucleosomes. Instead proteins, such as Hmo1 in yeast (and UBF in mammals), loop the rRNA genes, which stabilizes the nucleosome-free state [131]. However, histones have been suggested to be involved in active RNA pol I transcription also in yeast cells, in particular at the initiation step, at the promoter [132,133]. As the rRNA genes are highly transcribed and environmental stimuli that activate ribosomal transcription work directly on the RNA pol I transcription machinery, the requirement for the regulation of chromatin may not be as important as in RNA pol II transcription. Nevertheless, the intergenic spacers between active genes are organized with nucleosomes, so even if elongation occurs in a nucleosome-free surrounding, initiation of transcription occurs in a chromatin environment.

In mammalian-, *Xenopus-*, and *Drosophila* cells, Chromatin immunoprecipitation (ChIP) experiments documented that histones with different modifications are present at active RNA pol I genes, and are particularly enriched in H4–Ac and H3–Ac modifications. This observation suggests that, in contrast to yeast rRNA, nucleosomes or at least assemblies of histones are present on the rRNA genes (for examples refer to Refs. [83,98,134–137]. Modified histones have been found both at the promoters and in the transcribed regions, but the histone modifications at the promoters have gained most attention. Histone H4–Ac in the promoter is thus implicated in the activation of transcription [1,81,138,139]. The modifications associated with rRNA gene expression can be deposited by factors that bind the pol I transcription machinery. An example is represented by PHD finger protein 8 (PHF8), which activates rRNA transcription by demethylating H3K9me1/2 and by recruiting H3K4-methyltransferase activity [134]. Histone-modifying activities may also be recruited by regulatory transcription factors, such as c-MYC, Rb, and p53, which are known to bind to the promoter of RNA pol I genes [5,33,140,141], apart from regulating RNA pol II- and RNA pol III genes. In RNA pol II transcription, these factors influence the local chromatin environment at promoters by recruiting histone-modifying enzymes, such asHATs. It is thus possible that they function in a similar manner also during RNA pol I transcription. Histone-modifying enzymes—such as the HATs p300, GCN5, PCAF, and MOF—have indeed been found to bind to the active rRNA genes [79,98,135]. The c-MYC protein, for example, recruits the TRRAP complex with the HATs GCN5 and TIP60 during RNA pol II transcription [142]. It binds to the promoter and gene body of the RNA pol I loci [143,144], and its binding correlates with the level of histone acetylation at these sites. In addition to its function in regulating histone modifications, c-MYC may affect rRNA expression also by other mechanisms. For example, c-MYC is involved in loop formation between upstream- and downstream rDNA sequences at active rRNA genes to enhance transcription [86,145–147]. Finally, a novel rRNA specific histone modification, glutamine methylation of histone H2A, has recently been implicated in RNA pol I elongation in yeast [148]. The methylase responsible for the modification has turned out to be the nucleolar protein, Nop1, in yeast and its ortholog fibrillarin in mammalian cells and in plants [148,149]. Fibrillarin is an abundant nucleolar RNA methyltransferase, which is also involved in the processing of rRNA as part of snoRNP [150] in many species [151]. Although the role of glutamine methylation of H2A in mammalian cells has not been elucidated, it has been reported to inhibit the binding of the chromatin remodeling factor FACT to rRNA genes in yeast.

10.4.5 Activating Chromatin Remodeling Complexes: FACT, CSB, and B-WICH

Chromatin remodeling complexes, which change the chromatin structure by using ATP hydrolysis, are also recruited to the RNA pol I genes (for review refer to Ref. [152]). The FACT complex, which is involved in moving histones during the transcriptional elongation phase of RNA pol II transcription [153], has been shown to bind to the rRNA genes in mammalian cells and to increase the transcription of rRNA genes [154]. End-targeting proteomics of isolated chromatin fragments (ePICh) of rRNA promoters also identified the FACT subunits SSRP1 and SPT16 bound to rRNA promoters [155]. Furthermore, FACT subunits bind nucleolar histone H1 directly [156,157]. The binding studies did not distinguish between active- and inactive rRNA genes, but histone H1 is most likely associated with transcriptionally inactive gene copies. The fact that FACT facilitates transcriptional elongation in mammalian cells [154] suggests that it is present either on active rRNA copies or it is associated with poised rRNA genes. H1 is proposed to serve as a hub that interacts with factors bound to either active rRNA genes (UBF and nucleolin) or to inactive rRNA genes (HDACs and HP1) [157]. On the other hand, subunits of the NoRC complex, such as TIP5/BAZ2A or SNF2h, were not found to be associated with histone H1. Nor were they identified as proteins on the promoter fragment, which might be due to technical limitations of the methods and their inability to capture transient interactions [155,157].

Another chromatin remodeling factor found at the RNA pol I genes is the Cockayne syndrome protein B (CSB) factor, involved in DNA repair and in transcriptional regulation [158]. CSB is recruited to the rRNA promoters by TTF-I, where it activates transcription (Fig. 10.2) [159,160]. The interaction between TTF-I and CSB positions CSB at the promoter and allows it to interact with the NuRD complex and the promoter nucleosome [75]. CSB possesses ATPase activity and it has been suggested to be the chromatin remodeling that factor releases the poised state by moving the promoter nucleosome into the active position [72,73,100,160]. This is supported by the findings that while most of the CSB proteins associate with the active nucleosome at the promoter, a small fraction of it binds to the poised "repressed" nucleosomes [75]. CSB is then responsible for transcriptional activation by recruiting the histone methyltransferase G9a to gene bodies, and the HAT PCAF to the promoters [160,161]. While the recruitment of PCAF leads to histone acetylation at the promoters and to the facilitation of transcription initiation [161], binding of G9a to rRNA gene bodies, and the consequent methylation of H3K9me2/3, promotes the elongation process [75]. An interesting regulator of transcriptional activity of rRNA genes is TTF-I, which is able to exert context-dependent and even opposing effects on RNA transcription. On one hand, prevention of TTF-I binding to the T_0 position by nucleolin has been shown to maintain gene activity by inhibiting the TTF-I-mediated recruitment of NoRC [58]. On the other hand, the binding of TTF-I at the promoter and at the end of the rRNA genes is required for the activating loop to form. Furthermore, how TTF-I balances between activation of transcription by recruiting CSB or reducing transcription by recruiting NoRC is not known. Several mechanisms, not yet elucidated, appear to be operating, possibly by posttranslational modifications of TTF-1 in response to external cues.

CSB has been isolated as a loosely associated partner of the B-WICH complex, an additional chromatin remodeling complex regulating RNA pol I transcription. B-WICH is an extended version of the WICH complex [162] and comprises William syndrome transcription factor (WSTF/BAZ1B), SNF2h, and nuclear myosin 1 (NM1) in addition to a number of RNA-binding proteins and CSB [163,164]. WSTF associates with the promoter and the transcribed regions of rRNA genes. It is required for the binding of the histone acetyltransferases p300, PCAF, and GCN5, and for the subsequent acetylation of histone H3 at the promoter, which activates rRNA transcription [135]. However, no or very weak interactions between these HAT enzymes and the core subunits of B-WICH could be detected, with the exception for the association of NM1 with PCAF [165]. As NM1 binds both to SNF2h and the pol I machinery via its association with the transcription initiation factor TIF-1A, it emerges as an important structural component at the rRNA promoters, which connects chromatin remodeling to

RNA pol I transcription [166–168]. In addition, NM1 binds to actin, which may regulate the chromatin remodeling activity of B-WICH, as the binding of actin and SNF2h to NM1 are mutually exclusive [165]. Interestingly, β-actin$^{-/-}$ mouse embryonic fibroblasts have less TTF-1 loading at the rRNA promoters and form less accessible chromatin structure at the T_0 site. Although WSTF and SNF2h are still associated with the rRNA promoters in these cells, the association of NM1 is reduced. This suggests that the association between B-WICH and actin dictates whether chromatin remodeling takes place at certain sites [169]. In human cells, B-WICH is responsible for rendering a 200-bp region at the rRNA promoter more accessible for transcription [135]. This region corresponds to the promoter nucleosome and it is possible therefore that it is the B-WICH complex, instead of CSB, that may remodel the nucleosome to an active position. This is supported by the findings that siRNA-mediated silencing of WSTF blocks binding of RRN3/TIF-1A and RNA pol I to rRNA promoters (Rolika et al., unpublished results). It is not known, however, how B-WICH is targeted to the ribosomal genes, although its interaction with CSB could link it to the TTF-I-regulated network. Surprisingly, WSTF or SNF2h are not found among the proteins bound to the rRNA promoters by ePICh, similarly to TIP5 and CSB [155]. This may indicate that these chromatin remodeling complexes interact very transiently with the rRNA promoters or associate at different positions than examined in these experiments. Nor are WSTF or SNF2h associated directly with histone H1 [157]. The amount of WSTF, however, is higher in the nucleolus of histone H1–knock out cells than in those cells with normal H1 levels. It is possible therefore that WSTF occupies the place of H1 as the B-WICH complex remodels chromatin to maintain an active state. The targeting of B-WICH is likely regulated by transcription factors, such as c-MYC, as both c-MYC and WSTF are induced and bind to the rRNA promoters in hypertrophic muscle [170]. In growing cells, WSTF stays at the promoter of rRNA genes during mitosis, suggesting that it is required early in the activation of rRNA genes on exit from mitosis [165]. Importantly, WSTF and SNF2h have been shown to interact with the deacetylase SIRT7 [171], which activates ribosomal transcription [172] by deacetylating the RNA pol I subunit, PAF53 [173], and also stays at the rDNA during mitosis [174]. The acetylation acts as a switch, in which PAF53 is acetylated by CBP to inhibit transcription and deacetylated by SIRT7 to activate transcription by increasing the association of RNA pol I with the rRNA genes. SIRT7 binds to UBF and RNA pol I in an RNA-dependent manner, and is recruited to the rRNA genes by the 47/45S rRNA during glucose refeeding (Fig. 10.3) [173]. WSTF also binds the 47/45S rRNA [164], but WSTF remains associated to the promoter also when transcription is inhibited by actinomycin D [165], suggesting that the association with the promoter is mediated by protein–protein interactions. In summary, the main mechanism by which B-WICH activates transcription is by remodeling the chromatin of rRNA promoters and facilitating subsequent histone acetylation to enable binding of factors required for pol I transcription. Furthermore, its action may also involve deacetylation of RNA pol I through its interaction with SIRT7.

10.4.6 Other Chromatin Remodeling Complexes in rRNA Transcription

Other chromatin remodeling complexes have also been associated with RNA pol I transcription. The PIH1 protein, for example, interferes with TIP5 binding at the RNA pol I promoter and recruits BRG1-containing SWI/SNF complex to activate transcription on glucose stimulation [175]. PIH1 thus remodels the promoter nucleosome with the help of SWI/SNF to activate rRNA transcription. SWI/SNF has also been associated with activating RNA pol I transcription in yeast [176].

10.5 NONCODING RNAS IN THE STRUCTURAL AND FUNCTIONAL INTEGRITY OF THE NUCLEOLUS

ncRNAs play important roles in the structure and the many functions of the nucleolus. The stability of the nucleolus is promoted by the DJ-ncRNA transcribed from the distal flanking region of the rDNA and was proposed to anchor rDNAs to perinucleolar heterochromatin [27]. Moreover, ncRNAs transcribed from intronic Alu repeats by pol II were also shown to be required for the integrity of the nucleolus and for efficient pol I–mediated transcription [28]. Apart from having a structural function, ncRNAs can affect rRNA transcription by recruiting proteins that modify chromatin. As mentioned earlier, the pRNA [85,96] is complementary to rDNA promoters and affects chromatin states to silence rRNA genes in *trans* by binding to the NoRC complex [103]. Furthermore, the antisense transcript *PAPAS* recruits Suv4-20h2 to the rRNA genes (Fig. 10.3), which is responsible for the deposition of the silent epigenetic mark H4K20me3 [105]. The sense 47/45S rRNA transcripts recruit the activating deacetylase SIRT7 [173] and inactivate the silencing complex eNoSC [120]. Recently it was found that Argonaut also is recruited to the rRNA genes by micro-RNA targeting the sense transcript, and this attenuates transcription [177]. Moreover, ncRNAs have also been shown to play important roles in the maintenance of genome integrity and rDNA copy number. In *S. cerevisiae*, two divergent pol II transcripts (IGS1-F and IGS1-R) are thus produced from the IGS [178–182]. One of them (IGS1-R) is regulated by degradation by the exosome and is proposed to affect chromatin structure and/or DNA repair at a replication fork barrier [178,179,181,183]. IGS1-F, on the other hand, has been shown to increase the probability of unequal recombination. A model was proposed in which the effects of IGS1-F are mediated by the transcription-induced displacement of cohesin at replication forks, reducing sister chromatid cohesion at the rDNA IGS1, and thereby promoting unequal recombination [178]. The ncRNAs might also be required to enable recombination to take place during meiosis, and should be tightly regulated in other cell cycle phases [184]. Dysregulated transcription thus leads to detrimental recombination events, resulting in loss of copy number and formation of extrachromosomal ribosomal circles [181,184]. The expression of these ncRNAs is affected by the local chromatin structure and is downregulated by histone deacetylation and chromatin remodeling by ISW1 and ISW2 [185,186]. The architectural protein, NHP6, and deacetylases thus inhibit RNA pol II ncRNA transcription and consequent recombination, whereas acetylation of H3K16 increases ncRNA expression and recombination [187,188].

Ribosomal transcription and the integrity of the nucleolus are sensitive not only to environmental changes, but also to stress signals that arrest transcription and disintegrate the nucleolus [30,189]. A group of RNA pol I transcripts originating from various positions in the IGS [190] has been implicated in various types of stress responses, such as heat shock and acidosis, which induce stress responses to maintain cellular homeostasis [191]. The ncRNAs from the IGS have been suggested to be part of a molecular network to immobilize certain proteins to avoid acute cellular collapse [192,193]. For example, the IGS16 and IGS22 are induced by heat shock and bind HSP70; IGS28 is induced by acidosis and binds von Hippel Lindau (VHL) protein; the DNA methyltransferase DNMT1, and the DNA polymerase subunit POLD1. The sequestering and immobilization of stress proteins by ncRNAs also induce reversible morphological changes to the nucleolus [193,194]. Interestingly, transcriptional stress induced by a block in rRNA transcription can lead to the accumulation of proteins, such as the MDM2, at certain regions within the IGS, even though no ncRNA has been isolated from these regions [190]. Furthermore, a genome-wide study uncovered that boundaries of rDNA-similar sequences located on NOR(−) human chromosomes tend to be marked by the location of stress stimuli–inducible loci of the IGS. These sequences are located sometimes near Alu sequences and in the vicinity of protein-coding genes, raising the question whether ncRNAs from the IGS might also regulate protein-coding genes [195]. In addition to ncRNAs transcribed from the rDNA loci, the nucleolus contains many RNA species, such as snoRNAs, which originate outside of the nucleolus. SnoRNAs are part of snoRNPs U3, U8, and U12, which further process the rRNAs. Recent sequencing of small-sized RNAs in the nucleolus showed that snoRNA-derived small RNAs (sdRNA) accumulate in the nucleolus and are bound to fibrillarin [196]. The sdRNAs can be found in different cellular compartments and have new regulatory roles in translation and gene expression [197]. In addition to the RNAs, about 1400 proteins enter the nucleolus [198], which can take on yet unknown functions.

10.6 RNA POL III TRANSCRIPTION

The forth RNA in the eukaryotic ribosomes is the 5S rRNA, produced from a 120-bp gene transcribed by RNA pol III. RNA pol III genes, such as the 5S rRNA, tRNAs, and snU6, are short, ranging from 70 to 300 bp. This poses a challenge with respect to the analysis of chromatin states at these loci due to the resolution of the commonly used assays, such as ChIP. The 5S rRNA genes in yeast are located in between the RNA pol I genes in the nucleolus, whereas those in mammalian cells are found in clusters on different chromosomes, which can be preferentially localized in the interior of the nucleus or in the perinucleolar space (reviewed in Refs. [4,6]). The tRNA genes are scattered throughout the genome in both yeast and human cells, although the number of genes is higher in the human genome, which also harbors tRNA-like pseudogenes. Predictions of tRNAs in the human genome have uncovered about 500 RNA pol III genes, only half of which are occupied by RNA pol III, as determined by ChIP-sequencing experiments [199].

RNA pol III genes display a variety of promoter architectures. The 5S rRNA genes have thus internal promoters that require the binding of several auxiliary factors, such as TFIIIA, TFIIIB (a multiprotein initiation factor), and TFIIIC, which recognize specific DNA sequences. TFIIIA binds to the internal element, whereas TFIIIC is recruited to the so-called A- and B-box *cis* elements, and TFIIIB is recruited to the genes by TFIIIC (for review see Refs. [4,6]). The tRNA genes lack the internal element and only rely on TFIIIB and TFIIIC for gene expression. The RNA pol III genes have a precise termination site, which is determined by the RNA pol III and *cis*-acting sequences on both strands [200]. As these genes are small, they are regarded as nucleosome-free [201–209], with both nucleosome-positioning sequences and the RNA pol III machinery acting as factors responsible for the opening and the maintenance of this state. Furthermore, TFIIIC, which has two subunits with HAT activity [210,211], is required for chromatin alterations in the vicinity of pol III genes.

The transcription of the 5SrRNA genes is regulated in response to the same external cues and by similar key transcription factors, such as c-MYC, RB, and p53, as the transcription of RNA pol I genes [6,212]. Similarly to RNA pol I genes, these transcription factors bind to the RNA pol III machinery, in particular to BRF1 or BRF2 present in TFIIIB. BRF1 is required for the transcription of 5S rRNA and tRNA genes, whereas BRF2 is necessary for the transcription of snU6- types of RNA pol III genes [5,213]. BRF1 and BRF2 are often dysregulated in different cancers and promote cell growth [5,213]. Signaling pathways, such as the MAP kinase and mTOR pathways, also target TFIIIB [214–216]. The mTOR pathway induces the binding of the negative regulator MAF1 to TFIIIC and RNA pol III, which inhibits RNA pol III transcription [217–220]. MAF1 thus inhibits energy expenditure by regulating 5S rRNA- and tRNA transcription. In line with this function, MAF$^{-/-}$ mice are resistant to obesity [221]. The levels of the ribosomal proteins RPL5 and RPL11 also regulate the amount of 5S rRNA produced, both by binding to the to the 5S rRNA genes and by regulating c-MYC [222,223]. Thus, external stimuli regulate transcription by either promoting the assembly of the RNA pol III machinery on its target genes or by inhibiting RNA pol III. Finally, although the RNA pol III genes are devoid of nucleosomes, the surrounding nucleosomes are acetylated [201,206,208,224]. It has been suggested that these histone modifications make these genes accessible for the recruitment of TFIIIB and subsequently for RNA pol III [206].

Several genome-wide studies have shown that histone modifications around RNA pol III genes resemble those found at RNA pol II genes [205–207]. Methylation of H3K4 (H3K4me1–3), but not the silencing modification H3K27me3, is associated with active RNA pol III genes [205]. Whether histone modifications surrounding the RNA pol III genes are the result of RNA pol II transcription, and whether the RNA pol III machinery is also responsible for the recruitment of histone-modifying enzymes are still open questions. Although not all, many of the active RNA pol III genes are situated in the vicinity of RNA pol II genes. The genes located in isolation are also surrounded by active histone modifications, suggesting that these marks are set by transcription factors linked with RNA pol III transcription. In addition to the c-MYC sites (E boxes), AP-1 sites binding to c-JUN and c-FOS have also been found at RNA pol III genes [208]. Moreover, these RNA pol II–regulatory factors are also found to bind in the vicinity of RNA pol III genes [208,216]. The regulation of pol III genes by c-MYC involves the recruitment of the TRRAP complex and the subsequent acetylation of histone H3. However, the TRRAP complex involved in RNA pol III transcription only harbors GCN5, and not TIP60, that acts only in RNA pol II transcription [225]. PCAF is also absent from the 5S rRNA- and 7SL genes [226]. Nevertheless, histone modifications around the RNA pol III genes most likely affect transcription by rendering the chromatin region more or less accessible to regulatory factors, similarly to RNA pol II transcription. Chromatin remodeling complexes that play roles in RNA pol II transcription have also been found to associate with RNA pol III genes, further emphasizing the similarity in these processes.

10.7 CHROMATIN REMODELING COMPLEXES AFFECTING RNA POL III TRANSCRIPTION

In yeast, the chromatin remodeling complex RSC has been shown to alter the chromatin structure at RNA pol III genes [227,228]. The ISWI complexes ISW1 and ISW2 as well as FACT have also been associated with RNA pol III transcription in yeast [229,230]. While ISW1 and ISW2 are required to remodel or evict the upstream flanking nucleosome of tRNA genes, the RSC is recruited to remodel the downstream nucleosome upon stress signals [230]. Furthermore, the FACT complex exchanges H2A to H2Az to facilitate the transcription of tRNA genes and to activate poised RNA pols in response to stress [229]. The chromatin remodeling factors ISW1 and 2, RSC, and FACT were found to be associated with RNA pol III on chromatin using chromatin affinity purifications, suggesting that these factors indeed are present at the RNA pol III genes in exponentially growing yeast cells [35].

Whether or not chromatin remodeling complexes operate on RNA pol III genes also in mammalian cells is less established. Although FACT proteins associate with RNA pol III genes and affect their transcription [154], their mechanism of action is not clear. As FACT subunits bind to all types of RNA pol III genes independently of gene architecture, FACT may function as a histone exchanger also in RNA pol III transcription, similarly to its proposed role in RNA pol II transcription. The B-WICH complex, which is also associated with RNA pol I, is required for the transcription of 5S rRNA- and 7SL RNA genes (forms RNA in the signal recognition particle) [164]. The complex remodels the chromatin at specific sites outside of the pol III genes, including a region harboring a c-MYC site between the genes in the human 5S rRNA cluster [226]. Chromatin remodeling facilitates the binding of the c-MYC/MAX/MXD complex, RNA pol III and its cofactors to these sites, and promotes 5S- and 7SL rRNA transcription [226]. Association of RNA pol II to RNA pol III genes is also common, usually binding at 300 bp upstream of the genes [206]. Whether RNA pol III genes require RNA pol II transcription is not fully elucidated. Although the snU6 genes require RNA pol II transcription [231], this has not been observed in the case of other genes. Instead, RNA pol II may contribute to the patterns of histone modifications, which might indirectly affect RNA pol III transcription.

10.8   RNA POL III AND TRANSCRIPTION OF SHORT INTERSPERSED REPEAT ELEMENTS

RNA pol III DNA sites have been associated also with other functions than transcription of classical RNA pol III genes. RNA pol III thus also transcribes some viral genes and short interspersed repeat elements (SINEs) and SINE-derived transcription units (for review refer to Refs. [232,233]). These SINEs are nonautonomous retrotransposons, the 5′-terminal heads of which are derived from the RNA pol III genes tRNA, 5S or the 7SL genes, and contain A- and B-boxes in their promoters (reviewed in Refs. [234,235]). The mouse genome contains two main families of SINEs, namely the B1 SINEs, derived from tRNA, and B2 SINEs, derived from 7SL genes, which together occupy 5% of the mouse genome. In primates, 10% of the genome is composed of Alu sequences, which are derived from the 7SL genes. Alu sequences have been proposed to have several functions in the human genome, such as promoting recombination and influencing 3D nuclear architecture by changing the 3D chromatin structure and by mediating chromatin fiber interactions [233,234,236,237]. Genome-wide studies of the binding of the RNA pol III machinery in human cells showed that these factors occupy several sites close to the human Alu sequences [206,207,209], suggesting that some sites are transcribed by RNA pol III. The transcription of RNA pol III–Alu sequences is activated upon stress and viral infections [238–240], and may induce a cytokine response by the inflammasomes [241]. The inflammasome is comprised of intracellular pattern recognition receptors and caspases, and these activate the proinflammatory cytokines IL-1β and IL-18 for secretion [242]. In addition, Alu RNAs influence gene expression, both at the transcriptional- and the translational levels [243–245]. A large proportion of the Alu sequences is found in introns and is transcribed by RNA pol II (reviewed in Refs. [232,246]). These Alu RNAs also contribute to the regulation of nuclear functions. Many of the intronic Alu sequences are inverted repeats, which are often A-to-I edited and, as a consequence, retained in the nucleus [247]. Although the majority of the Alu sequences is inserted into introns, they may sometimes be spliced into the mRNAs and function as exons [237]. This may lead to nonfunctioning mRNAs with premature stop codons. The RNA pol III–Alu sequences are expressed at a lower level than the RNA pol II–transcribed Alu sequences, but are proposed to be responsible for the retrotransposition that still occurs in the human genome. RNA pol II–Alu RNAs, on the other hand, are proposed to affect gene transcription and translation [234,237].

10.9 RNA POL III GENES, BOUNDARY ELEMENTS, AND INSULATOR FUNCTION

RNA pol III genes and promoters have been linked with the domain organization of chromatin structure and the proper maintenance of gene expression domains. RNA pol III promoters and tRNA genes have thus been found at the boundaries between active regions and silent, heterochromatic regions in *S. cerevisiae*, and they were shown to be important in separating these domains [248–251]. These sites have now been associated with boundary-element activity, because they prevent the spreading of heterochromatin into actively transcribed regions [252–254]. Efficient boundary elements harbor at least the B-box *cis* element of the promoter and resemble tRNA genes. Interestingly, tRNA genes can also separate heterochromatin domains from active domains in *Schizosaccharomyces pombe* [255,256]. In yeast, flanking regions around the tRNA genes seem to distinguish the function of the regulatory element, and determine whether a tRNA gene is transcribed or has boundary function [257,258]. Genome-wide analyses in *S. cerevisiae* [259,260] and *S. pombe* [261] have shown that many RNA pol III sites bind only part of the RNA pol III machinery, namely the TFIIIC. These extratranscriptional TFIIICs sites (ETCs) only have a B-box and act as boundary elements only if multiples copies of these sites are present [261,262]. Early studies have shown that TFIIIB can also be associated to boundary elements [257], although most of the sites bind only TFCIIIC. Furthermore, the ETCs are essentially nucleosome free [203,258,263–266], suggesting that these sites change the underlying nucleosome pattern. The binding of the TFIIIC at these sites is preceded by histone acetylation and chromatin remodeling by RSC and Isw2 (reviewed in Ref. [252]). Furthermore, the chromatin remodeling protein Fun30 (Fft3) localizes to tRNA genes and maintains the heterochromatin mark H3K9me2, thereby limiting the spreading of active marks in *S. pombe* [267,268].

In mammalian- and *Drosophila* cells, genome-wide analyses have identified TFCIIIC-only sites, ETCs, and similarly to yeast, these have been associated with boundary function [206,208,224,269,270]. This has led to the proposal that tDNA/ETCs can serve as evolutionary-conserved boundary elements, working with chromatin architectural proteins to generate topological barriers between chromatin domains with different levels of transcriptional activity [270,271]. For example, ETCs are required to separate Polycomb domains, rich in H3K27me3, from active chromatin domains [224,254].

The ETC clusters in higher eukaryotes also block interactions between enhancers and promoters if placed in between the two [270], and thereby function as insulators. Insulator activity in metazoans are linked to chromatin architectural proteins, such as CTCF, cohesin, and condensin [124,272]. ETC sites often localize in the vicinity to CTCF sites in mammalian cells [206,273] and map close to CTCF and other insulator proteins in *Drosophila* [270]. Moreover, TFIIIC also interacts with cohesin in *S. cerevisiae* [248,274]. SINEs derived from RNA pol III genes can also work as insulators if present in many copies at their locations [233,235]. The barrier and the enhancer-blocking activities of tRNA/ETCs may be caused by a network of chromatin loops organized by factors binding to these elements, as has been suggested for other insulator proteins [270,271,275].

10.10 RIBOSOMAL TRANSCRIPTION AND DISEASES

Despite the fact that the ribosome production is required in all cells, dysregulation of ribosomal biogenesis has emerged as an important contributor to several diseases. Ribosomal transcription is often perturbed in cancer, caused by oncogenes, such as c-MYC, and the loss of tumor suppressor functions that control rRNA expression, such as Rb and p53, [5,276,277]. The RNA pol I- and RNA pol III transcription factors, such as BRF1 and BRF2, are also dysregulated in cancer, and contribute to the imbalance in ribosomal biogenesis. In recent years, ribosomal transcription and the nucleolus emerged as important targets for cancer drugs [26,278]. Dysregulated ribosomal biogenesis is also found in many developmental disorders. Ribosomopathies manifest themselves in defects in a few specific tissues, mostly in impaired function of hematopoietic cells or in the development of craniofacial-, limb-, heart- and urinotract abnormalities (reviewed in Refs. [276,279–283]). The underlying causes include mutations in proteins involved in ribosomal biogenesis and translation, which can affect components of the ribosomes as well as proteins involved in ribosome biogenesis (examples shown in Table 10.1). The defects have been attributed to ribosomal proteins or ribosome- associated proteins in Diamond–Blackfan anemia, Schwachman–Diamond syndrome, and 5q syndrome [282,284]. SnoRNP and ribosomal assembly factors, such as cirhin, are affected in Pader–Willi syndrome [279,285]. It has been suggested that the tissue specificity

Table 10.1 Syndromes That are Caused by an Impaired Ribosome Biogenesis

Syndrome	Mutated Gene	Proposed Function	Clinical Features
Treacher–Collin syndrome	Treacle (TFCO1 gene)	Ribosomal transcription	Craniofacial abnormities and anemia
Diamond–Blackfan anemia syndrome	RPS19, but also RPS24, RPS17, RPS26, RPL5, and RPL11	Ribosome biogenesis	Craniofacial abnormalities and macrocytic anemia
Schwachmann–Diamond syndrome	SBDS (7q11.21)	Ribosome biogenesis	Aplastic anemia and hematological abnormalities
5q syndrome	RPS14	Ribosome biogenesis	Macrocytic anemia
Dyskeratosis cognenita	DKC1 at chromosome Xq28	Telomere and rRNA processing	Short telomeres
Pader–Willi syndrome	SNORD116 (15q11.2–q13)	rRNA processing	Facial disorder and obesity
North American Indian childhood cirrhosis	hUTP4/Cirhin	Ribosome biogenesis	Cirrhosis and portal hypertension

For more details see [276] and [281].

observed in these patients is caused by the existence of tissue-specific biogenesis factors, and defects in any of those could result in the specific features present in ribosomopathies. The Treacle protein transcribed from the TCOF1 gene is associated with RNA pol I transcription and has been found mutated in many patients with Treacher–Collins syndrome (reviewed in Ref. [286]). As ePICT analysis shows that the Treacle protein binds to the rRNA promoters [155], it is likely involved in rRNA transcription. Moreover, it has been implicated in the downregulation of rRNA transcription in response to DNA damage [287]. Treacher–Collins syndrome patients exhibit craniofacial abnormalities, which is also seen in patients with other ribosomopathies. Haploinsufficient expression of the Treacle protein has been shown to lead to p53-mediated apoptosis in neural crest cells, which are the cells involved in craniofacial development [286]. Some individuals with Treacher–Collins syndrome have heterozygous mutations in POLR1C- and POL1C, subunits of RNA pol I and III, [288,289], which further substantiates that defects in ribosomal transcription is underlying disorders, such as Treacher–Collins syndrome. Craniofacial characteristic traits are also observed in William syndrome, which is caused by haploinsufficiency for a 1.5-Mb region on chromosome 7 comprising up to 20 genes [290]. WSTF/BAZ1B is one of the deleted genes, and its expression pattern during development suggests that it is involved in neural crest and craniofacial development [291–293]. As WSTF is a component of the B-WICH complex, important in rRNA transcription, William syndrome could be considered to be a ribosomopathy. WSTF is, however, involved also in processes outside of rRNA transcription, and it was recently shown that haploinsufficiency of WSTF/BAZ2A leads to dysregulation of neurodevelopmental pathways [294]. As the ribosome biogenesis is a fundamental process, the impact of the dysfunction of individual gene products is difficult to predict, in particular which tissues will be affected. Thus the list of rare diseases and syndromes that may be caused by impaired ribosome biogenesis is constantly growing.

References

[1] McStay B, Grummt I. The epigenetics of rRNA genes: from molecular to chromosome biology. Annu Rev Cell Dev Biol 2008;24:131–57.

[2] Drygin D, Rice WG, Grummt I. The RNA polymerase I transcription machinery: an emerging target for the treatment of cancer. Annu Rev Pharmacol Toxicol 2010;50:131–56.

[3] Schneider DA. RNA polymerase I activity is regulated at multiple steps in the transcription cycle: recent insights into factors that influence transcription elongation. Gene 2012;493(2):176–84.

[4] White RJ. Transcription by RNA polymerase III: more complex that we thought. Nature Rev 2011;12:459–63.

[5] White RJ. RNA polymerases I and III, non-coding RNAs and cancer. Trends Genet 2008;24(12):622–9.

[6] Moir RD, Willis IM. Regulation of pol III transcription by nutrient and stress signaling pathways. Biochim Biophys Acta 2012;1829(3–4):361–75.

[7] Boisvert FM, van Koningsbruggen S, Navascues J, Lamond AI. The multifunctional nucleolus. Nat Rev Mol Cell Biol 2007;8:574–85.

[8] Sirri V, Urcuqui-Inchima S, Roussel P, Hernandez-Verdun D. Nucleolus: the fascinating nuclear body. Histochem Cell Biol 2008;129(1):13–31.

[9] Pederson T. The nucleolus. Cold Spring Harb Perspect Biol 2010;3:1–5.

[10] Stults DM, Killen MW, Pierce HH, Pierce AJ. Genomic architecture and inheritance of human ribosomal RNA gene clusters. Genome Res 2008;18(1):13–8.

[11] Kobayashi T. Ribosomal RNA gene repeats, their stability and cellular senescence. Proc Jpn Acad Ser B Phys Biol Sci 2014;90(4):119–29.

[12] O'Sullivan AC, Sullivan GJ, McStay B. UBF binding in vivo is not restricted to regulatory sequences within the vertebrate ribosomal DNA repeat. Mol Cell Biol 2002;22:657–68.

[13] Esponda P, Giménez-Martín G. Ultrastructural morphology of the nucleolar organizing regions. J Ultrastruct Res 1972;39(5):509–19.

[14] Miller OL Jr, Beatty BR. Visualization of nucleolar genes. Science 1969;164:955–7.

[15] Conconi A, Widmer RM, Koller T, Sogo JM. Two different chromatin structures coexist in ribosomal RNA genes throughout the cell cycle. Cell 1989;57(5):753–61.

[16] Heliot L, Kaplan H, Lucas L, Klein C, Beorchia A, Doco-Fenzy M, Menager M, Thiry M, O'Donohue MF, Ploton D. Electron tomography of metaphase nucleolar organizer regions: evidence for a twistedloop organization. Mol Biol Cell 1997;8:2196–9.

[17] Prieto JL, McStay B. Psedo-NORs: a novel model for studying nucleoli. Biochim Biophys Acta 2008;1783(11):2116–23.

[18] Farley KI, Surovtseva Y, Merkel J, Baserga SJ. Determinants of mammalian nucleolar architecture. Chromosoma 2015;124(3):323–31.

[19] Mais C, Wright JE, Prieto J, Raggett SL, McStay B. UBF-binding site arrays form pseudo-NORs and sequester the RNA polymerase I transcription machinery. Genes Dev 2005;19:50–64.

[20] Grob A, Colleran C, McStay B. Construction of synthetic nucleoli in human cells reveals how a major functional nuclear domain is formed and propagated through cell division. Genes Dev 2014;28:220–30.

[21] Roussel P, Andre C, Masson C, Geraud G, Hernandez VD. Localization of the RNA polymerase I transcription factor hUBF during the cell cycle. J Cell Sci 1993;104:327–37.

[22] Hernandez-Verdun D. Assembly and disassambly of the nucleolus during the cell cycle. Nucleus 2011;2:189–94.

[23] Grob A, McStay. Construction of synthetic nucleoli and what it tells us about propagation of sub-nuclear domains through cell division. Cell Cycle 2014;13(16):2501–8.

[24] Merz K, Hondele M, Goetze H, Gmelch K, Stoeckl U, Griesenbeck J. Actively transcribed rRNA genes in *S. cerevisiae* are organised in a special chromatin associated with the high-mobility group protein Hmo1 and are largely devoid of histone molecules. Genes Dev 2008;22(9):1190–204.

[25] Weipoltshammer K, Schöfer C. Morphology of nuclear transcription. Histochem Cell Biol 2016;145:343–58.

[26] Quin JE, Devlin JR, Cameron D, Hannan KM, Pearson RB, Hannan RD. Targeting the nucleolus for cancer intervention. Biochim Biophys Acta 2014;1842(6):802–16.

[27] Floutsakou I, Agrawal S, Nguyen TT, Seoighe C, Ganley AR, McStay B. The shared genomic architecture of human nucleolar organizer regions. Genome Res 2013;23(12):2003–12.

[28] Caudron-Herger M, Pankert T, Seiler J, Németh A, Voit R, Grummt I, Rippe K. Alu element-containing RNAs maintain nucleolar structure, function. EMBO J 2015;34(22):2758–67.

[29] Stimpson KM, Sullivan LL, Kuo ME, Sullivan BA. Nucleolar organization, ribosomal DNA array stability, and acrocentric chromosome integrity are linked to telomere function. PLoS One 2014;9(3):e92432.

[30] Grummt I. The nucleolus-guardian of cellular homeostasis and genome integrity. Chromosoma 2013;122(6):487–97.

[31] Stepanchick A, Zhi H, Cavanaugh AH, Rothblum K, Schneider DA, Rothblum LI. DNA binding by the ribosomal DNA transcription factor rrn3 is essential for ribosomal DNA transcription. J Biol Chem 2013;288(13):9135–44.

[32] Kusnadi EP, Hannan KM, Hicks RJ, Hannan RD, Pearson RB, Kang J. Regulation of rDNA transcription in response to growth factors, nutrients and energy. Gene 2015;556(1):27–34.

[33] Woods SJ, Hannan KM, Pearson RB, Hannan RD. The nucleolus as a fundamental regulator of the p53 response and a new target for cancer therapy. Biochim Biophys Acta 2015;1849(7):821–9.

[34] Davis WJ, Lehmann PZ, LI W. Nuclear PI3K signaling in cell growth and tumorigenesis. Front Cell Dev Biol 2015;3:24.

[35] Nguyen NT, Saguez C, Conesa C, Lefebvre O, Acker J. Identification of proteins associated with RNA polymerse III using a modified tandem chromatin affinity purification. Gene 2015;556(1):51–60.

[36] Russell J, Zomerdijk JC. The RNA polymerase machinery. Biochem Soc Symp 2006;(73):203–16.

[37] Grummt I. Life on a planet of its own: regulation of RNA polymerase I transcription in the nucleolus. Genes Dev 2003;17:1691–702.

[38] Moss T, Stefanovsky VY. At the center of eukaryotic life. Cell 2002;109:545–8.

[39] Friedrich JK, Panov KI, Cabart P, Russell J, Zomerdijk JC. TBP-TAF complex SL1 directs RNA polymerase I pre-initiation complex formation and stabilizes upstream binding factor at the rDNA promoter. J Biol Chem 2005;280(33):29551–8.

[40] Panov KI, Friedrich JK, Russell J, Zomerdijk JC. UBF activates RNA polymerase I transcription by stimulating promoter escape. EMBO J 2006;25(14):3310–22.

[41] Drakas R, Tu X, Baserga R. Control of cell size through phosphorylation of upstream binding factor 1 by nuclear phosphatidylinositol 3-kinase. Proc Natl Acad Sci Acad Sci USA 2004;101(25):9272–6.

[42] Hannan KM, Brandenburger Y, Jenkins A, Sharkey K, Cavanaugh A, Rothblum L, Moss T, Poortinga G, McArthur GA, Pearson RB, et al. Mol Cell Biol 2003;23:8862–77.

[43] Stefanovsky VY, Pelletier G, Hannan R, Gagnon-Kugler T, Rothblum LI, Moss T. An immediate response of ribosomal transcription to growth factor stimulation in mammals is mediated by ERK phosphorylation of UBF. Mol Cell 2001;8:1063–73.

[44] Panova TB, Panov KI, Russell J, Zomerdijk JC. Casein kinase 2 associates with initiation-competent RNA polymerase I and has multiple roles in ribosomal DNA transcription. Mol Cell Biol 2006;26(16):5957–68.

[45] Mayer C, Zhao J, Yuan X, Grummt I. mTOR-dependent activation of the transcription factor TIF-IA links rRNA synthesis to nutrient availability. Genes Dev 2004;18:423–34.

[46] Zhao J, Yuan X, Frödin M, Grummt I. ERK-dependent phosphorylation of the transcription initiation factor TIF-IA is required for RNA polymerase I transcription and cell growth. Mol Cell 2003;11(2):405–13.

[47] Bierhoff H, Dundr M, Michels AA, Grummt I. Phosphorylation by casein kinase 2 facilitates rRNA gene transcription by promoting dissociation of TIF-IA from elongating RNA polymerase I. Mol Cell Biol 2008;28(16):4988–98.

[48] Mayer C, Bierhoff SH., Grummt SI. The nucleolus as a stress sensor: JNK2 inactivates the transcription factor TIF-IA and downregulates rRNA synthesis. Genes Dev 2005;19(8):933–41.

[49] Pelletier G, Stefanovsky VY, Faubladier M, Hirschler-Laszkiewicz I, Savard J, Rothblum LI, Côté J, Moss T. Competetive recruitment of CBP and Rb-HDAC regulates UBF acetylation and ribosomal transcription. Mol Cell 2000;6(5):1059–66.

[50] Nguyen leXT, Mitchell BS. Akt activation enhances ribosomal RNA synthesis through casein kinas II and TIF-IA. Proc Natl Acad Sci USA 2013;110(51):20681–6.

[51] Muth V, Nadaud S, Grummt I, Voit R. Acetylation of TAF(I)68, a subunit of TIF-IB/SL1, activates RNA polymerase I transcription. EMBO J 2001;20(6):1353–62.

[52] Richardson LA, Reed BJ, Charette JM, Freed EF, Fredrickson EK, Locke MN, Baserga SJ, Gardner RG. A conserved deubiquitinating enzyme controls cell growth by regulating RNA polymerase I stability. Cell Rep 2012;2(2):372–85.

[53] Fátyol K, Grummt I. Proteiasomal ATPases are associated with rDNA: the ubiquitin proteasome system plays a direct role in RNA polymerase I transcription. Biochim Biophys Acta 2008;1779(12):850–9.

[54] Zhao Z, Dammert MA, Grummt I, Bierhoff H. lncRNA-induced nucleosome repositioning reinforces transcriptional repression of rRNA genes upon hypotonic stress. Cell Rep 2016;14(8):1876–82.

[55] Mongelard F, Bouvet P. Nucleolin: a multiFACeTed protein. Trends Cell Biol 2007;17(2):80–6.

[56] Durut N, Sáez-Vásquez J. Nucleolin: dual roles in rDNA chromatin transcription. Gene 2015;556(1):7–12.

[57] McKeown PC, Shaw PJ. Chromatin: linking structure and function in the nucleolus. Chromosoma 2009;118(1):11–23.

[58] Cong R, Das S, Ugrinova I, Kumar S, Mongelard F, Wong J, Bouvet P. Interaction of nucleolin with ribosomal RNA genes and its role in RNA polymerase I transcription. Nucl Acids Res 2012;40(19):9441–54.

[59] Gaume X, Monier K, Argoul F, Mongelard F, Bouvet P. In vivo study of the histone chaperone activity of nucleolin by FRAP. Biochem Res Int 2011;187624.

[60] Hisaoka M, Ueshima S, Murano K, Nagata K, Okuwaki M. Regulation of nucleolar chromatin by B23/nucleophosmin jointly depends upon its RNA binding activity and transcription factor UBF. Mol Cell Biol 2010;30(20):4952–64.

[61] Sanij E, Hannan RD. The role of UBF in regulating the structure and dynamics of transcriptionally active rDNA chromatin. Epigenetics 2009;4(6):374–82.

[62] Sanij E, Diesch J, Lesmana A, Poortinga G, Hein N, Lidgerwood G, Cameron DP, Ellul J, Goodall GJ, Wong LH, Dhillon AS, Hamdane N, Rothblum LI, Pearson RB, Haviv I, Moss T, Hannan RD. A novel role for the Pol I transcription factor UBFT in maintaining genome stability through the regulation of highly transcribed Pol II genes. Genome Res 2015;25(2):201–12.

[63] Kermekchiev M, Workman JL, Pikaard CS. Nucleosome binding by the polymerase I transactivator upstream binding factor displaces linker histone H1. Mol Cell Biol 1997;17: 5833–42.

[64] Sanij E, Poortinga G, Sharkey K, Hung S, Holloway TP, Quin J, Robb E, Wong LH, Thomas WG, Stefanovsky V, et al. UBF levels determine the number of active ribosomal RNA genes in mammals. J Cell Biol 2008;183:1259–74.

[65] Diesch J, Hannan RD, Sanij E. Genome wide mapping of UBF binding-sites in mouse and human cell lines. Genom Data 2015;3:103–5.

[66] Ugrinova I, Monier K, Ivaldi C, Thiry M, StorckS, Mongelard F, Bouvet P. Inactivation of nucleolin leads to nucleolar disruption, cell cycle arrest and defects in centrosome duplication. BMC Mol Biol 2007;8:66.

[67] Angelov D, Bondarenko VA, Almagro S, Menoni H, Mongelard F, Hans F, Mietton F, Studitsky VM, Hamiche A, Dimitrov S, et al. Nucleolin is a histone chaperone with FACT-like activity and assists remodeling of nucleosomes. EMBO J 2006;25:1669–79.

[68] Erard MS, BelenguerP, Caizergues-Ferrer M, Pantaloni A, Amalric F. A major nucleolar protein, nucleolin, induces chromatin decondensation by binding to histone H1. Eur J Biochem 1988;175:525–30.

[69] Wittner M, Hamperl S, Stockl U, Seufert W, Tschochner H, Milkereit P, Griesenbeck J. Establishment and maintenance of alternative chromatin states at a multicopy gene locus. Cell 2011;145:543–54.

[70] Albert B, Colleran C, Leger-Silvestre I, Berger AB, Dez C, Normand C, Perez-Fernandez J, McStay B, Gadal O. Structure-function analysis of Hmo1 unveils an ancestral organization of HMG-Box factors involved in ribosomal DNA transcription from yeast to human. Nucleic Acids Res 2013;41:10135–49.

[71] Li J, Längst G, Grummt I. NoRC-dependent nucleosome positioning silences rRNA genes. EMBO J 2006;25:5735–41.

[72] Felle M, Exler JH, Merkl R, Dachauer K, Brehm A, Grummt I, Längst G. DNA sequence encoded repression of rRNA gene transcription in chromatin. Nucleic Acids Res 2010;38(16): 5304–14.

[73] Manelyte L, Strohner R, Gross T, Längst G. Chromatin targeting signals, nucleosome positioning mechanism and non-coding RNA—mediated regulation of the chromatin remodelling complex NoRC. PLoS Genet 2014;10(3):e1004157.

[74] Gagnon-Kugler T, Langlois F, Stefanovsky V, Lessard F, Moss T. Loss of human ribosomal gene CpG methylation enhances cryptic RNA polymerase II transcription and disrupts ribosomal RNA processing. Mol Cell 2009;35:414–25.

[75] Xie W, et al. The chromatin remodeling complex NuRD establishes the poised state of rRNA genes characterized by bivalent histone modifications and altered nucleosome positions. Proc Natl Acad Sci USA 2012;109:8161–6.

[76] Feng Q, Zhang. The NuRD complex: linking histone modification to nucleosome remodeling. Curr Top Microbiol Immunol 2003;274:269–90.

[77] Poortinga G, Wall M, Sanij E, Siwicki K, Ellul J, Brown D, Holloway TP, Hannan RD, McArthur GA. C-MYC co-ordinately regulates ribosomal gene chromatin remodelling and Pol I availability during granulocyte differentiation. Nucleic Acids Res 2011;39(8):3267–81.

[78] Hamdane N, Stefanovsky VY, Tremblay MG, Németh A, Paquet E, Lessard F, Sanij E, Hannan R, Moss T. Conditional inactivation of upstream binding factor reveals its epigenetic functions and the existence of a somatic nucleolar precursor body. PLoS Genet 2014;10(8):e1004505.

[79] Stefanovsky V, Moss T. Regulation of rRNA synthesis in human and mouse cells is not determined by changes in active gene count. Cell Cycle 2006;5:735–9.

[80] Strohner R, Nemeth A, Jansa P, Hofmann-Rohrer U, Santoro R, Längst G, Grummt I. NoRC—a novel member of mammalian ISWI-containing chromatin remodeling machines. EMBO J 2001;20:4892–900.

[81] Grummt I, Längst G. Epigenetic control of RNA polymerase I transcription in mammalian cells. Biochim Biophys Acta 2013;1829(3–4):393–404.

[82] Németh A, Strohner R, Grummt I, Längst G. The chromatin remodeling complex NoRC and TTF-1 cooperate in the regulation of the mammalian rRNA genes in vivo. Nucleic Acids Res 2004;32(14):4091–9.

[83] Zhou Y, Grummt I. The PHD finger/bromodomain of NoRC interacts with acetylated histone H4K16 and is sufficient for rDNA silencing. Curr Biol 2005;15:1434–8. 9.

[84] Tallant C, Valentini E, Fedorov O, Overvoorde L, Ferguson FM, Filippakopoulos P, Svergun DI, Knapp S, Ciulli A. Molecular basis of histone tail recognitions by human TIP5 PHD finger and bromodomain of the chromatin complex NoRC. Structure 2015;23(1):80–92.

[85] Mayer C, Neubert M, Grummt I. The structure of NoRC-associated RNA is crucial for targeting the chromatin remodelling complex NoRC to the nucleolus. EMBO Rep 2008;9:774–80.

[86] Nemeth A, Guibert S, Tiwari VK, Ohlsson R, Längst G. Epigenetic regulation of TTF-I-mediated promoter-terminator interactions of rRNA genes. EMBO J 2008;27:1255–65.

[87] Diermeier SD, Németh A, Rehli M, Grummt I, Längst G. Chromatin-specific regulation of mammalian rDNA transcription by clustered TTF-I binding sites. PLoS Genet 2013;9(9):e1003786.

[88] Schmitz KM, Mayer C, Postepska A, Grummt I. Interaction of noncoding RNA with the rDNA promoter mediates recruitment of DNMT3b and silencing of rRNA genes. Genes Dev 2010;24:2264–9.

[89] Lessard F, Morin F, Ivanchuk S, Langlois F, Stefanovsky V, Rutka J, Moss T. The ARF tumor suppressor controls ribosome biogenesis by regulating the RNA polymerase I transcription factor TTF-I. Mol Cell 2010;38(4):539–50.

[90] Lessard F, Stefanovsky V, Tremblay MG, Moss T. The cellular abundance of the essential transcription termination factor TTF-I regulates ribosome biogenesis and is determined by MDM2 ubiquitinylation. Nucleic Acids Res 2012;40(12):5357–67.

[91] Khan A, Giri S, Wang Y, Chakraborty A, Ghosh AK, Anantharaman A, Aggarwal V, Sathyan KM, Ha T, Prasanth KV, Prasanth SG. BEND3 represses rDNA transcription by stabilizing a NoRC component via USP21 deubiquitinase. Proc Natl Acad Sci USA 2015;112(27):8338–43.

[92] Chakrabarti R, Sanyal S, Ghosh A, Bhar K, Das C, Siddhanta A. Phosphatidylinositol-4-phosphatate 5-kinase modulates ribosomal RNA gene silencing through its interaction with histone H3 lysine 9 trimethylation and heterochromatin protein HP1-α. J Biol Chem 2015;290(34):20893–903.

[93] Gillies J, Hickey CM, Su D, Wu Z, Peng J, Hochstrasser M. SUMO pathway modulation of regulatory protein binding at the ribosomal DNA locus in *Saccharomyces cerevisiae*. Genetics 2016;202(4):1377–94.

[94] Kuhn A, Grummt I. A novel promoter in the mouse rDNA spacer is active in vivo and in vitro. EMBO J 1987;6:3487–92.

[95] Januszyk K, Lima CD. The eukaryotic RNA exosome. Curr Opin Struct Biol 2014;24:132–40.

[96] Mayer C, Schmitz KM, Li J, Grummt I, Santoro R. Intergenic transcripts regulate the epigenetic state of rRNA genes. Mol Cell 2006;22:351–61.

[97] Anosova I, Melnik S, Tripsianes K, Kateb F, Grummt I, Sattler M. A novel RNA binding surface of the TAM domain of TIP5/BAZ2A mediates epigenetic regulation of rRNA genes. Nucleic Acids Res 2015;43(10):5208–20.

[98] Zhou Y, Schmitz KM, Mayer C, Yuan X, Akhtar A, Grummt I. Reversible acetylation of the chromatin remodelling complex NoRC is required for non-coding RNA-dependent silencing. Nat Cell Biol 2009;11(8):1010–6.

[99] Wehner S, Dörrich AK, Ciba P, Wilde A, Marz M. pRNA: NoRC-associated RNA of rRNA operons. RNA Biol 2014;11(1):3–9.

[100] Erdel F, Krug J, Längst G, Rippe K. Targeting chromatin remodelers: signal and search mechanisms. Biochim Biophys Acta 2011;1809(9):497–508.

[101] Santoro R, Li J, Grummt I. The nucleolar remodelling complex NoRC mediates heterochromatin formation and silencing of ribosomal gene transcription. Nat Genet 2002;32:393–6.

[102] Zhou Y, Santoro R, Grummt I. The chromatin remodeling complex NoRC targets HDAC1 to the ribosomal gene promoter and represses RNA polymerase I transcription. EMBO J 2002;21:4632–40.

[103] Santoro R, Schmitz KM, Sandoval J, Grummt I. Intergenic transcripts originating from a subclass of ribosomal DNA repeats silence ribosomal RNA genes in trans. EMBO Rep 2010;11(1):52–8.

[104] Guetg C, Santoro R. Noncoding RNAs link PARP1 to heterochromatin. Cell Cycle 2012;11:2217–8.

[105] Bierhoff H, Schmitz K, Maass F, Ye J, Grummt I. Noncoding transcripts in sense and antisense orientation regulate the epigenetic state of ribosomal RNA genes. Cold Spring Harb Symp Quant Biol 2010;75:357–64.

[106] Bierhoff H, Postepska-Igielska A, Grummt I. Noisy silence: non-coding RNA and heterochromatin formation at repetitive elements. Epigenetics 2014;9(1):53–61.

[107] Zhao Z, Dammert MA, Hoppe S, Bierhoff H, Grummt I. Heat shock represses rRNA synthesis by inactivation of TIF-1A and lncRNA-dependent changes in nucleosome positioning. Nucleic Acids Res 2016;. Epub ahead of print.

[108] Zhao Z, Dammert MA, Grummt I, Bierhoff H. lncRNA-induced nucleosome repositioning reinforces transcriptional repression of rRNA genes upon hypotonic stress. Cell Rep 2016;14(8):1876–82.

[109] Hwang YJ, Han D, Kim KY, Min SJ, Kowall NW, Yang L, Lee J, Kim Y, Ryu H. ESET methylates UBF at K232/254 and regulated nucleolar heterochromatin plasticity and rDNA transcription. Nucleic Acids Res 2014;42(3):1628–43.

[110] Guetg C, Lienemann P, Sirri V, Grummt I, Hernandez-Verdun D, Hottiger MO, Fussenegger M, Santoro R. The NoRC complex mediates the heterochromatin formation and stability of silent rRNA genes and cetromeric repeats. EMBO J 2010;29(13):2135–46.

[111] Nguyen leXT, Raval A, Garcia JS, Mitchell BS. Regulation of ribosomal gene expression in cancer. J Cell Physiol 2015;230(6):1181–8.

[112] Huang J, Moazed D. Association of the RENT complex with non-transcribed and coding regions of rDNA and a regional requirement for the replication fork block protein Fob1 in rDNA. Genes Dev 2003;17:2162–76.

[113] Ha CW, Sung MK, Huh WK. Nsi1 plays a significat role in the silencing of ribosomal DNA in *Saccharomyces cerevisiae*. Nucleic Acids Res 2012;40(11):4892–903.

[114] Hannan A, Abraham NM, Goyal S, Jamir I, Priyakumar UD, Mishra K. Sumoylation of Sir2 differentially regulates transcriptional silencing in yeast. Nucleic Acids Res 2015;43(21):10213–26.

[115] Lewinska A, Miedziak B, Kulak K, Molon M, Wnuk M. Links between nucleolar activity, rDNA stability, aneuploidy and chronical aging in the yeast *Saccharomyces cerevisiae*. Biogerontology 2014;15(3):289–316.

[116] Kan WK, Kim Yh, Kan HA, Kwon KS, Kim JY. Sir2 phosphorylation through cAMP-PKA and CK2 signaling inhibits the lifespan extension activity of Sir2 in yeast. eLife 2015;4:e09709.

[117] Zillner K, Komatsu J, Filarsky K, Kalepu R, Bensimon A, Németh A. Active human nucleolar organizer regions are interspersed with inactive rDNA repeats in normal and tumor cells. Epigenomics 2015;7(3):363–78.

[118] Hoppe S, Bierhoff H, Cado I, Weber A, Tiebe M, Grummt I, Voit R. AMP-activated protein kinase adapts rRNA synthesis to cellular energy supply. Proc Natl Acad Sci USA 2009;106(42):17781–6.

[119] Murayama A, Ohmori K, Fujimura A, Minami H, Yasuzawa-Tanaka K, Kuroda T, Oie S, Daitoku H, Okuwaki M, Nagata K, Fukamizu A, Kimura K, Shimizu T, Yanagisawa J. Epigenetic control of rDNA loci in response to intracellular energy status. Cell 2008;133(4):627–39.

[120] Yang L, Song T, Chen L, Kabra N, Zheng H, Koomen J, Seto E, Chen. Regulation of SirT1-nucleomethylin binding by rRNA coordinates ribosome biogenesis with nutrient availability. J Mol Cell Biol 2013;33(19):3835–48.

[121] Song T, Yang L, Kabra N, Chen L, Koomen J, Haura EB, Chen J. The NAD+ synthesis enzyme nicotinamide mononucleotide adenylyltransferase (NMNAT1) regulates ribosomal RNA transcription. J Biol Chem 2013;288(29):20908–17.

[122] Voit R, Seiler J, Grummt I. Cooperative actions of Cdk1/cyclin B and SIRT1 is required for mitotic repression of rRNA synthesis. PLoS Genet 2015;11(5):e1005246.

[123] Bruderer M, Richards RG, Alini M, Stoddart MJ. Role and regulation of RUNX2 in osteogenesis. Eur Cell Mater 2014;28:269–86.

[124] Ali SA, Dobson JR, Lian JB, Stein JL, van Wijnen AJ, Zaidi SK, Stein GS. A RUNX2–HDAC1 corepressor complex regulates rRNA gene expression by modulating UBF acetylation. J Cell Sci 2012;125(11):2732–9.

[125] Ling T, Xie W, Luo M, Shen M, Zhu Q, Zong L, Zhou T, Gu J, Lu Z, Zhang F, Tao W. CHD4/NuRD maintains demethylation state of rDNA promoters through inhibiting the expression of the rDNA methyltransferase recruiter TIP5. Biochem Biophys Res Commun 2013;437(1):101–7.

[126] Charton R, Guintini L, Peyresaubes F, Conconi A. Repair of UV induced DNA lesions in ribosomal gene chromatin and the role of "odd" RNA polymerases (I and III) DNA Repair 2015;36:49–58.

[127] Hamperl S, Wittner M, Babl V, Perez-Fernandez J, Tschochner H, Griesenbeck J. Chromatin states at ribosomal DNA loci. Biochim Biophys Acta 2013;1829(3–4):405–17.

[128] Derenzini M, Pasquinelli G, O'Donohue MF, Ploton D, Thiry M. Structural and functional organization of ribosomal genes within the mammalian cell nucleolus. J Histochem Cytochem 2006;54(2):131–45.

[129] Trendelenburg MF, Zatsepina OV, Waschek T, Schlegel W, Tröster H, Rudolph D, Schmahl G, Spring H. Multiparameter microscopic analysis of nucleolar structure and ribosomal gene transcription. Histochem Cell Biol 1996;106(2):167–92.

[130] Griesenbeck J, Wittner M, Charton R, Conconi A. Chromatin endogenous cleavage and psoralen crosslinking assay to analyse rRNA gene chromatin in vivo. Methods Mol Biol 2012;809:291–301.

[131] Murugesapillai D, McCauley MJ, Huo R, Nelson Holte MH, Stepanyants A, Maher LJ 3rd, Israeloff NE, Williams MC. DNA bridging and looping by HMO1 provides a mechanism for stabilizing nucleosome-free chromatin. Nucleic Acids Res 2014;42(14):8996–9004.

[132] Keener J, Dodd JA, Lalo D, Nomura M. Histones H3 and H4 are components of upstream activation factor required for the high-level transcription of yeast rDNA by RNA polymerase I. Proc Natl Acad Sci USA 1997;94(25):13458–62.

[133] Tongaonkar P, French SL, Oakes ML, Vu L, Schneider DA, Beyer AL, Nomura M. Histones are required for transcription of yeast rRNA genes by RNA polymerase I. Proc Natl Acad Sci USA 2005;102(29):10129–34.

[134] Feng W, Yonezawa M, Ye J, Jenuwein T, Grummt I. PHF8 activates transcription of rRNA genes through H3K4me3 binding and H3K9me1/2 demethylation. Nat Struct Mol Biol 2010;445–50.

[135] Vintermist A, Böhm S, Sadeghifar F, Louvet E, Mansén A, Percipalle P, Östlund Farrants AK. The chromatin remodelling complex B-WICH changes the chromatin structure and recruits histone acetyl-transferase to activate rRNA genes. PLoS One 2011;6(4):e19184.

[136] Chen H, Fan M, Pfeffer LM, Laribee RN. The histone H3 lysine 56 acetylation pathway is regulated by target of rapamycin (TOR) singaling and functions directly in ribosomal RNA biogenesis. Nucleic Acids Res 2012;40(14):6534–46.

[137] Johnson JM, French SL, Osheim YN, Li M, Hall L, Beyer AL, Smith Js. Rpd3 and spt16-mediated nucleosome assembly and transcriptional regulation on yeast ribosomal DNA genes. Mol Cell Biol 2013;33:2748–59.

[138] Birch JL, Zomerdijk JC. Structure and function of ribosomal RNA gene chromatin. Biochem Soc Trans 2008;36:619–24.

[139] Brown SE, Szyf M. Dynamic epigenetic states of ribosomal RNA promoters during the cell cycle. Cell Cycle 2008;7(3):382–90.

[140] Chan JC, Hannan KM, Riddell K, Ng PY, Peck A, Lee RS, Hung S, Astle MV, Bywater M, Wall M, Poortinga G, Jastrzebski K, Sheppard KE, Hemmings BA, Hall MN, Johnstone RW, McArthur GA, Hannan RD, Pearson RB. AKT promotes rRNA synthesis and cooperates with c-MYC to stimulate ribosome biogenesis in cancer. Sci Signal 2011;4(188):ra56.

[141] Pickard AJ, Bierbach U. The cell's nucleolus: an emerging target for chemotherapeutic intervention. Chem Med Chem 2013;8(9):1441–9.

[142] McMahon SB, Wood MA, Cole MD. The essential cofactor TRRAP recruits the histone acetyltransferase hGCN5 to c-Myc. Mol Cell Biol 2000;20(2):556–62.

[143] Grandori C, Gomez-Roman N, Felton-Edkins ZA, Ngouenet C, Galloway DA, Eisenman RN, White RJ. c-Myc binds to human ribosomal DNA and stimulates transcription of rRNA genes by RNA polymerase I. Nat Cell Biol 2005;7(3):311–8.

[144] Arabi A, Wu S, Ridderstråle K, Bierhoff H, Shiue C, Fatyol K, Fahlén S, Hydbring P, Söderberg O, Grummt I, Larsson LG, Wright AP. c-Myc associates with ribosomal DNA and activates RNA polymerase I transcription. Nat Cell Biol 2005;7(3):303–10.

[145] Shiue CN, Berkson RG, Wright AP. c-Myc induces changes in higher order rDNA structure on stimulation of quiescent cells. Oncogene 2009;28(16):1833–42.

[146] Shiue CN, Nematollahi-Mahani A, Wright AP. Myc-induced anchorage of the rDNA IGS region to nucleolar matrix modulates growth-stimulated changes in higher-order rDNA architecture. Nucleic Acids Res 2014;42(9):5505–17.

[147] Denissov S, Lessard F, Mayer C, Stefanovsky V, van Driel M, Grummt I, Moss T, Stunnenberg HG. A model for the topology of active ribosomal RNA genes. EMBO Rep 2011;12(3):231–7.

[148] Tessarz P, Santos-Rosa H, Robson SC, Sylvestersen KB, Nelson CJ, Nielsen ML, Kouzarides T. Glutamine methylation in histone H2A is an RNA-polymerase-I-dedicated modification. Nature 2014;505(7484):564–8.

[149] Loza-Muller L, Rodríguez-Corona U, Sobol M, Rodríguez-Zapata LC, Hozak P, Castano E. Fibrillarin methylates H2A in RNA polymerase I trans-active promoters in *Brassica oleracea*. Front Plant Sci 2015;6:976.

[150] Tollervey D, Lehtonen H, Jansen R, Kern H, Hurt EC. Temperature-sensitive mutations demonstrate roles for yeast fibrillarin in pre-rRNA processing, pre-rRNA methylation, and ribosome assembly. Cell 1993;72:443–57.

[151] Rodriguez-Corona U, Sobol M, Rodriguez-Zapata LC, Hozak P, Castano E. Fibrillarin from Archaea to human. Biol Cell 2015;107(6):159–74.

[152] Narlikar GJ, Sundaramoorthy R, Owen-Hughes T. Mecahnisms and functions of ATP-dependent chromatin-remodeling enzyme. Cell 2013;154(3):490–503.

[153] Orphanides G, Wu WH, Lane WS, Hampsey M, Reinberg D. The chromatin-specific transcription elongation factor FACT comprises human SPT16 and SSRP1 proteins. Nature 1999;400(6741):284–8.

[154] Birch JL, Tan BC, Panov KI, Panova TB, Andersen JS, Owen-HughesTA, Russell J, Lee SC, Zomerdijk JC. FACT facilitates chromatin transcription by RNA polymerases I and III. EMBO J 2009;28:854–65.

[155] Ide S, Dejardin J. End-targeting proteomics of isolated chromatin segments of a mammalian ribosomal RNA gene promoter. Nat Commun 2015;6:6674.

[156] Kalashnikova AA, Winkler DD, McBryant SJ, Henderson RK, Herman JA, DeLuca JG, Luger K, Prenni JE, Hansen JC. Linker histone H1.0 interacts with an extensive network of proteins found in the nucleolus. Nucleic Acids Res 2013;41(7):4026–35.

[157] Szerlong HJ, Herman JA, Krause CM, DeLuca JG, Skoultchi A, Winger QA, Prenni JE, Hansen JC. Proteomic characterization of the nucleolar linker histone H1 interaction network. J Mol Biol 2015;427(11):2056–71.

[158] Jeong J. The role of Cockayne syndrome protein B in transcription regulation. Genom Data 2014;2:302–4.

[159] Bradsher J, Auriol J, Proietti de Santis L, Iben S, Vonesch JL, Grummt I, Egly JM. CSB is a component of RNA pol I transcription. Mol Cell 2002;10:819–29.

[160] Yuan X, Feng W, Imhof A, Grummt I, Zhou Y. Activation of RNA polymerase I transcription by Cockayne syndrome group B protein and histone methyltransferase G9a. Mol Cell 2007;27:585–95.

[161] Shen M, Zhou T, Xie W, Ling T, Zhu Q, Zong L, Lyu G, Gao Q, Zhang F, Tao W. The chromatin remodeling factor CSB recruits histone acetyltransferase PCAF to rRNA gene promoters in active state for transcription initiation. PLoS One 2013;8(5):e62668.

[162] Bozhenok L, Wade PA, Varga-Weisz P. WSTF-ISWI chromatin remodeling complex targets heterochromatic replication foci. EMBO J 2002;21:2231–41.

[163] Percipalle P, Fomproix N, Cavellán E, Voit R, Reimer G, Krüger T, Thyberg J, Scheer U, Grummt I, Farrants AK. The chromatin remodelling complex WSTF-SNF2h interacts with nuclear myosin 1 and has a role in RNA polymerase I transcription. EMBO Rep 2006;7(5):525–30.

[164] Cavellán E, Asp P, Percipalle P, Farrants AK. The WSTF-SNF2h chromatin remodeling complex interacts with several nuclear proteins in transcription. J Biol Chem 2006;281(24):16264–71.

[165] Sarshad A, Sadeghifar F, Louvet E, Mori R, Böhm S, Al-Muzzaini B, Vintermist A, Fomproix N, Östlund AK, Percipalle P. Nuclear myosin 1c facilitates the chromatin modifications required to activte rRNA gene transcription and cell cycle progression. PLoS Genet 2013;9(3):e1003397.

[166] Fomproix N, Percipalle P. An actin-myosin complex on actively transcribing genes. Exp Cell Res 2004;294(1):140–8.

[167] Philimonenko VV, Zhao J, Iben S, Dingová H, Kyselá K, Kahle M, Zentgraf H, Hofmann WA, de Lanerolle P, Hozák P, Grummt I. Nuclear actin and myosin I are required for RNA polymerase I transcription. Nat Cell Biol 2004;6(12):1165–72.

[168] Ye J, Zhao J, Hoffmann-Rohrer U, Grummt I. Nuclear myosin I acts in concert with polymeric actin to drive RNA polymerase I transcription. Genes Dev 2008;22(3):322–30.

[169] Almuzzaini B, Sarshad AA, Rahmanto AS, Hansson ML, Von Euler A, Sangfelt O, Visa N, Farrants AK, Percipalle P. In β-actin knockouts, epigenetic reprogramming and rDNA transcription inactivation lead to growth and proliferation defects. FASEB J. 2016;30(8):2860–73.

[170] von Walden F, Casagrande V, Östlund Farrants AK, Nader GA. Mechanical loading induces the expression of a Pol I regulon at the onset of skeletal muscle hypertrophy. Am J Physiol Cell Physiol 2012;302(10):C1523–C15230.

[171] Tsai YC, Greco TM, Boonmee A, Miteva Y, Cristea IM. Functional proteomics establishes the interaction of SIRT7 with chromatin remodelling complexes and expands its role in regulation of RNA polymerase I transcription. Mol Cell Proteomics 2012;11(5):60–76.

[172] Ford E, Voit R, Liszt G, Magin C, Grummt I, Guarente L. Mammalian Sir2 homolog SIRT7 is an activator of RNA polymerase I transcription. Genes Dev 2006;20(9):1075–80.

[173] Chen S, Seiler J, Santiago-Reichelt M, Felbel K, Grummt I, Voit R. Repression o RNA polymerase I upon stress is caused by inhibition of RNA-dependent deacetylation of PAF53 by SIRT7. Mol Cell 2013;52(3):303–13.

[174] Grob A, Roussel P, Wright JE, McStay B, Hernandez-Verdun D, Sirri V. Involvement of SIRT7 in resumption of rDNA transcription at the exit from mitosis. J Cell Sci 2009;122 (Pt. 4):489–98.

[175] Zhai N, Zhao ZL, Cheng MB, Di YW, Yan HX, Cao CY, Dai H, Zhang Y, Shen YF. Human PIH1 associates with histone H4 to mediate the glucose-dependent enhancement of pre-rRNA synthesis. J Mol Cell Biol 2012;4(4):231–41.

[176] Zhang Y, Anderson SJ, French SL, Sikes ML, Viktorovskaya OV, Huband J, Holcomb K, Hartman JL 4th, Beyer AL, Schneider DA. The SWI/SNF chromatin remodelling complex influences transcription by RNA polymerase I in *Saccharomyces cerevisiae*. PLoS One 2013;8(2):e56793.

[177] Atwood BL, Woolnough JL, Lefevre GM, Ribeiro MS, Felsenfeld G, Giles KE. Human Argonaute 2 is tehthered to ribosomal RNA throught microRNA interactions. J Biol Chem 2016 [Epub ahead of print].

[178] Houseley J, Kotovic K, El Hage A, Tollervey D. Trf4 targets ncRNAs from telomeric and rDNA spacer regions and functions in rDNA copy number control. EMBO J 2007;26(24):4996–5006.

[179] Milligan L, Decourty L, Saveanu C, Rappsilber J, Ceulemans H, Jacquier A, Tollervey D. A yeast ecosome cofactor, Mpp6, functions in RNA surveillance and in the degradation of noncoding RNA transcripts. Mol Cell Biol 2008;28(17):5446–57.

[180] Vasiljeva L, Kim M, Terzi N, Soares LM, Buratowski S. Transcription termination and RNA degradation contribute to silencing of RNA polymerase II transcription within heterochromatin. Mol Cell 2008;29:313–23.

[181] Kobayashi T, Ganley ARD. Recombination regulation by transcription-induced cohesin dissociation in rDNA repeats. Science 2005;309:1581–4.

[182] Caudy AA, Pikaard CS. *Xenopus* ribosomal RNA gene intergenic spacer elements conferring transcriptional enhancement and nucleolar dominance-like competition in oocytes. J Biol Chem 2002;277(35):31577–84.

[183] Steinmetz EJ, Warren CL, Kuehner JN, Panbehi B, Ansari AZ, Brow DA. Genome-wide distribution of yeast RNA polymerase II and its control by Sen1 helicase. Mol Cell 2006;24:735–46.

[184] Ide S, Miyazaki T, Maki H, Kobayashi T. Abundance of ribosomal RNA gene copies maintains genome integrity. Science 2010;327:693–6.

[185] Mueller JE, Bryk M. Isw1 acts independently of the Isw1a and Isw1b complexes in regulating transcriptional silencing at the ribosomal DNA locus in *Saccharomyces cerevisiae*. J Mol Biol 2007;371:1–10.

[186] Mueller JE, Li C, Bryk M. Isw2 regulates gene silencing at the ribosomal DNA locus in *Saccharomyces cerevisiae*. Biochem Biophys Res Commun 2007;361:1017–21.

[187] Cesarini E, Mariotti FR, Cioci F, Camilloni G. RNA polymerase I transcription silences noncoding RNAs at the ribosomal DNA locus in *Saccharomyces cerevisiae*. Eukaryot Cell 2010;9:325–35.

[188] Cesarini E, D'Alfonso A, Camilloni G. H4K16 acetylation affects recombination and nc RNA transcription at rDNA in *Saccharomyces cerevisiae*. Mol Biol Cell 2012;23(14):2770–81.

[189] Ruggero D. Revisiting the nucleolus: from marker to dynamic integrator of cancer signaling. Sci Signal 2012;5(241):pe38.

[190] Audas TE, Jacob MD, Lee S. Immobilization of proteins in the nucleolus by ribosomal intergenetic spacer noncoding RNA. Mol Cell 2012;45(2):147–57.

[191] James A, Wang Y, Raje H, Rosby R, DiMario P. Nucleolar stress with and without p53. Nucleus 2014;5(5):402–26.

[192] Audas TE, Jacob MD, Lee S. The nucleolar detention pathway: a cellular strategy for regulating molecular networks. Cell Cycle 2012;11(11):2059–62.

[193] Jacob MD, Audas TE, Uniacke J, Trinkle-Mulcahy L, Lee S. Environmental cues induce a long noncoding RNA-dependent remodelling of the nucleolus. Mol Biol Cell 2013;24(18):2943–53.

[194] Audas TE, Lee S. Stressing out over long noncoding RNA. Biochim Biophys Acta 2016;1859(1): 184–191.

[195] Kupriyanova NS, Netchvolodov KK, Sadova AA, Cherepanova MD, Ryskov AP. Non-canonical ribosomal DNA segments in the human genome, and nucleoli functioning. Gene 2015;572(2):237–42.

[196] Bai B, Yegnasubramanian S, Wheelan SJ, Laiho M. RNA-Seq of the nucleolus reveals abundant SNORD44-derived small RNAs. PLoS One 2014;9(9):e107519.

[197] Falaleeva M, Stamm S. Processing of snoRNAs as a new source of regulatory non-coding RNAs: snoRNA fragments form a new class of functional RNAs. Bioessays 2013;35(1):46–54.

[198] Ahmad Y, Boisvert FM, Gregor P, Cobley A, Lamond AI. NOPdb: nucleolar proteome database—2008 update. Nucleic Acids Res 2009;37:D181–4.

[199] Alla RK, Cairns BR. RNA polymerase III transcriptomes in human embryonic stem cells and induced pluripotent stem cells and relationships with pluripotency transcription factors. PLoS One 2014;9(1):e85648.

[200] Arimbasseri AG, Maraia RJ. A high density of cis-information terminated RNA polymerase III on a two-rail track. RNA Biol 2015;1–6.

[201] Morse RH, Roth SY, Simpson RT. A transcriptionally active tRNA gene interferes with nucleosome positioning in vivo. Mol Cell Biol 1992;12(9):4015–25.

[202] Lee CK, Shibata Y, Rao B, Strahl BD, Lieb JD. Evidence for nucleosome depletion at active regulatory regions genome-wide. Nat Genet 2004;36(8):900–5.

[203] Mavrich TN, Ioshikhes IP, Venters BJ, Jiang C, Tomsho LP, Qi J, Schuster SC, Albert I, Pugh BF. A barrier nucleosome model for statistical positioning of nucleosomes throughout the yeast genome. Genome Res 2008;18(7):1073–83.

[204] Schones DE, Cui K, Cuddapah S, Roh TY, Barski A, Wang Z, Wei G, Zhao K. Dynamic regulation of nucleosome positioning in the human genome. Cell 2008;132(5):887–98.

[205] Barski A, Chepelev I, Liko D, Cuddapah S, Fleming AB, Birch J, Cui K, White RJ, Zhao K. Pol II and its associated epigenetic marks are present at Pol III-transcribed noncoding RNA genes. Nat Struct Mol Biol 2010;17:629–34.

[206] Moqtaderi Z, Wang J, Raha D, White RJ, SnyderM, Weng Z, Struhl K. Genomic binding profiles of functionally distinct RNA polymerase III transcription complexes in human cells. Nat Struct Mol Biol 2010;17:635–6340.

[207] Oler AJ, All RK, Roberts DN, Wong A, Hollenhorst PC, Chandler KJ, Cassiday PA, Nelson A, Hagedorn CH, Graves BJ, Cairns BR. Human RNA polymerase III transcriptomes and relationships to Pol II promoter chromatin and enhancer-binding factors. Nat Struct Mol Biol 2010;17:620–8.

[208] Raha D, Wang Z, Moqtaderi Z, Wu L, Zhong G, Gerstein M, Struhl K, Snyder M. Close association of RNA polymerase II and many transcription factors with Pol III genes. Proc Natl Acad Sci USA 2010;107:3639–44.

[209] Canella D, Praz V, Reina JH, Cousin P, Hernandez N. Defining the RNA polymerase III transcriptome: genome-wide localization of the RNA polymerase III transcription machinery in human cells. Genome Res 2010;20(6):710–21.

[210] Kundu TK, Wang Z, Roeder RG. Human TFIIIC relieves chromatin-mediated repression of RNA polymerase III transcription and contains an intrinsic histone acetyltransferase activity. Mol Cell Biol 1999;19:1605–15.

[211] Hsieh YJ, Kundu TK, Wang Z, Kovelman R, Roeder RG. The TFIIIC90 subunit of TFIIIC interacts with multiple components of the RNA polymerase III machinery and contains a histone-specific acetyltransferase activity. Mol Cell Biol 1999;19(11):7697–704.

[212] Kenneth NS, White RJ. Regulation by c-Myc of ncRNA expression. Curr Opin Genet Dev 2009;19(1):38–43.

[213] Cabarcas S, Schramm L. RNA polymerase III transcription in cancer: the BRF2 connection. Mol Cancer 2011;10:47.

[214] Johnston IM, Allison SJ, Morton JP, Schramm L, Scott PH, White RJ. CK2 forms a stable complex with TFIIIB and activates RNA polymerase III transcription in human cells. Mol Cell Biol 2002;22:3757–68.

[215] Felton-Edkins ZA, FairleyJA, GrahamEL, Johnston IM, White RJ, Scott PH. The Mitogen-Activated Protein (MAP) kinase ERK induces tRNA synthesis by phosphorylating TFIIIB. EMBO J 2003;22:2422–32.

[216] Gomez-Roman N, Grandori C, Eisenman RN, White RJ. Direct activation of RNA polymerase III transcription by c-Myc. Nature 2003;421:290–4.

[217] Kantidakis T, Ramsbottom BA, Birch JL, Dowding SN, White RJ. mTOR associates with TFIIIC, is found at tRNA and 5S rRNA genes, and targets their repressor Maf1. Proc Natl Acad Sci USA 2010;107(26):11823–8.

[218] Boguta M. Maf1, a general negative regulator of RNA polymerase III in yeast. Biochim Biophys Acta 2013;1829(3–4):376–84.

[219] Moir RD, Willis IM. Regulating maf1 expression and its expanding biological functions. PLoS Genet 2015;11(1):e1004896.

[220] Wei Y, Zheng XS. Maf1 regulation: a model of signal transduction inside the nucleus. Nucleus 2010;1(2):162–5.

[221] Bonhoure N, Byrnes A, Moir RD, Hodroj W, Preitner F, Praz V, Marcelin G, Chua SC Jr, Martinez-Lopez N, Singh R, Moullan N, Auwerx J, Willemin G, Shah H, Hartil K, Vaitheesvaran B, Kurland I, Hernandez N, Willis IM. Loss of the RNA polymerse III repressor MAF1 confers obesity resistance. Genes Dev 2015;29(9):934–47.

[222] Challagundla KB, Sun XX, Zhang X, DeVine T, Zhang Q, Sears RC, Dai MS. Ribosomal protein L11 recruits miR-24/miRISC to repress c-Myc expression in response to ribosomal stress. Mol Cell Biol 2011;31:4007–21.

[223] Dai MS, Sun XX, Lu H. Ribosomal protein L11 associates with c-Myc at 5 S rRNA and tRNA genes and regulates their expression. J Biol Chem 2010;285(17):12587–94.

[224] Raab JR, Chiu J, Zhu J, Katzman S, Kurukuti S, Wade PA, Haussler D, Kamakaka RT. Human tRNA genes function as chromatin insulators. EMBO J 2012;31(2):330–50.

[225] Kenneth NS, Ramsbottom BA, Gomez-Roman N, Marshall L, Cole PA, White RJ. TRRAP and GCN5 are used by c-Myc to activate RNA polymerase III transcription. Proc Natl Acad Sci USA 2007;104:14917–22.

[226] Sadeghifar F, Böhm S, Vintermist A, Östlund Farrants AK. The B-WICH chromatin remodelling complex regulates RNA polymerase III transcription by promoting Max-dependent c-Myc binding. Nucleic Acids Res 2015;43(9):4477–90.

[227] Parnell TJ, Huff JT, Cairns BR. RSC regulates nucleosome positioning at Pol II genes and density at Pol III genes. EMBO J 2008;27:100–10.

[228] Arimbasseri AG, Bhargava P. Chromatin structure and expression of a gene transcribed by RNA polymerase III are independent of H2A.Z. deposit. Mol Cell Biol 2008;(8):2598–607.

[229] Mahapatra S, Dewari PS, Bhardwaj A, Bhargava P. Yeast H2A.Z, FACT complex and RSC regulate transcription of tRNA gene through differential dynamics of flanking nucleosomes. Nucleic Acids Res 2011;39:4023–34.

[230] Kumar Y, Bhargava P. A unique nucleosome arrangement, maintained actively by chromatin remodels facilitates transcription of yeast tRNA genes. BMC Genomics 2013;14:402.

[231] Listerman I, Bledau AS, Grishina I, Neugebauer KM. Extragenic accumulation of RNA polymerase II enhances transcription by RNA polymerase III. PLoS Genet 2007;3:e212.

[232] Wang C, Huang S. Nuclear function of Alus. Nucleus 2014;5(2):131–7.

[233] Lunyak VV, Atallah M. Genomic relationship between SINE retrotransposons, Pol III-Pol II transcription and chromatin organization: the journey form junk to jewel. Biochem Cell Biol 2011;89(5):495–504.

[234] Dridi S. Alu mobile elements: from junk DNA to genomic gems. Scientifica 2012;2012:545328.

[235] Orioli A, Pascali C, Pagano A, Teichmann M, Dieci G. RNA polymerase III transcription control elements: themes and variations. Gene 2011;493(2):185–94.

[236] Cowley M, Oakey RJ. Transposable elements re-wire and fine-tune the transcriptome. PLoS Genet 2013;9:e1003234.

[237] Ule J. Alu elements: at the crossroads between disease and evolution. Biochem Soc Trans 2013;41(6):1532–5.

[238] Liu WM, Chu WM, Choudary PV, Schmid CW. Cell stress and translational inhibitors transiently increases the abundance of mammalian SINE transcripts. Nucleic Acids Res 1995;23:1758–65.

[239] Panning B, Smiley JR. Activation of RNA polymerase III transcription of human Alu elements by herpes simplex virus. Virology 1994;202:408–17.

[240] Panning B, Smiley JR. Activation of RNA polymerase III transcription of human Alu repetitive elements by adenovirus type 5: requirement for the E1b 58-kilodalton protein and the products of E4 open reading frames 3 and 6. Mol Cell Biol 1993;13:3231–44.

[241] Tarallo V, Hirano Y, Gelfand BD. DICER1 loss and Alu RNA induce age-related macular degeneration via the NLRP3 inflammasome and MyD88. Cell 2012;149:847–59.

[242] Broz P, Dixit VM. Inflammasomes: mechanism of assembly, regulation and signalling. Nat Rev Immunol 2016;16(7):407–20.

[243] Mariner PD, Walters RD, Espinoza CA, Drullinger LF, Wagner SD, Kugel JF, Goodrich JA. Human Alu RNA is a modular transacting repressor of mRNA transcription during heat shock. Mol Cell 2008;29(4):499–509.

[244] Yakovchuk P, Goodrich JA, Kugel JF. B2 RNA and Alu RNA repress transcription by disrupting contacts between RNA polymerase II and promoter DNA within assembled complexes. Proc Natl Acad Sci USA 2009;106:5569–74.

[245] Rubin CM, Kimura RH, Schmid CW. Selective stimulation of translational expression by Alu RNA. Nucleic Acids Res 2002;30:3253–61.

[246] Häsler J, Strub K. Alu RNP and Alu RNA regulate translation initiation in vitro. Nucleic Acids Res 2006;34(8):2374–85.

[247] Chen LL, Carmichael GG. Altered nuclear retention of mRNAs containing inverted repeats in human embryonic stem cells: functional role of a nuclear noncoding RNA. Molecular Cell 2009;35(4):467–78.

[248] Donze D, Adams CR, Rine J, Kamakaka RT. The boundaries of the silenced HMR domain in *Saccharimyces cerevisiae*. Genes Dev 1999;13(6):698–708.

[249] Donze D, Kamakaka RT. Braking the silence: how heterochromatic gene repression is stopped in its tracks. Bioessays 2002;24(4):344–9.

[250] Simms TA, Dugas SL, Gremillion JC, Ibos ME, Dandurand MN, Toliver TT, Edwards DJ, Donze D. TFIIIC binding sites function as both heterochromatin barriers and chromatin insulators in *Saccharomyces cerevisiae*. Eukaryot Cell 2008;7(12):2078–86.

[251] Valenzuela L, Dhillon N, Kamakaka RT. Transcription independent insulation at TFIIIC-dependent insulators. Genetics 2009;183(1):131–48.

[252] Kirkland JG, Raab JR, Kamakaka RT. TFIIIC bound DNA elements in nuclear organization and insulation. Biochim Biophys Acta 2013;1829(3–4):418–24.

[253] Donze D. Extra-transcriptional functions of RNA Polymerase III complexes: TFIIIC as a potential global chromatin bookmark. Gene 2012;493(2):169–75.

[254] Van Bortle K, Ramos E, Takenaka N, Yang J, Wahi JE, Corces VG. *Drosphila* CTCF tandemly aligns with other insulator proteins at the borders of H3K27me3 domains. Genome Res 2012;22(11):2176–87.

[255] Scott KC, Merrett SL, Willard HF. A heterochromatin barrier partitions the fission yeast centromere into discrete chromatin domains. Curr Biol 2006;16:119–29.

[256] Scott KC, White CV, Willard HF. An RNA polymerase III-dependent heterochromatin barrier at fission yeast centromere 1. PLoS One 2007;2(10):e1099.

[257] Donze D, Kamakaka RT. RNA polymerase III and RNA polymerase II promoter complexes are heterochromatin barriers in *Saccharomyces cerevisiae*. EMBO J 2001;20(3):520–31.

[258] Dhillon N, Raab J, Guzzo J, Szyjka SJ, Gangadharan S, Aparicio OM, Andrews B, Kamakaka RT. DNA polymerase epsilon, acetylases and remodellers cooperate to form a specialized chromatin structure at a tRNA insulator. EMBO J 2009;28:2583–600.

[259] Roberts DN, Stewart AJ, Huff JT, Cairns BR. The RNA polymerase III transcriptome revealed by genome-wide localization and activity-occupancy relationships. Proc Natl Acad Sci USA 2003;100(25):14695–700.

[260] Moqtaderi Z, Struhl K. Genome-wide occupancy profile of the RNA polymerase III machinery in *Saccharomyces cerevisiae* reveals loci with incomplete transcription complexes. Mol Cell Biol 2004;24(10):4118–27.

[261] Noma K, Cam HP, Maraia RJ, Grewal SI. A role for TFIIIC transcription factor complex in genome organization. Cell 2006;125:859–72.

[262] Wallrath LL, Geyer PK. TFIIIC boxes in the genome. Cell 2006;125(5):829–31.

[263] Jansen A, Verstrepen KJ. Nucleosome positioning in *Saccharomyces cerevisiae*. Microbiol Mol Biol Rev 2011;75(2):301–20.

[264] Xu Z, Wei W, Gagneur J, Perocchi F, Clauder-Münster S, Camblong J, Guffanti E, Stutz F, Huber W, Steinmetz LM. Bidirectional promoters generate pervasive transcription in yeast. Nature 2009;457(7232):1033–7.

[265] Jiang C, Pugh BF. A compiled and systematic reference map of nucleosome positions across the *Saccharomyces cerevisiae* genome. Genome Biol 2009;10:R109.

[266] Xu Z, Wei W, Gagneur J, Perocchi F, Clauder-Münster S, Camblong J, Guffanti E, Stutz F, Huber W, Steinmetz LM. Bidirectional promoters generate pervasive transcription in yeast. Nature 2009;457(7232):1033–7.

[267] Strålfors A, Walfridsson J, Bhuiyan H, Ekwall K. The FUN30 chromatin remodeler, Fft3, protects centromeric and subtelomeric domains from euchromatin formation. PLoS Genet 2011;7(3):e1001334.

[268] Steglich B, Strålfors A, Khorosjutina O, Persson J, Smialowska A, Javerzat JP, Ekwall K. The Fun30 chromatin remodeler Fft3 controls nuclear organization and chromatin structure of insulators and subtelomeres in fission yeast. PLoS Genet 2015;11(3):e1005101.

[269] Pascali C, Teichmann M. RNA polymerase III transcription- regulated by chromatin structure and regulator of nuclear chromatin organization. Subcell Biochem 2013;61:261–87.

[270] Van Bortle K, Nichols MH, Li L, Ong CT, Takenaka N, Qin ZS, Corces VG. Insulator function and topological domain border strength scale with architectural protein occupancy. Genome Biol 2014;15(6):R82.

[271] Van Bortle K, Corces VG. tRNA insulators and the emerging role of TFIIIC in genome organization. Transcription 2012;3(6):277–84.

[272] Barkess G, West AG. Chromatin insulator elements: establishing barriers to ser heterochromatin-boundaries. Epigenomics 2012;4(1):67–80.

[273] Carrière L, Graziani S, Alibert O, Ghavi-Helm Y, Boussouar F, Humbertclaude H, Jounier S, Aude JC, Keime C, Murvai J, Foglio M, Gut M, Gut I, Lathrop M, Soutourina J, Gérard M, Werner M. Genomic binding of Pol III transcription machinery and relationship with TFIIS transcription factor distribution in mouse embryonic stem cells. Nucleic Acids Res 2012;40(1):270–83.

[274] D'Ambrosio C, Schmidt CK, Katou Y, Kelly G, Itoh T, Shirahige K, Uhlmann F. Identification of cis-acting sites for condensin loading onto budding yeast chromosomes. Genes Dev 2008;22(16):2215–27.

[275] Comet I, Schuettengruber B, Sexton T, Cavalli G. A chromatin insulator driving three-dimensional Polycomb response element (PRE) contacts and Polycomb association with the chromatin fiber. Proc Natl Acad Sci USA 2011;108(6):2294–9.

[276] Shenoy N, Kessel R, Bhagat TD, Bhattacharyya S, Yu Y, McMahon C, Verma A. Alterations in the ribosomal machinery in cancer and hematologic disorders. J Hematol Oncol 2012;5:32.

[277] Yu F, Shen X, Fan L, Yu Z. Analysis of histone modifications at human ribosomal DNA in liver cancer cells. Sci Rep 2015;5:18100.

[278] Poortinga G, Quinn LM, Hannan RD. Targeting RNA polymerase I to treat MYC-driven cancer. Oncogene 2015;34(4):403–12.

[279] Sondalle SB, Baserga SJ. Human diseases of the SSU processome. Biochim Biophys Acta 2014;1842(6):758–64.

[280] Freed EF, Bleichert F, Dutca LM, Baserga SJ. When ribosomes go bad: diseases of ribosome biogenesis. Mol BioSyst 2010;6:481–93.

[281] Nakhoul H, Ke J, Zhou X, Liao W, Zeng SX, Lu H. Ribosomopathies: mechanisms of disease. Clin Med Insights Blood Disord 2014;7:7–16.

[282] Yelick PC, Trainor PA. Ribosomopathies: global process, tissue specific defects. Rare Dis 2015;3(1):e1025185.

[283] Hannan KM, Sanij E, Rothblum LI, Hannan RD, Pearson RB. Dysregulation of RNA polymerase I transcription during disease. Biochim Biophys Acta 2013;1829(3–4):342–60.

[284] Narla A, Ebert BL. Ribosomopathies: human disorders of ribosome dysfunction. Blood 2010;115:3196–205.

[285] Cassidy SB, Schwartz S, Miller JL, Driscoll DJ. Prader-Willi syndrome. Genet Med 2012;14(1):10–26.

[286] Kadakia S, Helman SN, Badhey AK, Saman M, Ducic Y. Treacher Collins syndrome: the genetics of a craniofacial desease. Int J Pediatr Otorhinolaryngol 2014;78(6):893–8.

[287] Larsen DH, Hari F, Clapperton JA, Gwerder M, Gutsche K, Altmeyer M, Jungmichel S, Toledo LI, Fink D, Rask MB, Grøfte M, Lukas C, Nielsen ML, Smerdon SJ, Lukas J, Stucki M. The NBS1–Treacle complex controls ribosomal RNA transcription in response to DNA damage. Nat Cell Biol 2014;16(8):792–803.

[288] Dauwerse JG, Dixon J, Seland S, Ruivenkamp CA, van Haeringen A, Hoefsloot LH, Peters DI, Boers AC, Daumer-Haas C, Maiwald R, Zweier C, Kerr B, Cobo AM, Toral JF, Hoogeboom AJ, Lohmann DR, Hehr U, Dixon MJ, Breuning MH, Wieczorek D. Mutations in genes encoding subunits of RNA polymerases I and III cause Treacher Collins syndrome. Nat. Genet 2011;43:20–2.

[289] Schaefer E, Collet C, Genevieve D, Vincent M, Lohmann DR, Sanchez E, Bolender C, Eliot MM, Nürnberg G, Passos-Bueno MR, et al. Autosomal recessive POLR1D mutation with decrease of TCOF1 mRNA is responsible for Treacher Collins syndrome. Genet Med 2014;16(9):720–4.

[290] Nikitina EA, Medvedeva AV, Zakharov GA, Savvateeva-Popova EV. Williams syndrome as a model for elucidation of the pathway genes- the brain-cognigive functions: genetics and epigenetics. Acta Naturae 2014;6(1):9–22.

[291] Cus R, Maurus D, Kühl M. Cloning and developmental expression of WSTF during *Xenopus* laevis embryogenesis. Gene Expr Patterns 2005;6(4):340–6.

[292] Ashe A, Morgan DK, Whitelaw NC, Bruxner TJ, Vickaryous NK, Cox LL, Butterfield NC, Wicking C, Blewitt ME, Wilkins SJ, Anderson GJ, Cox TC, Whitelaw E. A genome-wide screen for modifiers of transgene variegation identifies genes with critical roles in development. Genome Biol 2008;9(12):R182.

[293] Barnett C, Yazgan O, Kuo HC, Malakar S, Thomas T, Fitzgerald A, Harbour W, Henry JJ, Krebs JE. Williams syndrome transcription factor is critical for neural crest cell function in *Xenopus* laevis. Mech Dev 2012;129(9–12):324–38.

[294] Lalli MA, Jang J, Park JC, Wang Y, Guzman E, Zhou H, Audouard M, Bridges D, Tovar KR, Papuc SM, Tutulan-Cunita AC, Huang Y, Budisteanu M, Arghir A, Kosik KS. Haploinsufficiency of BAZ contributes to Williams syndrome through transcriptional dysregulation of neurodevelopmental pathways. Hum Mol Genet 2016;25(1):1294–306.

CHAPTER 11

Chromatin Dynamics and DNA Repair

P. Agarwal, K.M. Miller

Department of Molecular Biosciences, Institute for Molecular and Cellular Biology, University of Texas at Austin, Austin, TX, United States

11.1 INTRODUCTION

The stability and accurate duplication of genetic information is vital for cellular and organismal homeostasis. DNA in the nucleus of eukaryotic cells is packaged into chromatin, which organizes DNA into the volume of the nucleus, and regulates the accessibility and the use of the genetic material (Fig. 11.1). DNA is under constant threat of being damaged by reactions involving metabolic by-products, DNA processing enzymes and errors that occur during DNA synthesis. In addition, UV radiation, environmental agents, and medical treatments including radiotherapy and current chemotherapeutics also act as exogenous DNA damaging agents [1,2]. DNA damage must be repaired rapidly and precisely to ensure the protection and maintenance of the genome (Fig. 11.1). DNA damage also threatens the epigenome, changes in which can alter how genetic information is accessed and expressed. To combat these genetic and epigenetic risks, cells employ DNA damage response (DDR) pathways that function to detect the damaged DNA lesion within chromatin, signal its presence to promote the appropriate response, and finally to repair the lesion to maintain the correct sequence of the DNA within its appropriate chromatin environment (Fig. 11.1, [1,2]).

Determinants of chromatin dynamics that influence DNA repair include histone posttranslational modifications (PTMs), histone variants, chaperones, and chromatin remodeling complexes; as well as the factors that "read," "write," and "erase" histone marks (Fig. 11.2, [3–7]). DNA damage occurs within these chromatin environments, and current research has revealed an essential interplay between chromatin and the DDR that is crucial for repairing damaged DNA across the chromatin landscape to ensure genome stability.

Here we present our current understanding of how chromatin dynamics orchestrates DNA repair. We limit our discussion to eukaryotic model systems

CONTENTS

Chromatin Regulation and Dynamics. http://dx.doi.org/10.1016/B978-0-12-803395-1.00011-3

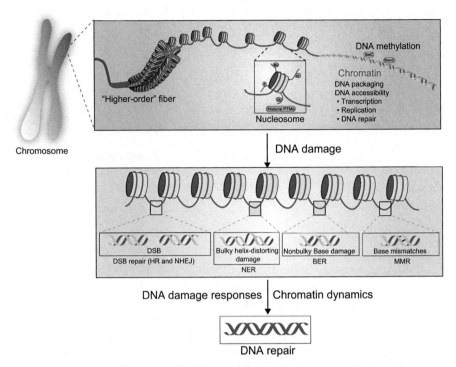

FIGURE 11.1 Nuclear DNA in eukaryotes is organized into nucleosomes and packaged in higher-order chromatin structure that regulates several DNA-based processes including DNA repair. DNA damage occurs within chromatin where specialized DNA repair pathways detect, signal and promote an appropriate DNA-damage repair pathway in order to maintain genome integrity. DNA damage response pathways *(DDR)* cooperate with various chromatin pathways to orchestrate DNA repair within the context of chromatin. *DSB*, DNA double-strand break; *BER*, base excision repair; *MMR*, mismatch repair; *NER*, nucleotide excision repair.

with a focus on yeast and mammalian cell systems. We first introduce the major DNA repair pathways and how they operate within chromatin, with a focus on nucleotide excision repair (NER) and DNA double-strand break repair (DSBR). We highlight how DNA repair is coordinated with transcription, and describe how nuclear dynamics, including chromosome mobility, impact DNA repair. We also cover emerging techniques that are advancing our ability to study how chromatin dynamics regulates DNA repair including genome editing, nucleosome dynamics, site-specific and inducible DNA damage induction systems, and next-generation sequencing strategies. We will not discuss chromatin dynamics during DNA replication, the cell cycle or the role that chromatin–remodeling complexes play to promote DNA repair since these topics will be covered in other chapters of the book.

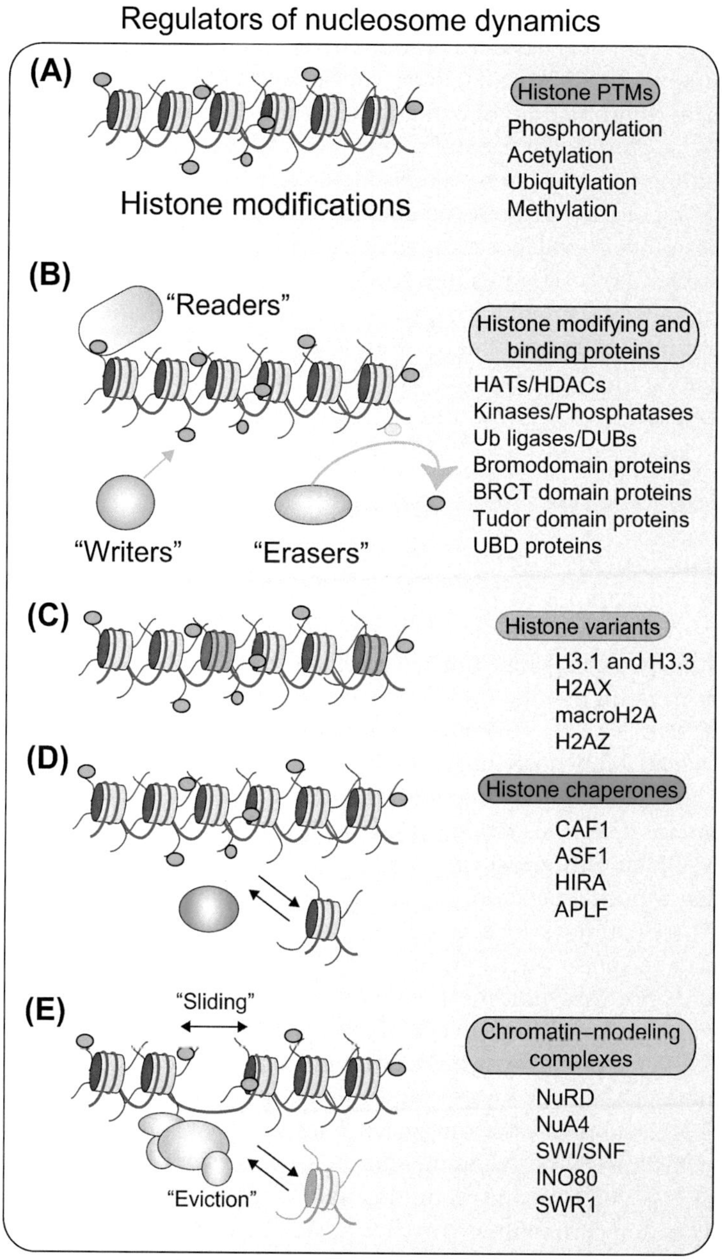

Regulators of nucleosome dynamics

FIGURE 11.2 Illustration of various pathways involved in chromatin dynamics and DNA repair.
(A) Histone modifications. (B) Histone PTM modifying and binding proteins that "write," "erase," and
"read" chromatin marks. (C) Histone variants. (D) Histone chaperones, and (E) Chromatin remodeling
complexes that regulate nucleosome and chromatin dynamics through activities including sliding and
eviction of nucleosomes. A limited set of examples for each category is provided.

Chromatin alterations occurring in diseases including cancer are known to influence DNA repair. This has been especially well-established in the context of cancer treatments, as DNA damaging agents are used as frontline therapies to treat cancer, and chromatin-based mechanisms have been identified as mediators of these responses. For example, cancer cells resistant to DNA damaging agents including cisplatin can be reversed to a drug-sensitive state through the reprogramming of chromatin by histone deacetylase inhibitors [8]. These and other studies provide evidence that epigenetic drugs can be used in the treatment of diseases [1,9–11]. Thus, the study of chromatin dynamics and DNA repair is not only fundamental for understanding genome maintenance, but it is also becoming an area of research with potential to provide novel insights into modulating the epigenome for therapeutic interventions to treat diseases including cancer.

11.2 CHROMATIN DYNAMICS AND DNA REPAIR

11.2.1 Nucleotide Excision Repair

Metabolites and by-products (ROS-generated cyclopurines, nitrosamines), environmental chemicals, UV radiation [cyclobutane pyrimidine dimers and 6-4 photoproducts (6,4-PPs)], and anticancer DNA intrastrand cross-linking agents (e.g., cisplatin) result in bulky DNA adducts and lesions that distort the helical structure of DNA [12,13]. NER detects and repairs these lesions within chromatin (Fig. 11.1). NER functions through two distinct subpathways, global genome (GG)-NER and transcription-coupled (TC)-NER, which differ in the initial step of DNA damage detection. While the GG-NER pathway acts within the entire genome, TC-NER is only activated when RNA polymerase II (RNAP II) is stalled during transcription elongation due to the presence of a template strand lesion [14]. These pathways are essential for human health, as defective GG-NER pathway leads to cancer predisposition. TC-NER defects cause several diseases including UV sensitive syndrome, premature aging conditions including Cockayne syndrome, and mutations in any of seven NER genes, XP-A through XP-G, give rise to a hereditary disease xeroderma pigmentosum, which is characterized by extreme UV sensitivity and cancer predisposition [13]. After lesion detection, the NER subpathways converge and involve the coordinated action of 20–30 proteins for DNA unwinding and incision/excision around the lesion, gap-filling DNA-synthesis, and strand ligation to restore the DNA [12,13,15]. Accessibility within chromatin is crucial for the NER machinery to function. Pioneering studies using human fibroblasts showed increased nuclease sensitivity in chromatin regions undergoing NER after UV damage. These results provided the first evidence that chromatin rearrangements occur in response to DNA damage and led to the "access-repair-restore" model [3,16,17]. This model set forth the basic notion that chromatin dynamics was essential for DNA repair.

NER is more efficient in naked DNA compared to chromatin, indicating that chromatin is a barrier for NER [18–20]. Chromatin remodeling complexes, histone chaperones, and histone modifications modulate NER within the chromatin environment. The ATP-dependent chromatin remodeling complexes, SWI/SNF and ISW2, act on UV-damaged nucleosomes to aid in DNA damage recognition and stimulate repair by photolyases in vitro and in yeast cells [18–20]. The human SWI/SNF component SNF5 interacts with the UV damage recognition factor XPC to facilitate DDR kinase ATM recruitment to lesions in order to increase the accessibility of the NER machinery to DNA and promote repair [21]. TC-NER also modulates spliceosome activity through a pathway involving ATM, resulting in damage-induced alternative pre-RNA splicing in response to UV damage [22]. BRG1 within the SWI/SNF complex facilitates UV-induced chromatin relaxation and stabilization of XPC to stimulate the recruitment of NER factors XPG and PCNA to UV-induced lesions [23]. These findings illustrate the importance of chromatin remodeling during NER.

Several histone PTMs have been reported to regulate NER. Histone acetylation was the first DNA damage-related chromatin modification to be identified and occurs after UV irradiation to promote NER [24]. UV-induced acetylation of H3K9 and H4K16 are well studied in the context of NER and occur prior to repair, suggesting that these acetylations may facilitate repair protein accessibility to the damage [25]. Indeed, inactivation of human acetyltransferase GCN5 decreases recruitment of NER factors to damaged DNA [26]. These studies highlight how acetylation facilitates the access of NER machinery to lesions in chromatin.

Histone modifications in addition to acetylation also participate in NER. For instance, H3K79 methylation is involved in GG-NER [25]. UV damage also induces NER-dependent H2A monoubiquitylation and occurs after incision of the damaged region [27]. The UV-DDB E3 ubiquitin ligase complex plays a key role in organizing chromatin during GG-NER initiation by ubiquitylating core histones in order to activate repair by histone displacement [12]. These findings demonstrate how histone PTMs modulate chromatin dynamics upstream of NER initiation. Chromatin modifications could serve as a platform to recruit chromatin remodeling complexes to evict, remodel or slide nucleosomes at sites of DNA lesions. These activities would pave the way for NER damage recognition factors to locate the lesion and initiate NER.

The involvement of complex chromatin dynamics in the NER pathway is further indicated by the discovery that histone exchange occurs at UV damaged sites. For instance, accelerated H2A and H2B histone exchange at sites of TC-NER (mediated by the histone chaperone FACT complex) stimulates repair and regulates transcription [28]. Moreover, new histone H3.1 deposition by histone chaperones CAF1 and ASF1, as well as H3.3 variant incorporation by HIRA occurs at UV damage site [29,30]. These findings highlight dynamic

Table 11.1 Summary of Chromatin Factors Involved in DNA Repair That are Highlighted in the Text

DNA Repair Pathway	Histone PTM	Histone Chaperone	Histone Variant
DSBR	H2A/X S139p		
	H2A/X K13/15-Ub		
	H2A/X K118/119-Ub		
	H2BK120-Ub	CAF-1	
	H3K9me2/3	ASF1	H2AX
	H3K27me3	FACT	macroH2A1
	H4K5/K8/K12/K16Ac	Anp32e	H2AZ
	H3K9/K14/K18/K23/K56Ac		
	H2AX K5/K36Ac		
	H3K79me		
	H4K20me2		
NER	H3K9Ac		
	H4K16Ac	CAF-1	
	H3K9me3	ASF1	H2AX
	H3K79me1/2/3	HIRA	H3.3
	H2AK119-Ub	FACT	
	H2AX S139p		
BER	Not determined	Not determined	Not determined
MMR	H3 Acetylation	CAF-1	Not determined
	H3K36me3		

chromatin reorganizations in response to DNA damage that promote NER (Table 11.1). Additional studies on the interplay between chromatin and NER across the genome and epigenome will likely reveal mechanistic insights into NER repair.

11.2.2   DNA Double-Strand Break Repair
11.2.2.1   Chromatin and DSBR

A DNA double-strand break (DSB) is a precarious type of lesion, as both DNA strands are severed (Fig. 11.1). Broken DNA ends can be degraded or ligated to other breaks incorrectly, resulting in alterations to the genetic material. DSBs are repaired principally by homologous recombination (HR) or nonhomologous end-joining (NHEJ) [31]. HR uses a template to faithfully repair the DSB, while NHEJ repairs the break by ligating the broken ends together. In mammalian cells, NHEJ is active throughout the cell cycle, including S/G2 when HR occurs [32,33]. Therefore, the choice for engaging a particular DSB repair pathway has been an important question in the field [34]. It is now widely accepted that chromatin orchestrates DNA repair reactions including how DSB repair pathway choice is achieved. Identifying how DNA repair reactions occur within the context of chromatin is fundamental for mechanistically understanding DNA repair in cells [3].

The presence of a DSB results in alterations to the chromatin surrounding the break that promote DNA damage signaling, activation of the DDR and ultimately repair of the break (Fig. 11.1). Histone PTMs and histone variants play critical roles in DSB repair [35]. Histone PTMs function to alter the chromatin structure to facilitate repair, and also provide molecular DSB recognition signals for PTM-binding proteins. DSBs activate DDR kinases including ATM and DNA-PK that phosphorylate the C-terminus of the histone variant H2AX (called γH2AX) in up to a megabase of chromatin surrounding a DSB [36]. γH2AX forms microscopically visible nuclear foci that bolster the focal accumulation of DDR factors to the break sites [37,38]. The DNA damage protein MDC1 is a phospho-binding protein that binds γH2AX and mediates the recruitment of additional DDR factors to DSBs [35]. In addition to phosphorylation signaling, several E3 ubiquitin ligases including RNF8, RNF168, BRCA1, RING1B, and BMI1 are recruited to DSBs to ubiquitylate the surrounding chromatin [39,40]. For example, H2A/H2AX is ubiquitylated by RNF168 on Lys-13/15 [41–43], and by RING1B/BMI1 on Lys-118/119 [44–47]. These modifications mediate the chromatin association of DDR factors including 53BP1, which binds to RNF168-dependent H2A-Ub and H4K20me2 through ubiquitin- and methyl-binding domains [43]. Recruitment of 53BP1 to damaged chromatin promotes DNA repair by NHEJ. DSB recruitment of over half of all histone acetyltransferases and histone deacetylases in mammalian cells supports the notion that acetylation signaling plays important roles in chromatin responses to DSBs [7]. For example, acetylation of H4K16 at DSBs by the TIP60 HAT promotes BRCA1 binding to facilitate HR, while deacetylation of this mark by HDAC1/2 promotes 53BP1 binding and NHEJ [48,49]. Along with the recruitment of acetylation "writers" and "erasers," one-third of acetylation readers (i.e., bromodomain proteins) relocalize in response to DNA damage [50]. Finally, the histone variants macroH2A1 and H2AZ, along with the H2AZ histone chaperone ANP32E, have been shown to be critical players in orchestrating DSB repair [51–53]. Thus, these examples illustrate how DDR chromatin-binding proteins along with site-specific histone PTMs and histone variants mediate critical signaling events that promote sensing and repair of DSBs in chromatin (Table 11.1).

In addition to PTMs and histone variants, other chromatin components influence DSB repair. In response to DSBs, DDR kinases, including ATM, also phosphorylate nonhistone and chromatin components that mediate chromatin responses to DSBs. ATM phosphorylates the transcriptional corepressor KAP1 in response to DSBs, leading to the decompaction of chromatin that is believed to facilitate DDR signaling and allow access of the DSB repair machinery to the damage [54]. This may be particularly important in heterochromatic regions containing DSBs due to the compact packaging of the chromatin, a potential barrier to DSB processing and repair [55]. Decompaction of chromatin in the absence of DNA damage also activates ATM, suggesting an important relationship between chromatin structure and the DDR [56]. It is clear that DSB repair relies on different mechanisms in euchromatin and heterochromatin due to the presence of different chromatin structures, factors that reside in these different chromatin environments, and the activities occurring in these regions including transcription. It is important to note that unlike budding yeast, mammalian cells have heterochromatin that is enriched in H3K9me3 and DNA methylation (i.e., constitutive heterochromatin), and histone H2A ubiquitylation (H2A-Ub) and H3K27me3 (i.e., facultative heterochromatin) [57,58]. Thus, differences between DSB repair mechanisms will occur due to the presence of different chromatin states and the pathways that regulate them, making simple comparisons between chromatin-based DSB repair pathways in different organisms complex [3–6].

11.2.2.2 Transcriptional Repression by DSBs

Transitions between euchromatin and heterochromatin during DSB repair also impact transcription. DSBs can thus signal the repression of transcription in its vicinity [59–61], most likely to avoid interference between transcription and repair machineries. Several mechanisms have been identified by which the DDR regulates transcription following DNA damage. The DDR kinases ATM and DNA-PK are required to repress transcription following DSB formation [59,61]. ATM activation has been shown to be regulated by the HAT TIP60, which itself is activated by the heterochromatin associated mark H3K9me3 [62]. Interestingly, H3K9me3 is induced at DSBs and TIP60 is required to repress transcription following DNA damage [50,63]. These findings help explain how TIP60 can be activated in transcriptionally active chromatin, which normally has low levels of H3K9me3. Acetylation of chromatin by TIP60 also results in the recruitment of the bromodomain protein ZMYND8 and the NuRD chromatin–remodeling complex, which deacetylates chromatin to repress transcription and promote HR [50]. Thus, TIP60 plays several key roles in modifying chromatin to both promote DSB repair and repress transcription. Facultative heterochromatin also participates in DSB repair. Polycomb repressive complexes 1 and 2 (i.e., PRC1 and PRC2) thus associate with DSBs and induce the repressive histone marks H2A-Ub and H3K27me3 [60,64]. One role of the PRC1 complex in DSB repair is to interact with the elongating RNAPII complex, which recruits PRC1 to sites of active transcription within damaged chromatin. This allows for the selective deposition of H2A-Ub around the DSB that represses transcription to facilitate NHEJ [64]. In summary, several chromatin-based DDR mechanisms have evolved to coordinate transcription with DSB repair.

11.2.2.3 Role of Regulated DSBs in Transcription

Although transcription can be negatively regulated by DSBs, it appears that DNA damage can also play a positive role during transcription. Rosenfeld and coworkers identified several scenarios where DNA damage induced by topoisomerases I or II (TOPI or TOPII) at enhancers or promoters, respectively, regulates transcription [65,66]. Formation of DSBs by TOPII was shown to activate PARP1 resulting in the exchange of histone H1 with HMGB1 followed by gene activation. Conversely, inhibition of TOPII or PARP1 activity blocked transcriptional induction. These studies are particularly interesting given that TOPI/II inhibitors are used therapeutically to eliminate cancer cells. Identifying whether these compounds regulate transcription in addition to creating cytotoxic DNA damage as part of their mechanism of action deserves consideration. Activation of transcription by estrogen results in demethylation of histones, including H3K9me2, a process that creates hydrogen peroxide as a by-product [67]. Hydrogen peroxide can oxidize DNA resulting in DNA damage. Interestingly, this reaction was shown to promote transcription through recruiting DNA repair proteins and TOPIIβ, which altered chromatin structure to enhance transcription [67]. In addition to transcriptional regulation by DNA damage, posttranscriptional processes also appear to play important roles in DSB repair. Noncoding RNAs have been identified in the vicinity of DSBs and play a role in their repair. Indeed, impairment of the RNA interference pathway involving DICER and DROSHA results in defective DSB repair [68]. These studies illustrate the complex relationship between transcription and the DDR. In some situations, DNA damage represses transcription, while in other scenarios it activates transcription. There are likely many layers of regulation and signals that act combinatorially and in a context-specific manner to promote the appropriate DDR pathway. Further studies are needed to integrate these findings into a clear molecular and genome-wide view of how transcription and the DDR cooperate to maintain genome function and stability.

11.2.3 Base Excision Repair

The reactive nature of DNA results in 10,000–20,000 individually damaged bases per mammalian genome per cell every day [69]. Spontaneous damage to DNA bases resulting from cellular metabolism, for example, reactive oxygen species (ROS), include oxidation, deamination, methylation and loss of the base entirely to create abasic sites (i.e., apurinic/pyrimidinic AP sites) [70]. These single-nucleotide, nonhelical distorting, damaged bases are repaired mainly by base excision repair (BER) [71,72]. BER must recognize base damages, including those contained within nucleosome structures, and repair them (Fig. 11.1). This process, called short patch BER (SP-BER), first involves recognition and removal of the damaged base by a DNA glycosylase. The damaged site is then processed by the endonuclease APE1 to promote base addition through template-directed synthesis by DNA polymerase beta that functionally interacts with XRCC1. After base addition, DNA Ligase III seals the nick. Single strand breaks resulting from BER intermediates or other DNA damaging reactions can use a subset of SP-BER factors to promote long Patch BER (LP-BER). In LP-BER, the single strand break is recognized by PARP1/XRCC1 complex, and ~2–10 bp are excised and replaced through the combined action of several LP-BER associated factors including PCNA, Pol δ/ε, FEN1, and DNA Ligase I/III. These reactions must occur within diverse chromatin environments, and some studies have already started to address how BER functions within chromatin [20,73].

The wrapping of DNA around the nucleosome creates a potential barrier for BER to recognize and repair damaged bases. However, transient "DNA unwrapping" events have been observed between DNA and histone octamers suggesting that nucleosomes contain intrinsic instability [74], which could allow DNA sampling by glycosylases within the nucleosome. In a similar fashion, other chromatin components, including DNA binding proteins and chromatin-associated RNAs, might likewise display transient interactions with DNA to allow BER processes to take place in chromatin. Studies of BER within nucleosomes have shown that damaged bases are repaired less efficiently within the nucleosome compared to free DNA [75,76]. The placement of the damaged base on either the "inner-facing" or "outer-facing" DNA strand in relationship to the histone octamer also governs repair efficiency. Indeed, lesions placed on the "outer-facing" DNA display increased repair kinetics compared to "inner-facing" base lesions [76]. These observations are thought to represent the increased occlusion of "inner-facing" base lesions due to the sterical hindrance between the histone octamer and BER factors. The final ligation step by the Ligase IIIα–XRCC1 complex disrupts the nucleosome due to its requirement for binding completely around the DNA substrate, a process that must break histone–DNA contacts [77]. These results obtained in vitro are supported by data obtained in cells showing that damaged bases are preferentially repaired in euchromatin versus heterochromatin regions [78].

Chromatin dynamics is therefore likely to participate in BER. For example, ATP-dependent chromatin remodeling complexes, including SWI/SNF and ISW1/2 have been shown to facilitate BER [20,73]. The ability of these complexes to remodel and position nucleosomes could help to orchestrate BER activities within chromatin. Chromatin dynamics regulating euchromatin and heterochromatin states also utilize histone modifications and histone variants, which have the potential to influence BER in chromatin. Finally, BER enzymes accumulate on damaged chromatin to facilitate repair, suggesting that chromatin also aids in localizing these factors to sites of repair [79]. In agreement with this idea, chromatin modifying enzymes involved in BER, such as the poly (ADP-ribose) polymerase (PARP) enzymes, localize to chromatin upon base damage [80,81]. Modification of chromatin, including poly (ADP) ribosylation, could support the recruitment of BER factors or alter chromatin structure to promote BER. Future studies identifying how BER manages to repair thousands of damaged bases within diverse chromatin environments will surely offer new insights into this essential DNA repair pathway.

11.2.4 Mismatch Repair

The faithful duplication of the genome and epigenome into two identical daughter cells is fundamental for preserving the integrity and regulation of genetic material. Although replication utilizes DNA polymerases with proof-reading capabilities, errors involving the incorrect copying of bases resulting in base-base mismatches and small insertion-deletion mismatches still occur frequently. Prokaryotic and eukaryotic cells contain a genome maintenance pathway called mismatch repair (MMR), which primarily resolves mismatch base errors in a strand-specific manner [82,83]. MMR discriminates between the parental and daughter DNA strands so that the repair event replaces the base complementary to the parental "correct" strand. Mismatches are recognized by either MutSα (MSH2–MSH6) or MutSβ (MSH2–MSH3) complexes. These complexes help load the MutLα and replication associated factors including PCNA, the exonuclease EXO1, Pol δ and Ligase I, which collectively remove the mismatch and synthesize the correct base(s) to repair the mismatched error(s) [84]. MMR increases the "error-free" rate of replication ~threefold, and defects in MMR are associated with increased mutation rates and cancer. A noncanonical MMR pathway also exists, which corrects mismatches outside of DNA replication [85]. MMR is also involved in genetic recombination where it suppresses recombination between homologous sequences. Collectively, these MMR pathways are vital for the maintenance of genome integrity.

How MMR occurs within chromatin is poorly understood despite that recent studies have started to address this question. Although MMR reactions have been reconstituted in vitro, the addition of nucleosomes onto DNA templates inhibits MMR, suggesting that MMR must overcome this barrier to ensure repair [86]. Consistent with this idea, EXO1, an important nuclease in MMR, is inhibited by nucleosomes [87]. Both nucleosome assembly and MMR occur during replication, suggesting that these processes are coordinated. MMR proteins inhibit nucleosome assembly, providing one potential mechanism to ensure that nucleosomes do not interfere with repair reactions. An interaction between MutSα and the chromatin assembly factor CAF-1 has thus been identified in human cells, which could explain how MMR regulates chromatin deposition to ensure unimpeded repair by MMR [88]. If the MMR machinery does encounter a nucleosome, one possible mechanism to overcome nucleosome barriers is through chromatin remodeling. Interestingly, the human MSH2–MSH6 complex has been reported to contain chromatin-remodeling activity, as this heterodimer can disassemble nucleosomes in vitro [89]. The nucleosome-remodeling activity of MSH2–MSH6 is enhanced by acetylation of H3, suggesting that histone modifications that reduce DNA–histone contacts can facilitate MMR. Consistent with this idea, H3K56Ac, a histone mark known to reduce histone–DNA contacts [90], participates in MMR to reduce spontaneous DNA lesion formation in yeast [91]. These studies provide important evidence for the role of chromatin dynamics in aiding MMR. Histone modifications in addition to acetylation have also been shown to be involved in MMR. The MutSα recognition factor MSH6 thus contains a Pro-Trp-Trp-Pro (PWWP) domain, a chromatin reader domain known to bind to methylated lysines or arginines. The PWWP domain on MSH6 was shown to bind to SETD2-generated H3K-36me3 in human cells to facilitate chromatin binding of the MutSα complex to mismatched DNA lesions [92]. Human cells lacking SETD2 and H3K36me3 display a mutator phenotype, consistent with a role for H3K36me3 in MMR [86]. Additional studies investigating chromatin regulation by MMR are warranted. For example, although the human MutSα contains a histone modification reader domain, the MutSβ (MSH2–MSH3) complex does not [86,92], suggesting a different mode of chromatin interaction. Additionally, H3K36me3 is important in transcription regulation and DSB repair, which could also impact MMR [93–95]. These studies highlight the importance of chromatin dynamics in facilitating MMR, suggesting the need for additional studies to further illuminate how MMR occurs within chromatin (Table 11.1).

11.3   CHROMOSOME MOBILITY AND THE DDR

Chromatin organization can be influenced locally by the presence of DNA damage, which in turn can also affects chromosome dynamics and mobility. Early experiments analyzing chromosomes after UV irradiation in quiescent normal human fibroblasts thus detected the appearance of elongated chromosomes after UV damage, suggesting decompaction of the chromatin [96]. Similar results were observed after UVA laser treatment and ionizing radiation in mammalian cells [97]. Using energy-filtering transmission electron microscopy, ~40% chromatin decompaction was observed at damage sites. Although local chromatin structure was altered by DNA damage, real-time tracking of DSBs by fluorescent microscopy did not observe a general increase in the mobility of the damaged DNA [97]. In other studies that analyzed DSB mobility, including the use of α particle radiation-induced DNA damage or site-specific endonuclease induced DSB formation in yeast, mobility of DNA damage sites was observed [98,99]. Given that variable results were obtained from different experimental systems and organisms, this suggests that chromosome mobility following DNA damage is quite complex and could take place in a context-specific manner [100].

Increased chromosome mobility following DNA damage could provide a mechanism that promotes DNA repair. Analysis of whole chromosome movements following DNA damage in yeast revealed that DSB induction increased the mobility of a chromosome containing the DSB as well as the mobility of undamaged chromosomes under certain conditions [101,102]. Defects in factors that regulate chromosome movement in response to DSBs in yeast, including chromatin–remodeling complexes and DDR factors, also cause concomitant impairment of DNA repair, including HR [100]. These studies support the idea that chromosome mobility serves, at least in part, to promote homology search during HR. This also appears to be the case for recombination-prone telomeres found in ALT cancer cells. Increased mobility of ALT telomeres and uncapped telomeres undergoing fusion by NHEJ repair have been observed, suggesting that telomeres could represent chromosomal locations prone to mobilization when damaged or undergoing repair reactions [103,104]. Additional functions for chromosome mobility following DNA damage have also been proposed. For example, in yeast, DSBs occurring in the nucleolus are shuttled outside this nuclear compartment [105]. Defects in relocalization of nucleolar DSBs in yeast result in hyperrecombination between rDNA repeats and genome instability. In flies, DSBs located in heterochromatin are relocalized to euchromatic regions to facilitate repair by HR [106]. Therefore, mobilization of DSBs located in repetitive rDNA or heterochromatin repeats from one nuclear environment to another represents one strategy for regulating the spatial and temporal repair of DNA lesions within these genomic loci.

In addition to chromatin–remodeling complexes, regulation of histone acetylation also impacts chromosome movement in response to DSBs in human cells [107]. Therefore, chromatin dynamics are likely to impact both local and global aspects of DNA repair within chromatin, along chromosomes and in different nuclear settings. Identification of how 3D nuclear architecture cooperates with the modifications of the primary chromatin fiber in response to genotoxic stress to maintain the fidelity of the genome and epigenome will be an exciting and active area of future research.

11.4   SITE-SPECIFIC ANALYSIS OF DNA DAMAGE AND CHROMATIN

Several experimental systems have been employed to study the DDR in eukaryotic cells [108] (Fig. 11.3). These include the use of exogenous DNA damaging agents, such as ionizing radiation and UV, as well as radiomimmetic drugs (i.e., bleomycin) or TOPI/II inhibitors. Laser systems have also been used to generate DNA damage in cells, and have provided a wealth of information about the cell's response to DNA damage [109]. Endonucleases that cleave DNA recognition sequences within the genome have been widely used to study DNA damage. More recently, the identification of nucleases, including zinc-finger nucleases (ZFNs), TALENs and CRISPR-Cas9 nucleases, which can be designed to target virtually any sequence in the genome, have started to be used by researchers for DDR studies. The use of site-specific nuclease systems coupled with chromatin immunoprecipitation (ChIP) and sequencing have provided particularly powerful systems to study chromatin dynamics and DSB repair, although each system has its advantages and disadvantages for studying chromatin-based DDR mechanisms [55].

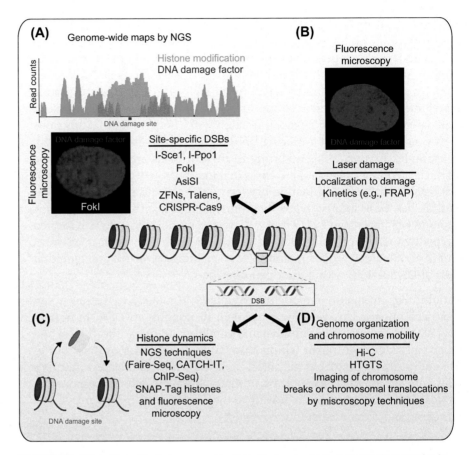

FIGURE 11.3 Selection of techniques used to study the involvement of chromatin dynamics in DNA repair pathways.
(A) Site-specific DNA double-strand break *(DSB)* systems. These systems have been combined with genome-wide analysis using next generation sequencing and fluorescence microscopy approaches to provide innovative approaches to study chromatin dynamics following DNA damage. (B) Laser damage coupled with fluorescence microscopy techniques have been used extensively to analyze dynamics of DNA repair and chromatin factors at DNA damage sites. (C) Techniques used to study histone dynamics. (D) Genome organization and chromosome mobility have been studied using a combination of approaches involving sequencing and microscopy.

The use of nucleases to study chromatin dynamics upon DSB formation was originally performed in yeast using the HO DSB system. HO endonuclease normally recognizes an 18 bp sequence in the mating-type (MAT) locus of *Saccharomyces cerevisiae*, which it cleaves to form a DSB. The formation of a DSB in the MAT locus facilitates recombination to promote mating type switching. However, the recognition sequence of the HO endonuclease can be used to create DSBs in other regions of the genome as well. For example, the HO

DSB system was used to study γH2AX formation surrounding a single DSB. This analysis used ChIP followed by PCR to identify the spatiotemporal formation of γH2AX surrounding the HO-induced DSB as well as the genetic requirements for γH2AX formation and spreading around the break [110]. This analysis revealed that γH2AX was increased in a region that spanned over 50 kb around the break, and required the DDR kinases Tel1p and Mec1p, the human ATM and ATR homologs, respectively. A plethora of studies have performed similar experiments in yeast to identify how a single HO-induced DSB alters chromatin dynamics and DSB repair, which has greatly advanced our understanding of these processes [4].

Engineered site-specific DSB systems have been used to study chromatin alterations surrounding a DSB in mammalian cells. The I-*SceI* endonuclease from *S. cerevisiae* is frequently used to study DSB repair and chromatin in mouse and human cells. An early example of the use of I-*SceI* was provided by Herceg and coworkers who inserted the I-*SceI* recognition sequence into the HPRT locus in embryonic stem (ES) cells [111]. Following ectopic expression of I-*SceI*, ChIP analysis with PCR was performed to analyze DDR factor occupancy and chromatin dynamics at an I-*SceI*-induced DSB. This study found that the HAT TIP60 localized to DSBs and regulated H2A and H4 acetylation surrounding the break [111]. Another rare-cutting nuclease, I-*PpoI*, has been used similarly in mammalian cells. Mammalian genomes only contain a few hundred I-*PpoI* recognition sequences, and transient expression of I-*PpoI* results in ~10% target recognition and cutting. Kastan and coworkers developed this system to create DSBs at defined, genomically encoded target sequences in mammalian cells [112]. With this system, histone loss dependent on ATM was detected surrounding the I-*PpoI* DSB. Greenberg and coworkers developed a site-specific DSB generating system to create DSBs near a reporter gene to analyze the effect of DSBs on transcription [59]. These site-specific DSBs were generated by expressing the nonspecific *FokI* nuclease domain fused to the lac repressor in cells where the lac operator was placed upstream of the reporter gene. This system allowed the recruitment of *FokI* to the reporter gene through LacI:LacO binding to generate site-specific DSBs (Fig. 11.3). Using this system, it was demonstrated that the ATM kinase represses transcription upon DSB formation through histone H2A-Ub [59]. These studies have provided insights into how DSBs alter chromatin dynamics to promote the DDR.

A limitation of several site-specific DSB systems is that only one to very few DSBs can be analyzed at a time, and the expression of the nuclease is difficult to control. These limitations have been overcome by the development of the DIvA (DSB inducible via *AsiSI*) system by Legube and coworkers. For DIvA, the *AsiSI* restriction enzyme was fused to an inducible estrogen receptor ligand-binding domain and an auxin-inducible degron [93]. This nuclease can thus be both "turned-on" and "turned-off". In addition, the *AsiSI* recognition sequences that are cleaved upon expression of *AsiSI* are relatively common in the mammalian genome allowing for several hundred *AsiSI*-induced DSBs to be analyzed simultaneously. This system represents the first inducible DSB system to be analyzed genome-wide by ChIP-Seq to provide high-resolution maps of chromatin marks, including γH2AX and H3K36me3, at multiple DSBs across the genome. Interestingly, by analyzing many DSBs, this study demonstrated that actively transcribed regions marked by SETD2-dependent H3K36me3 are preferentially repaired by HR [93]. To further support these findings, a TALEN-induced DSB was formed in an endogenous gene that was transcriptionally silent or active after induction by interferon. Remarkably, the same gene was mainly repaired by NHEJ if repressed or HR if transcriptionally active [93]. Thus, transcriptional status is a major determinant of DSB repair pathway choice. This study demonstrates the advantages of using genome-wide approaches to study chromatin and DSB repair. The implementation of these techniques to study DNA damaging agents across the genome is emerging, and will provide new insights into how genomic features and chromatin processes regulate DNA repair pathways throughout the diverse genome and epigenome landscapes (Fig. 11.3, [11]).

11.5   HISTONE DYNAMICS FOLLOWING DNA DAMAGE

The composition and organization of histone proteins within chromatin are regulated mainly by chromatin–remodeling complexes and histone chaperones (see also Chapter 4). Several methodologies have been applied to measure histone dynamics following DNA damage (Fig. 11.3). In addition to site-specific DNA damage systems that have identified histone exchange at DNA damage sites [3], other experimental systems provide opportunities to analyze DNA damage-induced histone responses on a broader cellular and genome wide scale. Techniques involving fluorescently tagged histones have provided single-cell analysis using fluorescence microscopy to determine histone dynamics at DNA damage sites. Upon DNA damage, including by UV irradiation, incorporation of newly synthesized H3.1 was detected at the damage site [29]. H3.1 was tagged and analyzed by indirect immunofluorescence. Directly tagging H2A and H2B with GFP also provided evidence of increased histone exchange on chromatin after UV irradiation [28]. These observations were obtained using FRAP analysis, where the GFP-tagged histone can be photobleached and its recovery measured and monitored by live-cell fluorescence microscopy. Building upon their earlier observations of H3.1 recruitment to UV damage sites, Almouzni and coworkers used SNAP-tag technology to analyze newly synthesized histones, and observed de novo H3.3 incorporation following UV damage that was dependent on the histone chaperone HIRA [30]. Further investigation of this phenomenon suggested that H3.3 incorporation is important for resuming transcription following repair of UV damage [30]. Thus, these studies demonstrate how reorganization of histones following UV damage acts to facilitate not only repair but also the restoration of normal chromatin function at sites of DNA damage.

Photoactivatable histones have also been used to analyze chromatin dynamics at DNA damage sites. Using photoactivated GFP-tagged H2B, laser-damaged chromatin was fluorescently marked by "turning on" the GFP signal where the laser damage occurred. In this study, increased expansion of the chromatin at the laser damage was observed [97]. Using a similar approach, another study identified a compaction phase of the damaged chromatin region, which was dependent on macroH2A1 [52]. Thus, DNA damage can cause both the compaction and decompaction of chromatin. Alteration of chromatin structure could be achieved through exchanging histones. Genome-wide techniques involving next-generation sequencing (NGS) that can analyze histone exchange (i.e., CATCH-IT) and nucleosome-free regions (i.e., FAIRE-seq) have been applied to study histone dynamics following DNA damage. Following treatment with the TOPII inhibitor doxorubicin, a drug that induces DNA damage and is used as an anticancer chemotherapeutic drug, an increase in histone eviction was observed [113,114]. The fate of the exchanged histones is unclear, although degradation of histones following DNA damage has been observed [115]. Remodeling of histones following DNA damage could play several roles in the DDR. Removal of histones will likely reshape damaged chromatin including a change in histone modifications and the reader proteins that bind to them. These changes could regulate DNA repair activities as well as chromatin functions within the damaged region, including transcription. These studies clearly demonstrate that DNA damage is intimately linked with its chromatin environment. Emerging evidence suggests that DDR factors and chromatin must cooperate to ensure that proper DNA repair takes place. Due to the plasticity of chromatin, chromatin-based DDR mechanisms must have evolved to function within variable chromatin environments to repair DNA. Genome-wide strategies to study chromatin dynamics in response to DNA damage should be employed to provide a better understanding of how DNA repair occurs across the genome within different chromatin environments. This is particularly important in biological scenarios where the epigenome has been modified, including in diseases, such as cancer. It is well-established that changes in the epigenome can have drastic effects on cellular homeostasis. We propose that these epigenetic changes will also likely have important consequences for DNA repair. These are important considerations given that many cancer treatments involve DNA damage, and cancer genomes often exhibit epigenetic defects as well. Understanding the importance of epigenetics and the DDR could help improve the use of DNA damaging agents and epigenetic drugs to treat diseases.

11.6 GENOME ORGANIZATION AND CHROMOSOMAL TRANSLOCATIONS

Genome instability is a hallmark of cancer and can be manifested by recurrent translocations between different damaged chromosome regions [116]. Complex genome rearrangements have traditionally been identified through cytogenetic methods, but the widespread use of NGS technologies has allowed for their high-throughput identification across many individual cancer genomes. Although translocations are prevalent in cancer genomes, the identification of the genomic environment that favors their production as well as the factors that promote their formation are poorly understood. Several innovative technologies have begun to provide penetrating new insights into how translocations occur in mammalian cells (Fig. 11.3).

Novel NGS-based techniques have begun to map and identify the spatial, three-dimensional (3D) architecture of the genome. For example, the Hi-C technique allows for the identification of chromatin interactions across the entire genome [117]. This technique has identified spatially compartmentalized chromatin domains that are likely to regulate all chromatin-based processes including gene regulation, replication, and DNA repair (see also Chapter 1). In a pioneering study of genome organization and translocations, 3D-spatial genome organization maps generated by Hi-C of mouse pro-B cells were compared to high-throughput genome-wide translocation sequencing (HTGTS) maps [118]. This study identified the spatial proximity of DSBs as one of the major determinants of translocation formation. Using I-*SceI* site-specific DSBs (see Section 11.4) as well as randomly formed, IR-induced DSBs; it was observed that both intra- and interchromosomal interactions are significantly associated with translocation frequency between break sites. Thus, both the formation of DSBs and the factors that regulate and maintain the spatial organization of the genome are likely contributors to translocations.

The antibody gene mutator, activation-induced cytidine deaminase (AID), can induce DSBs leading to translocations [119–121]. AID normally induces DNA damage including DSBs as part of the normal genome-rearrangement programs associated with class switch recombination and somatic hypermutation [122]. However, AID can induce DSBs also outside of genomic loci associated with these biological processes, an activity linked with translocation formation [119,120]. Using techniques including Hi-C, AID-mediated translocations were shown to be associated with super-enhancer regions within multiple cell types including B cells and mouse embryonic fibroblasts (MEFs) [123–125]. Super-enhancers represent highly transcribed regulatory regions that integrate signals from multiple pathways to control the expression of genes important for cell identity. Indeed, cancer cells contain active super-enhancers associated with oncogenes involved in tumorigenesis [126,127]. In the case of AID-mediated translocations, regulatory regions could act to recruit AID, perhaps through their transcriptional activity. The recruitment of AID results in DSB formation in these areas, leading to translocations within unique subsets of genes that are spatially proximal to active super-enhancers [124]. This work suggests that DSB formation and spatial proximity of genes can lead to AID-induced translocations, which could help explain why the translocations that occur frequently in cancer cells exhibit cell type-specific formation between different gene partners. These studies suggest that recurrent translocations likely take place within certain cells as a consequence of frequent DSB formation within genomic loci that have proximity to each other within the spatial organization of the genome.

The advent of high-throughput cell biology techniques that allow the visualization of individual DSB sites within living cells enable the tracking of translocation events in real-time, and have also provided new insights into how translocations are formed. The Misteli group has engineered a powerful system using multiple integrated I-*SceI* sites flanked by LacO/TetO operator arrays to induce and fluorescently mark damaged sites, which can be analyzed by live-cell fluorescence microcopy to study translocation formation [128]. This system was used in living cells to monitor the spatiotemporal parameters associated with translocation events occurring between two different break sites in independent chromosomal locations. Translocations were observed within hours of DSB formation in a cell-cycle independent manner [128]. In accordance with results obtained by sequencing techniques, spatial proximity also appeared to be an important factor in promoting translocations, because the majority of events observed in this system were between spatially proximal break sites. However, a small frequency of distal DSB sites did result in translocations, suggesting that chromosomal mobility could also influence these chromosomal rearrangements. The power of this technique was further demonstrated by the findings that inhibition of DDR factors altered the frequency of translocation formation. For example, DNA-PK inhibition increased translocation formation, while inhibition of MRN by the small molecule inhibitor Mirin reduced the frequency of translocations [128]. Thus, these results highlight the impressive range of questions that can be answered using microscopy-based techniques to study the biogenesis of translocations in cells [118,124,128]. These experimental systems will continue to provide critical insights into understanding how chromatin dynamics and DNA repair within the spatial organization of the genome influence chromosomal rearrangements that drive genome evolution within cancer genomes (Fig. 11.3).

11.7 CONCLUSIONS

Protection of genetic information is vital for avoiding mutations that can alter cellular homeostasis leading to diseases. Cells have evolved diverse DNA repair pathways to handle a wide variety of DNA damage lesions that can threaten the genome. A complete mechanistic understanding of DNA repair can only be achieved by delineating how these pathways are orchestrated within chromatin, given that the true in vivo substrate of the DDR are lesions occurring in chromatin. Here we have provided primary examples that demonstrate the involvement of chromatin in the maintenance of genome and epigenome integrity, which involve histone modifications, histone variants, chaperones, chromatin remodeling complexes, chromatin modifying and binding factors (Fig. 11.2). The fact that these chromatin factors participate in DNA repair firmly establishes the integral connection between chromatin dynamics and DNA repair. The development of innovative strategies to study these processes across the genome and in different chromatin environments has started to provide important new insights into this area of research (Fig. 11.3). It is vital to continue to interrogate and identify chromatin-based DDR pathways that function across the varied genome and epigenome to understand how these processes are capable of maintaining genome fidelity and function in the presence of DNA damage. Knowledge gained from these studies will illuminate how defects in these pathways result in diseases and provide insights into the use of DNA damaging agents and drugs targeting the epigenome for therapeutic benefit.

Glossary

CATCH-IT Covalent attachment of tags to capture histones and identify turnover. A method used to measure genome-wide nucleosome assembly, disassembly, and turnover using metabolically labeled, newly synthesized histones for affinity-based chromatin capture

ChIP-seq Chromatin immunoprecipitation followed by sequencing. A method used to identify interaction sites of chromatin-binding proteins with the genome. This technique uses specific antibodies to precipitate the protein of interest, which isolates bound DNA fragments that are analyzed by sequencing

CRISPR-Cas9, TALENs, and ZFNs Genome editing strategies that use engineered site-specific nucleases (TALENs and ZFNs) or RNA guided DNA endonucleases (CRISPR-Cas9) to induce targeted DNA damage to stimulate DNA repair mechanisms including NHEJ or HR at specific genomic locations to modify the genome

DNA damage response (DDR) A complex network of cellular pathways and signaling events that detect, direct, and promote repair of DNA lesions to maintain genetic integrity

Epigenome Combination of chemical changes to DNA, histone PTMs, and their interacting proteins that regulate heritable chromatin structure, genome accessibility, and function

FAIRE-seq Formaldehyde-assisted isolation of regulatory elements coupled with high-throughput sequencing. A molecular biology-based method for identifying DNA associated with regulatory activity, that is, DNA from nucleosome-free regions

FRAP Fluorescence recovery after photobleaching. A technique used to study the mobility and dynamics of fluorescently labeled proteins

Hi-C A technique based on Chromosome Conformation Capture (3C) to assess chromosome architecture and interactions genome-wide. DNA fragments generated from fixed cells are isolated and analyzed by high-throughput sequencing to determine genome organization

Next-Generation Sequencing NGS Also known as massively parallel sequencing, it is a post-Sanger sequencing technique that can sequence a single molecule of DNA rather than a sequence generated from multiple DNA templates, allowing identification of millions of DNA sequences from a single sample

SNAP-tag technology Protein labeling system that enables specific covalent attachment of any inert molecule including dyes, fluorophores, biotin or beads conjugated to guanine, or chloropyridine groups via a benzyl linker to a protein of interest

Abbreviations

ALT	Alternative lengthening of telomeres
AP sites	Apurinic/apyrimidinic sites
AsiSI	Restriction endonuclease A (Source: *Arthrobacter* species)
ATM	Ataxia telangiectasis mutated
BRCA1	Breast cancer type 1 susceptibility protein
BRCT	BRCA1 C terminus domain proteins (phospho-binding)
CATCH-IT	Covalent attachment of tagged histones to capture and identity turnover
ChIP	Chromatin immunoprecipitation
ChIP-Seq	Chromatin immunoprecipitation followed by sequencing
CRISPR	Clustered regularly interspaced short palindromic repeats
DDR	DNA damage response
DIvA	DSB inducible via AsiSI
DSB	DNA double-strand break
DUBs	Deubiquitylating enzymes
FAIRE-seq	Formaldehyde-assisted isolation of regulatory elements coupled with high-throughput sequencing
FokI	Bacterial type IIS endonuclease (Source: *Flavobacterium okeanokoites*)
FRAP	Fluorescence recovery after photobleaching
γH2AX	Phosphorylated H2AX
HO endonuclease	Homothallic switching endonuclease
HPRT locus	Hypoxanthine phosphoribosyltransferase locus
HTGTS	High-throughput genome wide translocation sequencing
I-PpoI	Group I intron encoded endonuclease (Source: *Physarum polycephalum*)
I-SceI	Intron encoded endonuclease I-SceI (Source: *Saccharomyces cerevisiae*)
LacI:LacO	Lac repressor/Lac operator system
LacO/TetO	LacO and TetO operators
NGS	Next generation sequencing
PCR	Polymerase chain reaction
PTMs	Posttranslational modifications
ROS	Reactive oxygen species
TALENs	Transcription activator-like effector nuclease
UBD proteins	Ubiquitin binding domain containing proteins
ZFNs	Zinc finger nucleases

References

[1] Jackson SP, Bartek J. The DNA-damage response in human biology and disease. Nature 2009;461(7267):1071–8.

[2] Ciccia A, Elledge SJ, The DNA. damage response: making it safe to play with knives. Mol Cell 2010;40(2):179–204.

[3] Soria G, Polo SE, Almouzni G. Prime, repair, restore: the active role of chromatin in the DNA damage response. Mol Cell 2012;46(6):722–34.

[4] Papamichos-Chronakis M, Peterson CL. Chromatin and the genome integrity network. Nat Rev Genet 2013;14(1):62–75.

[5] Price BD, D'andrea AD. Chromatin remodeling at DNA double-strand breaks. Cell 2013;152(6):1344–54.

[6] Smeenk G, van Attikum H. The chromatin response to DNA breaks: leaving a mark on genome integrity. Annu Rev Biochem 2013;82:55–80.

[7] Gong F, Miller KM. Mammalian DNA repair: HATs and HDACs make their mark through histone acetylation. Mutat Res 2013;750(1-2):23–30.

[8] Sharma SV, Lee DY, Li B, Quinlan MP, Takahashi F, Maheswaran S, et al. A chromatin-mediated reversible drug-tolerant state in cancer cell subpopulations. Cell 2010;141(1):69–80.

[9] Baylin SB, Jones PA. A decade of exploring the cancer epigenome—biological and translational implications. Nat Rev Cancer 2011;11(10):726–34.

[10] Helin K, Dhanak D. Chromatin proteins and modifications as drug targets. Nature 2013;502(7472):480–8.

[11] Rodriguez R, Miller KM. Unravelling the genomic targets of small molecules using high-throughput sequencing. Nat Rev Genet 2014;15(12):783–96.

[12] Scharer OD. Nucleotide excision repair in eukaryotes. Cold Spring Harb Perspect Biol 2013;5(10):a012609.

[13] Marteijn JA, Lans H, Vermeulen W, Hoeijmakers JH. Understanding nucleotide excision repair and its roles in cancer and ageing. Nat Rev Mol Cell Biol 2014;15(7):465–81.

[14] Svejstrup JQ. Mechanisms of transcription-coupled DNA repair. Nat Rev Mol Cell Biol 2002;3(1):21–9.

[15] Hoeijmakers JHJ. Human nucleotide excision repair syndromes: molecular clues to unexpected intricacies. Eur J Cancer. 1994;30A(13):1912–21.

[16] Smerdon MJ. DNA repair and the role of chromatin structure. Curr Opin Cell Biol 1991;3(3):422–8.

[17] Green CM, Almouzni G. When repair meets chromatin—first in series on chromatin dynamics. Embo Rep 2002;3(1):28–33.

[18] Gaillard H, Fitzgerald DJ, Smith CL, Peterson CL, Richmond TJ, Thoma F. Chromatin remodeling activities act on UV-damaged nucleosomes and modulate DNA damage accessibility to photolyase. J Biol Chem 2003;278(20):17655–63.

[19] Mone MJ, Bernas T, Dinant C, Goedvree FA, Manders EMM, Volker M, et al. In vivo dynamics of chromatin-associated complex formation in mammalian nucleotide excision repair. Proc Natl Acad Sci USA 2004;101(45):15933–7.

[20] Rodriguez Y, Hinz JM, Smerdon MJ. Accessing DNA damage in chromatin: preparing the chromatin landscape for base excision repair. DNA Repair 2015;32:113–9.

[21] Ray A, Mir SN, Wani G, Zhao Q, Battu A, Zhu QZ, et al. Human SNF5/INI1, a component of the human SWI/SNF chromatin remodeling complex, promotes nucleotide excision repair by influencing ATM recruitment and downstream H2AX phosphorylation. Mol Cell Biol 2009;29(23):6206–19.

[22] Tresini M, Warmerdam DO, Kolovos P, Snijder L, Vrouwe MG, Demmers JA, et al. The core spliceosome as target and effector of non-canonical ATM signalling. Nature 2015;523(7558):53–8.

[23] Zhao Q, Wang QE, Ray A, Wani G, Han CH, Milum K, et al. Modulation of nucleotide excision repair by mammalian SWI/SNF chromatin–remodeling complex. J Biol Chem 2009;284(44):30424–32.

[24] Ramanathan B, Smerdon MJ. Changes in nuclear protein acetylation in u.v.-damaged human cells. Carcinogenesis 1986;7(7):1087–94.

[25] Li S. Implication of posttranslational histone modifications in nucleotide excision repair. Int J Mol Sci 2012;13(10):12461–86.

[26] Guo R, Chen J, Mitchell DL, Johnson DG. GCN5 and E2F1 stimulate nucleotide excision repair by promoting H3K9 acetylation at sites of damage. Nucleic Acids Res 2011;39(4):1390–7.

[27] Bergink S, Salomons FA, Hoogstraten D, Groothuis TA, de Waard H, Wu J, et al. DNA damage triggers nucleotide excision repair-dependent monoubiquitylation of histone H2A. Genes Dev 2006;20(10):1343–52.

[28] Dinant C, Ampatziadis-Michailidis G, Lans H, Tresini M, Lagarou A, Grosbart M, et al. Enhanced chromatin dynamics by FACT promotes transcriptional restart after UV-induced DNA damage. Mol Cell 2013;51(4):469–79.

[29] Polo SE, Roche D, Almouzni G. New histone incorporation marks sites of UV repair in human cells. Cell 2006;127(3):481–93.

[30] Adam S, Polo SE, Almouzni G. Transcription recovery after DNA damage requires chromatin priming by the H3.3 histone chaperone HIRA. Cell 2013;155(1):94–106.

[31] Huertas P. DNA resection in eukaryotes: deciding how to fix the break. Nat Struct Mol Biol 2010;17(1):11–6.

[32] Beucher A, Birraux J, Tchouandong L, Barton O, Shibata A, Conrad S, et al. ATM and Artemis promote homologous recombination of radiation-induced DNA double-strand breaks in G2. EMBO J 2009;28(21):3413–27.

[33] Lieber MR. The mechanism of double-strand DNA break repair by the nonhomologous DNA end-joining pathway. Annu Rev Biochem 2010;79:181–211.

[34] Chapman JR, Taylor MR, Boulton SJ. Playing the end game: DNA double-strand break repair pathway choice. Mol Cell 2012;47(4):497–510.

[35] Miller KM, Jackson SP. Histone marks: repairing DNA breaks within the context of chromatin. Biochem Soc Trans 2012;40(2):370–6.

[36] Rogakou EP, Pilch DR, Orr AH, Ivanova VS, Bonner WM. DNA double-stranded breaks induce histone H2AX phosphorylation on serine 139. J Biol Chem 1998;273(10):5858–68.

[37] Rogakou EP, Boon C, Redon C, Bonner WM. Megabase chromatin domains involved in DNA double-strand breaks in vivo. J Cell Biol 1999;146(5):905–16.

[38] Paull TT, Rogakou EP, Yamazaki V, Kirchgessner CU, Gellert M, Bonner WM. A critical role for histone H2AX in recruitment of repair factors to nuclear foci after DNA damage. Curr Biol 2000;10(15):886–95.

[39] Jackson SP, Durocher D. Regulation of DNA damage responses by ubiquitin and SUMO. Mol Cell 2013;49(5):795–807.

[40] Panier S, Durocher D. Push back to respond better: regulatory inhibition of the DNA double-strand break response. Nat Rev Mol Cell Biol 2013;14(10):661–72.

[41] Mattiroli F, Vissers JH, van Dijk WJ, Ikpa P, Citterio E, Vermeulen W, et al. RNF168 ubiquitinates K13-15 on H2A/H2AX to drive DNA damage signaling. Cell 2012;150(6):1182–95.

[42] Gatti M, Pinato S, Maspero E, Soffientini P, Polo S, Penengo L. A novel ubiquitin mark at the N-terminal tail of histone H2As targeted by RNF168 ubiquitin ligase. Cell Cycle 2012;11(13):2538–44.

[43] Fradet-Turcotte A, Canny MD, Escribano-Diaz C, Orthwein A, Leung CC, Huang H, et al. 53BP1 is a reader of the DNA-damage-induced H2A Lys 15 ubiquitin mark. Nature 2013;499(7456):50–4.

[44] Facchino S, Abdouh M, Chatoo W, Bernier G. BMI1 confers radioresistance to normal and cancerous neural stem cells through recruitment of the DNA damage response machinery. J Neurosci 2010;30(30):10096–111.

[45] Ismail IH, Andrin C, McDonald D, Hendzel MJ. BMI1-mediated histone ubiquitylation promotes DNA double-strand break repair. J Cell Biol 2010;191(1):45–60.

[46] Chagraoui J, Hebert J, Girard S, Sauvageau G. An anticlastogenic function for the Polycomb Group gene Bmi1. Proc Natl Acad Sci USA 2011;108(13):5284–9.

[47] Ginjala V, Nacerddine K, Kulkarni A, Oza J, Hill SJ, Yao M, et al. BMI1 is recruited to DNA breaks and contributes to DNA damage-induced H2A ubiquitination and repair. Mol Cell Biol 2011;31(10):1972–82.

[48] Miller KM, Tjeertes JV, Coates J, Legube G, Polo SE, Britton S, et al. Human HDAC1 and HDAC2 function in the DNA-damage response to promote DNA nonhomologous end-joining. Nat Struct Mol Biol 2010;17(9):1144–51.

[49] Tang J, Cho NW, Cui G, Manion EM, Shanbhag NM, Botuyan MV, et al. Acetylation limits 53BP1 association with damaged chromatin to promote homologous recombination. Nat Struct Mol Biol 2013;20(3):317–25.

[50] Gong F, Chiu LY, Cox B, Aymard F, Clouaire T, Leung JW, et al. Screen identifies bromodomain protein ZMYND8 in chromatin recognition of transcription-associated DNA damage that promotes homologous recombination. Genes Dev 2015;29(2):197–211.

[51] Xu Y, Ayrapetov MK, Xu C, Gursoy-Yuzugullu O, Hu Y, Price BD. Histone H2A.Z controls a critical chromatin remodeling step required for DNA double-strand break repair. Mol Cell 2012;48(5):723–33.

[52] Khurana S, Kruhlak MJ, Kim J, Tran AD, Liu J, Nyswaner K, et al. A macrohistone variant links dynamic chromatin compaction to BRCA1-dependent genome maintenance. Cell Rep 2014;8(4):1049–62.

[53] Gursoy-Yuzugullu O, Ayrapetov MK, Price BD. Histone chaperone Anp32e removes H2A.Z from DNA double-strand breaks and promotes nucleosome reorganization and DNA repair. Proc Natl Acad Sci USA 2015;112(24):7507–12.

[54] Ziv Y, Bielopolski D, Galanty Y, Lukas C, Taya Y, Schultz DC, et al. Chromatin relaxation in response to DNA double-strand breaks is modulated by a novel ATM- and KAP-1 dependent pathway. Nat Cell Biol 2006;8(8):870–6.

[55] Lemaitre C, Soutoglou E. Double strand break (DSB) repair in heterochromatin and heterochromatin proteins in DSB repair. DNA Repair 2014;19:163–8.

[56] Bakkenist CJ, Kastan MB. DNA damage activates ATM through intermolecular autophosphorylation and dimer dissociation. Nature 2003;421(6922):499–506.

[57] Trojer P, Reinberg D. Facultative heterochromatin: Is there a distinctive molecular signature? Mol Cell 2007;28(1):1–13.

[58] Zhou VW, Goren A, Bernstein BE. Charting histone modifications and the functional organization of mammalian genomes. Nat Rev Genet 2011;12(1):7–18.

[59] Shanbhag NM, Rafalska-Metcalf IU, Balane-Bolivar C, Janicki SM, Greenberg RA. ATM-dependent chromatin changes silence transcription in cis to DNA double-strand breaks. Cell 2010;141(6):970–81.

[60] Chou DM, Adamson B, Dephoure NE, Tan X, Nottke AC, Hurov KE, et al. A chromatin localization screen reveals poly (ADP ribose)-regulated recruitment of the repressive polycomb and NuRD complexes to sites of DNA damage. Proc Natl Acad Sci USA 2010;107(43):18475–80.

[61] Pankotai T, Bonhomme C, Chen D, Soutoglou E. DNAPKcs-dependent arrest of RNA polymerase II transcription in the presence of DNA breaks. Nat Struct Mol Biol 2012;19(3):276–82.

[62] Sun Y, Jiang X, Price BD. Tip60: connecting chromatin to DNA damage signaling. Cell Cycle 2010;9(5):930–6.

[63] Ayrapetov MK, Gursoy-Yuzugullu O, Xu C, Xu Y, Price BD. DNA double-strand breaks promote methylation of histone H3 on lysine 9 and transient formation of repressive chromatin. Proc Natl Acad Sci USA 2014;111(25):9169–74.

[64] Ui A, Nagaura Y, Yasui A. Transcriptional elongation factor ENL phosphorylated by ATM recruits polycomb and switches off transcription for DSB repair. Mol Cell 2015;58(3):468–82.

[65] Ju BG, Lunyak VV, Perissi V, Garcia-Bassets I, Rose DW, Glass CK, et al. A topoisomerase IIbeta-mediated dsDNA break required for regulated transcription. Science 2006;312(5781):1798–802.

[66] Puc J, Kozbial P, Li W, Tan Y, Liu Z, Suter T, et al. Ligand-dependent enhancer activation regulated by topoisomerase-I activity. Cell 2015;160(3):367–80.

[67] Perillo B, Ombra MN, Bertoni A, Cuozzo C, Sacchetti S, Sasso A, et al. DNA oxidation as triggered by H3K9me2 demethylation drives estrogen-induced gene expression. Science 2008;319(5860):202–6.

[68] Francia S, Michelini F, Saxena A, Tang D, de Hoon M, Anelli V, et al. Site-specific DICER and DROSHA RNA products control the DNA-damage response. Nature 2012;488(7410):231–5.

[69] Lindahl T, Barnes DE. Repair of endogenous DNA damage. Cold Spring Harb Symp Quant Biol 2000;65:127–33.

[70] Lindahl T. Instability and decay of the primary structure of DNA. Nature 1993;362(6422):709–15.

[71] Krokan HE, Bjoras M. Base Excision Repair. Cold Spring Harb Perspec Biol 2013;5(4).

[72] Lindahl T, Barnes DE. Repair of endogenous DNA damage. Cold Spring Harb Symp Quant Biol 2000;65:127–33.

[73] Odell ID, Wallace SS, Pederson DS. Rules of engagement for base excision repair in chromatin. J Cell Physiol 2013;228(2):258–66.

[74] Li G, Levitus M, Bustamante C, Widom J. Rapid spontaneous accessibility of nucleosomal DNA. Nat Struct Mol Biol 2005;12(1):46–53.

[75] Nilsen H, Lindahl T, Verreault A. DNA base excision repair of uracil residues in reconstituted nucleosome core particles. EMBO J 2002;21(21):5943–52.

[76] Beard BC, Wilson SH, Smerdon MJ. Suppressed catalytic activity of base excision repair enzymes on rotationally positioned uracil in nucleosomes. Proc Natl Acad Sci USA 2003;100(13):7465–70.

[77] Odell ID, Barbour JE, Murphy DL, Della-Maria JA, Sweasy JB, Tomkinson AE, et al. Nucleosome disruption by DNA ligase III-XRCC1 promotes efficient base excision repair. Mol Cell Biol 2011;31(22):4623–32.

[78] Amouroux R, Campalans A, Epe B, Radicella JP. Oxidative stress triggers the preferential assembly of base excision repair complexes on open chromatin regions. Nucleic Acids Res 2010;38(9):2878–90.

[79] Rodriguez Y, Hinz JM, Smerdon MJ. Accessing DNA damage in chromatin: Preparing the chromatin landscape for base excision repair. DNA Repair 2015;32:113–9.

[80] Gassman NR, Wilson SH. Micro-irradiation tools to visualize base excision repair and single-strand break repair. DNA Repair 2015;31:52–63.

[81] Lan L, Nakajima S, Oohata Y, Takao M, Okano S, Masutani M, et al. In situ analysis of repair processes for oxidative DNA damage in mammalian cells. Proc Natl Acad Sci USA 2004;101(38):13738–43.

[82] Hsieh P, Yamane K. DNA mismatch repair: molecular mechanism, cancer, and ageing. Mech Ageing Dev 2008;129(7-8):391–407.

[83] Li GM. Mechanisms and functions of DNA mismatch repair. Cell Res 2008;18(1):85–98.

[84] Reyes GX, Schmidt TT, Kolodner RD, Hombauer H. New insights into the mechanism of DNA mismatch repair. Chromosoma 2015;124(4):443–62.

[85] Pena-Diaz J, Bregenhorn S, Ghodgaonkar M, Follonier C, Artola-Boran M, Castor D, et al. Noncanonical mismatch repair as a source of genomic instability in human cells. Mol Cell 2012;47(5):669–80.

[86] Li GM. New insights and challenges in mismatch repair: getting over the chromatin hurdle. DNA Repair 2014;19:48–54.

[87] Adkins NL, Niu H, Sung P, Peterson CL. Nucleosome dynamics regulates DNA processing. Nat Struct Mol Biol 2013;20(7):836–42.

[88] Schopf B, Bregenhorn S, Quivy JP, Kadyrov FA, Almouzni G, Jiricny J. Interplay between mismatch repair and chromatin assembly. Proc Natl Acad Sci USA 2012;109(6):1895–900.

[89] Javaid S, Manohar M, Punja N, Mooney A, Ottesen JJ, Poirier MG, et al. Nucleosome remodeling by hMSH2-hMSH6. Mol Cell 2009;36(6):1086–94.

[90] Neumann H, Hancock SM, Buning R, Routh A, Chapman L, Somers J, et al. A method for genetically installing site-specific acetylation in recombinant histones defines the effects of H3 K56 acetylation. Mol Cell 2009;36(1):153–63.

[91] Kadyrova LY, Mertz TM, Zhang Y, Northam MR, Sheng Z, Lobachev KS, et al. A reversible histone H3 acetylation cooperates with mismatch repair and replicative polymerases in maintaining genome stability. PLoS Genet 2013;9(10):e1003899.

[92] Li F, Mao G, Tong D, Huang J, Gu L, Yang W, et al. The histone mark H3K36me3 regulates human DNA mismatch repair through its interaction with MutSalpha. Cell 2013;153(3):590–600.

[93] Aymard F, Bugler B, Schmidt CK, Guillou E, Caron P, Briois S, et al. Transcriptionally active chromatin recruits homologous recombination at DNA double-strand breaks. Nat Struct Mole Biol 2014;21(4):366–74.

[94] Jha DK, Strahl BD. An RNA polymerase II-coupled function for histone H3K36 methylation in checkpoint activation and DSB repair. Nat Commun 2014;5:3965.

[95] Pfister SX, Ahrabi S, Zalmas LP, Sarkar S, Aymard F, Bachrati CZ, et al. SETD2-dependent histone H3K36 trimethylation is required for homologous recombination repair and genome stability. Cell Rep 2014;7(6):2006–18.

[96] Hittelman WN, Pollard M. Visualization of chromatin events associated with repair of ultraviolet light-induced damage by premature chromosome condensation. Carcinogenesis 1984;5(10):1277–85.

[97] Kruhlak MJ, Celeste A, Dellaire G, Fernandez-Capetillo O, Muller WG, McNally JG, et al. Changes in chromatin structure and mobility in living cells at sites of DNA double-strand breaks. J Cell Biol 2006;172(6):823–34.

[98] Aten JA, Stap J, Krawczyk PM, van Oven CH, Hoebe RA, Essers J, et al. Dynamics of DNA double-strand breaks revealed by clustering of damaged chromosome domains. Science 2004;303(5654):92–5.

[99] Lisby M, Mortensen UH, Rothstein R. Colocalization of multiple DNA double-strand breaks at a single Rad52 repair centre. Nat Cell Biol 2003;5(6):572–7.

[100] Dion V, Gasser SM. Chromatin movement in the maintenance of genome stability. Cell 2013;152(6):1355–64.

[101] Dion V, Kalck V, Horigome C, Towbin BD, Gasser SM. Increased mobility of double-strand breaks requires Mec1, Rad9 and the homologous recombination machinery. Nat Cell Biol 2012;14(5):502–9.

[102] Mine-Hattab J, Rothstein R. Increased chromosome mobility facilitates homology search during recombination. Nat Cell Biol 2012;14(5):510–7.

[103] Dimitrova N, Chen YC, Spector DL, de Lange T. 53BP1 promotes non-homologous end joining of telomeres by increasing chromatin mobility. Nature 2008;456(7221):524–8.

[104] Cho NW, Dilley RL, Lampson MA, Greenberg RA. Interchromosomal homology searches drive directional ALT telomere movement and synapsis. Cell 2014;159(1):108–21.

[105] Torres-Rosell J, Sunjevaric I, De Piccoli G, Sacher M, Eckert-Boulet N, Reid R, et al. The Smc5-Smc6 complex and SUMO modification of Rad52 regulates recombinational repair at the ribosomal gene locus. Nat Cell Biol 2007;9(8):923–31.

[106] Chiolo I, Minoda A, Colmenares SU, Polyzos A, Costes SV, Karpen GH. Double-strand breaks in heterochromatin move outside of a dynamic HP1a domain to complete recombinational repair. Cell 2011;144(5):732–44.

[107] Krawczyk PM, Borovski T, Stap J, Cijsouw T, ten Cate R, Medema JP, et al. Chromatin mobility is increased at sites of DNA double-strand breaks. J Cell Sci 2012;125(Pt 9):2127–33.

[108] Nagy Z, Soutoglou E. DNA repair: easy to visualize, difficult to elucidate. Trends Cell Biol 2009;19(11):617–29.

[109] Polo SE, Jackson SP. Dynamics of DNA damage response proteins at DNA breaks: a focus on protein modifications. Genes Dev 2011;25(5):409–33.

[110] Shroff R, Arbel-Eden A, Pilch D, Ira G, Bonner WM, Petrini JH, et al. Distribution and dynamics of chromatin modification induced by a defined DNA double-strand break. Curr Biol 2004;14(19):1703–11.

[111] Murr R, Loizou JI, Yang YG, Cuenin C, Li H, Wang ZQ, et al. Histone acetylation by Trrap-Tip60 modulates loading of repair proteins and repair of DNA double-strand breaks. Nat Cell Biol 2006;8(1):91–9.

[112] Berkovich E, Monnat RJ Jr, Kastan MB. Roles of ATM and NBS1 in chromatin structure modulation and DNA double-strand break repair. Nat Cell Biol 2007;9(6):683–90.

[113] Pang B, Qiao X, Janssen L, Velds A, Groothuis T, Kerkhoven R, et al. Drug-induced histone eviction from open chromatin contributes to the chemotherapeutic effects of doxorubicin. Nat Commun 2013;4:1908.

[114] Yang F, Kemp CJ, Henikoff S. Doxorubicin enhances nucleosome turnover around promoters. Curr Biol 2013;23(9):782–7.

[115] Qian MX, Pang Y, Liu CH, Haratake K, Du BY, Ji DY, et al. Acetylation-mediated proteasomal degradation of core histones during DNA repair and spermatogenesis. Cell 2013;153(5):1012–24.

[116] Bunting SF, Nussenzweig A. End-joining, translocations and cancer. Nat Rev Cancer 2013;13(7):443–54.

[117] Lieberman-Aiden E, van Berkum NL, Williams L, Imakaev M, Ragoczy T, Telling A, et al. Comprehensive mapping of long-range interactions reveals folding principles of the human genome. Science 2009;326(5950):289–93.

[118] Zhang Y, McCord RP, Ho YJ, Lajoie BR, Hildebrand DG, Simon AC, et al. Spatial organization of the mouse genome and its role in recurrent chromosomal translocations. Cell 2012;148(5):908–21.

[119] Chiarle R, Zhang Y, Frock RL, Lewis SM, Molinie B, Ho YJ, et al. Genome-wide translocation sequencing reveals mechanisms of chromosome breaks and rearrangements in B cells. Cell 2011;147(1):107–19.

[120] Hakim O, Resch W, Yamane A, Klein I, Kieffer-Kwon KR, Jankovic M, et al. DNA damage defines sites of recurrent chromosomal translocations in B lymphocytes. Nature 2012;484(7392):69–74.

[121] Ramiro AR, Jankovic M, Eisenreich T, Difilippantonio S, Chen-Kiang S, Muramatsu M, et al. AID is required for c-myc/IgH chromosome translocations in vivo. Cell 2004;118(4):431–8.

[122] Keim C, Kazadi D, Rothschild G, Basu U. Regulation of AID, the B-cell genome mutator. Genes Dev 2013;27(1):1–17.

[123] Klein IA, Resch W, Jankovic M, Oliveira T, Yamane A, Nakahashi H, et al. Translocation-capture sequencing reveals the extent and nature of chromosomal rearrangements in B lymphocytes. Cell 2011;147(1):95–106.

[124] Qian J, Wang Q, Dose M, Pruett N, Kieffer-Kwon KR, Resch W, et al. B cell super-enhancers and regulatory clusters recruit AID tumorigenic activity. Cell 2014;159(7):1524–37.

[125] Wang Q, Oliveira T, Jankovic M, Silva IT, Hakim O, Yao K, et al. Epigenetic targeting of activation-induced cytidine deaminase. Proc Natl Acad Sci USA 2014;111(52):18667–72.

[126] Hnisz D, Abraham BJ, Lee TI, Lau A, Saint-Andre V, Sigova AA, et al. Super-enhancers in the control of cell identity and disease. Cell 2013;155(4):934–47.

[127] Hnisz D, Schuijers J, Lin CY, Weintraub AS, Abraham BJ, Lee TI, et al. Convergence of developmental and oncogenic signaling pathways at transcriptional super-enhancers. Mol Cell 2015;58(2):362–70.

[128] Roukos V, Voss TC, Schmidt CK, Lee S, Wangsa D, Misteli T. Spatial dynamics of chromosome translocations in living cells. Science 2013;341(6146):660–4.

CHAPTER 12

Regulation of Centromeric Chromatin

D. Bade*, S. Erhardt**,†

*Hubrecht Institute, Uppsalalaan, Utrecht, The Netherlands
**ZMBH, DKFZ-ZMBH-Alliance, University of Heidelberg,
Im Neuenheimer Feld, Heidelberg, Germany
†Cell Networks Excellence Cluster, University of Heidelberg,
Im Neuenheimer Feld, Heidelberg, Germany

12.1 CENTROMERE—AN ESSENTIAL CHROMATIN DOMAIN FOR ACCURATE CHROMOSOME SEGREGATION

The accurate segregation of the genetic material during mitosis is crucial to prevent cellular aneuploidy and thus to maintain genomic integrity. Centromeres, which are structurally defined as the primary constriction on metaphase chromosomes, play an essential role in this function, as they delineate the site of kinetochore formation during mitosis. The kinetochore is a multiprotein complex containing the KMN network (named after the three major kinetochore subcomplexes Knl1, Mis12, and Ndc80) that connects the spindle microtubules to the sister chromatids. Hence it facilitates the biorientation on the metaphase plate, a prerequisite for correct chromosome segregation in anaphase [1]. Considering that centromeres are the functional foundation for kinetochore formation, the presence of exactly one centromeric region per chromosome is critical to prevent genomic instability: the gain of additional (neo)centromeres or the loss of centromeres can cause chromosome segregation errors that may result in a gain or loss of chromosomes (aneuploidy) and thus mitotic instability [2].

This pattern of one distinct centromere per chromosome holds true for most species; however, some species, for instance the nematode *Caenorhabditis elegans* and some plant species, have holocentric centromeres that are broadly distributed over the entire chromosome arms. In depth studies clarified that *C. elegans* counts ∼100 individual centromeric sites on each chromosome [3]. These centromeres colocalize with kinetochore components and mediate spindle microtubule attachments to multiple sites per sister chromatid during

CONTENTS

Chromatin Regulation and Dynamics. http://dx.doi.org/10.1016/B978-0-12-803395-1.00012-5

mitosis. At this time, the benefit and precise regulation of these holocentric chromosomes await discovery.

12.2 CENP-A—THE DETERMINANT OF CENTROMERE IDENTITY

Despite differences in complexity, centromere function is conserved among species and most share one common feature: the histone variant CENP-A replaces the canonical histone H3 in a subset of centromeric nucleosomes [4] (Fig. 12.1). The high abundance of CENP-A molecules at centromeres serves as a specific mark for the assembly of nascent CENP-A nucleosomes ensuring that the centromere is self-templating and thus stably maintained across the cell cycle. Even the positioning of CENP-A nucleosomes relative to H3 nucleosomes is inherited through cell divisions [5]. In line with this centromere inheritance model, artificially tethered ectopic *Drosophila* CENP-A (also known as CID) directs the recruitment of new CENP-A molecules to these sites [6].

The first evidence that centromere identity is regulated independently of the primary DNA sequence was presented in 1985: chromosomes harboring two

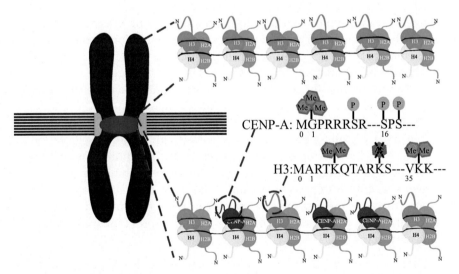

FIGURE 12.1 Centromeres (shown in red) are the functional foundation for kinetochores (shown in orange) that bind spindle microtubules (gray) during mitosis.
Centromeric chromatin is characterized by the presence of CENP-A nucleosomes adjacent to canonical nucleosomes. Furthermore, centromeric chromatin harbors distinct posttranslational modifications within the N-terminal tails of histone H3 and CENP-A itself. Whereas CENP-A is phosphorylated at several serine residues, H3 is dimethylated at lysine 4 and lysine 36 and lacks acetylation at lysine 9. *Me*, methylated; *P*, phosphorylated; *Ac*, acetylated.

identical arrays of centromeric DNA still displayed only one active centromere resulting in faithful chromosome segregation [7]. Around the same time, researchers identified CENP-A as a human autoantigen in patients suffering from the autoimmune disease scleroderma and subsequently observed that its absence correlates with inactive centromeres [8,9].

Sufficiency experiments conducted in flies clarified that *Drosophila* CENP-A is indeed the key factor in determining centromeric identity, as overexpression or alternatively targeted recruitment of the histone variant to ectopic sites results in the formation of functional ectopic kinetochores [6,10,11]. Similarly, targeted placing of CENP-A to noncentromeric sites causes the partial recruitment of kinetochore components, such as CENP-C and MIS18 in humans [12]. Consistent with its importance for centromere formation, CENP-A is an essential protein in any studied model system and its depletion prevents centromere and kinetochore formation [13–15]. For example, conditional depletion of a centromere in *Schizosaccharomyces pombe* leads to aneuploidy and mitotic cell death; however, some clones survive by spontaneously forming a neocentromere, a new active centromere. Interestingly, these neocentromeres are preferentially found close to the heterochromatin of telomeres, and, similarly to the endogenous centromere, are excluded from the gene-rich chromosome arms [16].

12.2.1 The Structural Features of CENP-A Nucleosomes

H3 and its variant CENP-A are structurally similar and have 48% sequence identity in humans [17]. The most conserved domain is the histone-fold domain (HFD) that is 62% identical [18]. Besides mediating the interaction of CENP-A with histone H4, the HFD of CENP-A is of utmost importance for centromere localization: it contains the CENP-A targeting domain (CATD), consisting of a loop (called L1) and an α-helix (named α-helix 2). Elegant experiments relying on a chimera of canonical H3 and the CATD of CENP-A clarified that this short stretch is sufficient for centromere targeting [19]. Interestingly, the H3^{CATD} chimera is able to compensate for CENP-A loss for a limited period of time [20]. Despite being able to compensate for CENP-A loss, the CATD domain is not the only segment of CENP-A that contributes to the recruitment of kinetochore components: a short hydrophobic stretch of six amino acids that is localized at the C-terminus of CENP-A has also been demonstrated to have this function [21–23].

Apart from influencing CENP-A localization, hydrogen/deuterium exchange mass spectrometry showed that the CATD makes CENP-A-containing nucleosomes more rigid and therefore structurally distinct [24]. In contrast to the HFD, the N-terminal tails of the different CENP-A homologs are very diverse and greatly vary in length (20 aa in *S. pombe* vs. 200 aa in *C. elegans*). The function of the N-termini of different CENP-A orthologs may therefore have diverse functions that are still not understood.

The composition of centromeric nucleosomes has been a matter of intense debate over the past years. The two prevailing models either favor a classical octameric nucleosome that, similarly to canonical nucleosomes, consists of two copies each of CENP-A, H4, H2A, and H2B, or a hemisome containing only one copy of each [25,26]. The idea of a centromeric hemisome came from isolating CENP-A/H4/H2A/H2B tetramers in chromatin immunoprecipitation experiments using cross-linked chromatin from *Drosophila* cells [27]. This observation was challenged soon afterward, as the residues required for cross linking are absent in the dimerization region of CENP-A, indicating that the observed tetramers might be an artifact of failed cross linking [28].

Interestingly, follow up experiments using atomic force microscopy experiments on human cells were interpreted as CENP-A nucleosomes that are only half as high as octameric H3 nucleosomes, supporting the previous findings of the hemisome [29]. Similarly, immuno-EM experiments with *Drosophila* CENP-A nucleosomes and centromere cleavage patterns obtained from yeast mapping experiments point toward a hemisome structure [29,30]. Moreover, CENP-A (also known as Cse4) nucleosomes of *Saccharomyces cerevisiae* induce positive supercoiling of DNA, a state incompatible with an octameric composition [31]. Nevertheless, most evidence point to an octamere as the predominant centromeric structure. Most importantly, Tachiwana et al. crystallized human centomeric nucleosomes-containing CENP-A and α-satellite DNA. This crystal structure reveals that CENP-A and canonical nucleosomes are structurally very similar octamers and only display small differences. For example, both types of nucleosomes wrap their DNA in a left-handed manner. Furthermore, it confirms the initial observation from MNase-protected DNA digests that CENP-A nucleosomes wrap 120 bp of DNA, a stretch of DNA too long to be wrapped by a hemisome [32–34]. Additionally, purified centromeric mononucleosomes from flies contain dimers of CENP-A and in vitro assembled CENP-A nucleosomes form octamers [35,36]. Lastly, photobleaching-assisted copy-number counting experiments also support an octameric CENP-A nucleosome in humans [37]. Although it seems that both models contradict and exclude each other, the reality may be in fact a combination of both scenarios: Krassovsky et al. suggested that CENP-A/Cse4 nucleosomes are dynamic and both forms may coexist in *S. cerevisiae* [38]. Alternatively, CENP-A nucleosome composition might alternate between a hemisome and an octamer in a cell cycle-dependent manner [39,40].

12.3   THE CHROMATIN ENVIRONMENT—AN IMPORTANT PLAYER IN CENTROMERE REGULATION

Besides the presence of CENP-A in a subset of centromeric nucleosomes [41], centromeric chromatin is characterized by a distinct pattern of posttranslational modifications (Fig. 12.1). Since centromeres are embedded into flanking heterochromatin, they were initially expected to display known heterochromatin-specific histone modifications, such as lysine 9 di- and trimethylation on the H3 nucleosomes (H3K9me2/3) that intersperse CENP-A nucleosomes. However, the pattern seems quite distinct from both hetero- and euchromatin. H3-containing nucleosomes adjacent to CENP-A nucleosomes carry the modifications H3K4me2 and H3K36me2, which is typically associated with active chromatin. At the same time, K9 and K36 on histone H3 are hypoacetylated, normally associated with inactive heterochromatin [42,43] (Fig. 12.1). Within this mixed chromatin state, H3K4me2 is noteworthy, as it is hardly present at centromeres in some species, such as *Zea mays*, or at vertebrate neocentromeres, but it is required for the transcription of underlying α-satellite DNA of endogenous human centromeres (see further) and for the assembly of newly synthesized CENP-A at human artificial centromeres [43–45].

Nevertheless, there seems to be a correlation between heterochromatin formation and neocentromere formation: neocentromere formation was reduced in mutants of the *S. pombe* H3K9 methyltransferase Clr4, suggesting that heterochromatin positively influences neocentromere formation [16]. Consistently, stable hotspots of CENP-A, after its overexpression in fly cells, are preferentially established at heterochromatin boundaries [11].

In addition to the distinct posttranslational modifications, the histone variant H2A.Z is also an integral component of a subset of centromeric and pericentric H3-containing nucleosomes. H2A.Z was suggested to function in the characteristic 3D organization of CENP-A nucleosomes that differs from canonical nucleosomes [46]. Accordingly, its depletion causes severe chromosome segregation defects in various organisms [46–48]. All in all, it can be concluded that the chromatin environment created by centromeric H3 nucleosomes and H2A.Z is a crucial part of centromere regulation.

12.3.1 Posttranslational Modifications on CENP-A

Like other histones, CENP-A is also subject to posttranslational modifications. Interestingly, whereas some modifications influence CENP-A stability, structure or positioning, the majority of them act on more downstream events, such as the recruitment of kinetochore components. In *S. cerevisiae*, the methylation on arginine 37 (R37) of CENP-A is required for the assembly of a complete kinetochore complex and consequently for proper chromosome segregation [49]. Similarly, the Aurora B-mediated mitotic phosphorylation of serine 7 within the N-terminal tail of human CENP-A is involved in the recruitment of the inner kinetochore protein CENP-C [50]. Unlike the previous modifications, serines 16 and 18 of CENP-A are already phosphorylated in a prenucleosomal context and directly impact CENP-A structure [51]. Recently, two different modifications were shown to have opposite effects on CENP-A deposition: whereas the CDK1 phosphorylation of serine 68 (S68) interferes with CENP-A binding to its loading factor Holliday junction–recognition protein (HJURP) and thus with its deposition to centromeric chromatin prior to mitotic exit (see further), the ubiquitylation of lysine 124 (K124) favors HJURP binding and is therefore essential for CENP-A loading onto chromatin [52,53]. Furthermore, ubiquitylation and the concomitant proteolysis of residual CENP-A helps to prevent its spreading into euchromatin in several organisms [54–57]. Lastly, ubiquitylation of CENP-A by the E3 ligase CUL3/RDX stabilizes CENP-A bound by its loading factor and thus is essential for centromere maintenance in *Drosophila melanogaster* [58]. Taken together, posttranslational modifications of CENP-A and H3-containing centromeric nucleosomes impact the formation of centromere chromatin, its regulation, and its function to allow faithful chromosome segregation.

12.4 CENTROMERIC DNA—SUPERFLUOUS OR NEEDED FOR CENTROMERE IDENTITY?

The underlying centromeric DNA is fast evolving and not conserved between species [59]. Nevertheless, centromeric regions are often characterized by the abundance of repetitive DNA sequences, such as α-satellite repeats in *Homo sapiens* [60].

As centromeres are defined by epigenetic means, the role of the underlying DNA is still not completely understood. Neocentromeres arise at previously noncentromeric sites [61], indicating that the sequence of the underlying DNA is of minor importance; however, evidence is accumulating that transcription at centromeric loci plays an important role in centromere maintenance (see further). As an exception, the centromere of *S. cerevisiae*, which classifies as a point centromere, is genetically determined by its 125 base-pair long centromeric DNA that consists of three conserved sequence elements (CDEI, CDEII, and CDEIII) on every chromosome. This stretch of DNA interacts with CENP-A/Cse4 and is necessary and sufficient to confer centromeric identity. *S. pombe*, many insects, plants and mammals, on the other hand, have regional centromeres covering up to 5 Mb of highly repetitive DNA with little to no sequence conservation even among different chromosomes of the same organism [62]. Human centromeric DNA consists, for example, of A/T rich 171-bp α-satellite repeats that build up chromosome-specific higher order repeats.

Besides being highly repetitive, human centromeric DNA also contains a 17-basepair motif, the CENP-B box, which together with the N-terminal tail of CENP-A recruits the centromeric protein B (CENP-B). Consistently, CENP-B levels drop by roughly 50% upon complete CENP-A depletion [63]. Although its precise role is still unclear, experiments with human artificial chromosomes suggest that CENP-B modulates the chromatin structure and thereby fulfills a dual role in facilitating the establishment of a new centromere, but also preventing the formation of multiple centromeres [64]. De novo centromere assembly on artificial chromosomes is more efficient in wild type mouse embryonic fibroblasts (MEFs) than in CENP-B-deficient cells, indicating that CENP-B enhances CENP-A recruitment. On the other hand, CENP-B prevents neocentromere formation on human α-satellite repeats stably integrated in mouse chromosomal DNA carrying an active endogenous centromere. This is possibly due to the heterochromatinization of the alphoid DNA, which is initiated by CENP-B. Consistently, MEFs of CENP-B knockout mice lack histone H3 lysine 9 trimethylation at these human α-satellite repeats [64]. Additionally, CENP-B binding to centromeres enhances the segregation fidelity possibly by interacting with and stabilizing CENP-C bound to centromeres [65]. Therefore, missegregation rates are higher in chromosomes lacking a functional CENP-B box within their centromeres, such as the human Y-chromosome.

Furthermore, CENP-B may also mark the minimal region required to maintain a functional centromere and thus protect the cell from centromere loss when CENP-A availability is limited [17]. Noteworthy, CENP-B is absent in many species, such as *D. melanogaster* or *C. elegans*, and is not required for neocentromere function in humans [66].

12.4.1 Centromeric Transcription and Its Role in Centromere Regulation

Centromeres, with the exception of rice centromeres, are generally gene poor. Nevertheless, transcripts originating from centromeric DNA repeats are described in various organisms [67–74]. In humans α-satellite repeats are actively transcribed by RNA polymerase II from late mitosis into early G1 [69,70]. Reducing the levels of these centromeric α-satellite transcripts by inhibiting RNA polymerase II results in an increase in segregation errors that is accompanied and probably caused by a loss of CENP-A and its loading factor HJURP from centromeres, although an indirect effect of transcription inhibition cannot be excluded [69,70].

In flies, RNA polymerase II also associates with centromeric chromatin during mitosis [67]. The *Drosophila* CENP-A loading factor CAL1 (see further) associates with RNA polymerase II and the chromatin remodeler FACT that facilitates transcription by nucleosome destabilization [68]. Experiments with an artificially created ectopic centromere revealed that CENP-A loading coincides with the transcription of the underlying DNA. Since the replenishment of CENP-A to ectopic centromeres depends on active transcription, an important function of transcription might be to open up and reorganize chromatin rather than to produce a specific transcript [68]. However, noncoding RNA transcripts originating from centromeric DNA are required for CENP-A loading, and it is therefore more likely that transcription fulfills a dual role at endogenous centromeres. Transcripts of satellite III repeats that cover megabases on the centromere of the fly X-chromosome localize *in trans* to all centromeres and physically bind to the inner kinetochore protein CENP-C. The depletion of satellite III repeats leads to erroneous chromosome segregation and reduced centromeric levels of CENP-A and CENP-C, supporting the view that noncoding RNA itself contributes to the epigenetic regulation of centromeres [67]. Comparable to the down-regulation of centromeric transcription, tremendously elevated levels (150 fold) of transcripts interfere with CENP-A loading onto human artificial chromosomes [75,75a]. In contrast, a more moderate increase (10-fold) of centromeric transcripts has no impact on CENP-A replenishment, indicating that variation to a certain extent is tolerated.

12.5   CENP-A LOADING AND THE MAINTENANCE OF CENTROMERIC CHROMATIN

When DNA is replicated in S phase, preexisting histones are distributed to both arms at the replication fork. At the same time, existing gaps are filled with nascent nucleosomes in a stepwise manner: at first histone chaperones and modifiers, such as chromatin assembly factor 1 (CAF1), load a H3–H4 tetramer onto DNA. Subsequently H2A–H2B dimers are deposited, resulting in the typical octamer (Fig. 12.2). Unlike these conventional nucleosomes, atypical nucleosomes, such as CENP-A-containing octamers, are replication-independently loaded onto chromatin in higher eukaryotes [76] (Fig. 12.2).

Interestingly, the exact timing of CENP-A replenishment differs between organisms and cell types. In humans, CENP-A incorporation occurs between telophase and the following G1 phase [77]. On the other hand, CENP-A is deposited during anaphase in fly embryos and during metaphase in fly tissue culture cells [78,79]. Meiotic timing seems again different, as CENP-A is assembled in prophase of meiosis I and after exit from meiosis II in spermatids of flies [80]. Despite these differences, the loading of all CENP-A variants relies on loading machineries distinct from the ones required for the replenishment of canonical nucleosomes. In addition, in all organisms, except *S. cerevisiae*, preexisting CENP-A molecules are distributed among the two daughter chromatids during the replication of centromeric DNA in mid to late S phase, and the emerging gaps are not filled with new CENP-A molecules until the next mitosis. Instead, histone H3.3 might function as a placeholder and fill these gaps until new CENP-A is loaded [81].

12.5.1   Priming of Centromeres

The replenishment of CENP-A nucleosomes is a multistep process where the centromere is first recognized and primed before CENP-A is loaded and finally maintained (Fig. 12.3). In humans, centromere licensing is assigned to the Mis18 complex, consisting of MIS18α, MIS18β, and their binding partner Mis18-binding

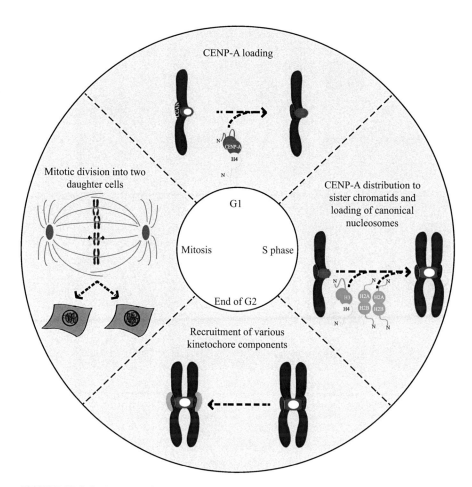

FIGURE 12.2 Centromeres throughout the human cell cycle.
(S phase) Preexisting nucleosomes are distributed to both arms of the replication fork. In parallel, canonical nucleosomes are replenished in a replication dependent manner. (G2 phase) At the end of G2, the kinetochore starts assembling on top of the centromeric chromatin. (Mitosis) The kinetochore binds to spindle microtubules and thereby allows the accurate segregation of chromosomes to the two daughter cells. (G1) The loading of CENP-A nucleosomes occurs after mitotic division and is therefore replication independent.

protein 1 (M18BP1). The Mis18 complex localizes to centromeres in telophase/G1, and its proper localization is a prerequisite for CENP-A deposition [82]. The kinetochore protein CENP-C thereby acts as a recruitment factor, as it directly interacts with the Mis18 complex through M18BP1 [83–85]. The time frame of M18BP1 localization to centromeres is defined by CDK1/2 kinase activities, which prevent its recruitment prior to mitosis [86,87]. Consistent with this finding, inhibition of CDK1/CDK2 causes premature M18BP1 localization and consequent

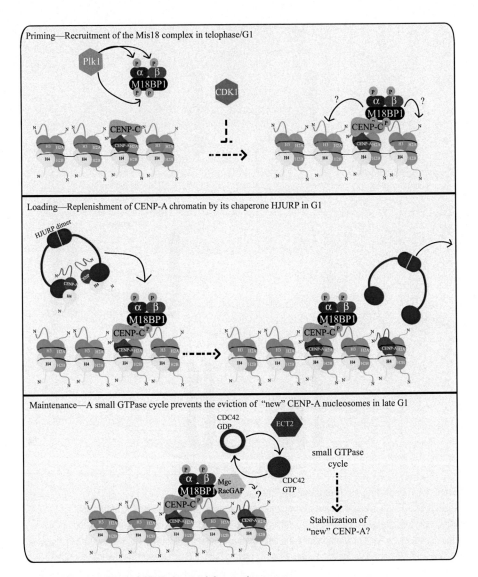

FIGURE 12.3 Loading of CENP-A-containing nucleosomes.

(Priming) The priming complex Mis18 binds to its centromeric anchor CENP-C in telophase. Whereas PLK1-dependent phosphorylations of Mis18 enhances its centromeric localization, CDK1-dependent phosphorylations counteract it. It remains unclear, whether Mis18 modifies surrounding chromatin. (Loading) The loading factor HJURP is recruited to centromeres in a Mis18-dependent manner. HJURP itself dimerizes, and each molecule interacts with one CENP-A/H4 dimer. HJURP loads CENP-A onto chromatin in G1 phase resulting in octameric CENP-A nucleosomes. (Maintenance) A GTPase cycle involving the small GTPase CDC42, the GTPase-activating protein MgcRacGAP and guanine nucleotide exchange factor Ect2 is required to prevent the eviction of newly assembled CENP-A chromatin. Up to date, it is unclear how these newly assembled nucleosomes are initially recognized and then stabilized. Additionally, CENP-C was shown to directly stabilize assembled CENP-A nucleosomes (see text for more details).

loading of CENP-A prior to mitosis [87]. Similarly, CDK1 phosphorylates CENP-A at serine 68, thereby abolishing its ability to bind its loading factor HJURP [53]. Hence, the down-regulation of CDK1/2 activity is crucial for CENP-A replenishment. In addition, upon mitotic exit, PLK1 activity plays an opposing role, promoting the centromeric localization of M18BP1 in G1 [86].

Although its exact mechanism of action is still not known, it is established that the Mis18 complex is exclusively required at the early stage of CENP-A loading, but is dispensable for CENP-A maintenance at centromeres. Based on the observation that treatment with the histone deacetylase inhibitor Trichostatin A suppresses the loss of CENP-A assembly upon Mis18 depletion, Mis18 was suggested to act through a mechanism involving histone acetylation [82]. In line with an epigenetic function of the Mis18 complex, MIS18α also has been suggested to recruit the DNA methyltransferases DNMT3A/3B, and its depletion affects the centromeric DNA methylation status, various histone modifications, and the level of noncoding RNA transcripts [88]. Alternatively, Mis18 might simply be the centromeric recruitment factor of the CENP-A chaperone HJURP, as artificial tethering of HJURP to a noncentromeric site bypasses the requirement of Mis18 for CENP-A deposition [84].

Mis18 was first described in *S. pombe*, where it fulfills a similar function as in humans [89]. Several components of the Mis18 complex are conserved in various organisms, such as the M18BP1 homologue Kinetochore Null 2 in *C. elegans* and *Arabidopsis thaliana*, but, interestingly, the whole complex seems absent in *D. melanogaster* and *S. cerevisiae* [90,91]. In *D. melanogaster*, the CENP-A-specific loading factor CAL1 was suggested to prime centromeres for CENP-A deposition, in analogy to the Mis18 complex in humans. This hypothesis is based on the observation that newly synthesized CAL1 is deposited shortly before the next round of CENP-A loading [79].

12.5.2 Loading of CENP-A

After centromere priming, CENP-A is loaded onto chromatin by chaperones (Fig. 12.3). These loading factors are specific for CENP-A and differ from the ones required for canonical nucleosome deposition in S phase. The chaperone is called HJURP in mammals and Scm3 in *S. pombe* and *S. cerevisiae* [92,93]. Although clear HJURP homologues are absent in some species, for instance in *C. elegans* and *D. melanogaster*, a functional analogue has been described in *Drosophila*, where CAL1 fulfills a very similar role [79,94,95].

In 2009, HJURP was identified as a component copurifying with prenucleosomal CENP-A complexes [92,93]. Its depletion causes a dramatic decrease in total CENP-A levels, and prevents the replenishment of newly synthesized CENP-A. The N-terminus of HJURP recognizes the CATD domain of CENP-A in the context of a CENP-A/H4 heterodimer [93]. Besides the CATD domain,

HJURP binds CENP-A residues adjacent to the CATD domain, and these additional contact points are a prerequisite for a stable HJURP/CENP-A/H4 complex [96]. As mentioned previously, the interaction between HJURP and CENP-A is abolished upon phosphorylation of CENP-A-S68 by CDK1, thereby restricting the interaction to a time point of low CDK1 activity [53]. As S68 resides outside of the CATD domain, CENP-A residues adjacent to the CATD seem to contribute to the timing of CENP-A assembly. Nevertheless, the H3^{CATD} chimera is deposited exclusively in G1, indicating that the CATD of CENP-A is sufficient to confer correct spatiotemporal regulation [97]. HJURP itself is recruited *via* the Mis18 complex and associates with centromeres exactly at the time of CENP-A loading, hence in late telophase and early G1 [84,92,93]. Similarly to the regulation of its interaction with CENP-A, HJURP localization is controlled by CDK1/2 activity: since the dephosphorylation of HJURP triggers its centromeric recruitment, its localization and CDK1/2 activity display a negative correlation [98]. Structural approaches clarified that HJURP interacts with a CENP-A/H4 heterodimer and simultaneously prevents the formation of CENP-A/H4 tetramers [99]. This observation raised the question, how an octameric nucleosome structure can then be achieved. One possible explanation is given by the fact that HJURP itself dimerizes *via* its C-terminus. This ability to dimerize is dispensable for its centromeric localization and crucial for the deposition of new CENP-A nucleosomes [100]. Hence, the soluble preassembly complex most likely consists of (HJURP/CENP-A/H4)$_2$ and concurrently delivers two CENP-A/H4 dimers to the centromeric chromatin, thereby supporting the formation of an octameric nucleosome. The subsequent loading of CENP-A depends on the ability of HJURP to directly bind to DNA, as DNA-binding mutants are unable to assemble CENP-A nucleosomes [98].

Elegant sufficiency experiments clarified that HJURP indeed acts as a CENP-A loading factor: the artificial tethering of HJURP to noncentromeric chromatin was sufficient to form functional centromeres. Likewise, HJURP specifically assembles CENP-A nucleosomes in vitro [84].

Although HJURP and its functional equivalent CAL1 do not share a common ancestor, they both influence centromere formation in a comparable manner [101]. First, CAL1, similarly to HJURP, interacts with CENP-A in a prenucleosomal context [79]. Second, artificially tethering CAL1 to a noncentromeric site results in ectopic kinetochore formation [35]. Third, CAL1 assembles CENP-A nucleosomes in vitro [35]. Fourth, depleting CAL1 also causes a decrease in total levels of CENP-A [95]. Taken together, organisms that lack clear HJURP homologues have developed comparable mechanisms to regulate CENP-A loading.

Besides the CENP-A-specific loading factor HJURP, members of the canonical chromatin assembly machinery affect centromeric CENP-A levels: RbAp46 and RbAp48 are both subunits of the CAF1 complex (see earlier); nevertheless, their depletion causes a decrease in centromeric CENP-A and total HJURP protein levels, and RbAp48 copurifies with HJURP [89,92]. Furthermore, MIS16, the orthologue of RbAp48 in *S. pombe,* is essential for CENP-A (also known as CNP1) localization [89]. These observations suggest that some components contribute to both canonical histone and CENP-A assembly. Possibly histone H4, which is shared by both nucleosome types, is the basis of competition between both pathways; however, whether these common components serve similar or different functions in these different loading mechanisms still needs to be determined.

12.5.3   Maintaining Centromeric CENP-A

Once CENP-A nucleosomes are assembled, their eviction is actively prevented, providing an epigenetic memory of centromere identity. Newly incorporated CENP-A nucleosomes were found to be stabilized by the GTPase activating protein MgcRacGAP (Fig. 12.3) that binds the licensing factor M18BP1 and specifically localizes to centromeres in late G_1. As shown by SNAP-tag pulse-chase experiments that allow the discrimination between nascent ("new") CENP-A nucleosomes and the ones already assembled in a previous cell cycle ("old"), depleting MgcRacGAP specifically causes a loss of nascent nucleosomes, whereas "old" CENP-A nucleosomes remain unaffected [102]. To date, it remains unclear how MgcRacGAP first recognizes and then protects the "new" CENP-A nucleosomes; however, besides MgcRacGAP, the guanine nucleotide exchange factor Ect2 and the small GTPases Cdc42 and Rac are required to maintain centromeric CENP-A, indicating that the stabilization of CENP-A nucleosome involves a small GTPase cycle [102].

Independently, CENP-C also physically stabilizes centromeric CENP-A nucleosomes: it changes the physical properties of the nucleosomes by altering their shape, thereby rigidifying the nucleosome structure overall. Accordingly, depleting CENP-C reduces the high stability of CENP-A at centromeres [103].

General chromatin remodelers have also been implicated in the stabilization of CENP-A chromatin: remodeling and spacing factor 1 (RSF1), a subunit of the ATP-dependent chromatin RSF complex, binds to and protects CENP-A chromatin [104]. Likewise, the Facilitates Chromatin Transcription (FACT) complex and the ATP-dependent remodeler CHD1 affect the loading and the maintenance of CENP-A nucleosomes [105].

12.6   THE CONSTITUTIVE CENTROMERE ASSOCIATED NETWORK (CCAN) AND ITS CONTRIBUTION TO CENTROMERE MAINTENANCE

Affinity purification experiments of CENP-A-containing chromatin resulted in the identification of a protein network constitutively associated with centromeric chromatin, the CCAN [106]. Besides assembling the kinetochore and regulating kinetochore-microtubule dynamics (reviewed in [107]), the CCAN functions in centromere maintenance. Based on protein-protein interaction and depletion studies, the CCAN components are divided into different subcomplexes, such as the CENP-H-I complex. Since the depletion of its components CENP-H, -I, -K, and -M affects centromeric CENP-A levels, the CENP-H-I complex might contribute to CENP-A replenishment [108]. Accordingly, CENP-H-containing complexes recruit the two remodelers FACT and CDH1 that affect centromere maintenance (see earlier) [105]. Additionally, artificially tethering CENP-I or CENP-C to an ectopic site in chicken DT40 cells that lack an endogenous centromere leads to the formation of a functional neocentromere. Due to the recruitment of CENP-A, these neocentromeres are self-maintaining and the initial targeting trigger is dispensable over time [109]. Since CENP-C functions as a kinetochore receptor for M18BP1, it influences an early step of the CENP-A loading pathway (see earlier). Consistently, its depletion disrupts the centromeric localization of HJURP and thus the assembly of CENP-A chromatin in *Xenopus laevis* [83]. Interestingly, CENP-C and HJURP were suggested to be direct-binding partners, and the recruitment of CENP-C to synthetic centromeres might require both CENP-A and HJURP [23]. Hence, by contributing to both CENP-A and CENP-C loading, HJURP possibly fulfills a dual role in centromere maintenance, and at the same time ensures its own centromeric recruitment during the next round of loading. Similarly, CAL1 and CENP-C form a stable complex in *D. melanogaster*, and they are mutually dependent for their centromeric localization, suggesting that CENP-C function is conserved [95]. This is especially interesting considering that the CCAN is largely absent in flies, and the centromere/kinetochore interface only consists of CENP-A, CAL1, and CENP-C.

Of note, the CCAN components CENP-T, -W, -S, and -X that form a heterotetrameric complex, and each contain a HFD and directly bind to centromeric DNA. Based on the observation that they display a similar DNA-binding surface as a canonical nucleosome, they were suggested to form a nucleosome-like structure at centromeres [110]; however, fluorescence recovery after photobleaching measurements on centromeres revealed that CENP-X and CENP-S exchange with very different kinetics, arguing against a stable tetrameric complex [111]. Although it is established that the CENP-S-containing complex is essential for outer kinetochore formation, it remains to be determined whether it contributes to the formation of the centromere-specific chromatin [112].

12.7 CENTROMERIC CHROMATIN AND ITS ROLE OUTSIDE OF CENTROMERES

3D imaging and thorough quantitative assessment of cellular CENP-A levels revealed that only a subset of cellular CENP-A nucleosomes localizes to centromeric sites in unperturbed cells [41]. Importantly, low levels of CENP-A molecules seem distributed throughout the chromatin. As active centromeres are only formed when a certain threshold of CENP-A molecules is reached, the question remains whether these low abundant ectopic CENP-A nucleosomes fulfill any function [41]. In contrast, most data support the hypothesis that the noncentromeric incorporation of CENP-A is actively counteracted and targeted for proteolysis in several organims [54–57]. Furthermore, cells may have developed mechanisms to prevent ectopic CENP-A incorporation. For instance, the histone-fold protein CHRAC14, a component of several chromatin remodeling complexes, directly binds to CENP-A, and prevents its ectopic incorporation to sites of DNA damage in flies [113]. Depleting CHRAC14 not only results in the accumulation of CENP-A clusters at sites of DNA lesions, but also causes severe mitotic defects [113]. Ectopic CENP-A loading needs to be studied in more detail in order to understand whether it can be beneficial for the cell under certain circumstances.

12.8   CENTROMERIC CHROMATIN AND ITS ROLE IN CANCER

Accurate chromosome segregation during mitosis is instrumental for preventing genomic instability, a major contributor to tumor formation and progression. Chromosomal instability is the main cause of aneuploidy, and chromosome gain or loss is found, for example, in 85% of colorectal cancers. As centromeres function as platforms for kinetochore formation, their tight regulation is crucial to avoid missegregations.

CENP-A protein levels are highly increased in many cancer tissues, including those derived from colorectal cancers [114,115]. Importantly, the levels of CENP-B and proliferating cell nuclear antigen, a factor required for DNA replication, were unchanged in these tissues, excluding increased proliferation rates as the driver of high CENP-A levels. Consistently, CENP-A clusters mislocalized to noncentromeric regions in these tumor tissues [114]. Follow-up studies using various colon cancer cell lines specified that ectopic CENP-A preferentially binds to accessible chromatin as found near gene promoters [115]. Interestingly, HJURP levels are not increased in these cell lines. Instead, ectopic CENP-A seems to utilize the histone H3.3 chaperone DAXX for ectopic chromatin incorporation and may form an atypical heterotypic tetramer containing both CENP-A/H4 and H3.3/H4 [115–117]. The resulting heterotypic octamers display unique physical properties, such as hyperstability [117]. As an effect, these atypical nucleosomes may alter the chromatin structure and impact DNA accessibility [116]. It remains to be determined, whether these ectopic CENP-A nucleosomes trigger neocentromere formation under certain circumstances, especially because they bind CENP-C in vitro [117]. Importantly, CENP-C is able to recruit the full kinetochore network in both humans and flies [118–120].

Independently, increased CENP-A levels seem to be the driving force of aneuploidy in colon cancer cells that lack the tumor suppressor retinoblastoma protein (pRb) [121]. These findings directly link CENP-A overexpression to chromosomal instability. Nevertheless, future research is needed to define the relevance of CENP-A overexpression in tumor progression.

12.9   PERSPECTIVES

Our understanding of the mechanisms regulating the dynamics of centromeric chromatin has greatly improved during the last few years. The identification of the loading factors and the crystallization of CENP-A nucleosomes only represent two of many milestones. Although the timing of CENP-A loading was intensively studied, it remains unclear how altered CENP-A loading affects cellular fitness/cell survival. It is well established now that CDK1 phosphorylation of CENP-A controls the timing of loading, however the functional impact and dynamics of many other posttranslational CENP-A modifications is still unclear. Similarly, the question whether centromeric nucleosomes are hemisomes or octamers, and whether their composition changes in a cell cycle-dependent manner remains to be ultimately clarified. Lastly, overexpressed CENP-A forms ectopic heterotypic nucleosomes that might be linked to aneuploidy in cancer cells. Hence, their impact on chromosomal instability and tumor progression needs to be addressed in the future to fully understand the significance of centromeric chromatin on chromosome segregation during development, health, and in disease.

Glossary

Centromere The chromatin region of a chromosome that mediates the connection to the microtubule spindle during mitosis and meiosis. Centromeres are defined by an enrichment of the histone variant CENP-A.

CENP-A A histone H3 variant that is enriched at centromeric chromatin and required for the correct assembly of the kinetochore and therefore the attachment to the microtubule spindle.

Epigenetics The study of heritable changes in cellular and physiological features that are not caused by changes in the primary DNA sequence.

Kinetochore A large multiprotein structure that forms at centromeric chromatin during mitosis and meiosis that mediates the attachment to the microtubule spindle.

Mitosis The phase in the cell cycle when duplicated chromosomes segregate into two daughter cells.

Abbreviations

CAF1	Chromatin assembly factor 1
CAL1	Chromosome alignment defect 1
CDK1	Cyclin-dependent kinase 1

CENP-A	Centromere protein A
CHRAC14	Chromatin accessibility complex 14
CID	Centromere identifier
Cse4	Chromosome segregation protein 4
Dnmt	DNA methyltransferase
HJURP	Holliday junction recognition protein
H3	Histone 3
CATD	CENP-A targeting domain
CCAN	Constitutive centromere-associated network
HFD	Histone-fold domain
PTM	Posttranslational modifications
ChIP	Chromatin immunoprecipitation
EM	Electron microscopy
C. elegans	*Caenorhabditis elegans*
D. melanogaster	*Drosophila melanogaster*
H. sapiens	*Homo sapiens*
S. pombe	*Schizosaccharomyces pombe*
S. cerevisiae	*Saccharomyces cerevisiae*
Z. mays	*Zea mays*
X. laevis	*Xenopus laevis*

Acknowledgments

We thank Lars Jansen, Aubry Miller, and Anne-Laure Pauleau for helpful comments on the manuscript and the Erhardt lab for discussions. Our research is funded by the Deutsche Forschungsgemeinschaft through the grants ER576 and EXC81 (CellNetworks) to SE and BA 5417/1-1 to DB.

References

[1] Cheeseman IM, Chappie JS, Wilson-Kubalek EM, Desai A. The conserved KMN network constitutes the core microtubule-binding site of the kinetochore. Cell 2006;127:983–97.

[2] Runge KW, Wellinger RJ, Zakian VA. Effects of excess centromeres and excess telomeres on chromosome loss rates. Mol Cell Biol 1991;11:2919–28.

[3] Steiner FA, Henikoff S. Holocentromeres are dispersed point centromeres localized at transcription factor hotspots. eLife 2014;3:e02025.

[4] Allshire RC, Karpen GH. Epigenetic regulation of centromeric chromatin: old dogs, new tricks? Nat Rev Genet 2008;9:923–37.

[5] Yao J, et al. Plasticity and epigenetic inheritance of centromere-specific histone H3 (CENP-A)-containing nucleosome positioning in the fission yeast. J Biol Chem 2013;288:19184–96.

[6] Mendiburo MJ, Padeken J, Fulop S, Schepers A, Heun P. *Drosophila* CENH3 is sufficient for centromere formation. Science 2011;334:686–90.

[7] Merry DE, Pathak S, Hsu TC, Brinkley BR. Anti-kinetochore antibodies: use as probes for inactive centromeres. Am J Hum Genet 1985;37:425–30.

[8] Earnshaw WC, Migeon BR. Three related centromere proteins are absent from the inactive centromere of a stable isodicentric chromosome. Chromosoma 1985;92:290–6.

[9] Earnshaw WC, Rothfield N. Identification of a family of human centromere proteins using autoimmune sera from patients with scleroderma. Chromosoma 1985;91:313–21.

[10] Heun P, et al. Mislocalization of the *Drosophila* centromere-specific histone CID promotes formation of functional ectopic kinetochores. Dev Cell 2006;10:303–15.

[11] Olszak AM, et al. Heterochromatin boundaries are hotspots for de novo kinetochore formation. Nat Cell Biol 2011;13:799–808.

[12] Gascoigne KE, et al. Induced ectopic kinetochore assembly bypasses the requirement for CENP-A nucleosomes. Cell 2011;145:410–22.

[13] Buchwitz BJ, Ahmad K, Moore LL, Roth MB, Henikoff S. A histone-H3-like protein in *C. elegans*. Nature 1999;401:547–8.

[14] Howman EV, et al. Early disruption of centromeric chromatin organization in centromere protein A (Cenpa) null mice. Proc Natl Acad Sci USA 2000;97:1148–53.

[15] Goshima G, Kiyomitsu T, Yoda K, Yanagida M. Human centromere chromatin protein hMis12, essential for equal segregation, is independent of CENP-A loading pathway. J Cell Biol 2003;160:25–39.

[16] Ishii K, et al. Heterochromatin integrity affects chromosome reorganization after centromere dysfunction. Science 2008;321:1088–91.

[17] Westhorpe FG, Straight AF. The centromere: epigenetic control of chromosome segregation during mitosis. Cold Spring Harb Perspect Biol 2015;7:a015818.

[18] Sullivan KF, Hechenberger M, Masri K. Human CENP-A contains a histone H3 related histone fold domain that is required for targeting to the centromere. J Cell Biol 1994;127:581–92.

[19] Black BE, et al. Structural determinants for generating centromeric chromatin. Nature 2004;430:578–82.

[20] Black BE, et al. Centromere identity maintained by nucleosomes assembled with histone H3 containing the CENP-A targeting domain. Mol Cell 2007;25:309–22.

[21] Guse A, Carroll CW, Moree B, Fuller CJ, Straight AF. In vitro centromere and kinetochore assembly on defined chromatin templates. Nature 2011;477:354–8.

[22] Logsdon GA, et al. Both tails and the centromere targeting domain of CENP-A are required for centromere establishment. J Cell Biol 2015;208:521–31.

[23] Tachiwana H, et al. HJURP involvement in de novo CenH3(CENP-A) and CENP-C recruitment. Cell Rep 2015;11:22–32.

[24] Black BE, Brock MA, Bedard S, Woods VL Jr, Cleveland DW. An epigenetic mark generated by the incorporation of CENP-A into centromeric nucleosomes. Proc Natl Acad Sci USA 2007;104:5008–13.

[25] Henikoff S, Furuyama T. The unconventional structure of centromeric nucleosomes. Chromosoma 2012;121:341–52.

[26] Quenet D, Dalal Y. The CENP-A nucleosome: a dynamic structure and role at the centromere. Chromosome Res 2012;20:465–79.

[27] Dalal Y, Wang H, Lindsay S, Henikoff S. Tetrameric structure of centromeric nucleosomes in interphase *Drosophila* cells. PLoS Biol 2007;5:e218.

[28] Black BE, Bassett EA. The histone variant CENP-A and centromere specification. Curr Opin Cell Biol 2008;20:91–100.

[29] Dimitriadis EK, Weber C, Gill RK, Diekmann S, Dalal Y. Tetrameric organization of vertebrate centromeric nucleosomes. Proc Natl Acad Sci USA 2010;107:20317–22.

[30] Henikoff S, et al. The budding yeast Centromere DNA Element II wraps a stable Cse4 hemisome in either orientation in vivo. eLife 2014;3:e01861.

[31] Furuyama T, Henikoff S. Centromeric nucleosomes induce positive DNA supercoils. Cell 2009;138:104–13.

[32] Tachiwana H, et al. Crystal structure of the human centromeric nucleosome containing CENP-A. Nature 2011;476:232–5.

[33] Dechassa ML, et al. Structure and Scm3-mediated assembly of budding yeast centromeric nucleosomes. Nat Commun 2011;2:313.

[34] Hasson D, et al. The octamer is the major form of CENP-A nucleosomes at human centromeres. Nat Struct Mol Biol 2013;20:687–95.

[35] Chen CC, et al. CAL1 is the *Drosophila* CENP-A assembly factor. J Cell Biol 2014;204:313–29.

[36] Zhang W, Colmenares SU, Karpen GH. Assembly of *Drosophila* centromeric nucleosomes requires CID dimerization. Mol Cell 2012;45:263–9.

[37] Padeganeh A, et al. Octameric CENP-A nucleosomes are present at human centromeres throughout the cell cycle. Curr Biol 2013;23:764–9.

[38] Krassovsky K, Henikoff JG, Henikoff S. Tripartite organization of centromeric chromatin in budding yeast. Proc Natl Acad Sci USA 2012;109:243–8.

[39] Bui M, et al. Cell-cycle-dependent structural transitions in the human CENP-A nucleosome in vivo. Cell 2012;150:317–26.

[40] Shivaraju M, et al. Cell-cycle-coupled structural oscillation of centromeric nucleosomes in yeast. Cell 2012;150:304–16.

[41] Bodor DL, et al. The quantitative architecture of centromeric chromatin. eLife 2014;3:e02137.

[42] Sullivan BA, Karpen GH. Centromeric chromatin exhibits a histone modification pattern that is distinct from both euchromatin and heterochromatin. Nat Struct Mol Biol 2004;11: 1076–83.

[43] Bergmann JH, et al. Epigenetic engineering shows H3K4me2 is required for HJURP targeting and CENP-A assembly on a synthetic human kinetochore. EMBO J 2011;30:328–40.

[44] Shang WH, et al. Chromosome engineering allows the efficient isolation of vertebrate neocentromeres. Dev Cell 2013;24:635–48.

[45] Alonso A, Hasson D, Cheung F, Warburton PE. A paucity of heterochromatin at functional human neocentromeres. Epigenetics Chromatin 2010;3:6.

[46] Greaves IK, Rangasamy D, Ridgway P, Tremethick DJ. H2A.Z. contributes to the unique 3D structure of the centromere. Proc Natl Acad Sci USA 2007;104:525–30.

[47] Rangasamy D, Greaves I, Tremethick DJ. RNA interference demonstrates a novel role for H2A.Z. in chromosome segregation. Nat Struct Mol Biol 2004;11:650–5.

[48] Kim HS, et al. An acetylated form of histone H2A.Z. regulates chromosome architecture in *Schizosaccharomyces pombe*. Nat Struct Mol Biol 2009;16:1286–93.

[49] Samel A, Cuomo A, Bonaldi T, Ehrenhofer-Murray AE. Methylation of CenH3 arginine 37 regulates kinetochore integrity and chromosome segregation. Proc Natl Acad Sci USA 2012;109:9029–34.

[50] Goutte-Gattat D, et al. Phosphorylation of the CENP-A amino-terminus in mitotic centromeric chromatin is required for kinetochore function. Proc Natl Acad Sci USA 2013;110:8579–84.

[51] Bailey AO, et al. Posttranslational modification of CENP-A influences the conformation of centromeric chromatin. Proc Natl Acad Sci USA 2013;110:11827–32.

[52] Niikura Y, et al. CENP-A K124 ubiquitylation is required for CENP-A deposition at the centromere. Dev Cell 2015;32:589–603.

[53] Yu Z, et al. Dynamic phosphorylation of CENP-A at Ser68 orchestrates its cell-cycle-dependent deposition at centromeres. Dev Cell 2015;32:68–81.

[54] Moreno-Moreno O, Torras-Llort M, Azorin F. Proteolysis restricts localization of CID, the centromere-specific histone H3 variant of *Drosophila*, to centromeres. Nucleic Acids Res 2006;34:6247–55.

[55] Moreno-Moreno O, Medina-Giro S, Torras-Llort M, Azorin F. The F box protein partner of paired regulates stability of *Drosophila* centromeric histone H3, CenH3(CID). Curr Biol 2011;21:1488–93.

[56] Hewawasam G, et al. Psh1 is an E3 ubiquitin ligase that targets the centromeric histone variant Cse4. Mol Cell 2010;40:444–54.

[57] Ranjitkar P, et al. An E3 ubiquitin ligase prevents ectopic localization of the centromeric histone H3 variant via the centromere targeting domain. Mol Cell 2010;40:455–64.

[58] Bade D, Pauleau AL, Wendler A, Erhardt S. The E3 ligase CUL3/RDX controls centromere maintenance by ubiquitylating and stabilizing CENP-A in a CAL1-dependent manner. Dev Cell 2014;28:508–19.

[59] Murphy WJ, et al. Dynamics of mammalian chromosome evolution inferred from multispecies comparative maps. Science 2005;309:613–7.

[60] Eichler EE. Repetitive conundrums of centromere structure and function. Hum Mol Genet 1999;8:151–5.

[61] Warburton PE. Chromosomal dynamics of human neocentromere formation. Chromosome Res 2004;12:617–26.

[62] Joglekar AP, et al. Molecular architecture of the kinetochore-microtubule attachment site is conserved between point and regional centromeres. J Cell Biol 2008;181:587–94.

[63] Fachinetti D, et al. A two-step mechanism for epigenetic specification of centromere identity and function. Nat Cell Biol 2013;15:1056–66.

[64] Okada T, et al. CENP-B controls centromere formation depending on the chromatin context. Cell 2007;131:1287–300.

[65] Fachinetti D, et al. DNA sequence-specific binding of CENP-B enhances the fidelity of human centromere function. Dev Cell 2015;33:314–27.

[66] Saffery R, et al. Human centromeres and neocentromeres show identical distribution patterns of >20 functionally important kinetochore-associated proteins. Hum Mol Genet 2000;9:175–85.

[67] Rosic S, Kohler F, Erhardt S. Repetitive centromeric satellite RNA is essential for kinetochore formation and cell division. J Cell Biol 2014;207:335–49.

[68] Chen CC, et al. Establishment of centromeric chromatin by the CENP-A assembly factor CAL1 requires FACT-mediated transcription. Dev Cell 2015;34:73–84.

[69] Quenet D, Dalal Y. A long non-coding RNA is required for targeting centromeric protein A to the human centromere. eLife 2014;3:e03254.

[70] Chan FL, et al. Active transcription and essential role of RNA polymerase II at the centromere during mitosis. Proc Natl Acad Sci USA 2012;109:1979–84.

[71] Li F, Sonbuchner L, Kyes SA, Epp C, Deitsch KW. Nuclear non-coding RNAs are transcribed from the centromeres of *Plasmodium falciparum* and are associated with centromeric chromatin. J Biol Chem 2008;283:5692–8.

[72] Ferri F, Bouzinba-Segard H, Velasco G, Hube F, Francastel C. Non-coding murine centromeric transcripts associate with and potentiate Aurora B kinase. Nucleic Acids Res 2009;37:5071–80.

[73] Choi ES, et al. Identification of noncoding transcripts from within CENP-A chromatin at fission yeast centromeres. J Biol Chem 2011;286:23600–7.

[74] Topp CN, Zhong CX, Dawe RK. Centromere-encoded RNAs are integral components of the maize kinetochore. Proc Natl Acad Sci USA 2004;101:15986–91.

[75] Bergmann JH, et al. Epigenetic engineering: histone H3K9 acetylation is compatible with kinetochore structure and function. J Cell Sci 2012;125:411–21.

[75a] Carone DM, Zhang C, Hall LE, Obergfell C, Carone BR, O'Neill MJ, O'Neill RJ. Hypermorphic expression of centromeric retroelement-encoded small RNAs impairs CENP-A loading. Chromosome Res 2013;21(1):49–62.

[76] Sullivan B, Karpen G. Centromere identity in *Drosophila* is not determined in vivo by replication timing. J Cell Biol 2001;154:683–90.

[77] Jansen LE, Black BE, Foltz DR, Cleveland DW. Propagation of centromeric chromatin requires exit from mitosis. J Cell Biol 2007;176:795–805.

[78] Schuh M, Lehner CF, Heidmann S. Incorporation of *Drosophila* CID/CENP-A and CENP-C into centromeres during early embryonic anaphase. Curr Biol 2007;17:237–43.

[79] Mellone BG, et al. Assembly of *Drosophila* centromeric chromatin proteins during mitosis. PLoS Genet 2011;7:e1002068.

[80] Dunleavy EM, et al. The cell cycle timing of centromeric chromatin assembly in *Drosophila* meiosis is distinct from mitosis yet requires CAL1 and CENP-C. PLoS Biol 2012;10:e1001460.

[81] Dunleavy EM, Almouzni G, Karpen GH. H3.3 is deposited at centromeres in S phase as a placeholder for newly assembled CENP-A in G(1) phase. Nucleus 2011;2:146–57.

[82] Fujita Y, et al. Priming of centromere for CENP-A recruitment by human hMis18alpha, hMis18beta, and M18BP1. Dev Cell 2007;12:17–30.

[83] Moree B, Meyer CB, Fuller CJ, Straight AF. CENP-C recruits M18BP1 to centromeres to promote CENP-A chromatin assembly. J Cell Biol 2011;194:855–71.

[84] Barnhart MC, et al. HJURP is a CENP-A chromatin assembly factor sufficient to form a functional de novo kinetochore. J Cell Biol 2011;194:229–43.

[85] Dambacher S, et al. CENP-C facilitates the recruitment of M18BP1 to centromeric chromatin. Nucleus 2012;3:101–10.

[86] McKinley KL, Cheeseman IM. Polo-like kinase 1 licenses CENP-A deposition at centromeres. Cell 2014;158:397–411.

[87] Silva MC, et al. Cdk activity couples epigenetic centromere inheritance to cell cycle progression. Dev Cell 2012;22:52–63.

[88] Kim IS, et al. Roles of Mis18alpha in epigenetic regulation of centromeric chromatin and CENP-A loading. Mol Cell 2012;46:260–73.

[89] Hayashi T, et al. Mis16 and Mis18 are required for CENP-A loading and histone deacetylation at centromeres. Cell 2004;118:715–29.

[90] Lermontova I, et al. Arabidopsis kinetochore null2 is an upstream component for centromeric histone H3 variant cenH3 deposition at centromeres. Plant Cell 2013;25:3389–404.

[91] Maddox PS, Hyndman F, Monen J, Oegema K, Desai A. Functional genomics identifies a Myb domain-containing protein family required for assembly of CENP-A chromatin. J Cell Biol 2007;176:757–63.

[92] Dunleavy EM, et al. HJURP is a cell-cycle-dependent maintenance and deposition factor of CENP-A at centromeres. Cell 2009;137:485–97.

[93] Foltz DR, et al. Centromere-specific assembly of CENP-a nucleosomes is mediated by HJURP. Cell 2009;137:472–84.

[94] Pauleau AL, Erhardt S. Centromere regulation: new players, new rules, new questions. Eur J Cell Biol 2011;90:805–10.

[95] Erhardt S, et al. Genome-wide analysis reveals a cell cycle-dependent mechanism controlling centromere propagation. J Cell Biol 2008;183:805–18.

[96] Bassett EA, et al. HJURP uses distinct CENP-A surfaces to recognize and to stabilize CENP-A/histone H4 for centromere assembly. Dev Cell 2012;22:749–62.

[97] Bodor DL, Valente LP, Mata JF, Black BE, Jansen LE. Assembly in G1 phase and long-term stability are unique intrinsic features of CENP-A nucleosomes. Mol Biol Cell 2013;24:923–32.

[98] Muller S, et al. Phosphorylation and DNA binding of HJURP determine its centromeric recruitment and function in CenH3(CENP-A) loading. Cell Rep 2014;8:190–203.

[99] Hu H, et al. Structure of a CENP-A-histone H4 heterodimer in complex with chaperone HJURP. Genes Dev 2011;25:901–6.

[100] Zasadzinska E, Barnhart-Dailey MC, Kuich PH, Foltz DR. Dimerization of the CENP-A assembly factor HJURP is required for centromeric nucleosome deposition. EMBO J 2013;32:2113–24.

[101] Phansalkar R, Lapierre P, Mellone BG. Evolutionary insights into the role of the essential centromere protein CAL1 in *Drosophila*. Chromosome Res 2012;20:493–504.

[102] Lagana A, et al. A small GTPase molecular switch regulates epigenetic centromere maintenance by stabilizing newly incorporated CENP-A. Nat Cell Biol 2010;12:1186–93.

[103] Falk SJ, et al. Chromosomes. CENP-C reshapes and stabilizes CENP-A nucleosomes at the centromere. Science 2015;348:699–703.

[104] Perpelescu M, Nozaki N, Obuse C, Yang H, Yoda K. Active establishment of centromeric CENP-A chromatin by RSF complex. J Cell Biol 2009;185:397–407.

[105] Okada M, Okawa K, Isobe T, Fukagawa T. CENP-H-containing complex facilitates centromere deposition of CENP-A in cooperation with FACT and CHD1. Mol Biol Cell 2009;20:3986–95.

[106] Foltz DR, et al. The human CENP-A centromeric nucleosome-associated complex. Nat Cell Biol 2006;8:458–69.

[107] McAinsh AD, Meraldi P. The CCAN complex: linking centromere specification to control of kinetochore-microtubule dynamics. Semin Cell Dev Biol 2011;22:946–52.

[108] Okada M, et al. The CENP-H-I complex is required for the efficient incorporation of newly synthesized CENP-A into centromeres. Nat Cell Biol 2006;8:446–57.

[109] Hori T, Shang WH, Takeuchi K, Fukagawa T. The CCAN recruits CENP-A to the centromere and forms the structural core for kinetochore assembly. J Cell Biol 2013;200:45–60.

[110] Nishino T, et al. CENP-T-W-S-X forms a unique centromeric chromatin structure with a histone-like fold. Cell 2012;148:487–501.

[111] Dornblut C, et al. A CENP-S/X complex assembles at the centromere in S and G2 phases of the human cell cycle. Open Biol 2014;4:130229.

[112] Amano M, et al. The CENP-S complex is essential for the stable assembly of outer kinetochore structure. J Cell Biol 2009;186:173–82.

[113] Mathew V, et al. The histone-fold protein CHRAC14 influences chromatin composition in response to DNA damage. Cell Rep 2014;7:321–30.

[114] Tomonaga T, et al. Overexpression and mistargeting of centromere protein-A in human primary colorectal cancer. Cancer Res 2003;63:3511–6.

[115] Athwal RK, et al. CENP-A nucleosomes localize to transcription factor hotspots and sub-telomeric sites in human cancer cells. Epigenetics Chromatin 2015;8:2.

[116] Lacoste N, et al. Mislocalization of the centromeric histone variant CenH3/CENP-A in human cells depends on the chaperone DAXX. Mol Cell 2014;53:631–44.

[117] Arimura Y, et al. Crystal structure and stable property of the cancer-associated heterotypic nucleosome containing CENP-A and H3. 3. Sci Rep 2014;4:7115.

[118] Screpanti E, et al. Direct binding of Cenp-C to the Mis12 complex joins the inner and outer kinetochore. Curr Biol 2011;21:391–8.

[119] Przewloka MR, et al. CENP-C is a structural platform for kinetochore assembly. Curr Biol 2011;21:399–405.

[120] Rago F, Gascoigne KE, Cheeseman IM. Distinct Organization and Regulation of the Outer Kinetochore KMN Network Downstream of CENP-C and CENP-T. Curr Biol 2015;25:671–7.

[121] Amato A, Schillaci T, Lentini L, Di Leonardo A. CENPA overexpression promotes genome instability in pRb-depleted human cells. Mol Cancer 2009;8:119.

CHAPTER 13

Telomere Maintenance in the Dynamic Nuclear Architecture

E. Micheli, A. Galati, A. Cicconi, S. Cacchione

Department of Biology and Biotechnology, Istituto Pasteur Italia - Fondazione Cenci Bolognetti, Sapienza University of Rome, Rome, Italy

13.1 INTRODUCTION

Switching from circular to linear chromosomes, eukaryotes had to solve two main problems. First, chromosome ends have to be distinguished from broken DNA ends to avoid processing by DNA repair enzymes, which is called the protection problem. Second, as DNA polymerases need a primer to start copying DNA, the removal of the terminal primer at the lagging strand leads to loss of genetic material at each replication cycle, which is called the replication problem. These problems found a solution with the advent of telomeres, protective structures located at the end of eukaryotic chromosomes. In most organisms, telomeres consist of short G-rich sequences repeated from tens to thousands fold depending on the species [1], ending in a G-rich single-stranded protrusion (G-tail or G-overhang). Protection is assured by the binding of specialized proteins [2]. Telomere maintenance is guaranteed by the action of telomerase, a ribonucleoprotein enzymatic complex that in most eukaryotes counteracts the erosion of chromosome ends by adding telomeric repeats at the 3′ termini [3].

In humans, telomeres play a pivotal role in several regulatory pathways that determine the cell fate. At birth, human telomeres are 10–15 kb long, consisting of thousands of TTAGGG repeats (Fig. 13.1) organized in a unique and compact chromatin and bound by the six-protein complex, shelterin [4]. In germinal and embryonic stem cells, telomere length is maintained by the activity of telomerase that adds TTAGGG repeats at the 3′ ends of chromosomes. By contrast, telomerase is inactive in somatic cells, and consequently telomeres shorten at each replication cycle till they reach a critical length that triggers a DNA damage response (DDR) pathway (see also Chapter 11) leading to permanent cell cycle arrest, a state known as replicative senescence [5].

A proper maintenance of telomeres is essential for life, and abnormal maintenance of telomere length can contribute to severe diseases. Anomalous

CONTENTS

325

Chromatin Regulation and Dynamics. http://dx.doi.org/10.1016/B978-0-12-803395-1.00013-7

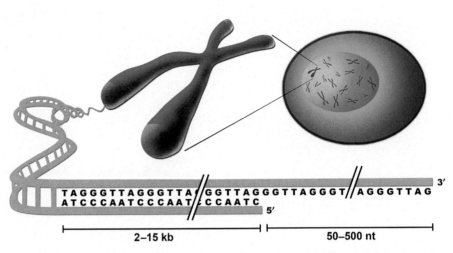

FIGURE 13.1 Representation of a metaphase chromosome with telomeres highlighted in *orange*. Human telomeric DNA is composed of thousands of tandem TTAGGG repeats, is very variable in length, and ends with a 50–500 nucleotides 3′-single-stranded overhang.

telomere shortening, for example, can lead to diseases such as dyskeratosis congenita and pulmonary fibrosis [6], whereas reactivation of telomerase allows most cancers to acquire the unlimited proliferative capacity necessary for their progression [7]. To fulfill their multiple functions, telomeres interplay with different nuclear environments and assume different structures. For example, telomeric chromatin structure needs to dynamically switch between closed and open conformation during the cell cycle to permit DNA replication and telomerase access in telomerase-positive cells.

In this chapter, we review what is known about telomere structure and telomeric chromatin organization, focusing on human telomeres. We will discuss how telomere structure changes on telomere shortening and during the steps leading to replicative senescence. Furthermore, we will examine the novel perspectives emerging from the analysis of the three-dimensional interactions of human telomeres.

13.2 TELOMERE MAINTENANCE

Linear chromosomes cannot be completely duplicated by conventional DNA polymerases, because they need an RNA primer to start DNA synthesis. After primer removal, a gap remains at the 5′ end of the lagging strand. To counteract the consequent loss of genetic material at any replication cycle, eukaryotes rely on telomerase, a ribonucleoproteic enzymatic complex conserved from yeast to humans (Fig. 13.2, see Section 13.2.1).

(A) G-rich strand synthesis by telomerase

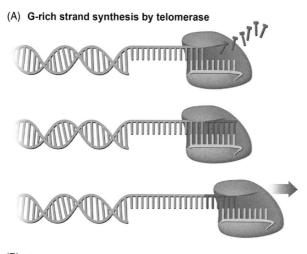

(B) C-rich strand synthesis by DNA polymerase

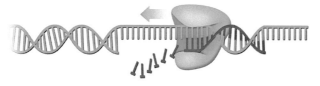

FIGURE 13.2 Scheme of telomerase elongation.

(A) The RNA subunit, telomerase RNA component, TER *(in yellow)* pairs with the 3′-overhang terminal nucleotides and acts as a template for the addition of a telomeric repeat to the 3′-OH, catalyzed by the protein component, telomerase reverse transcriptase, TERT *(in purple)*. Then, the telomerase moves toward the last newly synthesized nucleotides and starts a new elongation step. (B) The C-rich strand is synthesized by a DNA polymerase *(in green)*, in a canonical DNA synthesis reaction *(in red the RNA primer)*.

However, there are exceptions to telomerase-dependent telomere maintenance. The best characterized example of telomerase-independent telomere maintenance is found in *Drosophila*, an organism that lacks telomerase and G-rich repeats. *Drosophila* telomeres consist of three specialized retrotransposons (Het-A, TART, and TAHRE). Thus, the DNA sequences lost during the replication process are replaced by retrotransposition [8]. Other telomerase-free species have also been characterized (for a review see Ref. [9]). An interesting case is represented by the silkworm, *Bombyx mori*. In this species, telomere repeats are present and telomerase is expressed at very low levels; at the same time there are hundreds of copies of two retrotransposons that recognize the $(TTAGG)_n$ terminal DNA and transpose in the middle of the telomeric array. *B. mori* thus represents an intermediate situation between canonical telomere maintenance via telomerase and *Drosophila* transposition-based mechanism [10].

Finally, as an alternative to telomerase-mediated telomere elongation, telomeres might be maintained by recombination-based mechanisms. This phenomenon, termed alternative lengthening of telomeres (ALT), has been observed in 10–15% of human cancers (Section 13.2.2).

13.2.1   Telomerase

Telomerase is composed of two main subunits, the telomerase RNA component named TER (also known as TR or TERC) and the telomerase reverse transcriptase (TERT) protein, and a variable number of accessory factors [11]. Telomerase acts as a reverse transcriptase: a region within the RNA component anneals with the 3' end of chromosomes and provides the template for the addition of a telomeric repeat catalyzed by TERT (Fig. 13.2). Once a telomeric repeat is added, the telomerase moves on to the new 3' end and synthesizes a new repeat. Then the trimeric complex CST (CTC1–STN1–TEN1) binds the 3' end and favors the recruitment of a conventional DNA polymerase α/primase to fill in the C-rich complementary strand [12].

In humans, telomerase is active only in germinal tissues and in stem cells. In somatic cells, telomerase is repressed and, as a consequence, telomeres shorten at every cell cycle until they are not anymore able to form a protective structure [13]. As unprotected chromosome ends are not distinguishable from double-strand breaks (DSB), they activate a DDR signaling mediated by ataxia telangiectasia mutated (ATM) kinase or by ATM and Rad3-related (ATR) kinase. ATM or ATR phosphorylates nearby histone H2AX molecules producing γ-H2AX (phosphorylated Ser-139), which in turn triggers the recruitment of the signal transducer protein, 53BP1, to the exposed telomere. These events culminate in the activation of the cell cycle regulator p53 that mediates the cell cycle arrest [14]. In contrast to DDR in nontelomeric regions of the genome, DNA damage at telomeres is not repaired [15] and leads to a state of permanent cell cycle arrest termed replicative senescence. This is generally viewed as a tumor protection mechanism, because the limited number of divisions of somatic cells reduces the possibility of accumulating oncogenic mutations [16]. Furthermore, the inhibition of repair mechanisms acting on damaged telomeres impedes the chromosomal instability that would derive from telomeric fusion events [16].

Cells with defects in the p53 checkpoint can bypass the cell cycle block and continue dividing. In cycling cells, DNA repair activity at deprotected telomeres generates chromosome end-to-end fusions that result in cell death within a few cell generations, unless the cells acquire means to maintain telomeres, generally by reactivating telomerase. Thus, this process enables precancerous cells to acquire unlimited proliferating capacity and facilitates the transition into cancer. About 85–90% of cancers display an upregulated telomerase activity [5]; the remaining 10–15% are telomerase-negative and maintain telomeres through an alternative mechanism named ALT, based on homologous recombination (HR) [17].

13.2.2 Alternative Lengthening of Telomeres: Telomere Maintenance Without Telomerase

Several immortalized cell lines and, as noted above, 10–15% of tumors are telomerase-negative and yet maintain functional telomeres. This observation led to the assumption that an alternative way of telomere length maintenance must exist, termed ALT (for a review, see Refs. [17,18]). Telomeres of cells relying on ALT show several unusual features. For example, the telomere length is highly heterogeneous, ranging from almost undetectable telomeric sequences to tens of kilobases of telomeric repeats. Moreover, the size of individual telomeres can undergo rapid changes. Specific markers for ALT are the presence of extrachromosomal telomeric DNA, prevalently in a circular form (t-circles), and the formation of ALT-associated promyelocytic leukemia (PML) nuclear bodies (APBs). APBs are a subset of PML bodies that are functionally and structurally heterogeneous nuclear macromolecular structures likely involved in DDR and other nuclear processes. APBs contain chromosomal or extrachromosomal telomeric DNA, telomere-associated proteins, and proteins involved in homologous recombination. It is generally accepted that telomere maintenance in ALT cells is based on HR, even if the molecular details of the ALT mechanisms have not been resolved yet [18]. As HR at telomeres is repressed in normal cells and in telomerase-positive immortalized cells, ALT activation likely requires the inactivation of one or more factors that repress HR. For example, the protein α-thalassemia/mental retardation syndrome X-linked (ATRX) does not only inhibit HR, but is also able to repress ALT activity if transiently expressed in ALT-positive/ATRX-negative cells [19]. ATRX is a chromatin remodeler of the SWItch/sucrose nonfermentable complex (SWI/SNF) family that binds DNA tandem repeats both at telomeres and in euchromatin. ATRX also binds G-quadruplex structures in vitro [20], suggesting that it might play a role in resolving G-quadruplex structures forming at telomeres during replication (Section 13.3.1), thereby inhibiting replication fork stalling and HR. Together with the histone chaperone death domain–associated protein 6 (DAXX), it deposits the histone variant H3.3 at pericentric heterochromatic regions (see also Chapter 12) and at telomeres [21]. Moreover, in several ALT-positive tumors ATRX/DAXX and/or H3.3 are mutated, further supporting their role in the suppression of ALT [22]. A consequence of the failed deposition of H3.3 is the loss of heterochromatic marks at telomeres, but it is still unclear if this feature has a role in ALT establishment [21].

Another possible player in ALT establishment is TERRA (Section 13.4.4). TERRA levels are significantly increased in ALT cells, suggesting that TERRA might facilitate the process of HR [23].

13.3   TELOMERE STRUCTURE AND FUNCTION

In most eukaryotes, chromosomes termini consist of tandem repeats of short G-rich sequences (TTAGGG in vertebrates). Telomere length varies among species, ranging from tens of basepairs in ciliates to thousands of basepairs in higher eukaryotes. The G-rich strand extends to form a 3′ single-stranded overhang, the length of which is also species-specific. For example, in budding yeast (*Saccharomyces cerevisiae*) 3′ overhangs are short, 12–14 nt long sequences, whereas in mammals they can reach the length of a few hundred nucleotides [2]. The naked DNA component of every chromosome end is equivalent to a DSB. DSBs at non-telomeric sequences would activate DDR and cell cycle checkpoints to block the cell cycle progression, providing sufficient time to complete the repair mainly by HR or by nonhomologous end joining (NHEJ). If DNA damage is not repaired or too widespread, the fate of the cell will be the permanent cell cycle block or apoptosis. Natural chromosome ends assume a specific structure that prevents processing by DNA repair enzymatic complexes [16].

13.3.1   T-Loop and G-Quadruplex

The isolation of lasso-like structures visualized by electron microscopy [24] led to the telomerase-loop (t-loop) model of telomeres, in which the single-stranded 3′ overhang curves back to invade the upstream double-stranded telomeric DNA, forming a displacement loop (D-loop) at the point of insertion (Fig. 13.3A). The presence of t-loops in at least 25% of telomeres has been visualized directly in nuclei of mouse splenocytes by stochastic optical reconstruction microscopy, a super resolution fluorescence imaging method that allows imaging subcellular structure at near molecular scale definition [25]. T-loop formation is thermodynamically unfavorable: its stabilization requires

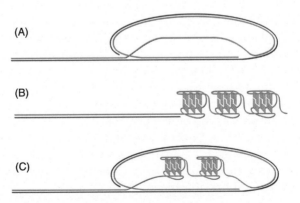

FIGURE 13.3 Three possible telomere capping configurations.
(A) Telomere-loop (T-loop) model; (B) G-quadruplex model; and (C) a possible model comprising both t-loop and G-quadruplex structures.

specific proteins, such as TRF2 [24,25], one of the components of the shelterin complex in mammals (Section 13.3.2).

Another structure that has been proposed as an alternative way of telomere capping is based on a four-stranded helical DNA conformation, known as G-quadruplex [26]. This structure derives from the folding of sequences containing four runs of three to four consecutive guanines, as it occurs in the 3′ single-stranded overhang of telomeric DNA (Fig. 13.3B). G-quadruplexes are very stable structures characterized by the stacking of G-quartets, planar cyclic arrangements of four guanines, held together by eight Hoogsteen hydrogen bonds and stabilized by the coordination of a central monovalent cation (typically potassium). G-quadruplexes form spontaneously in vitro at physiological ionic conditions, so it is not surprising that G-quadruplex structures have been found also in vivo [27]. Besides 3′ overhangs, G-quadruplex forms on DNA strand separation during replication, not only at telomeres but also in several sites in the genome [28]. The characterization of several helicases capable of binding and resolving G-quadruplexes supports the importance of these structures [29]. ATRX, involved also in repressing ALT establishment, is among the proteins that bind G-quadruplex (Section 13.2.2). An interesting study that provided evidence for the potential capping property of G-quadruplex has been carried out in budding yeast. When capping proteins are mutated, it has thus been demonstrated that G-quadruplex can form a rudimentary capping structure that protects telomeres [30].

T-loop- and G-quadruplex structures represent two possible solutions to the end-protection problem, which seem to coexist and/or dynamically convert into each other during different functional states of telomeres [31] (Fig. 13.3C). Interestingly, both structures need the activity of the RTEL1 helicase in order to be resolved and allow an efficient telomere replication [32].

13.3.2   The Shelterin Complex

T-loops, as well as any other protective structure telomeres may assume, need to be stabilized by the binding of specific proteins. In mammals, TTAGGG telomeric repeats are bound by the complex named shelterin [33], comprising six different protein factors. These components include telomeric repeat–binding factors 1 and 2 (TRF1 and TRF2), which bind to double-stranded telomeric DNA as homodimers, and protection of telomeres 1 (POT1) that recognizes single-stranded TTAGGG repeats, forming a heterodimer with TINT1/PIP1/PTOP1 (TPP1). Furthermore, repressor-activator protein 1 (RAP1) interacts with TRF2, and finally TRF1-interacting protein 2 (TIN2) bridges the complex by binding to TRF1, TRF2, and TPP1 (Fig. 13.4A).

Shelterin proteins display characteristic properties that distinguish them from other factors found at telomeres: they specifically localize at telomeres, they are present at telomeres throughout the cell cycle, and they have an exclusively

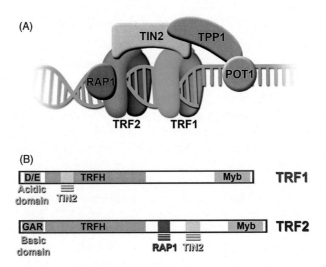

FIGURE 13.4 (A) Schematic representation of the shelterin complex assembled on terminal telomeric DNA. (B) Domain structures of telomeric repeat–binding factors, *TRF1* and *TRF2*; the regions of interaction with the other shelterin components are highlighted.

telomeric function. The localization of the shelterin complex relies on the ability of TRF1 and TRF2 to specifically bind double-stranded telomeric DNA [34,35]. These two proteins share a common domain architecture (Fig. 13.4B) containing a TRF homology (TRFH) domain that is responsible for their homodimerization, and a C-terminal SANT/Myb DNA-binding domain that allows the specific recognition of the double-stranded telomeric repeats. The dimerization and the binding domains are connected through an unstructured flexible hinge domain. The most divergent region is the N-terminal domain: while TRF2 has a basic Gly/Arg-rich (GAR) domain, TRF1 has an acidic N-terminal domain. The sequence-independent affinity of the GAR domain for Holliday junctions enables TRF2, but not TRF1, to promote strand invasion, and to form a t-loop structure if provided with the correct substrate in vitro [36]. Moreover, the GAR domain allows TRF2 to modify DNA topology and to condensate telomeric DNA, suggesting that this protein could have an important role in telomeric chromatin remodeling [36]. Furthermore, the TRF proteins interact with several factors, many of which are involved in DDR and DNA damage repair, in order to protect telomeres from fusion events (for a review, see Ref. [37]).

Finally, POT1 binds single-stranded telomeric DNA through its two OB-fold domains [38]. This protein, in its heterodimeric form with TPP1, is able to unfold G-quadruplex structures at the telomeric 3′ overhang [39], thus allowing telomerase to access the chromosome termini.

The protective role of the shelterin components has been intensively studied in mouse- and human cells [33,40]. TRF2 is the key protein in the formation and stabilization of t-loops [25], through its TRFH dimerization domain [41]. Deletion of shelterin complex components leads to the activation of two main DDR signaling pathways mediated by the ATM and ATR kinases [42]. ATM reacts to DSB and is inhibited by TRF2, whereas ATR is activated upon the binding of the protein RPA to single-stranded telomeric DNA deprived of POT1. The enrichment of γH2AX and 53BP1 at telomeres results in the formation of diagnostic foci known as telomere dysfunction–induced foci (TIFs). These events culminate in the activation of the cell cycle regulator p53, leading to either permanent cell cycle arrest or to cell death. The shelterin complex also acts as a platform for the recruitment of several factors, which are essential for telomeric DNA replication, telomere length regulation, and DDR repression [37]. Many of these factors are only transiently associated to telomeres, and all of them have known functions in other genomic sites. Among these, there are factors involved in the DDR, such as the MRE11–RAD50–NBS1 (MRN) complex that is activated by DSB, the Ku70/Ku80 heterodimer that mediates the NHEJ pathway, and ATM that interacts with TRF2. Paradoxically, all of these factors are essential for telomere functions in normal conditions, but in the presence of dysfunctional telomeres they trigger a DDR that will eventually lead to cell senescence or apoptosis.

13.4 TELOMERE CHROMATIN ORGANIZATION

Besides shelterin, telomeric DNA interacts with the histone octamer to form nucleosomes, as the rest of genomic DNA. Nucleosomal organization at telomeres is atypical [43]: Micrococcal nuclease (MNase) digestion of rat telomeric chromatin showed that nucleosome spacing is shorter than in the bulk of chromatin (Fig. 13.5 and Ref. [44]), an observation confirmed also in human cells (for a

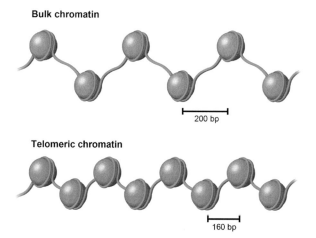

FIGURE 13.5 Schematic representation of nucleosome spacing of bulk and telomeric chromatin. The DNA linker between adjacent nucleosomes is about 40 bp shorter at telomeres.

review see Ref. [45]). Several in vitro assays demonstrated that telomeric nucleosomes are less stable [46] than other nucleosomes and can slide along telomeric DNA without help from ATP-remodeling enzymes [47]. These peculiarities are sequence-dependent and can be explained by that telomeric repeats (5–8 bp in most cases) are out of phase with the DNA helical repeat in the nucleosome (10.2 bp). As DNA folding around the histone octamer depends on sequence-dependent features, such as DNA curvability and flexibility [48,49], telomeric DNA requires more energy to form nucleosomes [50]. Moreover, telomeric DNA hosts a lot of nucleosomal isoenergetic positions spanning the same sequence [51], and the energy required to slide from one position to another is so low that nucleosomes can spontaneously move along DNA [47]. Also the short spacing between nucleosomes seems to be a sequence-dependent characteristics of telomeres. Using two different methods to reconstitute nucleosomal arrays in vitro, by salt dilution and by means of *Drosophila* embryonic extracts, it has been shown that telomeric DNA assembles in arrays with shortly spaced nucleosomes [52,53].

At present, the role played by nucleosomes in telomeric functions remains largely undefined. The relatively few papers [53–56] addressing these issues suggest that nucleosomes are involved in the proper assembly of the protective telomeric cap and in the regulated events leading to replicative senescence, but the mechanistic details are yet to be defined.

13.4.1 Interplay Between Shelterin and Nucleosomes

In eukaryotes, DNA-binding proteins interact with their recognition sequence in a nucleosomal context, that is to say that the histone octamer may often represent an obstacle to specific binding, more rarely an aid. This is true also for telomeric proteins. In lower eukaryotes, telomeres are so short that they often are nucleosome-free. In *Tetrahymena*, only 3–10% of the 200–500-bp long telomeres have nucleosomes [57]. In budding yeast, the double-stranded telomeric DNA (about 250–400 bp long) is covered by the protein Rap1, with the Ku complex binding to the very end; the 10–15-nt long 3′ tail is bound by the complex CST (for a review on yeast telomeres, see Ref. [58]). However, even if yeast telomeres are nucleosome-free, they are connected to the adjacent subtelomeric region via the silence information regulator 2–4 complex, Sir2–Sir3–Sir4 that bridges the telomeric protein Rap1 and subtelomeric nucleosomes, triggering the formation of a heterochromatic region (Section 13.4.2).

Studying the chromatin organization of telomeres in higher eukaryotes is technically demanding. Telomere length is heterogeneous even in the same cell, consisting of hundred to thousands of short repeats, which renders it difficult to assess whether nucleosomes occupy telomeres uniformly along their length, or whether the terminal and the proximal part of telomeres have a different nucleosomal organization. Nonetheless, MNase footprinting analyses on cultured cells indicate that mouse telomeres are organized in tightly spaced nucleosomes for most of their length and that this organization persists also till the very end of chromosomes [59]. This means that most telomeric repeats are wrapped around the histone octamer, limiting the DNA sequence available for the binding of telomeric proteins to the short linker between adjacent nucleosomes or to nucleosomal DNA exposed to the solvent. The accessibility of telomeric binding sites in a nucleosomal context to the human shelterin proteins TRF1 and TRF2 has been studied by using in vitro model systems [54,60]. From these studies it emerges that the two proteins are differently influenced by the presence of the histone octamer, despite the fact that they have the same affinity for naked telomeric DNA and share the same architecture. TRF1 binds efficiently also to binding sites on nucleosomal surface, although with a six-fold lower affinity than to naked DNA; furthermore, the affinity of TRF1 for naked DNA increases if there is a nucleosome adjacent to the binding site, as is the case at the linker DNA between consecutive nucleosomes. On the contrary, TRF2 binds to nucleosomal binding sites very poorly, about 150-fold lower than to naked DNA. Moreover, differently from TRF1, its affinity for TTAGGG-binding sites diminishes when adjacent to a nucleosome. Interestingly, binding assays using TRF2 mutants and tailless nucleosomes [54] indicate that the binding of TRF proteins to DNA in a nucleosomal context is modulated by interactions between histone tails and the N-terminal domains of TRF1 (rich of acidic residues) and TRF2 (rich of basic residues). This suggests that the scenario might change with a different pattern of histone modifications; for example, modifications that reduce the positive charge of histone tails, such as lysine acetylation, could increase the affinity of TRF2 for linker DNA and, at the same time, decrease the affinity of TRF1.

Analyzing the interplay between telomeric proteins and nucleosomes in vivo is challenging, and few analyses addressing this issue have been carried out so far. In mouse cells, the removal of shelterin proteins has no effect on nucleosomal organization of telomeres [42], not even in the very end of the telomere [59]. These findings indicate that at least mammalian telomeres have a strong nucleosomal backbone, on which shelterin proteins insert and organize the protecting structure. On the contrary, when TRF2 is overexpressed, nucleosome spacing is altered both in mouse keratinocytes [61] and in human cancer cells [53]. This concentration-dependent remodeling effect has been replicated also in vitro, using a model system of nucleosome array reconstitution based on *Drosophila* embryonic extracts. In this model, increasing TRF2 concentration caused an increase in nucleosome spacing [53], while higher concentrations of TRF1 resulted in complete deregulation of spacing between telomeric nucleosomes [54].

13.4.2 Epigenetic Status of Telomeres and the Discovery of Telomere Position Effect

Telomeres are generally considered heterochromatic regions. This is certainly true in budding yeast, where the Sir complex, recruited to telomeres by Rap1 and yKu, create a bridge with subtelomeric nucleosomes. Sir2 is a histone deacetylase (HDAC), whereas Sir3 and Sir4 bind the hypoacetylated N-terminal tails of H3 and H4. The Sir complex mediates the spread of heterochromatin toward the subtelomeric region generating the so-called telomere position effect (TPE). As a result, genes located in proximity of the telomere are silenced [62], and the extent of repression decreases at increasing distance from the end.

TPE has been found also in fission yeast, where heterochromatin formation at telomeric positions assumes a particular importance. In *Schizosaccharomyces pombe*, telomeres are protected by a shelterin-like complex and telomere maintenance is performed by telomerase. After deleting the gene encoding the catalytic moiety of telomerase, *trt*, some cells survive by circularizing chromosomes or by adopting a recombination-based way of maintaining telomeres. Rarely, different kind of linear telomerase-minus survivors have been isolated. These cells are capable of overcoming telomere loss by adopting an alternative mode to protect chromosome ends based on amplification and rearrangements of heterocromatic regions [63]. These survivors, named heterochromatin amplification-mediated and telomerase-independent (HAATI), still retain the 3' overhang–binding protein, POT1, and its interacting partner, Ccq1, that interacts also with the heterochromatic complex Snf2/HDAC-containing repressor complex (SHREC). These data suggest that the presence of heterochromatic regions at the chromosome ends allows the recruitment of POT1 via Ccq1, thus protecting chromosome ends even in the absence of telomeric sequences. These telomeres resemble *Drosophila* chromosome ends, lacking a specific telomeric sequence but protected by proteins that bind telomeres in a sequence-independent fashion [8].

TPE was first discovered in *Drosophila*, with seminal experiments demonstrating that insertions of the *white* reporter gene by transposition in the genome of *white*⁻ flies leads to the inactivation of the gene when inserted near centromeric heterochromatin or near the telomere-associated sequences (TAS) in the subtelomeric region [64]. Spreading of *Drosophila* TPE depends on the binding of heterochromatin protein 1 (HP1) to histone H3 trimethylated at lysine 9 (H3K9me3), which then recruits the histone methyltransferase (HMTase), SU(VAR)-3.9, propagating the heterocromatic signal [65].

Chromatin status at telomeres in other higher eukaryotes seems less defined, with the presence of both euchromatic and heterochromatic marks (for a review, see Ref. [43]). However, several lines of evidence indicate that the formation of a heterochromatic structure is essential for a correct telomere function in higher eukaryotes. In mouse, telomeres are enriched in heterochromatic marks, such as H3K9me3 that recruits HP1 [56]. The importance of a heterochromatic structure for mammalian telomere stability is supported also by indirect observations. For example, in human cells the loss of SIRT6, a

NAD$^+$-dependent deacetylase belonging to the same family as the yeast Sir2 and specific for the lysine residues 9 and 56 of the histone H3 (H3K9 and H3K56), leads to dysfunctional telomeres, including a significant increase in telomere fusion events [66].

13.4.3 Telomere Position Effect in Mammals and 3D Genome Folding

Due to genome complexity and telomere structure, studying TPE in detail in mammalian cells is technically more difficult than in yeast. However, it is generally accepted that genes close to telomeres are silenced due to the spreading of heterochromatin, as it happens in *S. cerevisiae*. Similarly to what found in yeast, genes artificially placed near telomeres in mammalian cells are transcriptionally silenced [67,68]. Moreover, the strength of the silencing is inversely correlated with the distance from the end, and can reach about 100 kb [69]. Until recently, only very few human genes have been reported to be negatively regulated by TPE. Such genes include the interferon-stimulated gene 15 (*ISG15*) located at about a megabase from the telomere. Hence, in human primary fibroblasts, the expression of *ISG15* increases when telomeres shorten, suggesting that this will cause also a decrease in heterochromatin spreading. Moreover, when telomeric heterochromatin is destabilized by inhibiting the synthesis of the histone deacetylase SIRT6, the telomere-mediated silencing of a luciferase reporter gene located close to a telomere and that of the *ISG15* gene are greatly reduced [70]. Another human gene likely regulated by TPE is double homeobox 4 (*DUX4*), which is located at about 50 kb from the end of chromosome 4q [71] and is expressed in a manner that is inversely proportional to telomere length. As the overexpression of DUX4 contributes to the development of facioscapulohumeral muscular dystrophy (FSHD), its regulation by TPE may explain the late onset of the disease, because the expression of *DUX4* is upregulated only after telomere shortening has taken place.

However, the spreading of heterochromatin along the linear chromatin fiber cannot always satisfactorily explain the mechanism of TPE. First, the *ISG15* gene is located one megabase away from the end, 10-fold further away from the telomere than the maximum distance compatible with TPE found using a reporter gene [69]. Second, the transcription of several genes positioned between *ISG15* and the telomere did not show dependence from telomere length. Recently, TPE in human cells has been studied at the genome-scale level using a variant of the HiC (chromosome conformation capture followed by high-throughput sequencing) method [72]. This technique allows the identification of chromosome sites in the nucleus that are in close enough physical proximity to be cross-linked by formaldehyde. Robin et al. adapted this method to obtain a map of telomeric and subtelomeric interactions occurring in the distal 20 Mb of human chromosome 6p. They found several physical interactions between the telomere and genes located megabases upstream, and validated these contacts by DNA fluorescence in situ hybridization (DNA FISH). From a global gene expression analysis it emerges that more than 140 genes located within 10 Mb from the end of the chromosome are affected by variations in telomere length, well before telomeres reach a critical short length that could trigger the DDR. These long-range interactions do not depend on heterochromatin spreading, as in conventional TPE, but instead involve chromatin looping that is influenced by telomere length and chromatin structure. The authors named this phenomenon as TPE over long distances (TPE-OLD) (Fig. 13.6). Although the mechanism underlying TPE-OLD is undefined, chromatin immunoprecipitation (ChIP) analysis of three genes regulated by TPE-OLD from three different chromosome ends showed an enrichment of TRF2 and the repressive H3K9me3 modifications at the promoters, indicating that silencing might be caused by the direct interaction between these genes and the repressive chromatin located at the telomeres. The analysis of transcription in single chromosome ends in different cell types showed a high variability of response to telomere shortening, with some genes being upregulated, others downregulated, yet another group of genes unaffected by variations in telomere length. These results suggest that regulation of TPE-OLD is not ruled by a simple mechanism, but

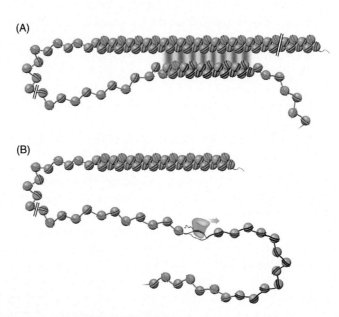

FIGURE 13.6 TPE over long distances (TPE-OLD) model.

Telomeric DNA is represented in *blue*, genomic DNA in *green* and the target gene in *red*. (A) A chromatin loop brings the target gene near the telomeric region. Telomeric heterochromatin spreads to the target gene by a still unknown mechanism, resulting in gene silencing. (B) Upon telomere shortening, the target gene separates from the telomeric region, resulting in gene derepression.

involves several levels of control that directly or indirectly affect the probability of interaction between gene promoters and telomeres, as well as the outcome of such events on the chromatin state of the genes involved.

Recently published data are consistent with a multifaceted regulation of TPE-OLD [73]. Wright and coworkers used the HiC method to characterize a chromatin loop at the 4q35 locus involving the sorbin and SH3 domain–containing protein 2 gene (*SORBS2*), encoding a skeletal muscle protein. This gene is highly expressed in FSHD myoblasts with short telomeres, but not in normal myoblasts or FSHD myoblasts with long telomeres. The increased expression of *SORBS2* is associated with a more relaxed chromatin organization of the genomic region comprising this gene. Moreover, the expression of other genes in the same locus is not altered in FSHD myoblasts with short telomeres, indicating that TPE-OLD regulation depends on several variables other than telomere length.

13.4.4 TERRA Transcription

Despite the fact that telomeres constitute heterochromatic regions, they are not transcriptionally inert. One of the most important findings of the last ten years in the telomere field is that telomeres are actively transcribed [74]. Starting from subtelomeric promoters, telomere transcription generates long G-rich noncoding RNAs named telomeric repeat–containing RNA (TERRA). TERRA molecules are heterogeneous in length, ranging from about 100 nt to several kilobases in mammals, with TERRA length increasing with telomere length. TERRA is transcribed mainly by RNA polymerase II, starting from promoters located on subtelomeres [75] and is widely conserved among eukaryotes. In humans, TERRA levels are high in G1 and decrease in S phase, increasing again in G2. Several telomeric functions have been associated to TERRA, including telomere replication, telomere length regulation, telomere DDR, and also telomere chromatin changes (for a updated review, see Ref. [76]). Furthermore, it has been proposed that TERRA mediates the RPA/POT1 switch at the single-stranded end of telomeres after replication [77] and that it plays a role in ALT establishment [23].

Transcription of human TERRA is inversely related to telomere length due to a negative feedback mechanism [78]. Long telomeres negatively regulate TERRA expression, which in turn causes telomere shortening. This is probably a consequence of the role played by TERRA in the establishment of heterochromatin at telomeres. TERRA thus promotes the increase in H3K9me3 levels at telomeres by recruiting the SUV39H1 methyltransferase and HP1α; conversely, by silencing the expression of the corresponding RNAs, it has been shown that TERRA transcription is negatively regulated by SUV39H1 and HP1α. Also the DNMT1/3B DNA methyltransferases negatively regulate TERRA transcription by methylating CpG-rich repeats present in about half of the TERRA promoters in humans. Another 61-bp element that regulates TERRA expression is located upstream of CpG repeats, and is bound by CCCTC-binding factor (CTCF) and the cohesin component radiation-sensitive 21 (RAD21). While CTCF [79] is known as an insulator protein and is involved in separating topologically associated domains (TADs) in chromosomes (Section 13.5), RAD21 is implicated in sister chromatid cohesion in mitosis [80] and chromatin loop formation between enhancers and promoters [79]. At TERRA promoters, CTCF and RAD21 act as positive regulators of transcription [81]. Finally, an important negative regulator of TERRA synthesis is TRF2 through its homodimerization domain [82].

13.5 TELOMERE CHROMATIN DYNAMICS DURING THE CELL CYCLE

The apparently stable and closed structure of telomeres that protects chromosome ends from nucleases and repair enzymes has to dynamically change its organization during the cell cycle. Hence, telomeres disrupt their protective structure to allow DNA replication in the S phase. Following t-loop resolution and the detachment of POT1/TPP1, single-stranded telomeric DNA is bound by replication protein A (RPA) [83] to promote replication. As RPA is also involved in DDR by recruiting the ATR kinase to sites with a single-stranded protrusion, RPA needs to be replaced by POT1 after each round of replication in order to avoid DDR signaling and to promote telomere capping. The POT1–RPA and RPA–POT1 switch thus plays a key role in the maintenance of genomic integrity [77].

The maintenance of 3' overhangs is a conserved feature of telomeres and is essential for the reassembly of a correctly capped structure after DNA replication, for the formation of the t-loop structure, and for priming telomerase elongation. As leading strand DNA replication results in a blunt end, the generation of telomeric 3' overhangs requires the concerted action of potentially damaging exonuclease activities for the resection of 5' ends [84]. After DNA replication, TRF2 recruits the Apollo/SNM1B nuclease at leading-end telomeres, which starts 3' overhang formation by degrading the 5' end. Then, exonuclease I (ExoI) resects 5' ends from both leading- and lagging-end telomeres in S/G2. Overresection by ExoI is finally corrected through the recruitment by POT1 of the CST complex, which in turn facilitates the loading at telomeres of the DNA polα–primase complex to partially fill in the C-rich strands [84].

Also other members of the DNA repair machinery, such as, MRE11, NBS, and ATM are present at telomeres from late S to G2 phase [85]. The presence of these factors at telomeres may simply reflect the exposure of chromosome ends due to the opening of the protecting structure. However, several lines of evidence indicate that DDR and DNA repair factors actively contribute to the correct formation of a capped telomere structure after DNA replication [83,85,86]. Importantly, the exposed telomere induces a noncanonical DDR, in which MRN and ATM are recruited but none of the downstream proteins of the pathway (such as Chk2 and p53) are activated as a consequence of the protective activity of TRF2 [87]; therefore, cell cycle arrest is not triggered by telomere exposure during S phase [88]. What emerges from these studies is that although chromosome termini are perceived as DNA damage during DNA replication, this is part of a controlled pathway that leads to telomere processing and to the reconstitution of a protective structure.

13.6   TELOMERE SHORTENING AND DEPROTECTION

Telomere-dependent replicative senescence is characteristic of organisms that do not express telomerase in the soma [89] and involves a finite number of replications with gradual telomere shortening, leading to DDR activation and cell cycle arrest. Interestingly, telomerase activity seems inversely correlated with body mass. Telomerase is expressed in somatic tissues of rodent species with small body mass and is repressed in somatic cells of species with large body mass, as a result of the evolution of a mechanism able to potentially reduce cancer risk [90].

Failure to establish replicative senescence via escape from cell cycle checkpoints and the activation of telomerase-dependent or -independent telomere length maintenance mechanisms thus constitute important steps toward malignancy. Much attention has been paid therefore to clarify the mechanisms that induce the activation of DDR at short telomeres, promote cell cycle arrest and induce entry into replicative senescence. A model [91] based on experimental evidence describes erosion of telomere protection as a three-step process (see Fig. 13.7). The first state is represented by fully protected telomeres, in which chromosome ends are shielded from DDR signaling and DNA repair, likely by assuming the t-loop conformation. The second, intermediate state is characterized by disruption of the closed conformation, in which DDR is activated, but DNA repair is inhibited. The third state features uncapped telomeres, with no residual protection from DNA repair, resulting in telomere fusions.

Most research on telomere deprotection focused on telomere uncapping upon the depletion of individual shelterin proteins, in particular TRF2. These studies allowed dissecting the functions of the shelterin components and revealed the prominent and complex role played by TRF2 in end protection (for a review see Ref. [92]). TRF2 prevents DDR activation and inhibits NHEJ via two distinct mechanisms. Whereas the TRFH homodimerization domain of TRF2 represses the ATM pathway of the DDR [87], inhibition of NHEJ and consequent telomere fusions are regulated by an independent region situated on the hinge domain of TRF2. This domain, named iDDR, is responsible for inhibiting 53BP1 localization to telomeres, which is a necessary step for the activation of NHEJ. These properties of TRF2 have been integrated with the three-state model of telomere erosion. Hence, in the closed state TRF2 is

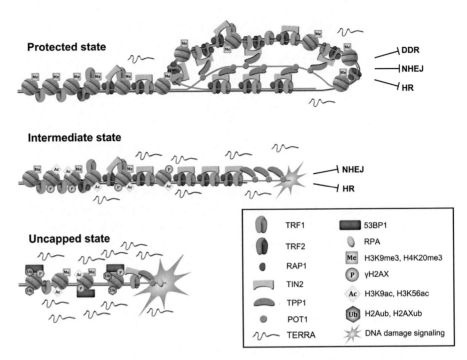

FIGURE 13.7 Schematic representation of the three telomeric states model, including shelterin protein localization and histone modifications during the different states.

Protected state: The telomere is in a closed configuration mediated by the shelterin complex. Telomeric chromatin is in a compact heterochromatic state, lysines 9 and 27 of histone H3 are trimethylated. This conformation inhibits DNA damage signaling and DNA repair at telomeres. *Intermediate state:* Telomere shortening leads to disruption of the closed conformation. Telomere is recognized as a damage site by the DNA damage response *(DDR)* machinery, resulting in replicative senescence, but it retains enough shelterin proteins to block nonhomologous end joining *(NHEJ)* and homologous recombination *(HR)* pathways. Telomeric chromatin is less condensed, as a result of the decrease in heterochromatic marks and is enriched in phosphorylated histones H2AX. *Uncapped state:* Inactivation of growth arrest checkpoint (p53) leads to excessive telomere shortening, resulting in a fully uncapped conformation. Loss of telomeric repeat–binding factor 2 (TRF2) and protection of telomeres 1 (POT1) leads to activation of DNA repair pathways, resulting in telomeric fusions. Telomeric chromatin is characterized by the presence of phosphorylated H2AX and ubiquitylated H2A and H2AX, markers of dysfunctional telomeres.

abundant at telomeres and mediates the formation of protective structures, such as the t-loop. When telomeres shorten, the reduced quantity of TRF2 is not able to protect against ATM signaling anymore, but still confers protection against DNA repair. In this state, DDR signaling advances the cell toward replicative senescence, unless the p53/RB checkpoints are disabled. Escape from the checkpoints enables cells to continue dividing, which leads to further telomere shortening, reduction in shelterin (TRF2) levels at telomeres and thus loss of protection against fusions.

13.7   TELOMERIC CHROMATIN DYNAMICS ON THE PATH TO REPLICATIVE SENESCENCE

Experimentally induced deprotection by shelterin component depletion, however, does not completely recapitulate the steps leading to replicative senescence. When replicative senescence is brought about by long-term cell culture/replication, telomere shortening–induced activation of DDR is preceded by extensive chromatin changes both in structure and in composition [43,56]. Several studies documented a decrease in heterochromatic marks (mainly H3K9me3 and H4K20me3) and, at the same time, an increase in euchromatic marks (H3K9ac) on telomere shortening. The role these epigenetic changes play in the destabilization of the closed telomeric structure and in the subsequent DDR response remains, however, unexplored.

A consequence of telomere erosion and related changes in telomeric chromatin structure is the reduction of classical TPE. Telomere shortening thus leads to an increased expression of reporter genes located in subtelomeric positions [69], which presumably reflects decreased telomeric heterochromatin spreading. In addition, shortening of telomeres and a decrease in heterochromatin marks have also been shown to alter the frequency of interactions between telomeric and subtelomeric regions, leading to expression changes at genes regulated by TPE-OLD [72]. As the expression of TPE-OLD-regulated genes is modulated by reduction in telomere length long before telomere erosion elicits the DDR, this mechanism might mediate aging related chromatin changes [72]. Finally, telomere shortening and altered chromatin structure might also affect TERRA expression, which is negatively regulated by heterocromatic marks [78] and plays important roles in the response to DNA damage.

Further telomere shortening leads to the destabilization of the shelterin complex and activates DDR signaling with consequent additional changes in the composition and organization of telomeric chromatin. As opposed to telomere deprotection triggered by deprivation of TRF2 or POT1, the destabilization of the closed capped structure at short telomeres does not depend on shortage of shelterin proteins [93]. Instead, destabilization is induced by the impaired formation of a protective closed structure (such as the t-loop) due to the short length of telomeric DNA. The remaining TRF2 and the other shelterin proteins bound to telomeres repress the DNA repair pathways, and direct the cell toward a safe entry into replicative senescence.

13.8   TELOMERES IN THE 3D SPACE OF THE NUCLEUS

Despite the different chromatin organization among species, telomeres seem to have a nonrandom localization in the 3D space of the nucleus. In budding yeast telomeres are associated with the nuclear envelope via the protein Mps3 that belongs to the SUN family and forms clusters mediated by the yKu complex and the Sir2/Sir3/Sir4 complex [94]. Interestingly, they seem to share several binding proteins and the nuclear envelope localization with DSBs that are recruited to repair centers near the nuclear pores [95]. In human cells, telomeres are tethered to the nuclear envelope during the nuclear assembly after mitosis, probably via interaction between RAP1 and the nuclear envelope protein SUN1 [96]. Several lines of evidence indicate that anchoring of telomeres to the nuclear envelope is mediated by A-type lamins, the main component of the nuclear lamina, and that this localization is important for correct telomere maintenance [97]. Many aspects related to the role of dynamic telomere anchoring to the nuclear envelope remain, however, to be explored.

A peculiar clustering characterizes ALT telomeres. It is thus known that a subset of ALT telomeres clusters in APBs [98]. Gathering chromosome ends seems important for ALT telomere maintenance that is based on HR. It has been recently shown that inducing a DSB response at ALT telomeres leads to a long-range movement of telomeres and to an increased clustering [99]. This phenomenon requires the machinery responsible for synapsis formation between homologous chromosomes during meiosis and allows ALT telomeres to cover micron distances and reach another telomere.

A tempting hypothesis is that telomere clustering in ALT cells and meiotic recombination might share at least some molecular mechanisms. The subnuclear localization of telomeres might potentially affect other fundamental functions that deserve further investigation, such as, nuclear organization, regulation of cell division, telomere maintenance, and expression of genes close to telomeres.

13.9 CONCLUSIONS

Telomeres are essential structures of the eukaryotic linear chromosomes. Starting from their basic functions—protecting chromosome ends from being recognized as DNA damage and assuring the complete replication of the genome—telomeres evolved together with the increasing complexity of organisms assuming new tasks that are far from being elucidated. In humans, telomere maintenance is strictly regulated, with the full conservation of the original telomere length only in the germline. In somatic tissues, telomere length is a measure of the remaining proliferative capacity of the cell, the so-called molecular clock. In a more general view, telomere length is seen as an indicator of individual life expectance, such as a molecular lifeline, responsive to stress and to life habits [100]. In long-lived organisms, such as humans, telomeres are a key structure both for aging and for cancer research.

At birth, human telomeres are 10–15 kb long, organized in a tight hetero-chromatic structure. This length is maintained in embryonic stem cells and in the germ line. In stem cells, telomerase is active but its action is not sufficient to impede a slow shortening of telomeres. In somatic cells, telomerase is in-active; the consequent progressive telomere erosion at every replication cycle conducts to eliciting DDR and to replicative senescence.

The vast majority of research on telomere shortening has been concentrated on the deprotection of the end-capping structure that activates the DNA damage pathway. The specific role of shelterin proteins has been deeply investigated, and many molecular interactions have been clarified. However, several questions are still left unsolved. Despite the efforts, there is neither precise definition for the "critical length" of the telomere that triggers the DNA damage signaling nor for the protective structure that is disrupted. Also the roles played by nucleosomal organization and by chromatin modifications are elusive. Telomere erosion seems to induce decompaction of telomeric heterocromatin [56], as suggested by the decrease of heterocromatic marks. It is not clear if these changes are a con-sequence of telomere deprotection or of the gradual telomere shortening. More-over, it would be interesting to investigate whether this altered chromatin envi-ronment also modulates the binding of shelterin proteins, analogously to what was found in vitro [54]. Finally, future research might clarify the role of TERRA in regulating the structure of telomeric chromatin and in ALT establishment.

The application of genome-wide techniques proved to be very useful in explor-ing the three-dimensional interactions of telomeric chromatin [72], revealing a new level of telomere length–dependent gene regulation, TPE-OLD, which affects genes distant from the telomere. TPE-OLD implies that the expression of a category of genes depends on telomere length; hence, it depends on how many cycles of replication the cell underwent. The findings of Robin et al. raise a series of issues to afford in the near future. First of all, more accurate analyses could assess how diffuse and important this phenomenon is at the cellular and the organismal level. Moreover, an exciting goal will be to decipher the mechanisms controlling nuclear three-dimensional organization and long-range regulation of gene expression mediated by the telomere, a phenomenon that is still largely unknown.

The lack of information on some important topics of telomere biology partly re-flects technical difficulties. As an example, powerful methods, such as ChIP and its genome-wide derivation, ChIP-seq, had to face the extreme heterogeneity of telo-mere lengths even in a single cell and did not give so far the wealth of informa-tion obtained for other parts of the genome. In order to deeply analyze the struc-ture and the functions of telomeric chromatin, it will be critical to develop more accurate methods of investigation that possibly could target single telomeres.

Glossary

G-quadruplex Four-stranded helical conformation of DNA, deriving from the folding of sequences containing four runs of three to four consecutive guanines (G); this structure is characterized by the stacking of G-quartets, a planar cyclic arrangement of four guanines, held together by eight Hoogsteen hydrogen bonds and stabilized by the coordination of a central cation (typically potassium).

Hoogsteen hydrogen bonds Noncanonical base pairing of nucleic acid that takes advantage of the N7 position of a purine, instead of the typical N1, as occurs in a Watson–Crick base pair.

OB-fold domain Oligomer-binding (OB)-fold domain. A structural motif that confers capability of binding to oligonucleotides or oligosaccharides. It is composed of a five-stranded, Greek key β-barrel capped by an α-helix located between the third and the fourth strands. OB-fold-containing proteins usually bind single-stranded DNA with two or more OB-fold domains at a time.

PML nuclear bodies Nuclear bodies organized by the PML protein, in which several different proteins involved in the fine-tuning of several nuclear processes localize. In ALT-positive cells, which rely on homologous recombination to maintain telomere length, ALT-associated PML bodies (APBs) contain several DNA double-strand break repair and homologous recombination factors and are associated with actively replicating telomeres during S/G2 phase of the cell cycle.

Replicative senescence Permanent block of the cell cycle in somatic cells that reached their maximum number of divisions. Due to the telomere erosion, after several replication cycles, telomeres reach a length that triggers a DNA damage response, resulting in the activation of the p53 pathway and subsequent cell cycle block. This is thought to be a tumor-preventing mechanism, as somatic cells can only replicate a limited number of times.

Shelterin The protein complex associated to mammalian telomeres, responsible for their stability and protection and essential for telomere function. It comprises six protein factors: TRF1 and TRF2 bind as homodimers double-stranded telomeric DNA; POT1 recognizes single-stranded TTAGGG repeats, forming an heterodimer with TPP1; RAP1 interacts with TRF2; TIN2 bridges the complex binding TRF1, TRF2, and TPP1.

Telomerase The ribonucleoproteic enzymatic complex composed of two main subunits, an RNA component, named TER (also known as TR or TERC) and the telomerase reverse transcriptase (TERT) protein, plus a variable number of accessory factors.

Telomere The nucleoprotein complex that protects the end of eukaryotic chromosomes. Telomeric DNA is characterized by short G-rich tandem repeated sequences; most of telomeric DNA is double-stranded, but the terminal region is characterized by the extension of the G-rich single strand, ending with a 3′-OH extremity, called 3′ overhang.

Telomeric repeats The tandem arrays of a short G-rich sequence found in telomeric DNA (TTAGGG in vertebrates).

t-loop Model for telomeric conformation, in which the single-stranded 3′ overhang curves back to invade the upstream double-stranded telomeric DNA, forming a displacement loop (D-loop) at the point of insertion.

Abbreviations

ALT	Alternative lengthening of telomeres
APB	ALT-associated promyelocitic leukemia (PML) nuclear body

ATM	Ataxia telangiectasia mutated kinase
ATR	ATM and Rad3-related kinase
ATRX	α-Thalassemia/mental retardation syndrome X-linked
ChIP	Chromatin immunoprecipitation
CST	CTC1-STN1-TEN1
CTCF	CCCTC-binding factor
DAXX	Death domain–associated protein 6
DDR	DNA damage response
D-loop	Displacement loop
DSB	Double-strand DNA break
GAR	Gly/Arg-rich domain
γ-H2AX	Phosphorylated H2AX on Ser-139
HDAC	Histone deacetylase
HP1	Heterochromatin protein 1
HR	Homologous recombination
ISG15	Interferon-stimulated gene 15
MNase	Micrococcal nuclease
MRN	MRE11/RAD50/NBS1 complex
NHEJ	Nonhomologous end joining
PML	Promyelocytic leukemia
POT1	Protection of telomeres 1
Rap1	Repressor-activator protein 1
SIR	Silent information regulator complex
SWI/SNF	SWItch/sucrose nonfermentable complex
TAD	Topologically associated domain
TAS	Telomere-associated sequences
TER (or TR or TERC)	Telomerase RNA component
TERRA	Telomeric repeat–containing RNA
TERT	Telomerase reverse transcriptase
TIF	Telomere dysfunction–induced focus
TIN2	TRF1-interacting protein 2
T-loop	Telomere loop
TPE	Telomere position effect
TPE-OLD	TPE over long distances
TPP1	TINT1/PIP1/PTOP1 complex
TRF1	Telomeric repeat–binding factor 1
TRF2	Telomeric repeat–binding factor 2
TRFH	TRF homology domain

References

[1] Blackburn EH. Structure and function of telomeres. Nature 1991;350(6319):569–73.

[2] Jain D, Cooper JP. Telomeric strategies: means to an end. Annu Rev Genet 2010;44:243–69.

[3] Greider CW, Blackburn EH. Identification of a specific telomere terminal transferase activity in *Tetrahymena* extracts. Cell 1985;43(2 Pt. 1):405–13.

[4] de Lange T. Shelterin: the protein complex that shapes and safeguards human telomeres. Genes Dev 2005;19(18):2100–10.

[5] Shay JW, Wright WE. Senescence and immortalization: role of telomeres and telomerase. Carcinogenesis 2005;26(5):867–74.

[6] Armanios M, Blackburn EH. The telomere syndromes. Nat Rev Genet 2012;13(10):693–704.

[7] Shay JW, Bacchetti S. A survey of telomerase activity in human cancer. Eur J Cancer 1997;33(5):787–91.

[8] Raffa GD, Cenci G, Ciapponi L, Gatti M. Organization and evolution of *Drosophila* terminin: similarities and differences between *Drosophila* and human telomeres. Front Oncol 2013;3:112.

[9] Mason JM, Randall TA, Capkova Frydrychova R. Telomerase lost? Chromosoma 2016;125(1):65–73.

[10] Fujiwara H, Osanai M, Matsumoto T, Kojima KK. Telomere-specific non-LTR retrotransposons and telomere maintenance in the silkworm, *Bombyx mori*. Chromosome Res 2005;13(5):455–67.

[11] Schmidt JC, Cech TR. Human telomerase: biogenesis, trafficking, recruitment, and activation. Genes Dev 2015;29(11):1095–105.

[12] Chen LY, Lingner J. CST for the grand finale of telomere replication. Nucleus 2013;4(4):277–82.

[13] Harley CB, Futcher AB, Greider CW. Telomeres shorten during ageing of human fibroblasts. Nature 1990;345(6274):458–60.

[14] Chapman JR, Taylor MR, Boulton SJ. Playing the end game: DNA double-strand break repair pathway choice. Mol Cell 2012;47(4):497–510.

[15] Rossiello F, Herbig U, Longhese MP, Fumagalli M, d'Adda di Fagagna F. Irreparable telomeric DNA damage and persistent DDR signalling as a shared causative mechanism of cellular senescence and ageing. Curr Opin Genet Dev 2014;26:89–95.

[16] Doksani Y, de Lange T. The role of double-strand break repair pathways at functional and dysfunctional telomeres. Cold Spring Harb Perspect Biol 2014;6(12):a016576.

[17] Cesare AJ, Reddel RR. Alternative lengthening of telomeres: models, mechanisms and implications. Nat Rev Genet 2010;11(5):319–30.

[18] Conomos D, Pickett HA, Reddel RR. Alternative lengthening of telomeres: remodeling the telomere architecture. Front Oncol 2013;3:27.

[19] Napier CE, Huschtscha LI, Harvey A, Bower K, Noble JR, Hendrickson EA, et al. ATRX represses alternative lengthening of telomeres. Oncotarget 2015;6(18):16543–58.

[20] Law MJ, Lower KM, Voon HP, Hughes JR, Garrick D, Viprakasit V, et al. ATR-X syndrome protein targets tandem repeats and influences allele-specific expression in a size-dependent manner. Cell 2010;143(3):367–78.

[21] Voon HP, Hughes JR, Rode C, De La Rosa-Velazquez IA, Jenuwein T, Feil R, et al. ATRX plays a key role in maintaining silencing at interstitial heterochromatic loci and imprinted genes. Cell Rep 2015;11(3):405–18.

[22] Schwartzentruber J, Korshunov A, Liu XY, Jones DT, Pfaff E, Jacob K, et al. Driver mutations in histone H3.3 and chromatin remodelling genes in paediatric glioblastoma. Nature 2012;482(7384):226–31.

[23] Arora R, Azzalin CM. Telomere elongation chooses TERRA ALTernatives. RNA Biol 2015;12(9):938–41.

[24] Griffith JD, Comeau L, Rosenfield S, Stansel RM, Bianchi A, Moss H, et al. Mammalian telomeres end in a large duplex loop. Cell 1999;97(4):503–14.

[25] Doksani Y, Wu JY, de Lange T, Zhuang X. Super-resolution fluorescence imaging of telomeres reveals TRF2-dependent T-loop formation. Cell 2013;155(2):345–56.

[26] Sen D, Gilbert W. Formation of parallel four-stranded complexes by guanine-rich motifs in DNA and its implications for meiosis. Nature 1988;334(6180):364–6.

[27] Biffi G, Tannahill D, McCafferty J, Balasubramanian S. Quantitative visualization of DNA G-quadruplex structures in human cells. Nat Chem 2013;5(3):182–6.

[28] Bochman ML, Paeschke K, Zakian VA. DNA secondary structures: stability and function of G-quadruplex structures. Nat Rev Genet 2012;13(11):770–80.

[29] Murat P, Balasubramanian S. Existence and consequences of G-quadruplex structures in DNA. Curr Opin Genet Dev 2014;25:22–9.

[30] Smith JS, Chen Q, Yatsunyk LA, Nicoludis JM, Garcia MS, Kranaster R, et al. Rudimentary G-quadruplex-based telomere capping in *Saccharomyces cerevisiae*. Nat Struct Mol Biol 2011;18(4):478–85.

[31] Maizels N, Gray LT. The G4 genome. PLoS Genet 2013;9(4):e1003468.

[32] Vannier JB, Pavicic-Kaltenbrunner V, Petalcorin MI, Ding H, Boulton SJ. RTEL1 dismantles T loops and counteracts telomeric G4-DNA to maintain telomere integrity. Cell 2012;149(4):795–806.

[33] Palm W, de Lange T. How shelterin protects mammalian telomeres. Annu Rev Genet 2008;42:301–34.

[34] Bianchi A, Smith S, Chong L, Elias P, de Lange T. TRF1 is a dimer and bends telomeric DNA. EMBO J 1997;16(7):1785–94.

[35] Bilaud T, Brun C, Ancelin K, Koering CE, Laroche T, Gilson E. Telomeric localization of TRF2, a novel human telobox protein. Nat Genet 1997;17(2):236–9.

[36] Poulet A, Pisano S, Faivre-Moskalenko C, Pei B, Tauran Y, Haftek-Terreau Z, et al. The N-terminal domains of TRF1 and TRF2 regulate their ability to condense telomeric DNA. NucleicAcids Res 2012;40(6):2566–76.

[37] Diotti R, Loayza D. Shelterin complex and associated factors at human telomeres. Nucleus 2011;2(2):119–35.

[38] Baumann P, Cech TR. Pot1, the putative telomere end-binding protein in fission yeast and humans. Science 2001;292(5519):1171–5.

[39] Hwang H, Buncher N, Opresko PL, Myong S. POT1-TPP1 regulates telomeric overhang structural dynamics. Structure 2012;20(11):1872–80.

[40] Bae NS, Baumann P. A RAP1/TRF2 complex inhibits nonhomologous end-joining at human telomeric DNA ends. Mol Cell 2007;26(3):323–34.

[41] Poulet A, Buisson R, Faivre-Moskalenko C, Koelblen M, Amiard S, Montel F, et al. TRF2 promotes, remodels and protects telomeric Holliday junctions. EMBO J 2009;28(6):641–51.

[42] Sfeir A, de Lange T. Removal of shelterin reveals the telomere end-protection problem. Science 2012;336(6081):593–7.

[43] Galati A, Micheli E, Cacchione S. Chromatin structure in telomere dynamics. Front Oncol 2013;3:46.

[44] Makarov VL, Lejnine S, Bedoyan J, Langmore JP. Nucleosomal organization of telomere-specific chromatin in rat. Cell 1993;73(4):775–87.

[45] Pisano S, Galati A, Cacchione S. Telomeric nucleosomes: forgotten players at chromosome ends. Cell Mol Life Sci 2008;65(22):3553–63.

[46] Cacchione S, Cerone MA, Savino M. In vitro low propensity to form nucleosomes of four telomeric sequences. FEBS Lett 1997;400(1):37–41.

[47] Pisano S, Marchioni E, Galati A, Mechelli R, Savino M, Cacchione S. Telomeric nucleosomes are intrinsically mobile. J Mol Biol 2007;369(5):1153–62.

[48] Shrader TE, Crothers DM. Effects of DNA sequence and histone-histone interactions on nucleosome placement. J Mol Biol 1990;216(1):69–84.

[49] Anselmi C, Bocchinfuso G, De Santis P, Savino M, Scipioni A. Dual role of DNA intrinsic curvature and flexibility in determining nucleosome stability. J Mol Biol 1999;286(5):1293–301.

[50] Filesi I, Cacchione S, De Santis P, Rossetti L, Savino M. The main role of the sequence-dependent DNA elasticity in determining the free energy of nucleosome formation on telomeric DNAs. Biophys Chem 2000;83(3):223–37.

[51] Rossetti L, Cacchione S, Fua M, Savino M. Nucleosome assembly on telomeric sequences. Biochemistry 1998;37(19):6727–37.

[52] Pisano S, Pascucci E, Cacchione S, De Santis P, Savino M. AFM imaging and theoretical modeling studies of sequence-dependent nucleosome positioning. Biophys Chem 2006;124(2):81–9.

[53] Galati A, Magdinier F, Colasanti V, Bauwens S, Pinte S, Ricordy R, et al. TRF2 controls telomeric nucleosome organization in a cell cycle phase-dependent manner. PLoS One 2012;7(4):e34386.

[54] Galati A, Micheli E, Alicata C, Ingegnere T, Cicconi A, Pusch MC, et al. TRF1 and TRF2 binding to telomeres is modulated by nucleosomal organization. Nucleic Acids Res 2015;43(12):5824–37.

[55] O'Sullivan RJ, Kubicek S, Schreiber SL, Karlseder J. Reduced histone biosynthesis and chromatin changes arising from a damage signal at telomeres. Nat Struct Mol Biol 2010;17(10):1218–25.

[56] Blasco MA. The epigenetic regulation of mammalian telomeres. Nat Rev Genet 2007;8(4):299–309.

[57] Cohen P, Blackburn EH. Two types of telomeric chromatin in *Tetrahymena* thermophila. J Mol Biol 1998;280(3):327–44.

[58] Wellinger RJ, Zakian VA. Everything you ever wanted to know about *Saccharomyces cerevisiae* telomeres: beginning to end. Genetics 2012;191(4):1073–105.

[59] Wu P, de Lange T. No overt nucleosome eviction at deprotected telomeres. Mol Cell Biol 2008;28(18):5724–35.

[60] Galati A, Rossetti L, Pisano S, Chapman L, Rhodes D, Savino M, et al. The human telomeric protein TRF1 specifically recognizes nucleosomal binding sites and alters nucleosome structure. J Mol Biol 2006;360(2):377–85.

[61] Benetti R, Schoeftner S, Munoz P, Blasco MA. Role of TRF2 in the assembly of telomeric chromatin. Cell Cycle 2008;7(21):3461–8.

[62] Gottschling DE, Aparicio OM, Billington BL, Zakian VA. Position effect at *S. cerevisiae* telomeres: reversible repression of pol II transcription. Cell 1990;63(4):751–62.

[63] Jain D, Hebden AK, Nakamura TM, Miller KM, Cooper JP. HAATI survivors replace canonical telomeres with blocks of generic heterochromatin. Nature 2010;467(7312):223–7.

[64] Hazelrigg T, Levis R, Rubin GM. Transformation of white locus DNA in *Drosophila*: dosage compensation, zeste interaction, and position effects. Cell 1984;36(2):469–81.

[65] Perrini B, Piacentini L, Fanti L, Altieri F, Chichiarelli S, Berloco M, et al. HP1 controls telomere capping, telomere elongation, and telomere silencing by two different mechanisms in *Drosophila*. Mol Cell 2004;15(3):467–76.

[66] Michishita E, McCord RA, Berber E, Kioi M, Padilla-Nash H, Damian M, et al. SIRT6 is a histone H3 lysine 9 deacetylase that modulates telomeric chromatin. Nature 2008;452(7186):492–6.

[67] Baur JA, Zou Y, Shay JW, Wright WE. Telomere position effect in human cells. Science 2001;292(5524):2075–7.

[68] Koering CE, Pollice A, Zibella MP, Bauwens S, Puisieux A, Brunori M, et al. Human telomeric position effect is determined by chromosomal context and telomeric chromatin integrity. EMBO Rep 2002;3(11):1055–61.

[69] Kulkarni A, Zschenker O, Reynolds G, Miller D, Murnane JP. Effect of telomere proximity on telomere position effect, chromosome healing, and sensitivity to DNA double-strand breaks in a human tumor cell line. Mol Cell Biol 2010;30(3):578–89.

[70] Tennen RI, Bua DJ, Wright WE, Chua KF. SIRT6 is required for maintenance of telomere position effect in human cells. Nat Commun 2011;2:433.

[71] Stadler G, Rahimov F, King OD, Chen JC, Robin JD, Wagner KR, et al. Telomere position effect regulates DUX4 in human facioscapulohumeral muscular dystrophy. Nat Struct Mol Biol 2013;20(6):671–8.

[72] Robin JD, Ludlow AT, Batten K, Magdinier F, Stadler G, Wagner KR, et al. Telomere position effect: regulation of gene expression with progressive telomere shortening over long distances. Genes Dev 2014;28(22):2464–76.

[73] Robin JD, Ludlow AT, Batten K, Gaillard MC, Stadler G, Magdinier F, et al. SORBS2 transcription is activated by telomere position effect-over long distance upon telomere shortening in muscle cells from patients with facioscapulohumeral dystrophy. Genome Res 2015;25(12):1781–90.

[74] Azzalin CM, Reichenbach P, Khoriauli L, Giulotto E, Lingner J. Telomeric repeat containing RNA and RNA surveillance factors at mammalian chromosome ends. Science 2007;318(5851):798–801.

[75] Schoeftner S, Blasco MA. Developmentally regulated transcription of mammalian telomeres by DNA-dependent RNA polymerase II. Nat Cell Biol 2008;10(2):228–36.

[76] Azzalin CM, Lingner J. Telomere functions grounding on TERRA firma. Trends Cell Biol 2015;25(1):29–36.

[77] Flynn RL, Centore RC, O'Sullivan RJ, Rai R, Tse A, Songyang Z, et al. TERRA and hnRNPA1 orchestrate an RPA-to-POT1 switch on telomeric single-stranded DNA. Nature 2011;471(7339):532–6.

[78] Arnoult N, Van Beneden A, Decottignies A. Telomere length regulates TERRA levels through increased trimethylation of telomeric H3K9 and HP1alpha. Nat Struct Mol Biol 2012;19(9):948–56.

[79] Ong CT, Corces VG. CTCF: an architectural protein bridging genome topology and function. Nat Rev Genet 2014;15(4):234–46.

[80] Nasmyth K, Haering CH. Cohesin: its roles and mechanisms. Annu Rev Genet 2009;43:525–58.

[81] Deng Z, Wang Z, Stong N, Plasschaert R, Moczan A, Chen HS, et al. A role for CTCF and cohesin in subtelomere chromatin organization, TERRA transcription, and telomere end protection. EMBO J 2012;31(21):4165–78.

[82] Porro A, Feuerhahn S, Delafontaine J, Riethman H, Rougemont J, Lingner J. Functional characterization of the TERRA transcriptome at damaged telomeres. Nat Commun 2014;5:5379.

[83] Verdun RE, Karlseder J. The DNA damage machinery and homologous recombination pathway act consecutively to protect human telomeres. Cell 2006;127(4):709–20.

[84] Wu P, Takai H, de Lange T. Telomeric 3′ overhangs derive from resection by Exo1 and Apollo and fill-in by POT1b-associated CST. Cell 2012;150(1):39–52.

[85] Verdun RE, Crabbe L, Haggblom C, Karlseder J. Functional human telomeres are recognized as DNA damage in G2 of the cell cycle. Mol Cell 2005;20(4):551–61.

[86] Audry J, Maestroni L, Delagoutte E, Gauthier T, Nakamura TM, Gachet Y, et al. RPA prevents G-rich structure formation at lagging-strand telomeres to allow maintenance of chromosome ends. EMBO J 2015;34(14):1942–58.

[87] Okamoto K, Bartocci C, Ouzounov I, Diedrich JK, Yates JR, 3rd, Denchi EL. A two-step mechanism for TRF2-mediated chromosome-end protection. Nature 2013;494(7438):502–5.

[88] Burgess RC, Misteli T. Not all DDRs are created equal: non-canonical dna damage responses. Cell 2015;162(5):944–7.

[89] Henriques CM, Ferreira MG. Consequences of telomere shortening during lifespan. Curr Opin Cell Biol 2012;24(6):804–8.

[90] Seluanov A, Chen Z, Hine C, Sasahara TH, Ribeiro AA, Catania KC, et al. Telomerase activity coevolves with body mass not lifespan. Aging Cell 2007;6(1):45–52.

[91] Cesare AJ, Kaul Z, Cohen SB, Napier CE, Pickett HA, Neumann AA, et al. Spontaneous occurrence of telomeric DNA damage response in the absence of chromosome fusions. Nat Struct Mol Biol 2009;16(12):1244–51.

[92] Feuerhahn S, Chen LY, Luke B, Porro A. No DDRama at chromosome ends: TRF2 takes centre stage. Trends Biochem Sci 2015;40(5):275–85.

[93] Kaul Z, Cesare AJ, Huschtscha LI, Neumann AA, Reddel RR. Five dysfunctional telomeres predict onset of senescence in human cells. EMBO Rep 2012;13(1):52–9.

[94] Taddei A, Gasser SM. Multiple pathways for telomere tethering: functional implications of subnuclear position for heterochromatin formation. Biochim Biophys Acta 2004;1677(1–3):120–8.

[95] Marcomini I, Gasser SM. Nuclear organization in DNA end processing: telomeres vs double-strand breaks. DNA Repair 2015;32:134–40.

[96] Crabbe L, Cesare AJ, Kasuboski JM, Fitzpatrick JA, Karlseder J. Human telomeres are tethered to the nuclear envelope during postmitotic nuclear assembly. Cell Rep 2012;2(6):1521–9.

[97] Gonzalez-Suarez I, Redwood AB, Perkins SM, Vermolen B, Lichtensztejin D, Grotsky DA, et al. Novel roles for A-type lamins in telomere biology and the DNA damage response pathway. EMBO J 2009;28(16):2414–27.

[98] Draskovic I, Arnoult N, Steiner V, Bacchetti S, Lomonte P, Londono-Vallejo A. Probing PML body function in ALT cells reveals spatiotemporal requirements for telomere recombination. Proc Natl Acad Sci USA 2009;106(37):15726–31.

[99] Cho NW, Dilley RL, Lampson MA, Greenberg RA. Interchromosomal homology searches drive directional ALT telomere movement and synapsis. Cell 2014;159(1):108–21.

[100] Lin J, Epel E, Blackburn E. Telomeres and lifestyle factors: roles in cellular aging. Mutat Res 2012;730(1–2):85–9.

CHAPTER 14

Epigenetic Regulation of X-Chromosome Inactivation

M.E. Donohoe

*Burke Medical Research Institute, White Plains, NY, United States;
Department of Neuroscience, Department of Cell and Developmental Biology,
Brain Mind Research Institute, Weill Cornell Medical College,
New York, NY, United States*

14.1 X-CHROMOSOME INACTIVATION: A HISTORICAL BACKGROUND

In the late 1940s, Murray Barr and his graduate student Ewart Bertram described a cytological entity present in the nuclei of female, but not male cat motor neurons [1]. They later reported that this dark, condensed perinucleolar structure (what we now refer to as the "Barr body") was present in the nuclei of mouse and human female somatic cells. 10 years later, Susumu Ohno suggested that the Barr body in mammals was a condensed or heterochromatic female X-chromosome [2]. This same year, Russell described that female mice harboring only one X-chromosome (XO), rather than the normal two Xs, lack the Barr body, develop normally, and are fertile, [3]. In 1961, Mary Lyon published a remarkable *Nature* article entitled "Gene action in the X-chromosome of the mouse (*Mus musculus* L)", in which she proposed a hypothesis explaining these earlier findings of Barr, Ohno, and Russell from new perspectives [4]. Lyon's landmark paper posited that female nuclei have a condensed or inactive X-chromosome (Barr body), because female cells inactivate or silence one of their two X-chromosomes to balance the gene dosage with XY males [4]. Her theory is also known as the Lyon Hypothesis or Lyonization.

Lyon was studying coat color variegation in *Tabby* and *Mottled* mutant mice, and made the astute observation that females had a mottled coat color, whereas males had a solid coat coloring. She noted that these mice are analogous to the variegated coat color of tabby and calico female cats, and suggested that this was due to the random silencing of the orange and black color genes harbored on the X-chromosome to balance the gene dosage with XY males [4]. Today we know that the Lyon hypothesis is indeed true, and mammalian females balance their somatic gene dosage with XY by randomly silencing one of their two Xs by an epigenetic mechanism called X-chromosome inactivation (XCI).

CONTENTS

353

Chromatin Regulation and Dynamics. http://dx.doi.org/10.1016/B978-0-12-803395-1.00014-9

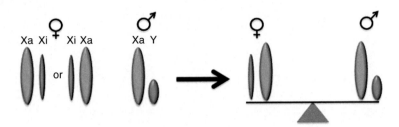

FIGURE 14.1 X-chromosome inactivation balances the gene dosage between XX females and XY males.

In the random form of X-chromosome inactivation *(XCI)* there is an equal possibility for the maternally or the paternally inherited X-chromosome (pink) to be silenced. Random XCI occurs in the soma and either of the two female X-chromosomes can be chosen to balance the gene dose to the single active X in males. The Y-chromosome is illustrated in blue. Xa, Active X-chromosome; Xi, inactive X-chromosome.

There are two forms of XCI in the mouse: imprinted and random [5]. The random form occurs in all somatic female cells with a mutually exclusive choice of the maternally (X^M) or paternally (X^P) inherited X-chromosome chosen for silencing (Fig. 14.1). Random XCI occurs about the time of epiblast (EPI) formation during mouse development, and results in all female somatic cells to be mosaic with either the X^P or X^M silenced. Remarkably, once an X has been chosen for inactivation, that very same X has this epigenetic fate following cellular division. The imprinted form of XCI selectively marks the paternally inherited X for silencing, and is regulated by chromatin marks laid down in the male germline [6,7]. This allele-specific silencing of the paternal X-chromosome precedes randome XCI, and it is analogous to parent-of-origin-specific genomic imprinting that occurs on the autosomes [8]. The imprinted form of XCI thus manifests in the mouse during very early embryogenesis, and is reversed in the inner cell mass of the blastocyst to pave the way for the subsequent establishment of random XCI. The extraembryonic tissues, however, continue to rely on the imprinted form of XCI [5]. Interestingly, female marsupials exclusively silence their paternal X-chromosome in all of the soma in contrast to the mouse [9]. The mammalian imprinted XCI is thus postulated to be a vestige of evolution [5]. Finally, whereas both the imprinted and random forms of XCI are established during mouse development, much less is known about the potential presence of imprinted XCI during human development [10].

The XCI cycle consists of dynamic inactivation and reactivation during mouse embryogenesis [10]. The X^P is preinactivated by a process called meiotic sex chromosome inactivation (MSCI), a result of the silencing of the unpaired X- and Y-chromosomes during the pachytene stage of male spermatogenesis [11–15]. After fertilization, the X^P undergoes silencing first at repetitive elements, which is followed by a genic silencing resulting in a near completely inactive paternal X-chromosome by the morula stage [16–18] (Fig. 14.2). These repetitive DNA elements consist of long interspersed elements (LINES)

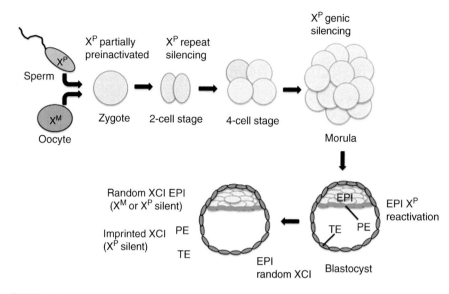

FIGURE 14.2 X-chromosome inactivation in early mouse development.
The paternal X-chromosome (X^P) is partially inactivated due to meiotic sex chromosome inactivation *(MSCI)* during spermatogenesis. The maternal X-chromosome (X^M) is active in the oocyte. Following fertilization, the female zygote maintains a partially preinactivated X^P. At the 2-cell stage 24 h after fertilization, the X^P repeats are silenced. By the morula stage (16 cells) the genic regions are silenced. At the blastocyst stage the XP is completely inactivated in the trophectoderm *(TE)* and the primitive endoderm *(PE)*. In contrast, the epiblast *(EPI)* undergoes a reactivation of the X^P. Random XCI occurs shortly thereafter, with the inactivation of the X^M or the X^P in the EPI. The extraembryonic tissues (TE and PE) maintain imprinted XCI with the exclusive silencing of the X^P.

and short interspersed elements (SINES). Interestingly, the X-chromosome has double the number of LINES and SINES than the autosomes. During the blastocyst stage [embryonic day 3.5 (E3.5)], a remarkable epigenetic reprogramming occurs in that the silent paternal X is reactivated in the EPI [19], the lineage that gives rise to the somatic cells of the embryo [20]. Notably, all repressive marks are erased from the X^P exclusively in the EPI [19]. Shortly thereafter, random XCI occurs in the EPI with a mutually exclusive choice of X^P versus X^M X-chromosome silencing. Interestingly the X^P-chromosome in the extraembryonic tissues [trophectoderm (TE) and primitive endoderm (PE)] resists reactivation and remains silent [10,21]. A second X reactivation occurs in the primordial germ cells, originating from the EPI, during their migration to the genital ridge at E7.0-E10.5 [22–27]. This second round of reactivation ensures the equal meiotic segregation of the X^M and X^P. The XCI cycle is then complete and recommences with silencing of the X^P during MSCI. Taken together, XCI is coupled with lineage segregation in the mouse embryo [10,28]. The stem cells derived from the EPI, TE, and PE lineages [embryonic stem cells (ESCs) [20], trophectoderm stem cells [7,29], and extraembryonic endoderm stem cells [30], respectively] faithfully recapitulate what is seen in the embryo, and can be used as a powerful tool to study XCI ex vivo [30–32].

14.2 LONG NONCODING RNAs AND XCI

Since Lyon's paper over 50 years ago, many labs sought to define the molecular elements that control XCI. It was determined that individuals with numerical variants of the X-chromosome had a counting mechanism, in which the X-to-autosome ratio is determined and XCI is induced in all but one X in a diploid nucleus [32–34]. This is known as the "$n-1$" rule in that XX and XXY individuals harbor one inactive X (Xi) or Barr body; whereas XY and XO individuals lack an Xi [35]. Adhering to the "$n-1$" rule, XXX individuals would exhibit two Barr bodies. A single region on the mouse X-chromosome, the *X inactivation center* (*Xic*), regulates the control of XCI including counting, the random choice of X (either the X^P or X^M X-chromosome) to be inactivated, and the initiation of silencing [32,35,36]. By cytogenetic mapping it was determined that the *Xic* control region is situated on a 1-Megabase (Mb) region of the human X-chromosome [37,38]. Subsequent studies narrowed the mouse *Xic* to 100-kilobase (kb) [39–41].

In 1991 Hunt Willard's lab successfully cloned the first gene in the human *XIC*, which they named *X-inactive-specific-transcript* (*XIST*) [37,42]. *XIST* is exclusively expressed from the inactive X-chromosome (Xi), and this RNA marks the Barr body. A similar analysis identified the mouse *Xist* gene that harbors eight exons encoding a 17-kb spliced and polyadenylated RNA with no significant open reading frames [43,44]. *XIST/Xist* expression can be visualized by RNA fluorescence in situ hybridization (FISH). RNA FISH analyses have thus uncovered that this long noncoding RNA (lncRNA) (lncRNAs are defined as noncoding RNA greater than 200 nucleotides in length) is upregulated during XCI, localizes in the nucleus, and "coats" the inactive X-chromosome forming an *Xist* cloud [44,45]. The genetic ablation of the mouse *Xist* promoter and the first exon in female mouse ESCs results in the continuous activity of this mutant X-chromosome, highlighting the functional importance of *Xist* in XCI [46]. A second deletion of *Xist* exons 1–7 in female mouse development confirmed these results, and also revealed the requirement for *Xist* in the establishment of imprinted XCI (the X^P X-chromosome is silenced) of the extraembryonic tissues [47]. Females that inherited the mutant *Xist* allele maternally thus developed normally, whereas females that inherited the mutant *Xist* allele paternally were growth retarded and had an early embryonic lethality due to the perturbation of the extraembryonic tissues carrying two active X chromosomes. Interestingly, distinct regions within the *Xist* RNA regulate its chromosomal localization and its effects on transcriptional silencing. While repeat A within *Xist* is an important element for XCI silencing, other domains of *Xist* are responsible for its chromosomal association and spreading on the Xi [48]. Collectively, *Xist* is required for the initiation of chromosomal-wide *cis*-silencing in both the random and the imprinted XCI early in mouse development [46,47].

In 1999, Jeannie Lee described *Tsix*, a 40-kb long noncoding RNA antisense to *Xist*. [49] *Tsix* acts in *cis* and is an antisense repressor of *Xist*, which is crucial for establishing the choice of inactivated X [41]. Targeted disruption of *Tsix* results in that allele to be chosen as the inactive X-chromosome [41,50]. *Tsix* has also been shown to have a primary role in maintaining random XCI [51]. Multiple mutations of *Tsix* have been generated and documented the importance of this lncRNA in XCI [52–58]. Moreover, the transcription of *Tsix*, rather than *Tsix* RNA itself, is thought to interfere with *Xist* transcription stopping its activity on the future active X [59,60]. *Tsix* is regulated by another noncoding RNA, *X-inactivation intergenic transcription element* (*Xite*), which is situated upstream of *Tsix* and harbors an enhancer on the future Xa that regulates the developmental-specific expression of *Tsix* [61,62].

Additional lncRNAs are present at the *Xic* (Fig. 14.3). For example, *Jpx* (also known as *Enox*) is a transacting lncRNA that activates *Xist via* a dose-dependent eviction of the chromatin insulator CTCF [63–66]. Another lncRNA within the

(A)

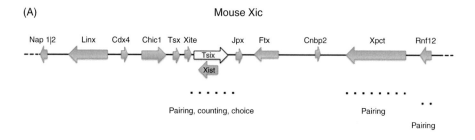

(B)

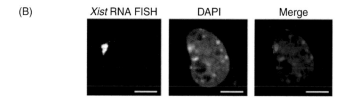

FIGURE 14.3 The mouse *X-inactivation center (Xic)* harbors many long noncoding RNAs (lncRNAs) including *Xist,* which is responsible for silencing and coats the inactive X-chromosome.

(A) The mouse *Xic* has a plethora of long noncoding RNAs *(lncRNAs)* including *Xist* (pink) and its antisense partner *Tsix* (yellow). The arrows indicate the direction of transcription. The dashed lines designate the regions responsible for X-chromosome pairing, counting, and choice. (B) The inactive female X-chromosome is coated with *Xist* RNA in differentiated female mouse embryonic stem cells. In the left hand panel, *Xist* RNA-Fluorescent in situ hybridization *(FISH)* shows that the *Xist* lncRNA coats the inactive X-chromosome (Xi). The middle panel marks the nucleus with 4′,6′-diamidino-2-phenylindole *(DAPI)* fluorescent staining. The Xi is enriched at a DAPI-dense heterochromatic region, as shown on the right-hand panel that merges the *Xist* RNA FISH (red) with the DAPI (blue) to highlight the Xi.

Xic is *Ftx* (named for its location *5′ to Xist*) [67]. When *Ftx* is deleted in male ESC, it alters the chromatin structure of the *Xic* and reduces the expression of *Xist* [67]. Interestingly, *Ftx* deletion did not affect XCI in female blastocyst [68]. *Testis-specific X-linked* (*Tsx*) is a putative lncRNA that, when knocked down, reduces *Tsix* and partially derepresses *Xist* expression [69–71]. Collectively, XCI is mediated by the interplay between numerous lncRNAs at the *Xic* (Fig. 14.3). The discovery of *Xist* thus sparked a new field of lncRNA regulation. Nearly 60–90% of the mammalian genome is transcriptionally active, and noncoding RNAs are important players in this pervasive expression [72–74], affecting gene expression at many loci during development and in diseases, such as cancer [75–79].

14.3 CHROMATIN MODIFICATIONS REGULATING X-CHROMOSOME INACTIVATION

The formation of the Barr body within the nucleus of female cells was the first description of facultative heterochromatin, representing a transcriptionally repressive environment that forms during differentiation [80]. Indeed, during XCI this extends to a chromosomal-wide silencing over 155-Mb on the Xi. The dynamics of XCI has been extensively studied in mouse development. The silencing of genes on the inactive X-chromosome is intimately linked to the presence of *Xist* lncRNA and its protein partners that not only promote silencing but also ensure the exclusive nuclear localization of the *Xist* RNA. For example, the scaffold protein SAF-A (hnRNPU) is required for the accumulation of *Xist* RNA on the Xi chromosome territory and for proper XCI in female ES cells [81]. Furthermore, the multifunctional transcription factor, YY1, with its ability to bind both DNA and RNA has also been implicated in the retention of *Xist* RNA on the Xi [82]. Since its discovery, numerous attempts have been made to isolate the proteins that interact with *Xist*. Apart form SAF-A (hnRNPU) and YY1, candidate approaches described, for example, several Polycomb–repressor complex (PRC) components and CTCF as *Xist*-binding partners [81–84]. Most recently, several groups have combined *Xist* lncRNA purification with high throughput quantitative mass spectrometry to identify a comprehensive *Xist*-interactome that includes chromatin remodeling and modification complexes, repressive–complexes, and matrix attachment factors, which collaborate to manifest and maintain the stable silencing of the Xi [85–87].

One of the earliest events in XCI is the depletion of RNA polymerase II (RNAP II) from the *Xist*-coated domain [15,88]. Proteins required for *Xist* RNA-mediated exclusion of RNAP II from the Xi and for silencing include SHARP that interacts with the SMRT corepressor, a factor that activates HDAC3, leading to histone deacetylation and consequent RNAP II exclusion [86]. The core of the Xi in both mouse and man thus excludes genes and mostly consists of repressed, nongenic sequences [88,89]. The gene-rich outer rim, on the other hand, includes genes that may escape silencing. Thus, the dynamic relocalization of genes into this inner repressed core emerging during XCI has been suggested to contribute to gene silencing and to the condensation of the Xi [88]. However, the spatial separation of RNAP II from the Xi is likely not sufficient for silencing, and XCI also involves locus-specific repression of regulatory elements [90].

The initiation of the maintenance phase of XCI is accompanied by epigenetic changes, such as the deposition of repressive posttranslational histone modifications (PTMs) and, at certain regions, eventually DNA methylation that collaborate to stably maintain silenced states [91]. The Xi is thus marked with the depletion of active histone PTMs, such as histone acetylation and histone H3 trimethylation at lysine 4 (H3K4me3) [92], and with the presence of several repression-associated PTMs that form distinct domains on the Xi. X-inactivated genes within the *Xist* chromatin domain, which contains the gene-dense repressed regions of the Xi, are silenced by Polycom-group complexes (PcGs) and by H4-Lysine-20-specific histone monomethyltransferase that generates H4K20me1 [90]. The Polycomb–repressive complex 1 (PRC1) catalyzes the ubiquitylation of histone H2A at lysine 119 (H2AK119ub1) [91,93,94], whereas the PRC2 complex catalyzes the trimethylation of histone 3 at lysine 27 (H3K27me3). These two PcG complexes become enriched at the Xi upon *Xist* RNA coating [93–96] to mediate the stability and heritability of silencing. The repressive histone variant MacroH2A also coats the Xi at regions marked by *Xist* RNA and H3K27me3, although its role in XCI is not completely understood [91]. Trimethylation of histone 3 at lysine 9 (H3K9me3), which represents a repressive histone modification characteristic of constitutive heterochromatin, is also involved in the stable repression of the Xi, and forms distinct chromatin domains that are spatially segregated from H3K27me3-marked regions. H3K-9me3-domains attract heterochromatin protein 1 (HP1) and are also enriched in repressive H4K20me3 modifications [97]. Moreover, the functional cross talk between these two types of heterochromatin (i.e., constitutive and facultative) as characterized by H3K9me3- and H3K27me3-enrichment was suggested to promote the stability of repressed states during cellular differentiation [98]. Taken together, the upregulation of *Xist* on the inactive X in XCI is accompanied by a number of histone PTMs and the recruitment of repressive heterochromatin proteins [91], which collaborate in the initiation and the remarkably stable maintenance of silenced states on the Xi in differentiated cells [87].

14.4 XCI STATUS IN PLURIPOTENCY AND DURING REPROGRAMMING

As discussed earlier, XCI status follows lineage segregation in the early mouse embryo [10]. Indeed, random XCI is induced upon cellular differentiation [10]. Undifferentiated mouse female ESCs with diploid number of chromosomes maintain both X-chromosomes in an active state [35]. In this case, both *Tsix* and *Xist* are expressed from each of the two female Xs. Upon differentiation, *Tsix* expression is downregulated, whereas *Xist* RNA is dramatically induced on the future inactive X [35]. XCI and differentiation are thus tightly connected, and inhibiting one process also interferes with the other [10]. The reversal of XCI takes place during epigenetic reprogramming in that conversion of female somatic cells to an induced pluripotent state is accompanied by the reactivation of the Xi, the reexpression of *Tsix* and *Xite*, and the disappearance of heterochromatic silent marks from the initially inactive X-chromosome. X-chromosome reactivation has been associated with the establishment of transcriptional networks maintaining pluripotency. Several pluripotent transcription factors (OCT4, REX, KLF4, c-MYC, and PRDM14) thus are implicated in the reactivation of the Xi [99,100–102]. The reactivation of the silent X during the conversion to the pluripotent state is paramount for understanding the molecular process of epigenetic reprogramming in regenerative medicine.

14.5 CHROMOSOMAL DYNAMICS AND SUBNUCLEAR LOCALIZATION OF XI

The dynamics of XCI is mediated by lncRNAs in *cis*, but there must be a communication between the two female X-chromosomes to ensure the mutually exclusive designation of active *versus* inactive X. Remarkably, the two X-chromosomes transiently communicate in the 3D nuclear space prior to XCI (during day 2–4 of female mouse embryonic stem cell differentiation) by forming a physical interaction at the *Xic* [103,104]. Live cell imaging has confirmed that homologous X pairing is completed within 45 min [105]. The importance of this process is underscored by the finding that failure to pair Xs during the initiation of XCI in female ESCs results in cell death [103]. This X chromosomal dance is mediated by active transcription and DNA sequences within the *Xic* including *Tsix*, *Xite*, and *Xpr* [103,104,106,107], as well as several proteins including the transcription factors CTCF and OCT4, and RNF12 [100,106,108]. RNF12 is an E3 ubiquitin ligase that is an XCI activator [109], and the maternal *Rnf12* is also required for imprinted XCI in mice [110].

The mechanism by which the *Xist* RNA spreads along the future inactive chromosome is under extensive investigation, and seems to be tightly linked with the 3D organization of the X-chromosome. Mary Lyon proposed that LINE elements on the X function as "way stations" propagating the spread of silencing and may aid facultative heterochromatin spreading [15,111,112]. Although evidence for this hypothesis in XCI is scant, the invention of novel technologies designed to map lncRNA occupancy and RNA--chromatin interactions have advanced our understanding of this process [113]. Recently, comparisons between high-resolution mapping of *Xist* RNA localization and 3D chromatin contact maps between the *Xist* locus and the rest of the X-chromosome have revealed that

Xist initially spreads along the Xi by searching in 3D [114,115]. Regions marked by early *Xist* localization were thus frequently in spatial proximity to the *Xist* locus, and were enriched in inactive genes, bordering actively transcribed, gene-rich regions [114,115]. Spreading of *Xist* RNA to transcribed regions required its A-repeat implicated in silencing, suggesting that active transcription might antagonize the presence of *Xist* RNA. The final step of the coating of the Xi by *Xist* RNA involves its variable spreading from gene-rich islands to intervening gene-poor domains, while being depleted form those genes that remain active (XCI escapers, see later) on the Xi [115].

Although XCI is generally considered to be a complete chromosomal-wide genic silencing on the Xi, there are genes that can thwart silencing. In humans, about 15% of the 1500 X-linked genes escape silencing [116]. In contrast, 3% of genes escape XCI in the mouse [90]. These escapers are frequently organized in clusters [116–118]. lncRNAs, boundary elements, paucity of DNA methylation, and chromatin insulators/3D genome organizers, such as CTCF, may shield these regions from the spread of heterochromatin [118–121]. Although LINE-1 element density is low in regions where X-chromosome genes escape silencing [112], future studies may reveal the molecular mechanism of XCI escape.

More recently, the Heard lab discovered that the promoters of *Xist* and *Tsix* reside in two distinct topologically associating domains (TADs) [122]. Genes that are harbored in the same TAD are likely to be coregulated during development due to frequent chromatin fiber contacts between enhancers and promoters localized within the same TAD. The localization of *Xist* and *Tsix* in different TADs might enable the differential regulation of these two lncRNAs during cell differentiation, with the prominent expression of *Tsix* in pluripotency and the upregulation of *Xist* upon cellular differentiation in female cells [122]. The structure of these TADs may be thus essential for generating asymmetries between the Xa and Xi by separating differential transcriptional territories [122]. After establishment of XCI, TAD structures are no longer detectable on the compact Xi by allele-specific Hi-C, a method designed to capture the all-to-all chromatin interactions in the nucleus [122]. Interestingly, emergence of the repressive compact structure of the Xi was paralleled by the eviction of cohesin, a 3D genome organizer, from Xa-specific cohesin binding sites within TADs as well as from TAD boundaries, to potentially interfere with enhancer–promoter interactions and the maintenance of TADs on the Xi. As the *Xist* RNA actively regulates the binding of cohesin components to the Xi, there seems to be a two-way relationship between *Xist* localization and 3D chromatin structures [87].

The higher-order chromatin compaction of the Xi is further stabilized by the SMCHD1–HBiX1 protein complex that mediates physical interactions between repressed, nonoverlapping domains of H3K9me2/3- or H3K27me3-enriched chromatin regions [123]. The combinatorial presence of H3K9me2 and H3K-27me3 histone marks also serves as platform for the recruitment of the CDYL protein early on during the XCI process, which not only reads the Xi but also recruits chromatin modifiers, such the H3K9me2 histone methyltransferase G9a, to propagate the repressed state [98]. Taken together, the Xi uses a number of mechanisms to compact and partition itself into particular 3D chromatin structures within the nucleus, which not only reflects but also contributes to the eastablishment and maintenance of active and silent epigenetic states along the chromosome.

Although the Barr body was originally described as a condensed structure that preferentially located in the perinucleolar territory of the nucleus, very little is known about how this subnuclear placement of the Xi might influence its 3D organization and contribute to the chromosome-wide repression. Interestingly, the stable maintenance of repressed chromatin states on the Xi has been suggested to depend on the *Xist* RNA-mediated localization of the Xi to a particular perinucleolar territory during S phase [124]. The loss of *Xist* RNA has thus been paralleled by the impaired localization of the Xi to the repressive environment of the perinucleolar ring, leading to the slow erosion of repressed states on the Xi over time. Although the dependence of the perinucleolar localization of Xi on *Xist* RNA complicates conclusions about the role of the perinucleolar space in silencing, this space has been proposed to harbor factors that enable the replication and the maintenance of repressive chromatin states, supporting the functional importance of 3D genome organisation in XCI [124].

14.6   X-LINKED DISEASES AND CANCER CONNECTIONS

Lyon's XCI hypothesis explained the non-Mendelian inheritance of sex- or X-linked diseases [125]. X-linked disorders manifest in males and not female carriers of diseases, such as Hemophilia A (a coagulopathy caused by a mutation in the gene that encodes the X-linked Factor VIII coagulation protein), red–green color blindness (present in approximately 4–6% of the normal male population as a result of mutations in the X-linked color sensitive cones), and *glucose-6-phosphate dehydrogenase* (*G6pdx*) gene deficiency (destruction of red blood cells in response to oxidative stress) [111,126–128]. Females heterozygous for X-linked mutations, on the other hand, display a somatic mosaicism due to random XCI.

Although the human X-chromosome contains an estimated 1098 genes (only about 4% of the genes in the human genome), there are approximately 168 X-linked diseases that can be attributed to mutations in the X chromosome. Many are associated with X-linked intellectual disability (formerly known as X-linked mental retardation). Fragile X syndrome is one of the most common causes of inherited intellectual disability and the leading monogenic cause of autism spectrum disorders. Whereas females are carriers, fragile X manifests in males due to the increased number of CGG DNA repeats in the X-linked Fragile X Mental Retardation 1 (*FMR1*) gene. While healthy individuals have only about 60 CGG repeats, the expansion of the FMR1 trinucleotide reiteration to greater than 200 repeats ultimately results in the epigenetic silencing of the *FMR1* gene, which leads to the down-regulation of its gene product, the Fragile X Mental Retardation Protein (FMRP) that is necessary for normal cognitive development [129,130]. Rett syndrome (RTT) is a neurodevelopmental disorder caused by a mutation in the X-linked methyl-CpG-binding protein-2 (*MeCP2*) gene [131]. Males with *MeCP2* mutation die within a year or so of birth; whereas females exhibit a deterioration of brain functions [132]. RTT symptoms include intellectual disability, autism, and developmental regression. Of note, neurons are particularly sensitive to the dosage of MeCP2, as not only does a loss of this protein cause symptoms, but a duplication of MeCP2 can also cause neuronal dysfunction [133]. Neuronal functions can be restored in a mouse model by supplying a proper MeCP2 dosage, suggesting a possible clinical therapy for RTT [133]. Understanding the role of MeCP2 in neuronal functions and the mechanism of its repression during XCI is important for devising future therapeutic approaches in RTT.

In addition to X-linked diseases, the X-chromosome is also important from the perspective of cancer development. Since its discovery over 70 years ago, the Barr body has been observed to disappear in breast and ovarian cancers [134–136]. This loss of the Barr body may be due to genetic loss or to an epigenetic instability of the repressed states on the Xi and transcriptional reactivation. Indeed, changes in DNA methylation, chromatin structure, noncoding RNAs, and nuclear organization all accompany tumorigenesis [137,138]. Although the functional consequences of the epigenetic instability on the Xi are not completely understood, several X-linked genes have been functionally implicated in cancer, including *AMER1/WT1* and *FOXP3* [139,140]. Interestingly, genetic ablation of *Xist* in adult female mice results in fully penetrant hematological malignancy suggesting that this noncoding RNA may have a role in tumor surveillance [141]. Recently, Edith Heard's lab reported that there is an epigenetic erosion of the Barr body in human breast cancers resulting in the reactivation of "cancer-specific escapee" genes from the inactive female X [142]. By genome-wide profiling of chromatin and transcription in breast cancer cells they observed an alteration in gene activity, heterochromatin pattern, and 3D nuclear organization on the Xi [142]. It remains to be determined whether the X-linked gene reactivation is the consequence of general epigenetic disruption observed in cancer cells, or it may also contribute to cancer initiation and/or progression. Taken together, gender differences in disease pathology are clinically relevant to XCI studies, and the human X contains a disproportionately high number of associated diseases.

14.7 FUTURE DIRECTIONS AND OPEN QUESTIONS

Since Mary Lyon's seminal article, numerous exciting discoveries have been made in the field of XCI. There remain, however, many open questions. The molecular mechanisms underlying XCI have for the most part been elucidated in mouse models (embryos and ESCs), but little is known about XCI in the human embryo. It is, for example, controversial whether the antisense lncRNA *TSIX* is present in humans. Also, it is not known whether the placental derived tissues in humans have the imprinted form of XCI. Further questions include what mechanisms allow selection for one X to be inactivated and the other to remain active. The signals and the mechanism of heterochromatin formation, and the cause and effect relationship between the *Xist* RNA and the acquisition of heterochromatic marks is also incompletely understood. Once an X-chromosome is chosen for inactivation, there is a remarkable epigenetic memory that maintains the very same X-chromosome silenced during mitosis. It will be thus important to determine how this epigenetic memory is mediated, and what regulates X reactivation in the mouse blastocyst. Interestingly, fifteen percent of the genes on the human X-chromosome and three percent of the genes on the mouse X-chromosome escape inactivation. It is still an enigma how XCI escape is mechanistically accomplished. Understanding this mechanism has clinical relevance, because our ability to selectively reactivate a silent allele on the Xi might prevent or reverse disorders, such as the Rett syndrome. These are but multiple open questions to be answered in future studies. Mary Lyon's seminal paper over 50 years sparked a new field: XCI. Research on XCI is not only important for our understanding of the molecular mechanisms regulating this crucial epigenetic process during development, but it has also huge implications for novel therapeutic approaches in a wide range of diseases.

Abbreviations

EPI	Epiblast
ESCs	Embryonic stem cells
lncRNAs	Long noncoding RNAs
LINES	Long interspersed elements
MSCI	Meiotic sex chromosome inactivation
PcGs	Polycomb–group complexes
PE	Primitive endoderm
PTMs	Posttranslational modifications
TE	Trophectoderm
Xa	Active X-chromosome
XCI	X-chromosome inactivation
Xi	Inactive X-chromosome
Xic	*X inactivation center*
Xist	X-inactive-specific-transcript
X^M	Maternal X-chromosome
X^P	Paternal X-chromosome

Acknowledgments

I thank my lab members for insightful discussions and comments on this review. Work in the Donohoe lab is supported by National Institute of Health 1R01MH 090267, the Burke Foundation, the Thomas and Agnes Carvel Foundation, and the Eisenberg Ahsen Foundation.

References

[1] Barr ML, Bertram EG. A morphological distinction between neurones of the male and female, and the behaviour of the nucleolar satellite during accelerated nucleoprotein synthesis. Nature 1949;163(4148):676.

[2] Ohno S, Kaplan WD, Kinosita R. Formation of the sex chromatin by a single X-chromosome in liver cells of Rattus norvegicus. Exp Cell Res 1959;18:415–8.

[3] Russell WL, Russell LB, Gower JS. Exceptional inheritance of a sex-linked gene in the mouse explained on the basis that the x/o sex-chromosome constitution is female. Proc Natl Acad Sci USA 1959;45(4):554–60.

[4] Lyon MF. Gene action in the X-chromosome of the mouse (*Mus musculus* L.). Nature 1961;190:372–3.

[5] Payer B, Lee JT, Namekawa SH. X-inactivation and X-reactivation: epigenetic hallmarks of mammalian reproduction and pluripotent stem cells. Hum Genet 2011;130(2):265–80.

[6] Lifschytz E, Lindsley DL. The role of X-chromosome inactivation during spermatogenesis (Drosophila-allocycly-chromosome evolution-male sterility-dosage compensation). Proc Natl AcadSci USA 1972;69(1):182–6.

[7] Takagi N, Sasaki M. Preferential inactivation of the paternally derived X chromosome in the extraembryonic membranes of the mouse. Nature 1975;256(5519):640–2.

[8] Lee JT, Bartolomei MS. X-inactivation, imprinting, and long noncoding RNAs in health and disease. Cell. 2013;152(6):1308–23.

[9] Sharman GB. Late DNA replication in the paternally derived X chromosome of female kangaroos. Nature 1971;230(5291):231–2.

[10] Monk M, Harper MI. Sequential X chromosome inactivation coupled with cellular differentiation in early mouse embryos. Nature 1979;281(5729):311–3.

[11] Turner JM, Mahadevaiah SK, Elliott DJ, Garchon HJ, Pehrson JR, Jaenisch R, et al. Meiotic sex chromosome inactivation in male mice with targeted disruptions of Xist. J Cell Sci 2002;115(Pt 21):4097–105.

[12] Namekawa SH, Park PJ, Zhang LF, Shima JE, McCarrey JR, Griswold MD, et al. Postmeiotic sex chromatin in the male germline of mice. Curr Biol 2006;16(7):660–7.

[13] Huynh KD, Lee JT. Imprinted X inactivation in eutherians: a model of gametic execution and zygotic relaxation. Curr Opin Cell Biol 2001;13(6):690–7.

[14] Huynh KD, Lee JT. Inheritance of a pre-inactivated paternal X chromosome in early mouse embryos. Nature 2003;426(6968):857–62.

[15] Okamoto I, Otte AP, Allis CD, Reinberg D, Heard E. Epigenetic dynamics of imprinted X inactivation during early mouse development. Science 2004;303(5658):644–9.

[16] Namekawa SH, Payer B, Huynh KD, Jaenisch R, Lee JT. Two-step imprinted X inactivation: repeat versus genic silencing in the mouse. Mol Cell Biol 2010;30(13):3187–205.

[17] Kalantry S, Purushothaman S, Bowen RB, Starmer J, Magnuson T. Evidence of Xist RNA-independent initiation of mouse imprinted X-chromosome inactivation. Nature 2009;460(7255):647–51.

[18] Okamoto I, Patrat C, Thepot D, Peynot N, Fauque P, Daniel N, et al. Eutherian mammals use diverse strategies to initiate X-chromosome inactivation during development. Nature 2011;472(7343):370–4.

[19] Mak W, Nesterova TB, de Napoles M, Appanah R, Yamanaka S, Otte AP, et al. Reactivation of the paternal X chromosome in early mouse embryos. Science 2004;303(5658).666–9.

[20] Evans MJ, Kaufman MH. Establishment in culture of pluripotential cells from mouse embryos. Nature 1981;292(5819):154–6.

[21] McLaren A, Monk M. X-chromosome activity in the germ cells of sex-reversed mouse embryos. J Reprod Fertil 1981;63(2):533–7.

[22] McMahon A, Fosten M, Monk M. Random X-chromosome inactivation in female primordial germ cells in the mouse. J Embryol Exp Morphol 1981;64:251–8.

[23] Monk M, McLaren A. X-chromosome activity in foetal germ cells of the mouse. J Embryol Exp Morphol 1981;63:75–84.

[24] Gartler SM, Liskay RM, Campbell BK, Sparkes R, Gant N. Evidence for two functional X chromosomes in human oocytes. Cell Differ 1972;1(4):215–8.

[25] Sugimoto M, Abe K. X chromosome reactivation initiates in nascent primordial germ cells in mice. PLoS Genet 2007;3(7):e116.

[26] de Napoles M, Nesterova T, Brockdorff N. Early loss of Xist RNA expression and inactive X chromosome associated chromatin modification in developing primordial germ cells. PloS One 2007;2(9):e860.

[27] Chuva de Sousa Lopes SM, Hayashi K, Shovlin TC, Mifsud W, Surani MA, McLaren A. X chromosome activity in mouse XX primordial germ cells. PLoS Genet 2008;4(2):e30.

[28] McMahon A, Fosten M, Monk M. X-chromosome inactivation mosaicism in the three germ layers and the germ line of the mouse embryo. J Embryol Exp Morphol 1983;74:207–20.

[29] Tanaka S, Kunath T, Hadjantonakis AK, Nagy A, Rossant J. Promotion of trophoblast stem cell proliferation by FGF4. Science 1998;282(5396):2072–5.

[30] Kunath T, Arnaud D, Uy GD, Okamoto I, Chureau C, Yamanaka Y, et al. Imprinted X-inactivation in extra-embryonic endoderm cell lines from mouse blastocysts. Development 2005;132(7):1649–61.

[31] Martin GR, Epstein CJ, Travis B, Tucker G, Yatziv S, Martin DW Jr, et al. X-chromosome inactivation during differentiation of female teratocarcinoma stem cells in vitro. Nature 1978;271(5643):329–33.

[32] Rastan S, Robertson EJ. X-chromosome deletions in embryo-derived (EK) cell lines associated with lack of X-chromosome inactivation. J Embryol Exp Morphol 1985;90:379–88.

[33] Lyon MF. Possible mechanisms of X chromosome inactivation. Nat New Biol 1971;232(34):229–32.

[34] Lyon MF. X-chromosome inactivation and developmental patterns in mammals. Biol Rev Camb Philos Soc 1972;47(1):1–35.

[35] Payer B, Lee JT. X chromosome dosage compensation: how mammals keep the balance. Annu Rev Genet 2008;42:733–72.

[36] Kay GF, Barton SC, Surani MA, Rastan S. Imprinting and X chromosome counting mechanisms determine Xist expression in early mouse development. Cell 1994;77(5):639–50.

[37] Brown CJ, Lafreniere RG, Powers VE, Sebastio G, Ballabio A, Pettigrew AL, et al. Localization of the X inactivation centre on the human X chromosome in Xq13. Nature 1991;349(6304):82–4.

[38] Migeon BR, Kazi E, Haisley-Royster C, Hu J, Reeves R, Call L, et al. Human X inactivation center induces random X chromosome inactivation in male transgenic mice. Genomics 1999;59(2):113–21.

[39] Lee JT, Strauss WM, Dausman JA, Jaenisch R. A 450 kb transgene displays properties of the mammalian X-inactivation center. Cell 1996;86(1):83–94.

[40] Heard E, Mongelard F, Arnaud D, Avner P. Xist yeast artificial chromosome transgenes function as X-inactivation centers only in multicopy arrays and not as single copies. Mol Cell Biol 1999;19(4):3156–66.

[41] Lee JT, Lu N, Han Y. Genetic analysis of the mouse X inactivation center defines an 80-kb multifunction domain. Proc Natl Acad Sci USA 1999;96(7):3836–41.

[42] Brown CJ, Ballabio A, Rupert JL, Lafreniere RG, Grompe M, Tonlorenzi R, et al. A gene from the region of the human X inactivation centre is expressed exclusively from the inactive X chromosome. Nature 1991;349(6304):38–44.

[43] Brockdorff N, Ashworth A, Kay GF, Cooper P, Smith S, McCabe VM, et al. Conservation of position and exclusive expression of mouse Xist from the inactive X chromosome. Nature 1991;351(6324):329–31.

[44] Brockdorff N, Ashworth A, Kay GF, McCabe VM, Norris DP, Cooper PJ, et al. The product of the mouse Xist gene is a 15 kb inactive X-specific transcript containing no conserved ORF and located in the nucleus. Cell 1992;71(3):515–26.

[45] Brown CJ, Hendrich BD, Rupert JL, Lafreniere RG, Xing Y, Lawrence J, et al. The human XIST gene: analysis of a 17 kb inactive X-specific RNA that contains conserved repeats and is highly localized within the nucleus. Cell 1992;71(3):527–42.

[46] Penny GD, Kay GF, Sheardown SA, Rastan S, Brockdorff N. Requirement for Xist in X chromosome inactivation. Nature 1996;379(6561):131–7.

[47] Marahrens Y, Panning B, Dausman J, Strauss W, Jaenisch R. Xist-deficient mice are defective in dosage compensation but not spermatogenesis. Genes Dev 1997;11(2):156–66.

[48] Wutz A, Rasmussen TP, Jaenisch R. Chromosomal silencing and localization are mediated by different domains of Xist RNA. Nat Genet 2002;30(2):167–74.

[49] Lee JT, Davidow LS, Warshawsky D. Tsix, a gene antisense to Xist at the X-inactivation centre. Nat Genet 1999;21(4):400–4.

[50] Lee JT. Disruption of imprinted X inactivation by parent-of-origin effects at Tsix. Cell 2000;103(1):17–27.

[51] Gayen S, Maclary E, Buttigieg E, Hinten M, Kalantry S. A primary role for the Tsix lncRNA in maintaining random X-chromosome inactivation. Cell Rep 2015;11(8):1251–65.

[52] Cohen DE, Davidow LS, Erwin JA, Xu N, Warshawsky D, Lee JT. The DXPas34 repeat regulates random and imprinted X inactivation. Dev Cell. 2007;12(1):57–71.

[53] Clerc P, Avner P. Role of the region 3′ to Xist exon 6 in the counting process of X-chromosome inactivation. Nat Genet 1998;19(3):249–53.

[54] Vigneau S, Augui S, Navarro P, Avner P, Clerc P. An essential role for the DXPas34 tandem repeat and Tsix transcription in the counting process of X chromosome inactivation. Proc Natl Acad Sci USA 2006;103(19):7390–5.

[55] Ohhata T, Hoki Y, Sasaki H, Sado T. Tsix-deficient X chromosome does not undergo inactivation in the embryonic lineage in males: implications for Tsix-independent silencing of Xist. Cytogenet Genome Res 2006;113(1-4):345–9.

[56] Sado T, Hoki Y, Sasaki H. Tsix silences Xist through modification of chromatin structure. Dev Cell 2005;9(1):159–65.

[57] Sado T, Hoki Y, Sasaki H. Tsix defective in splicing is competent to establish Xist silencing. Development 2006,133(24):4925–31.

[58] Sado T, Wang Z, Sasaki H, Li E. Regulation of imprinted X-chromosome inactivation in mice by Tsix. Development 2001;128(8):1275–86.

[59] Stavropoulos N, Lu N, Lee JT. A functional role for Tsix transcription in blocking Xist RNA accumulation but not in X-chromosome choice. Proc Natl Acad Sci USA 2001;98(18):10232–7.

[60] Luikenhuis S, Wutz A, Jaenisch R. Antisense transcription through the Xist locus mediates Tsix function in embryonic stem cells. Mol Cell Biol 2001;21(24):8512–20.

[61] Ogawa Y, Lee JT. Xite, X-inactivation intergenic transcription elements that regulate the probability of choice. Mol Cell 2003;11(3):731–43.

[62] Stavropoulos N, Rowntree RK, Lee JT. Identification of developmentally specific enhancers for Tsix in the regulation of X chromosome inactivation. Mol Cell Biol 2005;25(7):2757–69.

[63] Johnston CM, Newall AE, Brockdorff N, Nesterova TB. Enox, a novel gene that maps 10 kb upstream of Xist and partially escapes X inactivation. Genomics 2002;80(2):236–44.

[64] Chureau C, Prissette M, Bourdet A, Barbe V, Cattolico L, Jones L, et al. Comparative sequence analysis of the X-inactivation center region in mouse, human, and bovine. Genome Res 2002;12(6):894–908.

[65] Tian D, Sun S, Lee JT. The long noncoding RNA, Jpx, is a molecular switch for X chromosome inactivation. Cell 2010;143(3):390–403.

[66] Sun S, Del Rosario BC, Szanto A, Ogawa Y, Jeon Y, Lee JT. Jpx RNA activates Xist by evicting CTCF. Cell 2013;153(7):1537–51.

[67] Chureau C, Chantalat S, Romito A, Galvani A, Duret L, Avner P, et al. Ftx is a non-coding RNA which affects Xist expression and chromatin structure within the X-inactivation center region. Hum Mol Genet 2011;20(4):705–18.

[68] Soma M, Fujihara Y, Okabe M, Ishino F, Kobayashi S. Ftx is dispensable for imprinted X-chromosome inactivation in preimplantation mouse embryos. Sci Rep 2014;4:5181.

[69] Simmler MC, Cunningham DB, Clerc P, Vermat T, Caudron B, Cruaud C, et al. A 94 kb genomic sequence 3′ to the murine Xist gene reveals an AT rich region containing a new testis specific gene Tsx. Hum Mol Genet 1996;5(11):1713–26.

[70] Cunningham DB, Segretain D, Arnaud D, Rogner UC, Avner P. The mouse Tsx gene is expressed in Sertoli cells of the adult testis and transiently in premeiotic germ cells during puberty. Dev Biol 1998;204(2):345–60.

[71] Anguera MC, Ma W, Clift D, Namekawa S, Kelleher RJ 3rd, Lee JT. Tsx produces a long noncoding RNA and has general functions in the germline, stem cells, and brain. PLoS Genet 2011;7(9):e1002248.

[72] Carninci P, Kasukawa T, Katayama S, Gough J, Frith MC, Maeda N, et al. The transcriptional landscape of the mammalian genome. Science 2005;309(5740):1559–63.

[73] Kapranov P, Cheng J, Dike S, Nix DA, Duttagupta R, Willingham AT, et al. RNA maps reveal new RNA classes and a possible function for pervasive transcription. Science 2007;316(5830):1484–8.

[74] Ponting CP, Oliver PL, Reik W. Evolution and functions of long noncoding RNAs. Cell 2009;136(4):629–41.

[75] Guttman M, Donaghey J, Carey BW, Garber M, Grenier JK, Munson G, et al. lincRNAs act in the circuitry controlling pluripotency and differentiation. Nature 2011;477(7364):295–300.

[76] Orom UA, Derrien T, Beringer M, Gumireddy K, Gardini A, Bussotti G, et al. Long noncoding RNAs with enhancer-like function in human cells. Cell 2010;143(1):46–58.

[77] Wang KC, Yang YW, Liu B, Sanyal A, Corces-Zimmerman R, Chen Y, et al. A long noncoding RNA maintains active chromatin to coordinate homeotic gene expression. Nature 2011;472(7341):120–4.

[78] Rinn JL, Kertesz M, Wang JK, Squazzo SL, Xu X, Brugmann SA, et al. Functional demarcation of active and silent chromatin domains in human HOX loci by noncoding RNAs. Cell 2007;129(7):1311–23.

[79] Lee JT. Epigenetic regulation by long noncoding RNAs. Science 2012;338(6113):1435–9.

[80] Trojer P, Reinberg D. Facultative heterochromatin: is there a distinctive molecular signature? Mol Cell 2007;28(1):1–13.

[81] Hasegawa Y, Brockdorff N, Kawano S, Tsutui K, Tsutui K, Nakagawa S. The matrix protein hnRNP U is required for chromosomal localization of Xist RNA. Dev Cell 2010;19(3):469–76.

[82] Jeon Y, Lee JT. YY1 tethers Xist RNA to the inactive X nucleation center. Cell 2011;146(1):119–33.

[83] Zhao J, Sun BK, Erwin JA, Song JJ, Lee JT. Polycomb proteins targeted by a short repeat RNA to the mouse X chromosome. Science 2008;322(5902):750–6.

[84] Kung JT, Kesner B, An JY, Ahn JY, Cifuentes Rojas C, Colognori D, et al. Locus-specific targeting to the X chromosome revealed by the RNA interactome of CTCF. Mol Cell 2015;57(2):361–75.

[85] Chu C, Zhang QC, da Rocha ST, Flynn RA, Bharadwaj M, Calabrese JM, et al. Systematic discovery of Xist RNA binding proteins. Cell 2015;161(2):404–16.

[86] McHugh CA, Chen CK, Chow A, Surka CF, Tran C, McDonel P, et al. The Xist lncRNA interacts directly with SHARP to silence transcription through HDAC3. Nature 2015;521(7551): 232–6.

[87] Minajigi A, Froberg JE, Wei C, Sunwoo H, Kesner B, Colognori D, et al. Chromosomes. A comprehensive Xist interactome reveals cohesin repulsion and an RNA-directed chromosome conformation. Science 2015;349.(6245).

[88] Chaumeil J, Le Baccon P, Wutz A, Heard E. A novel role for Xist RNA in the formation of a repressive nuclear compartment into which genes are recruited when silenced. Genes Dev 2006;20(16):2223–37.

[89] Clemson CM, Hall LL, Byron M, McNeil J, Lawrence JB. The X chromosome is organized into a gene-rich outer rim and an internal core containing silenced nongenic sequences. Proc Natl Acad Sci USA 2006;103(20):7688–93.

[90] Calabrese JM, Sun W, Song L, Mugford JW, Williams L, Yee D, et al. Site-specific silencing of regulatory elements as a mechanism of X inactivation. Cell 2012;151(5):951–63.

[91] Wutz A. Gene silencing in X-chromosome inactivation: advances in understanding facultative heterochromatin formation. Nat Rev Genet 2011;12(8):542–53.

[92] Heard E, Rougeulle C, Arnaud D, Avner P, Allis CD, Spector DL. Methylation of histone H3 at Lys-9 is an early mark on the X chromosome during X inactivation. Cell 2001;107(6): 727–38.

[93] Plath K, Talbot D, Hamer KM, Otte AP, Yang TP, Jaenisch R, et al. Developmentally regulated alterations in Polycomb repressive complex 1 proteins on the inactive X chromosome. J Cell Biol 2004;167(6):1025–35.

[94] de Napoles M, Mermoud JE, Wakao R, Tang YA, Endoh M, Appanah R, et al. Polycomb group proteins Ring1A/B link ubiquitylation of histone H2A to heritable gene silencing and X inactivation. Dev Cell 2004;7(5):663–76.

[95] Silva J, Mak W, Zvetkova I, Appanah R, Nesterova TB, Webster Z, et al. Establishment of histone h3 methylation on the inactive X chromosome requires transient recruitment of Eed-Enx1 polycomb group complexes. Dev Cell 2003;4(4):481–95.

[96] Plath K, Fang J, Mlynarczyk-Evans SK, Cao R, Worringer KA, Wang H, et al. Role of histone H3 lysine 27 methylation in X inactivation. Science 2003;300(5616):131–5.

[97] Chadwick BP, Willard HF. Multiple spatially distinct types of facultative heterochromatin on the human inactive X chromosome. Proc Natl Acad Sci USA 2004;101(50):17450–5.

[98] Escamilla-Del-Arenal M, da Rocha ST, Spruijt CG, Masui O, Renaud O, Smits AH, et al. Cdyl, a new partner of the inactive X chromosome and potential reader of H3K27me3 and H3K9me2. Mol Cell Biol 2013;33(24):5005–20.

[99] Navarro P, Chambers I, Karwacki-Neisius V, Chureau C, Morey C, Rougeulle C, et al. Molecular coupling of Xist regulation and pluripotency. Science 2008;321(5896):1693–5.

[100] Donohoe ME, Silva SS, Pinter SF, Xu N, Lee JT. The pluripotency factor Oct4 interacts with Ctcf and also controls X-chromosome pairing and counting. Nature 2009;460(7251):128–32.

[101] Navarro P, Oldfield A, Legoupi J, Festuccia N, Dubois A, Attia M, et al. Molecular coupling of Tsix regulation and pluripotency. Nature 2010;468(7322):457–60.

[102] Payer B, Lee JT. Coupling of X-chromosome reactivation with the pluripotent stem cell state. RNA Biol 2014;11(7):798–807.

[103] Xu N, Tsai CL, Lee JT. Transient homologous chromosome pairing marks the onset of X inactivation. Science 2006;311(5764):1149–52.

[104] Bacher CP, Guggiari M, Brors B, Augui S, Clerc P, Avner P, et al. Transient colocalization of X-inactivation centres accompanies the initiation of X inactivation. Nat Cell Biol 2006;8(3):293–9.

[105] Masui O, Bonnet I, Le Baccon P, Brito I, Pollex T, Murphy N, et al. Live-cell chromosome dynamics and outcome of X chromosome pairing events during ES cell differentiation. Cell 2011;145(3):447–58.

[106] Xu N, Donohoe ME, Silva SS, Lee JT. Evidence that homologous X-chromosome pairing requires transcription and Ctcf protein. Nat Genet 2007;39(11):1390–6.

[107] Augui S, Filion GJ, Huart S, Nora E, Guggiari M, Maresca M, et al. Sensing X chromosome pairs before X inactivation via a novel X-pairing region of the Xic. Science 2007;318(5856):1632–6.

[108] Barakat TS, Loos F, van Staveren S, Myronova E, Ghazvini M, Grootegoed JA, et al. The transactivator RNF12 and cis-acting elements effectuate X chromosome inactivation independent of X-pairing. Mol Cell 2014;53(6):965–78.

[109] Jonkers I, Barakat TS, Achame EM, Monkhorst K, Kenter A, Rentmeester E, et al. RNF12 is an X-encoded dose-dependent activator of X chromosome inactivation. Cell 2009;139(5):999–1011.

[110] Shin J, Bossenz M, Chung Y, Ma H, Byron M, Taniguchi-Ishigaki N, et al. Maternal Rnf12/RLIM is required for imprinted X-chromosome inactivation in mice. Nature 2010;467(7318):977–81.

[111] Gartler SM, Riggs AD. Mammalian X-chromosome inactivation. Annu Rev Genet 1983;17:155–90.

[112] Chow JC, Ciaudo C, Fazzari MJ, Mise N, Servant N, Glass JL, et al. LINE-1 activity in facultative heterochromatin formation during X chromosome inactivation. Cell 2010;141(6):956–69.

[113] Chu C, Qu K, Zhong FL, Artandi SE, Chang HY. Genomic maps of long noncoding RNA occupancy reveal principles of RNA–chromatin interactions. Mol Cell 2011;44(4):667–78.

[114] Engreitz JM, Pandya-Jones A, McDonel P, Shishkin A, Sirokman K, Surka C, et al. The Xist lncRNA exploits three-dimensional genome architecture to spread across the X chromosome. Science 2013;341(6147):1237973.

[115] Simon MD, Pinter SF, Fang R, Sarma K, Rutenberg-Schoenberg M, Bowman SK, et al. High-resolution Xist binding maps reveal two-step spreading during X-chromosome inactivation. Nature 2013;504(7480):465–9.

[116] Carrel L, Willard HF. X-inactivation profile reveals extensive variability in X-linked gene expression in females. Nature 2005;434(7031):400–4.

[117] Carrel L, Park C, Tyekucheva S, Dunn J, Chiaromonte F, Makova KD. Genomic environment predicts expression patterns on the human inactive X chromosome. PLoS Genet 2006;2(9):e151.

[118] Li N, Carrel L. Escape from X chromosome inactivation is an intrinsic property of the Jarid1c locus. Proc Natl Acad Sci USA 2008;105(44):17055–60.

[119] Reinius B, Shi C, Hengshuo L, Sandhu KS, Radomska KJ, Rosen GD, et al. Female-biased expression of long non-coding RNAs in domains that escape X-inactivation in mouse. BMC Genom 2010;11:614.

[120] Horvath LM, Li N, Carrel L. Deletion of an X-inactivation boundary disrupts adjacent gene silencing. PLoS Genet 2013;9(11):e1003952.

[121] Filippova GN, Cheng MK, Moore JM, Truong JP, Hu YJ, Nguyen DK, et al. Boundaries between chromosomal domains of X inactivation and escape bind CTCF and lack CpG methylation during early development. Dev Cell 2005;8(1):31–42.

[122] Nora EP, Lajoie BR, Schulz EG, Giorgetti L, Okamoto I, Servant N, et al. Spatial partitioning of the regulatory landscape of the X-inactivation centre. Nature 2012;485(7398):381–5.

[123] Nozawa RS, Nagao K, Igami KT, Shibata S, Shirai N, Nozaki N, et al. Human inactive X chromosome is compacted through a PRC2-independent SMCHD1-HBiX1 pathway. Nat Struct Mol Biol 2013;20(5):566–73.

[124] Zhang LF, Huynh KD, Lee JT. Perinucleolar targeting of the inactive X during S phase: evidence for a role in the maintenance of silencing. Cell 2007;129(4):693–706.

[125] Lyon MF. Gene action in the X-chromosome of the mouse (*Mus musculus* L.). Nature 1961;190:372–3.

[126] Beutler E, Yeh M, Fairbanks VF. The normal human female as a mosaic of X-chromosome activity: studies using the gene for C-6-PD-deficiency as a marker. Proc Natl Acad Sci USA 1962;48:9–16.

[127] Linder D, Gartler SM. Glucose-6-phosphate dehydrogenase mosaicism: utilization as a cell marker in the study of leiomyomas. Science 1965;150(3692):67–9.

[128] Puck JM, Willard HF. X inactivation in females with X-linked disease. N Eng J Med 1998;338(5):325–8.

[129] Terracciano A, Chiurazzi P, Neri G. Fragile X syndrome. Am J Med Genet C 2005;137C(1):32–7.

[130] Usdin K, Kumari D. Repeat-mediated epigenetic dysregulation of the FMR1 gene in the fragile X-related disorders. Front Genet 2015;6:192.

[131] Amir RE, Van den Veyver IB, Wan M, Tran CQ, Francke U, Zoghbi HY. Rett syndrome is caused by mutations in X-linked MECP2, encoding methyl-CpG-binding protein 2. Nat Genet 1999;23(2):185–8.

[132] Chahrour M, Zoghbi HY. The story of Rett syndrome: from clinic to neurobiology. Neuron 2007;56(3):422–37.

[133] Cohen S, Greenberg ME. Communication between the synapse and the nucleus in neuronal development, plasticity, and disease. Annu Rev Cell Dev Biol 2008;24:183–209.

[134] Barr ML, Moore KL. Chromosomes, sex chromatin, and cancer. Proc Can Cancer Conf 1957;2:3–16.

[135] Perry M. Evaluation of breast tumour sex chromatin (Barr body) as an index of survival and response to pituitary ablation. Br J Surg 1972;59(9):731–4.

[136] Smethurst M, Bishun NP, Fernandez D, Allen J, Burn JI, Alaghband-Zadeh J, et al. Steroid hormone receptors and sex chromatin frequency in breast cancer. J Endocrinol Invest 1981;4(4):455–7.

[137] De Carvalho DD, Sharma S, You JS, Su SF, Taberlay PC, Kelly TK, et al. DNA methylation screening identifies driver epigenetic events of cancer cell survival. Cancer Cell 2012;21(5):655–67.

[138] Shen H, Laird PW. Interplay between the cancer genome and epigenome. Cell 2013;153(1):38–55.

[139] Rivera MN, Kim WJ, Wells J, Driscoll DR, Brannigan BW, Han M, et al. An X chromosome gene, WTX, is commonly inactivated in Wilms tumor. Science 2007;315(5812):642–5.

[140] Bennett CL, Christie J, Ramsdell F, Brunkow ME, Ferguson PJ, Whitesell L, et al. The immune dysregulation, polyendocrinopathy, enteropathy, X-linked syndrome (IPEX) is caused by mutations of FOXP3. Nat Genet 2001;27(1):20–1.

[141] Yildirim E, Kirby JE, Brown DE, Mercier FE, Sadreyev RI, Scadden DT, et al. Xist RNA is a potent suppressor of hematologic cancer in mice. Cell 2013;152(4):727–42.

[142] Chaligne R, Popova T, Mendoza-Parra MA, Saleem MA, Gentien D, Ban K, et al. The inactive X chromosome is epigenetically unstable and transcriptionally labile in breast cancer. Genome Res 2015;25(4):488–503.

CHAPTER 15

Interaction Between Cellular Metabolic States and Chromatin Dynamics

S.J. Linder, R. Mostoslavsky

Program in Biological and Biomedical Sciences, Harvard Medical School, Boston, MA, United States; Massachusetts General Hospital Cancer Center, Boston, MA, United States

15.1 INTRODUCTION

Cellular plasticity, in response to nutrient availability, is critical for proper cell function and survival. Cells have evolved to tightly link chromatin changes, and ultimately gene transcription, to changes in their environment by having chromatin-modifying factors dependent on metabolites for function. Chromatin dynamics, in response to metabolic cues and thus metabolite availability, have large implications for a range of processes, including gene transcription, cellular proliferation and differentiation, disease, and aging.

This chapter presents the importance of intermediary metabolism, and how its products are tightly linked to the activity of chromatin-modifying enzymes, how metabolites have a direct effect on epigenetics and disease, and finally how exposure to different states of nutrient availability can have lasting effects on chromatin, even with the potential to be passed on transgenerationally, placing subsequent generations at risk for disease.

15.2 INTERMEDIARY METABOLISM PRODUCTS AS SUBSTRATES OR COFACTORS FOR CHROMATIN MODIFICATIONS

Chromatin modifications can be both dynamic and stable in nature. Intriguingly, almost all enzymes that catalyze these modifications are dependent on metabolites as sources of substrates and/or cofactors for catalysis. These metabolites are all molecules in pathways of intermediary metabolism, including acetyl-CoA, S-adenosylmethionine (SAM), ATP, nicotinamide adenine dinucleotide (NAD^+), flavin adenine dinucleotide (FAD), alpha-ketoglutarate (alpha-KG), uridine diphosphate (UPD)-glucose, and more.

CONTENTS

Chromatin Regulation and Dynamics. http://dx.doi.org/10.1016/B978-0-12-803395-1.00015-0

While metabolite availability is subject to fluctuation under different metabolic states, the activity of these chromatin-modifying enzymes may change as well, as substrate and cofactor availability will ebb and flow. A number of landmark studies have proven this to be true with modifications, such as histone acetylation, as further described in detail, revealing a critical connection between cellular metabolism, chromatin dynamics, and ultimately gene transcription.

15.2.1 Acetyl-CoA and Histone Acetylation

Acetylation of lysine residues on the N-terminal tails of histone proteins is one of the most extensively studied histone posttranslational modifications (PTM). It is a mark that is associated with actively transcribed genes, as acetylation neutralizes the positive charge on lysine residues, enabling the loosening of the interaction between highly negatively charged DNA and the histone proteins [1]. This leads to the opening of chromatin, in turn allowing DNA-related processes, such as transcription, replication, and DNA recombination to occur. Additionally, this mark can act as a docking site, recruiting specific chromatin-associated proteins, such as those containing bromodomains.

The universal substrate for histone acetylation is acetyl-CoA, a major metabolite produced in both catabolic and anabolic pathways of intermediary metabolism [2,3]. Histone acetyl-transferases are dependent on intracellular levels of acetyl-CoA for function, revealing acetylation as a powerful example of how cellular metabolism is tightly linked to chromatin dynamics.

Acetyl-CoA is most well associated with its roles in fueling the TCA cycle in the mitochondria (Fig. 15.1), as well as acting as a building block for the production of macromolecules, such as lipids, cholesterol, and amino acids [1]. Under normal nutrient conditions, acetyl-CoA is primarily produced from pyruvate generated from the breakdown of glucose, otherwise known as glycolysis. Under fasting conditions, where glycogen stores have been depleted, cells utilize fatty acid beta-oxidation in order to generate sufficient acetyl-CoA to fuel the TCA cycle, maintain normal mitochondrial oxidative metabolism, and ultimately ATP synthesis [4]. Cells have evolved a number of other ways to maintain sufficient acetyl-CoA levels in the cell. Acetyl-CoA is also derived from acetate via acetyl-CoA synthase (ACSS) enzymes, with isoforms located in the mitochondrial, cytoplasmic, and nuclear compartments, which catalyze the generation of acetyl CoA from acetate, ATP, and CoA. Reductive carboxylation of alpha-KG, an intermediate in the TCA cycle, can also lead to acetyl-CoA production via the anaplerotic pathway [1].

Due to its polar and chemical nature, acetyl-CoA is unable to cross membranes, such as those found in the mitochondria. Therefore, acetyl-CoA leaves the mitochondria in the form of citrate, the product of condensation of oxaloacetate

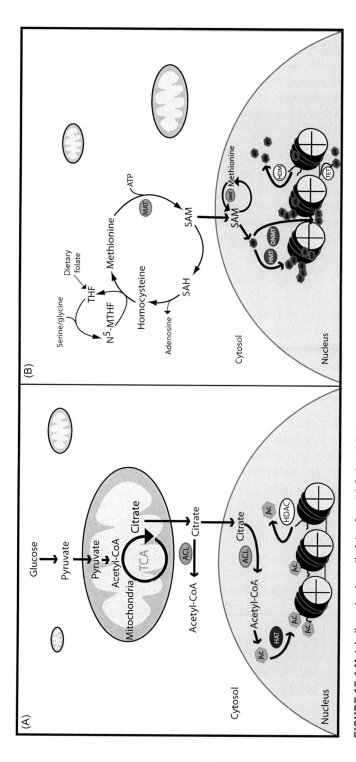

FIGURE 15.1 Metabolic and epigenetic fate of acetyl-CoA and SAM.

(A) In the cytosol, glucose is broken down to pyruvate, and subsequently acetyl-CoA within the mitochondria. Here, it is converted to citrate in the tricarboxylic acid cycle (TCA), which can then be transported out into the cytosol. Here, ATP citrate lyase (ACL) catalyzes the conversion of citrate back to acetyl-CoA. ACL is also in the nucleus, providing a source of acetyl-CoA that is used by histone acetyltransferases (HATs) to acetylate histone N-terminal tails. Acetyl groups are removed by the action of histone deacetylases (HDACs). (B) Methionine adenosyltransferase (MAT) catalyzes the biosynthesis of S-adenosylmethionine (SAM) through condensation of the amino acid methionine and ATP. MAT is located in both the cytosol and nucleus, and can potentially maintain nuclear pools of SAM. In the nucleus, SAM acts as a methyl donor for both histone methyl transferases (HMTs) and DNA methyltransferases (DNMTs). Methyl groups are removed from histones and DNA by histone demethylases (HDMs) and the ten-eleven translocation (TET) enzymes, respectively. After losing a methyl group, S-adenosylhomocysteine (SAH) is recycled back to methionine in a 5-methyl tetrahydrofolate (N^5-MTHF)-dependent reaction, where N^5-MTHF acts as the methyl donor.

and acetyl-CoA during the first step of the TCA cycle. When in excess, citrate is exported into the cytosol via a malate-citrate antiporter system. Here, it is converted back into oxaloacetate and acetyl-CoA in an ATP-driven reaction catalyzed by ATP citrate lyase (ACL) [1]. In the cytosol, acetyl-CoA acts as a carbon donor for anabolic reactions, as well as a substrate for protein PTMs. ACL is also found in the nucleus [5], ensuring the availability of acetyl-CoA in all compartments.

Availability of acetyl-CoA is dynamic, with its intracellular levels fluctuating under different states of nutrient availability. Histone acetylation itself is quite dynamic, with previous studies reporting a modification half-life of as short as 3 min [6]. Although the concept that fluctuations in intermediary metabolism could have a direct effect on gene transcription via epigenetic mechanisms was introduced several years ago [7,8], it was only until recently that proof of concept studies were explored.

Studies in yeast have proven particularly powerful in investigating the relationship between acetyl-CoA availability and histone acetylation. Microarray and ChIP-seq data from yeast supplemented with high acetyl-CoA or precursors of acetyl-CoA (glucose, galactose, ethanol) demonstrated hyperacetylation of histones at ~1000 growth-related genes, such as those coding for ribosomal RNA and protein translation machinery. Deficiency of the acetyltransferase GCN5 abrogated both hyperacetylation of H3 and H4 histones, as well as expression of these growth-related genes [9,10].

Studies in mammalian cells have displayed similar results, providing further evidence that intracellular acetyl-CoA availability has a direct effect on chromatin dynamics and thus transcriptional regulation. In a landmark study, Thompson and coworkers found that ACL, the enzyme capable of converting citrate to oxaloacetate and acetyl-CoA, is not only highly expressed within the nucleus of HeLa cells, but is also required for increased histone acetylation in response to stimulation with growth factors. Deletion of ACL resulted in a marked suppression in conversion of both nuclear and cytosolic acetyl-CoA, corresponding with a decrease in histone acetylation. Interestingly, the authors also observed a specific decrease in the expression of genes involved in the glycolytic pathway upon ACL deletion. This phenotype could be rescued by supplementing with high levels of acetate (5 mM) in culture, presumably via the conversion of acetate to acetyl-CoA by acetyl-CoA synthetase 2 (ACSS2) in the nucleus [11]. In addition to ACL and ACSS, the mitochondrial pyruvate dehydrogenase complex (PDC), which converts glycolysis-derived pyruvate into acetyl-CoA, was also recently identified to be active in the nucleus of mammalian cells. Nuclear PDC was found critical for nuclear acetyl-CoA production, and thus the maintenance of histone acetyl marks [12]. Presence of ACL, ACSS, and the PDC in the nucleus reveals the high level of regulation cells have evolved in order to maintain an acetyl-CoA pool within the nuclear compartment.

Taken together, these studies, along with many others, suggest that under high-energy conditions (and thus high acetyl-CoA availability), a more permissive chromatin configuration is achieved with increased histone acetylation. This allows for the selective increase in expression of genes regulating processes with high-energetic needs, such as those involved in growth and proliferation. How cells selectively control the expression of these specific genes remains poorly understood. Regardless, this relationship between acetyl-CoA and chromatin modification reveals how transcription can be regulated via metabolite availability through epigenetic mechanisms.

15.2.2 Histone Deacetylation and NAD⁺-Dependent Class III HDACs

In contrast to acetylation, deacetylation of histones by histone deacetylases (HDACs) is associated with condensed and compacted chromatin (heterochromatin), and ultimately transcriptional repression of genes [1]. Class III HDACs, otherwise known as sirtuins, are NAD⁺-dependent proteins whose activity is dependent upon the nutrient status of the cell. NAD⁺ is a redox cofactor involved in oxidative pathways, such as glycolysis, the TCA cycle, and fatty acid beta-oxidation. Unlike NAD⁺-independent Class I and Class II HDACs, which utilize zinc to catalyze the hydrolysis of acetyl groups on their acetylated substrates, sirtuins are dependent on NAD⁺ for their deacetylase function, and thus have the potential to act as nutrient sensors, tying histone deacetylation directly to cellular metabolic state.

Sirtuins use NAD⁺ as a cofactor to catalyze the removal of the acetyl group from their lysine targets, yielding O-acetyl-ADP-ribose (OAADPr), nicotinamide, and the deacetylated substrate as products [13,14]. While all seven mammalian sirtuins require NAD⁺ for deacetylase activity, each differs in their biological roles, dependent upon subcellular localization, tissue expression, and substrate specificity [15]. Nuclear sirtuins (SIRT1, 2, 6, and 7) all have HDAC activity, their substrates differing between specific histone–lysine residues. SIRT1 specifically deacetylates histone H3 at K9 and K14, at regions surrounding various genes [16]. SIRT2, on the other hand, plays an important role in global deacetylation of H4K16 during mitosis [17]. SIRT7 specifically deacetylates H3K18 [18], and SIRT6 H3K9 and H3K56, at a number of different loci [19–22].

Because all sirtuins rely on NAD⁺ as an essential cofactor, this begs the question; do intracellular fluctuations of NAD⁺, as a function of metabolic state, affect sirtuin HDAC activity? For years this concept has been up for debate, and still many questions go unanswered. But a body of evidence, most overwhelmingly for SIRT1, suggests that indeed NAD[+] levels do have a direct effect on sirtuin function, and thus tie nutrient status to the chromatin state of the cell.

Cells maintain intracellular NAD[+] pools by balancing two major mechanisms [15]. On one hand, NAD[+] is regulated by its rate of *production*, via biosynthesis from tryptophan or via salvage mechanisms, and the rate of its *consumption* by enzymes, such as sirtuins, PARPs or CD38. On the other hand, NAD[+] levels can also be regulated by *redox conversion* that serves to maintain the NAD[+]/NADH ratio. To date, no mitochondrial NAD[+] transporter has been identified in mammals, suggesting that mitochondrial and cytoplasmic pools are regulated individually [15]. While NAD[+] can move freely between the nucleus and cytosol through nuclear pores, it is possible that these two pools are differentially regulated as well, as individual isoforms of NAD[+] salvage pathway enzymes are located in both the cytoplasm and the nucleus [15].

Certain SIRT1 HDAC functions are regulated in a circadian manner, primarily through NAD[+] availability (see also Chapter 16). Studies in mice reveal intracellular NAD[+] levels to fluctuate in a circadian manner, as BMAL1 and CLOCK directly regulate the expression of nicotinamide phosphoribosyl transferase (NAMPT), the rate-limiting enzyme in the NAD[+] salvage pathway. When treated with a specific NAMPT inhibitor, FK866, NAD[+] oscillations were blunted, and the cyclic deacetylase activity of SIRT1 was lost [16].

Outside of circadian NAD[+] fluctuation, it is still unclear whether NAD[+] fluctuations due to normal metabolic states (such as feeding/fasting) are dramatic enough to have a direct effect on sirtuin activity. Limitations of previous conflicting studies include measurement of whole cell/whole tissue levels of NAD[+], rather than compartment-specific concentrations. Recent work has shown, however, that significant decreases in NAD[+], due to decreased NAD[+] salvage, in certain physiological/pathological conditions, such as aging and aging-induced type II diabetes have a direct effect on SIRT1 function. Nutritional supplementation with nicotinamide mononucleotide, a precursor of NAD[+], was able to reverse the phenotypes seen, by elevating intracellular concentrations of NAD[+], and enhancing SIRT1 deacetylase activity [23–25].

The deacetylase activity of sirtuins increases in times of nutrient stress. NAD[+] has been seen to increase upon calorie restriction (CR) [26], which could be attributed to an increase in AMP-activated protein kinase (AMPK) signaling. AMPK is stimulated in states of low energy, and one of its downstream targets includes NAMPT [27]. By stimulating AMPK signaling in CR, NAMPT expression increases, leading to an increase in cellular NAD[+] salvage. This increase in NAD[+] can then lead to an increase in sirtuin deacetylase activity. Class III HDACs, and their dependence on NAD[+], thus reveal another relationship where chromatin-modifying enzymes are directly dependent on metabolites in order to function.

15.2.3   SAM, Folate, and DNA/Histone Methylation

In contrast to histone acetylation, methylation can occur on both lysine and arginine residues on H3 and H4 N-terminal tails. This covalent modification does not affect charge, but rather acts as a docking site for proteins with specificity for this mark [1]. Methylation is associated with either transcriptional repression or activation, dependent upon, which residues are modified and the number of methyl groups added. DNA methylation, specifically addition of a methyl group to the 5-carbon of cytosines in CpG dinucleotides, is the most predominant epigenetic modification on DNA in vertebrates [28], and is associated with transcriptional repression (see also Chapter 3).

Methylation is directly linked to intermediary metabolism, with SAM acting as the universal methyl donor for all methyltransferases that methylate both histones and DNA (Fig. 15.1). SAM is a product of one-carbon metabolism, and is derived from the amino acid methionine and ATP by methionine adenosyltransferase (MAT) [29]. MAT was recently found in the nucleus [30], suggesting that SAM, just like acetyl-CoA, can be regulated locally within the nuclear compartment. After donating a methyl group, SAM becomes S-adenosylhomocysteine (SAH), which can be subsequently hydrolyzed to homocysteine and adenosine. Homocysteine is recycled back to methionine, accepting a methyl group from N^5-MTHF, a derivative of folate [1,4]. This recycling mechanism is critical for maintaining proper SAM/SAH ratio within the cell, as SAH is a powerful inhibitor of both DNA and histone methyltransferases.

As mentioned, tetrahydrofolate (THF), a derivative of dietary folate (vitamin B9), also acts as a methyl donor when methylated on its N5 atom (N^5-MTHF), but has a much lower transfer potential than SAM. Rather than serving as a methyl donor for histone/DNA methylation reactions, N^5-MTHF indirectly regulates methylation capability by playing critical roles in both the biosynthesis and the recycling of SAM [1].

Recent work in mouse embryonic stem cells (mESCs) revealed the importance of threonine as a source of SAM, in the maintenance of histone methylation throughout pluripotency. Highly proliferating mESCs express high levels of the enzyme threonine dehydrogenase (TDH), which catabolizes threonine to yield glycine and acetyl-CoA. Glycine is used by the mitochondrial glycine cleavage enzyme complex, which converts THF to N^5-MTHF. Inhibition of TDH in mESCs led to a dramatic increase in intracellular threonine and 5-aminoimidazole-4-carboxamide ribonucleotide, coupled with a drastic decrease in N^5-MTHF [31]. Subsequent work discovered that this drop in N^5-MTHF had a dramatic effect on specific histone methylation marks, impacting mESCs' ability to maintain pluripotency. As mentioned previously, N^5-MTHF plays a critical role in the regeneration of methionine from homocysteine, which can then be used to produce SAM. Threonine restriction of mESCs led to a dramatic, selective decrease in H3K4me2 and H3K4me3 [32]. Strikingly, these marks are part of a "bivalent" epigenetic state containing both activating (H3K4me3) and repressive (H3K27me3) histone modifications proposed to keep chromatin structure of developmentally important genes in their poised state [33]. Taken together, this data highlight the importance of threonine catabolism, and ultimately maintenance of SAM pools, to maintain epigenetic marks specific to mESC function [1]. How threonine restriction specifically affects H3K4me2 and H3K4me3 marks remains unclear.

While the threonine study exhibits some of the first evidence that an amino acid can influence histone methylation [32], direct evidence as to whether intracellular levels of SAM and SAH could have a direct effect on histone methylation remained unclear. Locasale and coworkers addressed this very question, establishing that methionine metabolism can directly affect methyltransferase activity by altering SAM levels, and thus methylation status [34]. By depleting the amino acid methionine, either in human cell culture or the diet of mice, drastic decreases in SAM, SAH, and the SAM/SAH ratio were observed. These decreases coincided with decreased histone methylation. Of the marks analyzed, the transcriptional activation-associated histone mark H3K4 tri-methylation (H3K4me3) was most consistently decreased. Intriguingly, decreases in H3K4me3 were enriched at the promoters of key one-carbon metabolism genes that play roles in utilizing SAM. This suggests a potential feedback mechanism, whereby cells can use histone methylation as a sensor to maintain one-carbon metabolism homeostasis. Changes in SAM, SAH, and histone methylation levels were reversed upon methionine supplementation [34]. This work provides clear evidence that methionine availability (and thus SAM and SAH), much like acetyl-CoA, has a direct effect on histone methylation status. Cellular dependence on methionine, threonine, SAM, and folate metabolites for maintenance of methylation marks reveals yet another link between intermediary metabolism, nutrition, and chromatin state [4].

15.2.4   Alpha-Ketoglutarate (2-Oxoglutarate)-Dependent Dioxygenases and Demethylation

Histone and DNA demethylation are processes facilitated by enzymes that, much like acetyltransferases, deacetylases, and methyltransferases, are reliant upon metabolites as their cofactors/substrates. In a landmark paper, Shi and coworkers discovered that histone methylation, just like acetylation, can be a dynamic and reversible process by identifying the first histone demethylase, LSD1 [35]. Removal of the histone methylation mark is facilitated through an oxidative reaction, using FAD as a cofactor [36]. FAD is produced in the mitochondria and like NAD^+, is a redox cofactor that exists in a balance between its oxidized and reduced forms ($FAD/FADH_2$). This dependence on FAD suggests that LSD1-mediated histone demethylation may also be dependent upon the energetic state of the cell.

In addition to LSD1, other demethylating enzymes were later discovered, including the Jumonji C (JmjC)-domain-containing proteins and the ten-eleven translocation (TET) proteins. While JmjC-domain-containing proteins facilitate the demethylation of histones, the TET proteins facilitate the oxidation and eventual demethylation of methylated cytosines [1]. These enzymes are dioxygenases dependent upon Fe^{2+} and alpha-KG (also known as 2-oxoglutarate) to facilitate oxidative removal of methyl marks [37].

Alpha-KG is an intermediate of the TCA cycle, produced from isocitrate in the mitochondria by isocitrate dehydrogenases 2 and 3 (IDH2/3). IDH1, a third paralog of this enzyme, resides in the cytosol and peroxisomes, offering an alternative source of alpha-KG in the cell. Other amino acids can act as sources of alpha-KG, including arginine, glutamine, histidine, and proline [1].

Fumarate and succinate, intermediates of the TCA cycle downstream of alpha-KG, can inhibit TET and JmjC-containing enzymes [38], suggesting that relative concentration of these metabolites may regulate TET/JmjC demethylase activity. While it has been reported that intracellular levels of alpha-KG are in the low millimolar range, substantially above the Km values of JmjC and TET proteins, these values were measured under in vitro conditions with purified proteins, lacking endogenous inhibitor molecules like fumarate and succinate, and without control for varying alpha-KG levels in different subcellular compartments [39,40]. Because of these limitations, it is still possible that changes in concentrations of metabolic agonists and antagonists could indeed affect demethylase activity. Indirect evidence suggests this may be the case, as in pathological conditions of altered metabolic enzyme activity/metabolite availability, the functions of these demethylating enzymes are affected (see Section 15.3).

15.2.5 Other Chromatin Modifications Dependent on Metabolites

In addition to acetylation and methylation, there are a number of other histone PTMs that rely upon metabolites, including but not limited to phosphorylation (ATP), GlycNAcylation (O-linked N-acetylglucosamine/O-GlcNAc), and a variety of acylations including the more recently identified crotonylation (crotonyl-CoA) [41]. These modifications are dependent upon metabolites produced in the cell and have the ability to be quite dynamic, depending on substrate availability. This section will discuss this dynamic nature, and how certain metabolites have the potential to out-compete others for particular PTM sites/residues.

Modification by O-GlycNAcylation is unique, as by nature it is more similar to phosphorylation, rather than classical protein glycosylation. Addition of O-GlcNAc is highly dynamic, and occurs on a variety of proteins within the cytosol and nucleus at serine and threonine residues, just like phosphorylation. The highly dynamic addition and removal of O-GlcNAc is facilitated by an opposing pair of enzymes: O-GlcNAc transferase (OGT) and O-GlcNAcase (OGA) [42], similar to kinases and phosphatases. O-GlcNAc has been identified on all four core histones, on residues that can also be alternatively modified by phosphorylation [43,44].

Recent work has revealed that changes in glucose levels have a direct effect on the rate of glycoslyation. O-GlcNAc is a product of the hexosamine biosynthetic pathway, a subsidiary pathway of glycolysis that generates a key metabolite, UDP-GlcNAc. In addition to glucose, glutamine plays an essential role in O-GlcNAc production as well, acting as a nitrogen donor [4,45]. Future studies comparing the rate of phosphorylation and O-GlcNAcylation of particular residues on histones, and how relative intracellular levels of ATP and O-GlcNAc affects these rates will likely reveal critical information on chromatin dynamics, and how differential modification of particular residues may affect chromatin structure and thereby gene transcription.

While distinct parallels can be drawn between phosphorylation and O-GlcNAcylation of serine and threonine residues on histone tails, crotonylation can be easily compared to acetylation, as both are acylation events that occur on lysine residues of histones. Tandem mass spectrometry proteomic analyses have identified other histone lysine acylations in addition to acetylation, including butyrylation, propionylation, 2-hydroxisobutyrylation, crotonylation, malonylation, and succinylation [46–49]. ChIP-seq analysis revealed that crotonylation, in addition to 2-hydroxisobutyrylation, is located on regulatory elements of actively transcribed regions of the genome [47,48]. These studies suggest that differential acylation of histones might play a distinct role in regulating chromatin dynamics, and ultimately transcriptional regulation. With the exception of crotonylation, the physiological role of the other acyl modifications remains to be determined.

Extensive investigation and characterization of in vitro crotonylation of histones revealed an important observation relating the availability of metabolites directly to histone modifications, which affects gene transcription [41]. Hypothesizing that availability of differential acyl groups may have an effect on histone modifications, Allis and coworkers found that by modulating intracellular levels of acetyl-CoA versus crotonyl-CoA, differential histone acylation was achieved. Addition of sodium crotonate to culture media of HeLa S3 cells caused an increase in panKCr levels on core histones, as well as increased gene expression of particular genes. This increase in gene expression correlated with an increase in histone crotonylation flanking the regulatory elements of these genes. Crotonylation was found to be dependent on p300, a classic histone acetyltransferase, revealing differential histone acylation capability of this enzyme. Intriguingly, knockdown of ACSS2 attenuated the observed increase in histone crotonylation when supplemented with sodium crotonate, implicating this enzyme in the production of crotonyl-CoA from crotonate, in addition to acetyl-CoA from acetate in the nucleus [41].

Knockdown of ACL strongly reduced H3K18ac (as seen in Ref. [11]), and increased H3K18Cr. This effect was reversed when acetate was supplemented in the media [41]. Taken together, these findings suggest that acetylation and crotonylation have the ability to compete for the same residues on histones. This calls into importance the relative intracellular crotonyl-CoA and acetyl-CoA levels, as availability of these two metabolites could have varying effects on gene expression. Indeed, crotonylation of regions flanking actively transcribed genes caused a more robust activation of genes, when compared to their acetylated form [41].

This study illustrates an elegant mechanism by which availability of certain metabolites directly translates into differential chromatin modification, corresponding to a distinct functional output. Importantly, the physiological relevance of this differential acylation event is unknown, as this work was conducted in cell culture. Yet these observations raise many questions, particularly when one directly compares acetylation to crotonylation. With both being acylation events that are facilitated by the same acyltransferase enzyme (p300), one can speculate that perhaps this modification may be removed by the same, or similar, enzymes, such as HDACs like the sirtuins. It has been established that sirtuins have deacylation capability of a number of acyl groups other than acetate [50,51], making nuclear sirtuins (SIRT1, 2, 6, 7) likely candidates for decrotonylation of histones. In this context, it is tempting to speculate that similarly to the competition between these two modifications, other yet to be defined modifications could also be competing for the same moieties, providing further layers of gene regulation by metabolites.

15.3 METABOLISM, CHROMATIN, AND DISEASE

More and more research has found that aberrant changes in chromatin play a powerful role in disease, in particular cancer. Recent work profiling the "cancer epigenome" identified a vast array of altered chromatin marks, which can play roles in regulating gene expression and genome stability [52]. As discussed in this chapter, a building body of evidence argues that changes in metabolism/activity of metabolic enzymes have the potential to facilitate changes in chromatin marks (Figs. 15.2 and 15.3). Strikingly, recent studies have suggested that particular metabolites, and changes in their relative concentration, can alter the properties of tumors, not only through energetics, but epigenetics

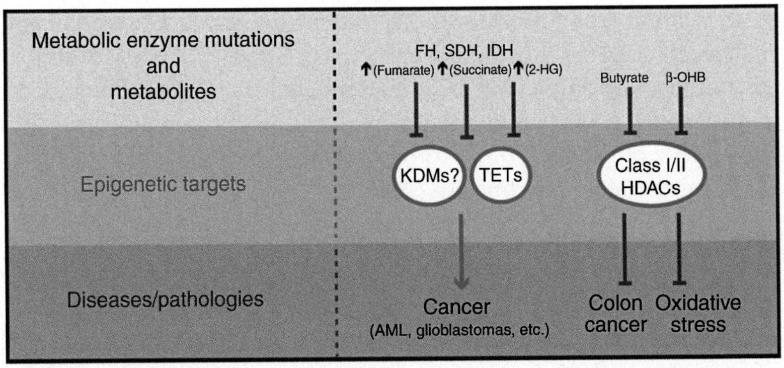

FIGURE 15.2 Relationship between metabolism, chromatin, and disease.
Mutations in the metabolic enzymes fumarate hydratase (FH), succinate dehydrogenase (SDH), and isocitrate dehydrogenase (IDH) lead to an accumulation of the metabolites fumarate, succinate, and 2-hydroxyglutarate (2-HG) within the cell. By competing with alpha-ketoglutarate (alpha-KG), these metabolites inhibit alpha-KG-dependent enzymes, such as the TETs and histone lysine demethylases (KDMs), altering both DNA and histone methylation patterns. This change in methylation can lead to a number of pathologies, including AML and glioblastoma. At sufficiently high levels, short-chain fatty acid butyrate and ketone body beta-hydroxybutyrate (β-OHB) can inhibit Class I and II HDAC deacetylase activity. Increases in butyrate and β-OHB have been found to protect against colon cancer and oxidative stress, respectively.

as well [38,53–56]. Metabolites harboring oncogenic and tumor suppressive roles have been identified, and will be discussed in the following section.

15.3.1 Metabolic and Chromatin-Modifying Enzyme Mutations, Metabolism and Cancer

Mutations in the TCA cycle enzymes IDH, succinate dehydrogenase (SDH), and fumarate hydratase (FH) have all been identified in a number of cancers [38,53,57,58]. These mutations lead to a buildup of metabolites that have the potential to prime cells for malignant transformation via inhibition of alpha-KG-dependent dioxygenases, altering the epigenetic landscape of cells. These mutations describe another side of metabolic regulation of epigenetics, where abnormal changes in metabolite concentration can play a role in disease (Fig. 15.2).

Hot spots for somatic mutations in IDH1 and 2 have been identified in a variety of cancers, including glioblastoma and acute myeloid leukemia (AML) [53]. These gain of function mutations cause IDH to have a neomorphic enzymatic capability, producing 2-hydroxyglutarate (2-HG) from alpha-KG, the normal product of IDH. 2-HG acts as a competitive inhibitor of alpha-KG in the cell, affecting the activity of alpha-KG-dependent enzymes, such as the dioxygenases which facilitate histone and DNA methylation (JmJC-containing demethylases and TET proteins) (Fig. 15.3). Tumors positive for this mutation exhibit an

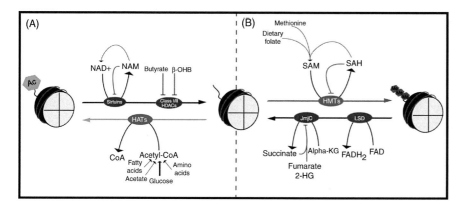

FIGURE 15.3 Dynamic acetylation/methylation of histones and key metabolites.
(A) HDACs facilitate the removal of acetyl groups from histone-lysine tails in nicotinamide adenine dinucleotide *(NAD)*+-independent (Class I/II) and NAD+-dependent (Sirtuins) manners. Class I/II HDAC enzymatic activity is inhibited by the short-chain fatty acid butyrate and ketone body β-OHB. Histone acetylation is catalyzed by HATs, which use the metabolite acetyl-CoA as a substrate. Acetyl-CoA is produced from a variety of metabolic means, including the breakdown of glucose, fatty acids, and amino acids and (B) methylation of histones is catalyzed by HMTs, using SAM as a methyl donor, producing SAH as a product. SAM pools are maintained through recycling of SAH, as well as biosynthesis from methionine and dietary folate. Histone demethylation is catalyzed by histone demethylases (HDMs) including jumonji-C *(JmjC)* domain-containing proteins, and lysine-specific demethylase 1 *(LSD1)*, which use alpha-KG and flavin adenine dinucleotide *(FAD)* as cofactors, respectively.

accumulation of this metabolite [57]. 2-HG has been termed an "oncometabolite," as it has been proposed that its ability to inhibit demethylase activity contributes to oncogenesis.

Inactivating mutations in SDH and FH have also been identified in a number of cancers, including familial paraganglioma and papillary renal cancers, respectively [58]. These mutations lead to an accumulation of succinate and fumarate which, as mentioned previously, can act as inhibitors of alpha-KG-dependent enzymes, such as the JmJC-containing demethylases and TET proteins [38].

Future work to identify the specific chromatin-modifying enzymes that are affected by these changes in metabolite concentration will prove critical for potential cancer therapies. In this context, histone demethylases KDM6A and KDM5C, as well as TET2 have been proposed as putative targets, as they have all been found to be inactivated in cancers [1]. A building body of evidence suggests that TET2 may be the likely candidate. IDH-mutant tumors exhibit a DNA hypermethylation phenotype, consistent with the loss of function of TET2. Mutually exclusive TET2 and IDH mutations have been identified in AML, consistent with the idea that 2-HG-mediated inhibition of demethylating enzymes can provide an alternative means of inactivating TET2 in cancer cells [54]. TET2 inactivation was found to be sufficient to promote hematopoietic stem cell self-renewal and block differentiation [54,55], suggesting that inhibition of TET2 by oncometabolites would facilitate the same effect, priming cells for transformation.

While metabolites like 2-HG, succinate, and fumarate can inhibit the activity of chromatin modifiers and thus contribute to oncogenesis, direct mutation of chromatin factors that regulate the expression of metabolic genes has also been found to contribute to cancer risk. The histone deacetylase SIRT6, which acts as a potent tumor suppressor by regulating glycolysis [59], is a powerful example. Most cancer cells adopt a unique metabolic signature in order to satisfy their energetic needs for rapid proliferation and growth. By increasing glucose uptake and decreasing oxidative metabolism, cancer cells utilize glycolytic intermediates to produce macromolecules that are used as building blocks during rapid cell division. This increase in glycolytic metabolism, despite oxygen availability, is termed "aerobic glycolysis," or the Warburg effect [60]. SIRT6 acts as a tumor suppressor, at least in the context of colorectal cancer, by regulating the Warburg effect. SIRT6 deacetylates H3K9 at the promoters of HIF1-alpha target genes, including those involved in glycolysis and glucose uptake, and represses their transcription. When SIRT6 is lost, H3K9 acetylation increases, corresponding with an increase in expression of these genes. This increase in glycolytic capacity allows for the transformation of cells and was observed to contribute to tumor number, size, and aggression in a model of colon cancer [59].

In the context of histone acetylation defects, recent studies have also demonstrated that in particular tumors, free acetate is an important metabolite for histone acetylation (Fig. 15.3). Here, the nucleocytosolic ACSS2, which converts acetate to acetyl-CoA, appears to act as a strong oncogene, providing a pool of acetyl-CoA that is used for both histone acetylation and lipid synthesis in tumor cells [56,61].

These examples shine a light on the clear connection between chromatin, metabolism and cancer, as mutations in metabolic enzymes, chromatin-modifying enzymes and altered metabolite pools all play a role in deregulating the epigenome of tumor cells.

15.3.2　Butyrate- and Beta-Hydroxybutyrate-Mediated HDAC Inhibition in Health and Disease

Butyrate and beta-hydroxybutyrate (β-OHB) are metabolites that serve both as carbon sources for energy, as well as functional metabolites with HDAC inhibitory capability (Fig. 15.3). Over the years evidence has accumulated showing that this regulatory function plays important protective roles against several diseases. This section will discuss these findings, offering yet another example where metabolites, through epigenetic means, can have direct consequences on cellular functions.

Butyrate is a short-chain fatty acid produced at high concentrations in the colon, via microbiota-mediated metabolism of dietary fiber. It is the main energy source of colonocytes, which oxidize butyrate to acetyl-CoA [62]. Over the years, nutritional studies have found that diets rich in fiber help prevent colitis and colon cancer in many organisms, including humans [63]. Until recently, reasons as to why this is the case have remained elusive.

Butyrate is a potent inhibitor of class I/II HDACs (Fig. 15.3), with a recorded IC_{50} of 90 µM in HT-29-colon carcinoma-derived cells [64]. Recent work in mice revealed that butyrate itself, via HDAC inhibition, can regulate transcriptional programs involved in proliferation of colon cells, acting as a protective agent against colonocyte transformation/hyperproliferation [65,66]. Butyrate levels exist in a gradient within the proximal colon, ranging from 5 mM in the epithelial wall, and diffusing down to 0.5 mM within the colon crypts. At such high levels (5 mM), butyrate inhibits the activity of class I/II HDACs, causing an increase in histone H3 lysine acetylation specifically at elements regulating transcriptional programs that *decrease* proliferation. In contrast, in vitro exposure to 10-fold less butyrate (0.5 mM), does not elicit this HDAC inhibitory effect. Instead, radio-tracer experiments in cells exposed to this lower concentration of butyrate revealed that acetyl-CoA derived from butyrate was used in the acetylation of histones associated with a different subset of genes, involved in *promoting* cell proliferation. Taken together, the results suggested a functional role for this butyrate gradient, arguing that butyrate diffusion from the epithelial wall down to the crypt allows for chromatin-mediated transcriptional control that supports quiescence of more differentiated luminal cells in the epithelium, while supporting active proliferation of progenitors within intestinal crypts [65]. This regulatory function of butyrate could act as a protective factor against colon malignancies, inhibiting aberrant hyperproliferation of the intestinal epithelium. In a follow-up study using a mouse model of colon cancer, it was determined that microbiota-derived butyrate from dietary fiber indeed had a tumor suppressive effect. Butyrate-mediated inhibition of HDAC activity led to an increase in histone acetylation at genes that drive apoptosis, decreasing tumor burden and grade [66]. These mouse studies provide evidence that increased dietary fiber protects against colon malignancies, and suggest that the mechanism of protection is related to metabolite-driven chromatin dynamics and transcriptional control. A diet rich in fiber, therefore could be used to prevent colon malignancies.

β-OHB, a ketone body with structural similarity to butyrate, also inhibits HDACs (Fig. 15.3). It has been found to have protective effects within the body, particularly against oxidative stress. Ketone bodies are produced in times of nutrient stress, when glucose levels are low [67]. These include conditions of fasting, CR, strenuous exercise, and ketogenic diets, which have the ability to increase serum β-OHB levels from the micromolar to millimolar range [4]. Ketone bodies serve as an alternative source of energy in peripheral tissues, particularly neurons, in order to maintain ATP production. Studies from over 10 years ago have found ketogenic diets to have a protective effect in models of degenerative neurological disorders, such as Alzheimer's and Parkinson's disease, by increasing neuronal survival [68]. Similar results were seen under states of CR, through a marked reduction in reactive oxygen species production within neurons [69].

How this functional metabolite facilitates the reduction in oxidative stress can be attributed to a number of mechanisms, including chromatin dynamics. For instance, in a kidney model, elevated levels of β-OHB was able to inhibit HDAC 1, 3, and 4, causing an increase in acetylation of H3K9 and H3K14 at key elements of the oxidative damage response. This increase in acetylation was associated with an increase in expression of these genes, including Foxo3a and metallothionein Mt2 [70]. Here, increased production of the ketone body β-OHB under states of nutrient stress plays a dual role: to produce an alternative energy source to maintain metabolic homeostasis, as well as a functional molecule that facilitates transcriptional programs associated with reducing oxidative stress under times of nutrient stress.

15.4 CROSSTALK BETWEEN METABOLISM AND CHROMATIN: A TWO-WAY STREET?

It is clear that metabolite availability can have a direct effect on the rate of chromatin modifications. Another intriguing question is whether the nucleus, with its wealth of PTMs, could act as a nutrient reservoir to support cellular metabolism in times of nutrient stress? The compelling idea, that crosstalk between chromatin and cellular metabolism can go both ways, has yet to be addressed within the field.

Two major questions have been formulated to tackle this complex issue: first, it will be essential to determine if the nucleus could act as a source of metabolites (reservoir), which are mobilized in order to maintain normal metabolic function/cell survival in times of metabolic need [15]. Second, it will be important to explore whether retention of metabolite moieties on chromatin could, under aberrant conditions of chromatin-modifying activity, affect the availability of metabolites needed to retain normal cellular metabolism [71].

DNA and histones carry an exceedingly high number of putative sites for acetylation and methylation, making the nucleus a highly attractive "metabolite sink" that could retain/provide metabolites in times of need [56,71]. While the putative turnover rate for acetyl groups as PTMs is around 10-fold slower than the rate of metabolic acetyl-CoA production, in times of metabolic stress, the nucleus could theoretically act as a buffer to maintain basal cellular metabolism [15].

In order to investigate this question, a number of factors must be addressed. First, the metabolic fate of these modifications, once removed, must be determined. It is known that acetate, the product of deacetylation of histones by class I/II HDACs, is converted back to acetyl-CoA, under the action of ACSS. Class III HDACs (sirtuins), on the other hand, remove acetyl groups from histones through an NAD^+-dependent reaction, yielding OAADPr as a product. It is currently unknown what role this product plays within the cell. An attractive hypothesis suggests that OAADPr, like acetyl-CoA, can act as an activated acetyl donor, playing a role in cellular metabolism [15].

A second major question pertains to the identification of the physiological stress or stresses that would trigger such a response. A putative stressor includes severe nutrient depletion, where acetyl-CoA production from nutrients, such as glucose and lipids are compromised. Interestingly, a recent study found that cells, under conditions of low pH, have the ability to mobilize acetate from histones, in order to regulate intracellular pH [72]. This mobilization was HDAC dependent, allowing cosecretion of acetate and protons in order to prevent acidification of the intracellular space [72]. This study provides a clear example where histone modifications can be actively removed, in times of stress, in order to maintain normal function of the cell. Future studies will help determine whether mobilization of metabolites from the nucleus indeed plays an important role in sustaining metabolic homeostasis.

15.5 NUTRIENT-INDUCED EPIGENETIC INHERITANCE

Up until now, this chapter has discussed highly dynamic chromatin changes that occur based on metabolic state and metabolite availability. In addition to the dynamic flexibility of chromatin and its PTMs, a growing body of evidence has amassed suggesting that stably inherited epigenetic modifications, accrued through environmental triggers, such as nutrition, occur and play an important role in human health and disease. Nutrient-induced epigenetic inheritance has been associated with a number of traits, including lifespan, diabetes, and obesity [4].

During gametogenesis the chromatin landscape is globally reprogrammed, removing changes that have accumulated over time [73]. This process is necessary for proper development, but some portions resist this reprogramming, and can be transmitted to the next generation(s) [74]. This inheritance of epigenetic marks can occur in two distinct fashions: intergenerational (inherited directly from parent to offspring) and transgenerational (inherited across multiple generations, where progeny display specific epigenetic signatures without exposure to the initial stimulus) [75]. While both phenomena are equally important and meaningful to the field of epigenetic inheritance, it is critical to distinguish between these two modes of inheritance in order to better understand the mechanisms behind these events.

The study of nutritional epigenetic inheritance has proven quite challenging over the years, as most studies are correlational with minimal mechanistic insight. Nevertheless, the field has been growing, with a number of proof-of-principle studies in model organisms providing evidence that transmission of disease risk by epigenetic mechanisms from parents to offspring over a number of generations does indeed occur. Epidemiological studies on particular human populations have suggested human relevance, stressing the importance of determining the clear mechanisms behind this process. It is currently believed that small RNAs, histone modifications, and DNA methylation may all play an important role [76].

Transgenerational epigenetic inheritance of longevity was first observed in 2011 by Brunet and coworkers, in *Caenorhabditis elegans*. Paternal deficiency in members of the H3K4me3 complex (WDR-5, ASH-2, and histone methyltransferase SET-2) caused an increase in longevity that was passed on to the F3 generation [77]. Deficiencies in these chromatin modifiers were accompanied by both epigenetic and transcriptional changes [77]. Interestingly, in 2014 a separate group observed that starvation induced an increased lifespan phenotype that lasted for 3 generations as well [78]. Worms starved for 6 consecutive days yielded F3 progeny with a 37% increase in lifespan relative to F3 offspring of normal-fed worms. The small RNA profiles of these longer lived F3 progeny were more similar to that of the starved F0 population, versus the fed F0 population. These small RNAs were enriched for targets involved in nutrient reservoir activity [78]. In contrast, worms fed a glucose-enriched diet yielded fewer progeny, and these progeny displayed a shorter lifespan phenotype relative to offspring of control-fed worms [79]. These studies reveal true transgenerational inheritance, as the stimulus-induced phenotype was inherited across multiple generations, even after the stimulus was removed. Whether or not the methylation phenotype and this nutrient-dependent phenotype are related has yet to be determined.

A number of rodent studies have also proven quite powerful, where both intergenerational and transgenerational inheritance of traits have been observed. Offspring of male mice fed a low-protein diet display a high-cholesterol phenotype, corresponding with changes in gene expression of particular lipid and cholesterol biosynthesis genes in the liver. This is associated with a modest change (20%) in DNA methylation at particular genes, such as *PPAR-alpha* [80]. In rats, chronic paternal feeding of high-fat diet (HFD) resulted in female offspring with impaired glucose tolerance and insulin sensitivity, due to altered expression of pancreatic islet genes associated with glucose metabolism [81].

While the former two studies display intergenerational transmission of traits, transgenerational inheritance has also been observed. Grandpaternal feeding with HFD caused an increase in adiposity of male and female offspring in the F2 generation (two generations after the HFD stimulus) [82]. A global reduction in DNA methylation, in addition to altered mRNA/miRNA expression in sperm, was associated with grandpaternal HFD feeding [82]. Insulin resistant male mice (through HFD and low-dose stretozotocin treatment) yield male F2 offspring with decreased insulin sensitivity and glucose tolerance. Changes in DNA methylation were consistently found at insulin signaling genes in the pancreatic islets of the F1 and F2 progeny [83]. These findings suggest a potential epigenetic mechanism explaining transgenerational transmission of this insulin-resistant phenotype. However, it remains to be determined whether such epigenetic changes play causative and mechanistic roles in these phenotypes.

Human epidemiological studies have also proven quite useful in studying events of under- or overnourishment, and how these events increase risk for metabolic disease in subsequent generations [84–86]. Data collected on those who experienced the Dutch famine of 1944 show that grandchildren of women exposed to famine during pregnancy exhibited increased neonatal adiposity. Additionally, offspring of prenatally undernourished fathers displayed increased obesity than those born from fathers who had not been undernourished prenatally [84,85]. Data collected on individuals in the Överkalix region of Sweden revealed that grandsons of grandfathers with access to excess food during prepubertal development had increased risk for cardiovascular disease and diabetes, as compared to those whose grandfathers were exposed to low food availability [86].

A major caveat of these studies and others like these is that they are strictly observational, and do not include any mechanistic link between disease risk and changes in chromatin. Additionally, other factors, such as postnatal nutrition, cannot be accounted for. A more recent study found that humans conceived under famine conditions display a decrease in DNA methylation at specific genes, including *IGF2*. This gene has been associated with increased risk for metabolic syndromes, such as dyslipidemia, insulin resistance, and obesity in later life [87]. Although, not mentioned in that study, it will be interesting to determine whether folate deficiency could account for such changes in DNA methylation. More studies aimed to identify the key nutritional, epigenetic, and genetic players in nutritional epigenetic inheritance are crucial toward a better understanding this phenomenon.

15.6 CONCLUSIONS, UNANSWERED QUESTIONS, AND FUTURE PERSPECTIVES

The cross talk between chromatin and cellular metabolism represents a unique mechanism that allows cells to rapidly adapt to their nutritional environment in order to make crucial cell fate decisions. By utilizing key metabolites, such as ATP, acetyl-CoA, SAM, NAD+, FAD and alpha-KG as substrates and cofactors, chromatin-modifying enzymes become dependent upon cellular metabolism, as their activity is susceptible to fluctuations in nutrient availability.

Studying the relationship between metabolism and chromatin dynamics is a rapidly growing field, with many questions still unanswered. Current work has shown that chromatin modifications, such as acetylation, can be quite dynamic depending upon substrate (metabolite) availability within the cell. Meticulous characterization of levels of these metabolites in particular subcellular compartments/tissues, how these levels fluctuate, and determination of whether these fluctuations are indeed limiting for enzyme activity in vivo, is of utmost importance. To address these questions, techniques, such as metabolomics and flux analysis with metabolic tracers, paired with genome-wide chromatin studies like ChIP-seq can be used to determine compartment-specific metabolite levels, and correlate these with chromatin state.

More generally, studies have yet to determine the direct effect changes in diet, such as CR, have on the chromatin landscape, and ultimately gene transcription. The decline in metabolic health has become a serious public health issue, particularly within the Western world. While the idea is still quite controversial, an interesting hypothesis has been brought to the forefront: could transgenerational inheritance of metabolic risk be a contributing factor to the growing metabolic syndrome/obesity epidemic? If this is the case, the increasing rate of obesity in the world may cause "epigenetic drift" (the progressive accumulation of transgenerational epigenetic changes), predisposing our future generations for metabolic disease [4,76]. Therefore, it is crucial to have a better understanding of how the diet and nutritional environment play a role in molding our epigenomes, and how this places individuals at risk for disease. New avenues of prevention and therapy could then be used, including certain dietary interventions before or during development. Particular epigenetic-modifying enzymes could also prove to be useful drug targets, as there are already histone modifier drugs currently in clinical trials for a number of diseases [88].

There is still much to be answered about the crosstalk between metabolism and chromatin dynamics, making it an actively growing field with many avenues to pursue. Further exploration into this compelling topic will help cement our understanding of this long-ignored relationship, with the potential to target it in order to prevent metabolic diseases and cancer.

Abbreviations

ACL	ATP citrate lyase
ACSS	Acetyl-CoA synthetase
Alpha-KG	Alpha-ketoglutarate
AML	Acute myeloid leukemia
AMP	Adenosine monophosphate
AMPK	AMP-activated protein kinase
ATP	Adenosine triphosphate
β-OHB	Beta-hydroxybutyrate
ChIP-seq	Chromatin immunoprecipitation-sequencing
CoA	Coenzyme A
CR	Calorie restriction
DNA	Deoxyribonucleic acid
FAD	Flavin adenine dinucleotide
FH	Fumarate hydratase
HAT	Histone acetyl-transferase
HDAC	Histone deacetylase
HFD	High-fat diet
HIF1-alpha	Hypoxia-inducible factor 1-alpha
2-HG	2-hydroxyglutarate
IDH (1,2,3)	Isocitrate dehydrogenase (1,2,3)
JmjC	Jumonji C
LSD1	Lysine-specific demethylase 1
MAT	Methionine adenosyltransferase
mESC	Mouse embryonic stem cell
miRNA	Micro-RNA
mRNA	Messenger RNA
NAD	Nicotinamide adenine dinucleotide
NAM	Nicotinamide
NAMPT	Nicotinamide phosphoribosyl transferase
NMN	Nicotinamide mononucleotide
OAADPr	O-acetyl-ADP-ribose
O-GlcNAc	O-linked N-acetylglucosamine
PDC	Pyruvate dehydrogenase complex
PPAR-alpha	Peroxisome proliferator-activated receptor alpha
PTM	Posttranslational modification
RNA	Ribonucleic acid
SAH	S-adenosylhomocysteine
SAM	S-adenosylmethionine
SDH	Succinate dehydrogenase
TCA	Tricarboxylic acid
TDH	Threonine dehydrogenase
TET	Ten-eleven translocation
THF	Tetrahydrofolate
UDP	Uridine diphosphate

Acknowledgments

We would like to thank the Mostoslavsky laboratory for invaluable discussions. Research in the Mostoslavsky laboratory is funded by the NIH. R.M. is a Kristin and Bob Higgins MGH Research Scholar.

References

[1] Kaelin WGJ, McKnight SL. Influence of metabolism on epigenetics and disease. Cell 2013;153(1):56–69.

[2] Lee KK, Workman JL. Histone acetyltransferase complexes: one size doesn't fit all. Nat Rev Mol Cell Biol 2007;(4):284–95.

[3] Shahbazian MD, Grunstein M. Functions of site-specific histone acetylation and deacetylation. Annu Rev Biochem 2007;76:75–100.

[4] Gut P, Verdin E. The nexus of chromatin regulation and intermediary metabolism. Nature 2013;502(7472):489–98.

[5] Wellen KE, Thompson CB. A two-way street: reciprocal regulation of metabolism and signalling. Nat Rev Mol Cell Biol 2012;13(4):270–6.

[6] Waterborg JH. Dynamics of histone acetylation in vivo. A function for acetylation turnover? Biochem Cell Biol 2002;80(3):363–78.

[7] Ladurner AG. Rheostat control of gene expression by metabolites. Mol Cell 2006;24(1):1–11.

[8] Shi Y, Shi Y. Metabolic enzymes and coenzymes in transcription—a direct link between metabolism and transcription? Trends Genet 2004;20(9):445–52.

[9] Tu BP, Kudlicki A, Rowicka M, McKnight SL. Logic of the yeast metabolic cycle: temporal compartmentalization of cellular processes. Science 2005;310(5751):1152–8.

[10] Cai L, Tu BP. On acetyl-CoA as a gauge of cellular metabolic state. Cold Spring Harb Symp Quant Biol 2011;76:195–202.

[11] Wellen KE, Hatzivassiliou G, Sachdeva UM, Bui TV, Cross JR, Thompson CB. ATP-citrate lyase links cellular metabolism to histone acetylation. Science 2009;324(5930):1076–80.

[12] Sutendra G, Kinnaird A, Dromparis P, Paulin R, Stenson TH, Haromy A, et al. A nuclear pyruvate dehydrogenase complex is important for the generation of acetyl-CoA and histone acetylation. Cell 2014;158(1):84–97.

[13] Denu JM. The Sir 2 family of protein deacetylases. Curr Opin Chem Biol 2005;9(5):431–40.

[14] Feldman JL, Dittenhafer-Reed KE, Denu JM. Sirtuin catalysis and regulation. J Biol Chem 2012;287(51):42419–27.

[15] Fan J, Krautkramer KA, Feldman JL, Denu JM. Metabolic regulation of histone post-translational modifications. ACS Chem Biol 2015;10(1):95–108.

[16] Nakahata Y, Sahar S, Astarita G, Kaluzova M, Sassone-Corsi P. Circadian control of the NAD+ salvage pathway by CLOCK-SIRT1. Science 2009;324(5927):654–7.

[17] Vaquero A, Scher MB, Lee DH, Sutton A, Cheng HL, Alt FW, et al. SirT2 is a histone deacetylase with preference for histone H4 Lys 16 during mitosis. Genes Dev 2006;20(10):1256–61.

[18] Barber MF, Michishita-Kioi E, Xi Y, Tasselli L, Kioi M, Moqtaderi Z, et al. SIRT7 links H3K18 deacetylation to maintenance of oncogenic transformation. Nature 2012;487(7405):114–8.

[19] Michishita E, McCord RA, Berber E, Kioi M, Padilla-Nash H, Damian M, et al. SIRT6 is a histone H3 lysine 9 deacetylase that modulates telomeric chromatin. Nature 2008;452(7186):492–6.

[20] Kawahara TL, Michishita E, Adler AS, Damian M, Berber E, Lin M, et al. SIRT6 links his tone H3 lysine 9 deacetylation to NF-kappaB-dependent gene expression and organismal life span. Cell 2009;136(1):62–74.

[21] Zhong L, D'Urso A, Toiber D, Sebastian C, Henry RE, Vadysirisack DD, et al. The histone deacetylase Sirt6 regulates glucose homeostasis via Hif1alpha. Cell 2010;140(2):280–93.

[22] Kugel S, Mostoslavsky R. Chromatin and beyond: the multitasking roles for SIRT6. Trends Biochem Sci 2014;39(2):72–81.

[23] Gomes AP, Price NL, Ling AJ, Moslehi JJ, Montgomery MK, Rajman L, et al. Declining NAD(+) induces a pseudohypoxic state disrupting nuclear-mitochondrial communication during aging. Cell 2013;155(7):1624–38.

[24] Yoshino J, Mills KF, Yoon MJ, Imai S. Nicotinamide mononucleotide, a key NAD(+) intermediate, treats the pathophysiology of diet- and age-induced diabetes in mice. Cell Metab 2011;14(4):528–36.

[25] Yoon MJ, Yoshida M, Johnson S, Takikawa A, Usui I, Tobe K, et al. SIRT1-mediated eNAMPT secretion from adipose tissue regulates hypothalamic NAD+ and function in mice. Cell Metab 2015;21(5):706–17.

[26] Canto C, Auwerx J. NAD+ as a signaling molecule modulating metabolism. Cold Spring Harb Symp Quant Biol 2011;76:291–8.

[27] Ho C, van der Veer E, Akawi O, Pickering JG. SIRT1 markedly extends replicative lifespan if the NAD+ salvage pathway is enhanced. FEBS Lett 2009;583(18):3081–5.

[28] Bird AP. Functions for DNA methylation in vertebrates. Cold Spring Harb Symp Quant Biol 1993;58:281–5.

[29] Sakata SF, Shelly LL, Ruppert S, Schutz G, Chou JY. Cloning and expression of murine S-adenosylmethionine synthetase. J Biol Chem 1993;268(19):13978–86.

[30] Reytor E, Perez-Miguelsanz J, Alvarez L, Perez-Sala D, Pajares MA. Conformational signals in the C-terminal domain of methionine adenosyltransferase I/III determine its nucleocytoplasmic distribution. FASEB J 2009;23(10):3347–60.

[31] Alexander PB, Wang J, McKnight SL. Targeted killing of a mammalian cell based upon its specialized metabolic state. Proc Natl Acad Sci USA 2011;108(38):15828–33.

[32] Shyh-Chang N, Locasale JW, Lyssiotis CA, Zheng Y, Teo RY, Ratanasirintrawoot S, et al. Influence of threonine metabolism on S-adenosylmethionine and histone methylation. Science 2013;339(6116):222–6.

[33] Bernstein BE, Mikkelsen TS, Xie X, Kamal M, Huebert DJ, Cuff J, et al. A bivalent chromatin structure marks key developmental genes in embryonic stem cells. Cell 2006;125(2):315–26.

[34] Mentch SJ, Mehrmohamadi M, Huang L, Liu X, Gupta D, Mattocks D, et al. Histone methylation dynamics and gene regulation occur through the sensing of one-carbon metabolism. Cell Metab 2015;22(5):861–73.

[35] Shi Y, Lan F, Matson C, Mulligan P, Whetstine JR, Cole PA, et al. Histone demethylation mediated by the nuclear amine oxidase homolog LSD1. Cell 2004;119(7):941–53.

[36] Forneris F, Binda C, Vanoni MA, Mattevi A, Battaglioli E. Histone demethylation catalysed by LSD1 is a flavin-dependent oxidative process. FEBS Lett 2005;579(10):2203–7.

[37] Loenarz C, Schofield CJ. Expanding chemical biology of 2-oxoglutarate oxygenases. Nat Chem Biol 2008;4(3):152–6.

[38] Xiao M, Yang H, Xu W, Ma S, Lin H, Zhu H, et al. Inhibition of alpha-KG-dependent histone and DNA demethylases by fumarate and succinate that are accumulated in mutations of FH and SDH tumor suppressors. Genes Dev 2012;26(12):1326–38.

[39] Chowdhury R, Yeoh KK, Tian YM, Hillringhaus L, Bagg EA, Rose NR, et al. The oncometabolite 2-hydroxyglutarate inhibits histone lysine demethylases. EMBO Rep 2011;12(5):463–9.

[40] Pritchard JB. Intracellular alpha-ketoglutarate controls the efficacy of renal organic anion transport. J Pharmacol Exp Ther 1995;274(3):1278–84.

[41] Sabari BR, Tang Z, Huang H, Yong-Gonzalez V, Molina H, Kong HE, et al. Intracellular crotonyl-CoA stimulates transcription through p300-catalyzed histone crotonylation. Mol Cell 2015;58(2):203–15.

[42] Hart GW, Housley MP, Slawson C. Cycling of O-linked beta-N-acetylglucosamine on nucleocytoplasmic proteins. Nature 2007;446(7139):1017–22.

[43] Sakabe K, Wang Z, Hart GW. Beta-N-acetylglucosamine (O-GlcNAc) is part of the histone code. Proc Natl Acad Sci USA 2010;107(46):19915–20.

[44] Fong JJ, Nguyen BL, Bridger R, Medrano EE, Wells L, Pan S, et al. beta-N-Acetylglucosamine (O-GlcNAc) is a novel regulator of mitosis-specific phosphorylations on histone H3. J Biol Chem 2012;287(15):12195–203.

[45] Hanover JA, Krause MW, Love DC. Bittersweet memories: linking metabolism to epigenetics through O-GlcNAcylation. Nat Rev Mol Cell Biol 2012;13(5):312–21.

[46] Chen Y, Sprung R, Tang Y, Ball H, Sangras B, Kim SC, et al. Lysine propionylation and butyrylation are novel post-translational modifications in histones. Mol Cell Proteomics 2007;6(5):812–9.

[47] Dai L, Peng C, Montellier E, Lu Z, Chen Y, Ishii H, et al. Lysine 2-hydroxyisobutyrylation is a widely distributed active histone mark. Nat Chem Biol 2014;10(5):365–70.

[48] Tan M, Luo H, Lee S, Jin F, Yang JS, Montellier E, et al. Identification of 67 histone marks and histone lysine crotonylation as a new type of histone modification. Cell 2011;146(6):1016–28.

[49] Xie Z, Dai J, Dai L, Tan M, Cheng Z, Wu Y, et al. Lysine succinylation and lysine malonylation in histones. Mol Cell Proteomics 2012;11(5):100–7.

[50] Feldman JL, Baeza J, Denu JM. Activation of the protein deacetylase SIRT6 by long-chain fatty acids and widespread deacylation by mammalian sirtuins. J Biol Chem 2013;288(43):31350–6.

[51] Hirschey MD, Zhao Y. Metabolic regulation by lysine malonylation, succinylation, and glutarylation. Mol Cell Proteomics 2015;14(9):2308–15.

[52] Baylin SB, Jones PA. A decade of exploring the cancer epigenome—biological and translational implications. Nat Rev Cancer 2011;11(10):726–34.

[53] Sturm D, Witt H, Hovestadt V, Khuong-Quang DA, Jones DT, Konermann C, et al. Hotspot mutations in H3F3A and IDH1 define distinct epigenetic and biological subgroups of glioblastoma. Cancer Cell 2012;22(4):425–37.

[54] Figueroa ME, Abdel-Wahab O, Lu C, Ward PS, Patel J, Shih A, et al. Leukemic IDH1 and IDH2 mutations result in a hypermethylation phenotype, disrupt TET2 function, and impair hematopoietic differentiation. Cancer Cell 2010;18(6):553–67.

[55] Turcan S, Rohle D, Goenka A, Walsh LA, Fang F, Yilmaz E, et al. IDH1 mutation is sufficient to establish the glioma hypermethylator phenotype. Nature 2012;483(7390):479–83.

[56] Comerford SA, Huang Z, Du X, Wang Y, Cai L, Witkiewicz AK, et al. Acetate dependence of tumors. Cell 2014;159(7):1591–602.

[57] Dang L, White DW, Gross S, Bennett BD, Bittinger MA, Driggers EM, et al. Cancer-associated IDH1 mutations produce 2-hydroxyglutarate. Nature 2009;462(7274):739–44.

[58] Kaelin WGJ. SDH5 mutations and familial paraganglioma: somewhere Warburg is smiling. Cancer Cell 2009;16(3):180–2.

[59] Sebastian C, Zwaans BM, Silberman DM, Gymrek M, Goren A, Zhong L, et al. The histone deacetylase SIRT6 is a tumor suppressor that controls cancer metabolism. Cell 2012;151(6):1185–99.

[60] Vander Heiden MG, Cantley LC, Thompson CB. Understanding the Warburg effect: the metabolic requirements of cell proliferation. Science 2009;324(5930):1029–33.

[61] Schug ZT, Peck B, Jones DT, Zhang Q, Grosskurth S, Alam IS, et al. Acetyl-CoA synthetase 2 promotes acetate utilization and maintains cancer cell growth under metabolic stress. Cancer Cell 2015;27(1):57–71.

[62] Roediger WE. Utilization of nutrients by isolated epithelial cells of the rat colon. Gastroenterology 1982;83(2):424–9.

[63] Kim YS, Milner JA. Dietary modulation of colon cancer risk. J Nutr 2007;137(11 Suppl.): 2576S–9S.

[64] Waldecker M, Kautenburger T, Daumann H, Busch C, Schrenk D. Inhibition of histone-deacetylase activity by short-chain fatty acids and some polyphenol metabolites formed in the colon. J Nutr Biochem 2008;19(9):587–93.

[65] Donohoe DR, Collins LB, Wali A, Bigler R, Sun W, Bultman SJ. The Warburg effect dictates the mechanism of butyrate-mediated histone acetylation and cell proliferation. Mol Cell 2012;48(4):612–26.

[66] Donohoe DR, Holley D, Collins LB, Montgomery SA, Whitmore AC, Hillhouse A, et al. A gnotobiotic mouse model demonstrates that dietary fiber protects against colorectal tumorigenesis in a microbiota- and butyrate-dependent manner. Cancer Discov 2014;4(12):1387–97.

[67] Cahill GFJ. Fuel metabolism in starvation. Annu Rev Nutr 2006;26:1–22.

[68] Kashiwaya Y, Takeshima T, Mori N, Nakashima K, Clarke K, Veech RL. D-beta-hydroxybutyrate protects neurons in models of Alzheimer's and Parkinson's disease. Proc Natl Acad Sci USA 2000;97(10):5440–4.

[69] Maalouf M, Rho JM, Mattson MP. The neuroprotective properties of calorie restriction, the ketogenic diet, and ketone bodies. Brain Res Rev 2009;59(2):293–315.

[70] Shimazu T, Hirschey MD, Newman J, He W, Shirakawa K, Le Moan N, et al. Suppression of oxidative stress by beta-hydroxybutyrate, an endogenous histone deacetylase inhibitor. Science 2013;339(6116):211–4.

[71] Martinez-Pastor B, Cosentino C, Mostoslavsky R. A tale of metabolites: the cross-talk between chromatin and energy metabolism. Cancer Discov 2013;3(5):497–501.

[72] McBrian MA, Behbahan IS, Ferrari R, Su T, Huang TW, Li K, et al. Histone acetylation regulates intracellular pH. Mol Cell 2013;49(2):310–21.

[73] Sasaki H, Matsui Y. Epigenetic events in mammalian germ-cell development: reprogramming and beyond. Nat Rev Genet 2008;9(2):129–40.

[74] Feil R, Fraga MF. Epigenetics and the environment: emerging patterns and implications. Nat Rev Genet 2011;13(2):97–109.

[75] Heard E, Martienssen RA. Transgenerational epigenetic inheritance: myths and mechanisms. Cell 2014;157(1):95–109.

[76] Stegemann R, Buchner DA. Transgenerational inheritance of metabolic disease. Semin Cell Dev Biol 2015;43:131–40.

[77] Greer EL, Maures TJ, Ucar D, Hauswirth AG, Mancini E, Lim JP, et al. Transgenerational epigenetic inheritance of longevity in *Caenorhabditis elegans*. Nature 2011;479(7373):365–71.

[78] Rechavi O, Houri-Ze'evi L, Anava S, Goh WS, Kerk SY, Hannon GJ, et al. Starvation-induced transgenerational inheritance of small RNAs in *C. elegans*. Cell 2014;158(2):277–87.

[79] Tauffenberger A, Parker JA. Heritable transmission of stress resistance by high dietary glucose in *Caenorhabditis elegans*. PLoS Genet 2014;10(5):e1004346.

[80] Carone BR, Fauquier L, Habib N, Shea JM, Hart CE, Li R, et al. Paternally induced transgenerational environmental reprogramming of metabolic gene expression in mammals. Cell 2010;143(7):1084–96.

[81] Ng SF, Lin RC, Laybutt DR, Barres R, Owens JA, Morris MJ. Chronic high-fat diet in fathers programs beta-cell dysfunction in female rat offspring. Nature 2010;467(7318):963–6.

[82] Fullston T, Ohlsson Teague EM, Palmer NO, DeBlasio MJ, Mitchell M, Corbett M, et al. Paternal obesity initiates metabolic disturbances in two generations of mice with incomplete penetrance to the F2 generation and alters the transcriptional profile of testis and sperm microRNA content. FASEB J 2013;27(10):4226–43.

[83] Wei Y, Yang CR, Wei YP, Zhao ZA, Hou Y, Schatten H, et al. Paternally induced transgenerational inheritance of susceptibility to diabetes in mammals. Proc Natl Acad Sci USA 2014;111(5):1873–8.

[84] Painter RC, Osmond C, Gluckman P, Hanson M, Phillips DI, Roseboom TJ. Transgenerational effects of prenatal exposure to the Dutch famine on neonatal adiposity and health in later life. BJOG 2008;115(10):1243–9.

[85] Veenendaal MV, Painter RC, de Rooij SR, Bossuyt PM, van der Post JA, Gluckman PD, et al. Transgenerational effects of prenatal exposure to the 1944-45 Dutch famine. BJOG 2013;120(5):548–53.

[86] Kaati G, Bygren LO, Pembrey M, Sjostrom M. Transgenerational response to nutrition, early life circumstances and longevity. Eur J Hum Genet 2007;15(7):784–90.

[87] Dominguez-Salas P, Cox SE, Prentice AM, Hennig BJ, Moore SE. Maternal nutritional status, C(1) metabolism and offspring DNA methylation: a review of current evidence in human subjects. Proc Nutr Soc 2012;71(1):154–65.

[88] Arrowsmith CH, Bountra C, Fish PV, Lee K, Schapira M. Epigenetic protein families: a new frontier for drug discovery. Nat Rev Drug Discov 2012;11(5):384–400.

CHAPTER 16

Circadian Plasticity of Chromatin States

W.J. Belden

Department of Animal Sciences, School of Environmental and Biological Sciences, Rutgers, The State University of New Jersey, New Brunswick, NJ, United States

16.1 THE CIRCADIAN CLOCK

The predominant mechanism underlying all circadian clocks is a mechanistically conserved transcriptional negative feedback loop that consists of heterodimeric transcriptional activators and a multisubunit repressor complex that inhibits clock gene expression (Fig. 16.1A–B). The transcriptional activator complex drives expression of the negative elements that feedback, with timed delays, to inhibit transcription. The transcriptional activators all contain PAS (*Per, Arnt,* and *Sim*) domains that allow them to dimerize while potentially serving as a redox sensors in order to drive the expression of negative elements [1–3]. These activator complexes consist of White collar 1 (WC-1) and WC-2 in *Neurospora* [4], CLOCK (Clk) and CYCLE (Cyc) in *Drosophila* [5–7], and CLOCK and BMAL1/MOP3 in mammals [8–11]. Once expressed and translated, the negative elements block transcription by inhibiting the positive elements in part by establishing a repressive chromatin template that prevents elongation. The negative elements are *frequency* (*frq*) and *frq-interacting RNA helicase* (*frh*) in *Neurospora* [12,13], *Period* (*Per*) and *Timeless* (*Tim*) in *Drosophila* [14,15], and *Per* and *Cryptochrome* (*Cry*) in mammals [16–19]. The circadian clock contains additional feedback loops, nested transcriptional networks, translational regulatory control, and an abundance of posttranslational regulatory mechanisms. These additional facets of the circadian clock are well-characterized in all the model systems, reviewed in *Methods in Enzymology* [20,21]. However, to fully understand the elegance of circadian clock, especially as it pertains to circadian chromatin, a brief summary of the mammalian clock is presented below.

In addition to driving expression of the *Period* (*Per1, 2,* and *3*) and *Cryptochrome* (*Cry1,* and *2*) genes whose proteins form the negative limb, CLOCK and BMAL1 (components of the positive limb) also drive the expression of *Rev-Erbα* and *Rora* that form an additional feedback loop (Fig. 16.1C). REV-ERBα and RORa have opposing effects on *Bmal1* expression; REV-ERBα negatively regulates

CONTENTS

Chromatin Regulation and Dynamics. http://dx.doi.org/10.1016/B978-0-12-803395-1.00016-2

(A)

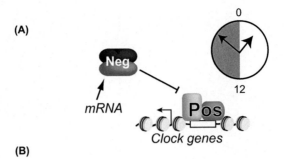

(B)

Negative elements	Postive elements
FRQ-FRH	WCC
PER-TIM	CLK-CYC
PER-CRY	CLOCK-BMAL1

(C)

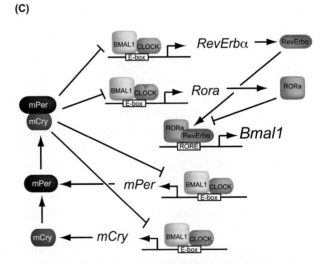

FIGURE 16.1 A schematic representation of circadian negative feedback.

(A) In its most simplified form, the circadian clock has a positive (Pos) transcriptional activating complex capable of initiating transcription at clock genes. Once the mRNA from the corresponding clock gene is expressed, the negative (Neg) inhibitory elements are translated and form a multimeric complex that inhibits the transcriptional activity of the positive elements. Time dependent delays generate *circa* 24 h rhythm in gene expression. (B) The table shows the positive and negative elements in *Neurospora, Drosophila,* and mammals in descending order. The acronyms are as follows: *FRQ,* frequency; *FRH,* FRQ-interacting RNA Helicase; *WCC,* white collar complex composed of WC-1 & WC-2; *PER,* period; *TIM,* timeless; *CLK,* clock; *CYC,* cycle; *CRY,* cryptochrome; *CLOCK,* circadian locomotor output cycles kaput; *BMAL1,* brain and muscle ARNT-like 1. (C) Schematic representation of the mammalian interconnected transcriptional feedback loops.

Bmal1 whereas RORa positively regulates *Bmal1* leading to a robust (but ancillary) rhythm in *Bmal1* expression. The nested transcriptional network is an indirect method of increasing the number of cyclic clock-controlled genes (*ccg*). This nested transcriptional network is formed when CLOCK:BMAL1 directly activate other transcription factors that are not part of the core clock causing indirect rhythmic expression in genes that are differentially phased relative to *Period* and *Cryptochrome*. An example of this is the circadian regulated transcription factor *c-Myc*. Rhythmic expression of c-MYC mediated by CLOCK:BMAL1 leads to rhythms in c-MYC target genes. The combination of interconnected positive and negative transcriptional feedback loops and nested transcriptional networks necessitate a role for chromatin regulation in clock function.

In organisms ranging from filamentous fungi to mammals, where DNA is packaged in a chromatin template, the clock transcriptional negative feedback loop entails a role for chromatin regulation in order to achieve rhythms in gene expression. Initially, circadian changes to chromatin were thought of as just a consequence, a necessary outcome of the underlying oscillations. However, in recent years it has become increasingly clear that a major function of the core components (both the activator and repressive complexes) is to recruit chromatin-remodeling and chromatin-modifying enzymes to facilitate the cycling between permissive and nonpermissive chromatin states necessary to regulate daily oscillations in gene expression. This chapter explores the past 15 years in circadian chromatin research and highlights the growing importance of the collaboration between the circadian clock and chromatin-modifying enzymes. Although our understanding of this process is fairly extensive, there are still many undiscovered enzymes, and the actual molecular mechanisms are not fully characterized, even though a unifying model of circadian clock-mediated chromatin transitions appears conserved from fungi to mammals.

16.2 CIRCADIAN CHROMATIN TRANSITIONS

To begin, one must first consider that nearly half of all protein coding genes in mammals are under circadian control [22] and that the core clock components are known to be associated with, or to recruit chromatin remodeling and modifying enzymes in *Neurospora, Drosophila,* and mammals [23–28]. In addition, both the positive and negative elements exhibit rhythmic associations with chromatin [29–31]. Based on these observations, it is easy to extrapolate that major portions of the genome exhibit oscillations in chromatin states as genes transition from expressed to repressed states and back again and again [31,32]. Although this concept is now readily apparent, this wasn't always the case. The first hint of circadian clock-regulated changes in genome-wide chromatin states came from early studies in the mouse suprachiasmatic nucleus (SCN), when it was shown by immunohistochemistry that histone

H3 serine 10 became phosphorylated (H3S10P) in response to light [33]. This study showing circadian H3S10P prompted a wave of subsequent research that has led to the identification of numerous histone modifications along with the corresponding histone modifying enzymes. These enzymes add and remove the modifications so that chromatin-remodeling enzymes can readjust the underlying chromatin to facilitate 24-h cycles in gene expression.

16.2.1 Acetylation and Deacetylation Rhythm

Early studies in circadian chromatin focused on histone acetylation (see also Chapter 2), due in part to the original finding that *Tetrahymena* GCN5 was a histone lysine acetyltransferase (KAT) and that acetylation accompanied transcriptional activation [34]. This finding spawned the discovery of rhythms in acetylation in *mPer1*, *mPer2*, and *Cry1* promoters with the peaks in acetylation occurring when the genes were actively transcribed [26,35]. The rhythm in acetylated histone H3 occurs as a result of the transcriptional activator CLOCK that recruits the ubiquitous KAT3B, p300, and CREB-binding protein (CBP) [26,35]. In addition, CLOCK has also been reported to have intrinsic in vitro histone acetyltransferase activity [36], but it is not clear if this activity is solely the result of CLOCK or due to coprecipitation with one of the many KATs. The robust rhythm in histone acetylation has led to the obvious hunt for the histone deacetylases (HDAC), and the list of HDACs now include HDAC1 & HDAC2 (part of both the SIN3 and NuRD complexes) [27,37,38] and the NAD$^+$-dependent deacetylase SIRT1 [39,40]. Multiple lines of evidence implicate HDAC1 and HDAC2 in clock function for rhythmic deacetylation of histones, firmly establishing them in clock-regulated gene expression [27,37,38]. However, the data pertaining to SIRT1 is significantly more complicated, in part due to alternative reports describing different acetylated targets including histone H3, BMAL1 [39], PER2 [40] and recently, MLL1, a histone methyltransferase important for the establishment of active chromatin states at enhancers and promoters [41]. Although the reports are disparate concerning the targets of SIRT1, its circadian expression and the circadian phenotypes that arise in the absence of SIRT1 [39,40], it seems definitive that SIRT1 is recruited to the CLOCK:BMAL1 complex [39,40] possibly via CLOCK [38], and it may assists in the acetylation rhythm of BMAL1, Histone H3 [39] and/or PER2 [40]. Interestingly, another sirtuin, SIRT6 is also involved in circadian metabolic output. Moreover, SIRT1 and SIRT6 assist in the control of distinct sets of circadian genes [42].

Collectively, data over the past decade has revealed that circadian clock genes undergo oscillation in histone acetylation that requires the addition of acetyl groups by KATs and their removal by HDACs. As pertains to SIRT1 and SIRT6, whether or not the levels of these proteins oscillate is still unresolved. However, the expression of the gene encoding nicotinamide phosphoribosyltransferase (*Nampt*) is under the direct control of CLOCK:BMAL1, and, as a consequence,

NAD$^+$ levels oscillate with the corresponding metabolic rhythm [43], paving the way for oscillation in the activity of SIRT1 and SIRT6 (see also Chapter 15).

Very little is known about acetylation and deacetylation rhythms in the *Neurospora* or *Drosophila* clocks. However, in *Neurospora* the GCN5 homolog (*ngf-1*) is needed for light-activated transcription at WCC target genes [44], and thus likely functions under circadian conditions as well. Unpublished results also indicate that the *Neurospora* HDAC, HDA-1, may function as the corresponding HDAC, and that the SIRT1 homolog is dispensable for rhythms in *Neurospora* (William J. Belden, unpublished observation). In *Drosophila*, acetylation is presumably needed for a normal rhythm, because BRAMHA subunits contain bromodomains that bind to acetylated histones. As discussed below, BRAMHA was recently shown to associate with CLK and CYC in *Drosophila* [28].

16.2.2 Methylation and Demethylation Rhythm

In addition to acetylation and deacetylation, the circadian clock guides rhythms in histone methylation and demethylation. Although there are some minor anomalies, a universal mechanism of dynamic facultative heterochromatin formation is beginning to emerge. One of the first reports showing rhythmic histone methylation suggested a role for Polycomb–repressive complex 2 (PRC2) in this process. During the repressive phase, the level of di- and trimethylation of histone H3 on lysine 27 (H3K27me2 & H3K27me3) thus increases at *mPer1* and *mPer2*, which is dependent on the Polycomb group protein KMT6/EZH2 [45]. EZH2 coprecipitates with CLOCK:BMAL1 in liver extracts over the entire circadian cycle, and siRNA against *Ezh2* affects the luciferase rhythm of both *Per2-luc* and *Bmal1-luc* reporters [45]. The implications of EZH2 and H3K27 methylation are, however, not fully understood, and more needs to be done to fully understand the role of PRC2 in the circadian clock and circadian output.

In 2006, a transformative report showed the presence of dimethylation at histone H3 lysine 9 (H3K9me2) and the binding of HP1α at the *Dbp* locus during the repressive phase of the circadian cycle [46]. H3K9me2 and HP1α rhythms were antiphasic to H3K9 acetylation and Histone H3 Lysine 4 trimethylation (H3K4me3) rhythm. Furthermore, the rhythm in facultative heterochromatin formation was dependent on CLOCK:BMAL1 binding, because in the absence of the E-box promoter element at the *Dbp* gene, H3K9me2 marks were prevalent and H3K9Ac marks were absent [46]. A rhythm in facultative heterochromatin formation, including binding of HP1, appears to be conserved mechanism between mammals, *Drosophila*, and *Neurospora*, although the HP1 isoform present may differ in mammals depending on the locus (Fig. 16.2). In mammals, KMT1 (Suv39h) and HP1γ are part of the PER–CRY complex and regulate the rhythm in H3K9me2, H3K9me3 levels and HP1γ binding at *Per1* E-box elements [25]. Absence of Suv39h results in shorter period length [25]. H3K9me3 and HP1 binding also occur at the central clock gene *frq* in *Neurospora* [47].

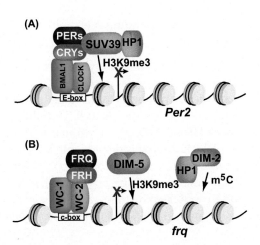

FIGURE 16.2 Circadian clock-regulated facultative heterochromatin.
A model illustrating circadian regulated facultative heterochromatin in (A) mammals and (B) *Neurospora*. In mammals, KMT1/SUV39 and HP1γ are part of the PER complex. Association of PER-CRY with CLOCK:BMAL1 guides KMT1/SUV39 to clock genes where it adds methyl groups to Histone H3 on Lysine 9. The chromodomain of HP1, which is also part of the PER complex, binds to H3K9me3. In *Neurospora*, a similar but slightly different model is proposed. Dicer-independent siRNAs (disiRNA) that arise from the *frq/qrf* sense/antisense pair are thus needed to direct KMT1/DIM-5-mediated H3K9me3 deposition. HP1, in a complex with the DNMT DIM-2, binds to H3K9me3 and guides DNA methylation. Whether or not KMT1/DIM-5 is associated with FRQ-FRH is currently unknown.

Loss of the KMT1 (DIM-5) in *Neurospora* predominantly manifests as a phase effect, but there is a minor period decrease as well [47]. Interestingly, there is an increase in circadian asexual spore production in the absence of H3K-9me3 in *Neurospora*, suggesting that, like *Dbp* in mammals, *ccg* also undergo rhythms in facultative heterochromatin formation, although this has yet to be tested directly. Evidence for rhythms in facultative heterochromatin also exist in *Drosophila*. There is a rhythm in H3K9me2, that is, antiphasic to H3K9Ac and H3K4me3 at *dper* [48], and unpublished proteomic approaches in *Drosophila* S2 cells indicate that suppressor of position-effect variegation [Su(var)] proteins associate with both CLK and dPER (Johanna Chiu and Isaac Edery, unpublished observation). The molecular mechanisms of KMT1 recruitment and the recruitment of the components of the negative limb to clock loci are still largely unexplored. An important observation in mice highlights that specific repression of *ccg* by general repressors can be achieved by the rhythmic targeted reconstitution of the histone deacetylase–NuRD complex [27]. Hence, one half of the NuRD complex associates with the CLOCK:BMAL1 complex, whereas the other half, which is necessary for its repressor function, is delivered to clock-controlled loci by the PER complex. Importantly, in *Neurospora* the *frq* natural antisense transcript, *qrf*, is required for efficient heterochromatin formation at *frq*, suggesting similarities to RNAi-mediated heterochromatin formation [49].

Given that all the *Per* genes in mammals have natural antisense transcripts, the mechanism for recruitment may also be conserved [31,32,49]. Despite this highly conserved mechanism of facultative heterochromatin formation at clock genes, the reversal of the heterochromatin is still under investigation.

As alluded to above, the rhythm in facultative heterochromatin at *Dbp* is accompanied by a rhythm in H3K4me3 [46]. Thus, similarly to H3K9me2/3, H3K4me3 is an oscillating circadian modification. However, in the case of H3K4me3, there appear to be substantial differences between *Neurospora* and mammals. CLOCK:BMAL1 interacts with the KMT2 MLL1 (mixed-lineage leukemia) that catalyzes the deposition of H3K4me3 at *Dbp* during the activation phase [50]. In *Mll1* knockout MEFs, circadian expression of *Dbp* is abolished [50]. Interestingly, MLL1 does not interact with CLOCKΔ19, which may provide insight into the phenotype associated with the *Clock*Δ19 mutation [50](i.e., a semidominant *Clock* allele generated by ENU mutagenesis containing an A to T transversion resulting in a *Clock* transcript that lacks exon 19 [51]). In addition, recent work indicates that MLL1 is acetylated and then deacetylated by SIRT1, and this regulates its methyltransferase activity [41].

H3K4me3 is also present at *frq* in *Neurospora*, and it is catalyzed by the KMT2 SET-1. Surprisingly, the role of H3K4me3 in *Neurospora* seems to be opposite to that of MLL1 in mammals. Instead of supporting activation, H3K4me3 thus appears to be needed for the repression of the clock gene *frq* in *Neurospora*. Moreover, KMT2 deletion (Δ*set1*) leads to a defect in negative feedback repression [52]. Moreover, SET1 is presumably required for subsequent heterochromatin formation at *frq*, because DNA methylation at *frq* is absent in Δ*set1* [52]. However, it appears that this central clock gene may be the exception to the widely-accepted notion that H3K4 tri-methylation is associated with active transcription, because other loci in *Neurospora* require KMT2/SET1 for expression [52]. An interesting caveat in this hypothesis was the discovery that WDR5, a subunit of the COMPASS (*Complex associated with Set1*) complex, is associated with PER1 [24,53]. Transient expression of WRD5 repressed CLOCK:BMAL1-mediated expression, and repression of WDR5 caused a reduction in both H3K4me3 and H3K9me3 levels [24]; a result entirely consistent with observations mentioned earlier on the *Neurospora* KMT2/SET-1. This raises the possibility that the H3K4me3 mark may have context dependent function at central clock genes, which is different from its effects at *ccg* like *Dbp*. Alternatively, the presence of mono-, di-, and trimethylated forms of H3K4 have different outcomes, complicating interpretations. Support for this second idea comes from two separate circadian genome-wide studies that examined H3K4 methylation states. One report found that peaks in H3K4me3 coincided with oscillating transcript levels [32], whereas another report found that H3K4me3 peaked during the repressive phase [31]. It is likely that the context-dependent nature of histone modifications, their underlying dynamics, and the potential for circadian bivalent domains may account for these differences.

The old adage "What goes up, must come down" holds true for the removal of histone methylation, and as it pertains to histone methylation, the rhythm necessitates demethylation. Despite the obvious requirement for circadian-regulated histone demethylation of H3K4me3 and H3K9me2/3, the lysine-specific demethylase (KDM) enzymes responsible have remained elusive. This is perhaps best illustrated by the lysine-specific demthylase 1 (KDM1/LSD1) and the JumonjiC and ARID domain-containing histone lysine demthylase 1a (KDM5A/JARID1a), both of which are needed for normal clock function, although neither of which appear to directly demethylate histones at clock genes [54,55]. KDM1/LSD1 is a dual-specificity histone demethylase that is capable of demethylating both H3K4 [56] and H3K9 [57] mono- and dimethylations. In the circadian system, KDM1/LSD1 is associated with CLOCK-BMAL1 and has rhythmic phosphorylation mediated by protein kinase Cα (PKCα) [55]. This phosphorylation is needed for efficient CLOCK:BMAL1 binding to *Per2* and *Dbp* E-boxes [55]. In mice expressing a site-specific KDM1/LSD1 mutant that replaced the phosphorylated serine (S112) with an alanine, the KDM1/LSD1 is not recruited to either *Per2* or *Dbp*, yet the levels of H3K4me2 and H3K9me2 mirror the WT rhythm, indicating that KDM1/LSD1 does not demethylate either modifications, or there is a compensatory mechanism that removes the methylation to maintain the rhythm [55]. There is, however, a significant decrease in H3K9Ac, which one might expect if H3K9 was not efficiently demethylated. KDM5A/JARID1a is also implicated in clock function. Like KDM1/LSD1, KDM5A/JARID1a also associates with CLOCK-BMAL1. *Kdm5a/Jarid1a* knockout (*Jarid1a*$^{-/-}$) MEFs display a shorter period and have a lower amplitude rhythm in *Per2* and *Cry1* expression [54]. Surprisingly, H3K4me3 levels are slightly increased in *Jarid1a*$^{-/-}$ MEFs, but H3K9Ac was reduced [54]. Further analysis revealed that KDM5A/JARID1a associates with HDAC-1, and this led the authors to conclude that KDM5A/JARID1a, in association with CLOCK:BMAL1, inhibits HDAC-1 activity, leading to the lower levels of H3K9Ac in *Jarid1a*$^{-/-}$ MEFs [54]. Hence, in both these KDM mutants, there is a decrease in the level of acetylation, but no defect in histone methylation rhythm can be detected, adding to the mystique of demethylation in circadian facultative heterochromatin dynamics.

16.2.3 Rhythmic Modifications During Transcriptional Elongation

In addition to the chromatin modifications and remodeling enzymes that regulate the transitions between permissive and nonpermissive chromatin states at the promoters of clock genes, there are rhythmic modifications that are routinely found in the gene body of actively transcribed genes. One such modification is H3K36me3, which is added by KMT3 (Set2) and was originally identified in *Saccharomyces cerevisiae* as a modification needed for repression [58]. Shortly thereafter, it was determined that KMT3/SET2 was associated with Pol II, and that the deposition of H3K36me3 coincided with transcriptional elongation [59 61]. These seemingly opposite functions are unified by the observation that H3K36me3 is a docking site for the histone deacetylase complex, Rpd3S, which deacetylates histones after Pol II elongation to suppress spurious transcription [62–64]. In the *Neurospora* clock, H3K36me3 is rhythmic at *frq*, and the peak appears to coincide with the peak in transcription, consistent with H3K36me3 occurring during elongation [65]. Moreover, SET2 is required for normal rhythms and suppresses *frq* [65]. Genome-wide studies in mammals revealed that H3K36me3 peaked at the tail end of the repression phase [31]. Unlike other circadian methyl-modifications, the K36-specific histone demethylase (KDM8/JMJD5) has been partially characterized in clock function. Analysis of microarray data in *Arabidopsis* revealed that *KDM8/JMJD5* cycled similarly to TOC1 (timing of *CAB1* expression), and *KDM8/JMJD5* nulls had a shorter rhythm in both *Arabidopsis* and mammals [66]. Even though H3K36me3 levels were never tested in this study, the *Arabidopsis* and mammalian KDM8/JMJD5 were interchangeable, strongly supporting the notion that it is required for clock gene-specific H3K36 demethylation [66].

16.2.4 ATP-Dependent Chromatin-Remodeling Enzymes in Circadian Transcription

The rhythm in histone modifications is acted upon by a variety of multisubunit complexes that interpret modifications either alone or in combination to ensure that the chromatin template is phased appropriately during transitions from permissive to nonpermissive states. Numerous ATP-dependent chromatin-remodeling enzymes and complexes that detect the underlying chromatin modifications are now well characterized (see also Chapter 4), although in most instances the mechanism of their binding to circadian chromatin has not been examined. A total of four separate ATP-dependent remodeling enzymes are known in *Neurospora*, two in *Drosophila* and one in mammals to affect the circadian rythm, and the results to date suggest a high degree of conservation (Fig. 16.3). However, as chromatin structural changes have been documented both at individual clock genes and on a genome wide scale, global remodeling events likely play important roles in circadian transcriptional regulation. In *Neurospora*, WCC is associated with the SWI/SNF ATP-dependent chromatin–remodeling complex, and SWI/SNF is needed for rhythmic *frq* expression [23]. WCC recruits SWI/SNF to the *frq* promoter and decreases nucleosome density at the clock-box (c-box) promoter element [23]. In *Drosophila*, the SWI/SNF BRAHMA complex plays an opposite role to SWI/SNF in *Neurospora*, but like SWI/SNF, it associates with the positive elements CLK and CYC. In biochemical–interaction studies in *Drosophila* S2 cells, BRAHMA was found to associate with CLK [28]. BRAHMA is localized to *dper* and *tim* over the entire circadian cycle, and it is needed to increase nucleosome occupancy [28]. In addition, BRAHMA regulates the association of CLK with *dper* independently of its ATP-dependent remodeling activity [28], suggesting a higher order function other than remodeling nucleosomes.

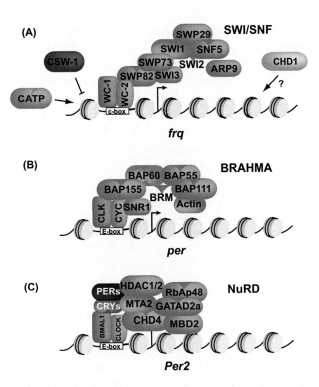

FIGURE 16.3 Chromatin remodeling in circadian clock-regulated gene expression.
The illustrations are schematic representations of the chromatin-remodeling enzymes/complexes in (A) *Neurospora*, (B) *Drosophila*, and (C) mammals. In *Neurospora*, SWI/SNF, CATP, CSW-1 and CHD-1 are all implicated in controlling chromatin structure at *frq*. SWI/SNF and CATP are both involved in creating a more transcriptionally permissive chromatin template, whereas CSW-1 is required for efficient negative feedback inhibition. Data suggest that CHD-1 is needed for activation. However, the role of CHD-1 remains enigmatic (highlighted by a question mark), because loss of CHD-1 causes constitutive heterochromatin causing potential pleiotropic effects. In *Drosophila*, BRAHMA is associated with CLK-CYC and binds to *per* and *tim* over the entire circadian cycle. BRAHMA creates more restrictive chromatin to inhibit *per* and *tim* expression, and also regulates the level of CLK binding independently of its remodeling activity. In mammals, there is a two-step mechanism for NuRD-mediated repression. During the activation phase, CHD4 and MTA2 are associated with CLOCK:BMAL1, and CHD4 promotes gene expression. During the repressive phase, PER-CRY, through an interaction with CLOCK:BMAL1, bring the remaining components of NuRD to *Per1* and *Per2* genes to achieve repression.

In addition, all model systems have chromo-domain helicase DNA binding proteins (CHD) implicated in circadian regulation. These are CHD1, Kismet, and CHD4/Mi-2 in *Neurospora, Drosophila*, and mammals respectively [27,67,68]. Both CHD1 in *Neurospora* and CHD4 in mammals are needed to promote transcription of the respective clock gene [27,67]. In the absence of CHD1 in *Neurospora*, DNA within the *frq* promoter is extensively methylated, indicating that loss of CHD1 causes constitutive heterochromatin and thus CHD1 may function in the reversal of heterochromatin [67]. This constitutive heterochromatic state may account for the reduced expression of *frq*. In mammals, CHD4 is associated with CLOCK:BMAL1 over the entire circadian cycle, and the PER complex acts as a bridge to deliver the remaining NuRD subunits to the appropriate circadian locus [27]. Once the remaining NuRD subunits are directed to the *Per* genes, corepression is achieved [27]. Of note, and highlighting potential similarities among the systems, NuRD contains a methyl DNA binding protein (MBD2) and *frq* is hypermethylated in *chd1* deletion strains. Thus it would be interesting to examine whether or not *Per* genes may exist in a constitutive heterochromatic state in the absence of CHD4. Support for this possibility can be found through the observation that Pol II levels at *Per1* and *Per2* are dramatically reduced in experiments using siRNA against *Chd4* [27], although H3K9me3 and HP1γ binding were not examined in this context.

In *Neurospora*, two additional ATP-dependent chromatin-remodeling enzymes have been implicated in clock function, Clockswitch (CSW-1), and Clock ATPase (CATP) [69]. CSW-1, whose homologs include *S.c.*Fun30, *S.p.*Ftf3, *m*ETL1, and *H.s.*SMARCAD1, remodels the nucleosome distal to the *frq* c-box element generating a condensed chromatin state that inhibits WC-2 binding. By all accounts, CSW-1 appears to counteract the activity of *Neurospora* SWI/SNF and CATP (see later). In the absence of CSW-1, the normal rhythmic binding of WC-2 is compromised, and WC-2 can be found at the c-box over the entire circadian cycle [70]. Studies on *S. cerevisiae* Fun30 (CSW-1 homolog) indicate that it exists as a homodimer and can remodel chromatin in vitro [71]. Interestingly, Fun30 appears to promote gene silencing at heterochromatic loci [72]. This observation is entirely consistent with the role of CSW-1 in negative feedback inhibition. Further supporting this hypothesis, a recent paper on SMARCAD1 indicates that it promotes silent heterochromatin formation by recruiting the H3K9 methyltransferase G9a [73]. CATP, on the other hand, seems to function analogous to SWI/SNF, and regulates nucleosome occupancy at the c-box [69]. In *catp* knockouts, the nucleosome density is increased at the c-box and there is less WC-2 bound [69]. In addition, there is a reduction in PolII association with *frq* in the transcribed region [69]. It is clear that a significant amount of work still needs to be done on the ATP-dependent remodeling enzymes to fully understand their role and mechanism of action in the circadian system.

Finally, the role of chromatin-remodeling enzymes in circadian regulation is further highlighted by the observed genome-wide changes to nucleosome occupancy during the circadian cycle. In both *Neurospora* and mammals, mononucleosome mapping of MNaseI treated chromatin has revealed that nucleosome occupancy decreases at regions surrounding WCC and CLOCK-BMAL1 binding sites in response to light or during the circadian cycle throughout the genome

[74,75]. These findings are consistent with early reports that documented MNaseI or DNaseI hypersensitive sites in circadian genes [26,70]. A model in 2007 (based on work done on the yeast Pho5 promoter) proposed that nucleosomes at circadian genes were first acetylated and then disassembled coinciding with binding of the general transcription factors (GTFs) [70,76,77]. The idea that nucleosomes are rearranged to facilitate transcription is not new. Studies showing that nucleosomes are remodeled to enable GTFs to access regulatory regions date back to the 1980s and possibly earlier [76–78], further emphasizing the requirement for the positive elements to instigate genome-wide circadian changes to chromatin structure. In this capacity, both the CLOCK:BMAL1 complex in mammals, and the WCC in *Neurospora*, serve as pioneering transcription factors that are required to initiate the nucleosome removal or repositioning [74,75].

16.3 GENOME-WIDE CHANGES IN 3D CHROMATIN ORGANIZATION

In addition to circadian-regulated repositioning/remodeling of individual nucleosomes, the clock appears to be involved in processes well-beyond mere regulation of rhythmic changes to the primary chromatin fiber. In fact, BMAL1 promotes the establishment of dynamic, long-range contacts between distant parts of the genome, suggesting that the circadian clock regulates oscillating changes in the three dimensional (3D) organization of nuclear structures [79]. Chromosomes within the genome are predominantly organized into globular domains of intrachromosomal contacts consisting of open and closed chromatin [80]. However, unexpected dynamics in genome organization was uncovered using *Dbp* as bait in chromosome conformation capture on a chip (4C) assays, which revealed time-dependent enrichment of chromatin contacts between active circadian genes in transcriptionally active foci of the nucleus [79]. The circadian dynamics of 3D genome organization has been further highlighted by the observations that mobility of clock genes between transcriptionally permissive and repressive subnuclear environments contributes to rhythms in facultative heterochromatin formation [81]. Importantly, the molecular components that regulate chromatin mobility between active and inactive compartments and help establish spatially distinct genomic structures on the circadian time scale are also implicated in clock function [81]. The genome organizers poly(ADP-ribose), Polymerase 1 (PARP-1), and CTCF thus exhibit rhythmic complex formation, and together promote the translocation of *ccg* to the repressive environment of the nuclear lamina, where they enter a heterochromatic state [81]. Loss of H3K9 methylation, or downregulation of either PARP1 or CTCF attenuated not only chromatin movement to the nuclear periphery, but also circadian transcription [81]. Collectively, the data point to a very dynamic genome that undergoes spatial and temporal structural changes on the local and global scale to facilitate circadian oscillations in transcription.

16.4 DNA-METHYLATION AND THE CIRCADIAN CLOCK

DNA methylation at clock genes was first observed in cancer lines [82–84], prompted by studies indicating that night-shift workers suffer a higher incidence of cancer relative to their day-shift counterparts [85,86]. The studies exploring *Per* DNA methylation were also based in part on the notion that *Per2* functioned as a tumor suppressor [87]. Even though the relevance of DNA methylation at clock genes is undefined, it is clear that the circadian clock has a role in establishing DNA methylation. Dynamic DNA methylation at a clock gene was first reported in *Neurospora* [67]. Characterization of the CHD1 deletion revealed that the *frq* locus was hypermethylated and that physiologic patterns of DNA methylation required functional FRQ-FRH, WC-dependent transcription and the *frq* natural antisense transcript *qrf* [67]. The requirement for *qrf* in establishing DNA methylation was confirmed when it was determined that regions encoding convergent transcripts in *Neurospora* all carry DNA methylation marks [88]. DNA methylation at *frq* is also dependent on H3K9me3 and HP1 [47]. The observation that the dynamic DNA methylation at *frq* is dependent on facultative heterochromatin raises some interesting caveats for mammals (see also Chapter 3). As mentioned earlier, facultative heterochromatin in clock regulation is conserved from *Neurospora* to mammals, and NuRD contains a methyl-DNA binding protein, MBD2. This raises the question, whether establishment of DNA methylation at *Per* genes requires Suv39 and HP1, and what are the consequences of CHD4 downregulation on DNA methylation. It is remarkable that when rodents are entrained to noncircadian day lengths, there are global alterations to DNA methylation relative to normal circadian days (11 h vs. 12 h light–dark cycle) [89]. This altered DNA methylation occurs despite the finding that overall DNA methylation does not change dramatically over the course of the day [32,89], once again a result entirely consistent with data generated in *Neurospora*. Moreover, potential rhythms in DNA methylation could be detected in postmortem analysis in humans [90]. These findings raise interesting questions pertaining to aging, the clock and age-related diseases. DNA methylation (and presumably the underlying chromatin modifications) and the clock change with age, and altered DNA methylation is a common finding in many age-related diseases like cancer and Alzheimer's disease [91]. This leads to the question whether or not age-related circadian decline is a consequence or the cause of the underlying changes to chromatin and DNA methylation. Based on results from the BMAL1 knockout mouse, which displays advanced aging phenotypes [92], and the work summarized in this chapter, it seems plausible that chromatin changes regulated by the circadian clock and age-related alterations in chromatin are self-reinforcing. As we age, chromatin becomes altered due to a declining circadian oscillator, which contributes to the propensity for disease progression.

16.5   CONCLUSIONS

The past decade and a half has revealed a growing importance for chromatin regulation in the circadian rhythm of eukaryotes, metazoans and mammals. Biochemical isolation of clock proteins in *Neurospora*, *Drosophila*, and mammals has uncovered that clock proteins are associated with chromatin-modifying and remodeling enzymes. Moreover, there are genome-wide rhythms in modifications and remodeling events, which collectively prove that one of the major functions of the circadian clock is to control genome-wide chromatin states so that clock genes and *ccg* exhibit exquisitely timed transitions from permissive to nonpremissive states. The unifying model in all systems is the need for plasticity in chromatin states that cycles from accessible euchromatin, capable of supporting transcription, to heterochromatin that silences clock genes and prevents expression. Many questions remain on how the heterochromatin is targeted and established. More specifically, is it a ncRNA-mediated process, as in the case with the *frequency* natural antisense transcript *qrf*, or is the mere association with the negative elements that guide the deposition of H3K9me3 and subsequent HP1 binding to loci activated by the positive elements? In addition, it is still unclear how the heterochromatin is reversed back to euchromatin state that supports transcription. These questions will most certainly be answered in the coming years.

Acknowledgments

Research in the Belden laboratory is supported by the National Institutes of Health (GM101378 and AA024330) and the National Institute of Food and Agriculture (NE1439).

References

[1] Reppert SM, Weaver DR. Coordination of circadian timing in mammals. Nature 2002;418(6901):935–41.

[2] Schibler U, Sassone-Corsi P. A web of circadian pacemakers. Cell 2002;111(7):919–22.

[3] Heintzen C, Liu Y. The *Neurospora crassa* circadian clock. Adv Genet 2007;58:25–66.

[4] Crosthwaite SK, Dunlap JC, Loros JJ. *Neurospora wc-1* and *wc-2*: transcription, photoresponses, and the origins of circadian rhythmicity. Science 1997;276(5313):763–9.

[5] Darlington TK, et al. Closing the circadian loop: CLOCK-induced transcription of its own inhibitors per and tim. Science 1998;280(5369):1599–603.

[6] Rutila JE, et al. CYCLE is a second bHLH-PAS clock protein essential for circadian rhythmicity and transcription of *Drosophila* period and timeless. Cell 1998;93(5):805–14.

[7] Allada R, White NE, So WV, Hall JC, Rosbash M. A mutant *Drosophila* homolog of mammalian Clock disrupts circadian rhythms and transcription of period and timeless. Cell 1998;93(5):791–804.

[8] Antoch MP, et al. Functional identification of the mouse circadian Clock gene by transgenic BAC rescue. Cell 1997;89(4):655–67.

[9] Gekakis N, et al. Role of the CLOCK protein in the mammalian circadian mechanism. Science 1998;280(5369):1564–9.

[10] Hogenesch JB, Gu YZ, Jain S, Bradfield CA. The basic-helix-loop-helix-PAS orphan MOP3 forms transcriptionally active complexes with circadian and hypoxia factors. Proc Natl Acad Sci USA 1998;95(10):5474–9.

[11] Bunger MK, et al. Mop3 is an essential component of the master circadian pacemaker in mammals. Cell 2000;103(7):1009–17.

[12] Aronson BD, Johnson KA, Dunlap JC. Circadian clock locus frequency: protein encoded by a single open reading frame defines period length and temperature compensation. Proc Natl Acad Sci USA 1994;91(16):7683–7.

[13] Cheng P, He Q, Wang L, Liu Y. Regulation of the *Neurospora* circadian clock by an RNA helicase. Genes Dev 2005;19(2):234–41.

[14] Hardin PE, Hall JC, Rosbash M. Feedback of the *Drosophila* period gene product on circadian cycling of its messenger RNA levels. Nature 1990;343(6258):536–40.

[15] Sehgal A, et al. Rhythmic expression of timeless: a basis for promoting circadian cycles in period gene autoregulation. Science 1995;270(5237):808–10.

[16] Shearman LP, Zylka MJ, Weaver DR, Kolakowski LF Jr, Reppert SM. Two period homologs: circadian expression and photic regulation in the suprachiasmatic nuclei. Neuron 1997;19(6):1261–9.

[17] Sun ZS, et al. RIGUI, a putative mammalian ortholog of the *Drosophila* period gene. Cell 1997;90(6):1003–11.

[18] Tei H, et al. Circadian oscillation of a mammalian homologue of the *Drosophila* period gene. Nature 1997;389(6650):512–6.

[19] Kume K, et al. mCRY1 and mCRY2 are essential components of the negative limb of the circadian clock feedback loop. Cell 1999;98(2):193–205.

[20] Sehgal A. Methods in Enzymology. Circadian rhythms and biological clocks, part B. Preface. Methods Enzymol 2015;552:xvii–xvii10.

[21] Sehgal A. Methods in Enzymology. Circadian rhythms and biological clocks, part A. Preface. Methods Enzymol 2015;551:xvii–xvii10.

[22] Zhang R, Lahens NF, Ballance HI, Hughes ME, Hogenesch JB. A circadian gene expression atlas in mammals: implications for biology and medicine. Proc Natl Acad Sci USA 2014;111(45):16219–24.

[23] Wang B, Kettenbach AN, Gerber SA, Loros JJ, Dunlap JC. *Neurospora* WC-1 recruits SWI/SNF to remodel frequency and initiate a circadian cycle. PLoS Genet 2014;10(9):e1004599.

[24] Brown SA, et al. PERIOD1-associated proteins modulate the negative limb of the mammalian circadian oscillator. Science 2005;308(5722):693–6.

[25] Duong HA, Weitz CJ. Temporal orchestration of repressive chromatin modifiers by circadian clock Period complexes. Nat Struct Mol Biol 2014;21(2):126–32.

[26] Etchegaray JP, Lee C, Wade PA, Reppert SM. Rhythmic histone acetylation underlies transcription in the mammalian circadian clock. Nature 2003;421(6919):177–82.

[27] Kim JY, Kwak PB, Weitz CJ. Specificity in circadian clock feedback from targeted reconstitution of the NuRD corepressor. Mol Cell 2014;56(6):738–48.

[28] Kwok RS, Li YH, Lei AJ, Edery I, Chiu JC. The catalytic and non-catalytic functions of the brahma chromatin-remodeling protein collaborate to fine-tune circadian transcription in *Drosophila*. PLoS Genet 2015;11(7):e1005307.

[29] Menet JS, Abruzzi KC, Desrochers J, Rodriguez J, Rosbash M. Dynamic PER repression mechanisms in the *Drosophila* circadian clock: from on-DNA to off-DNA. Genes Dev 2010;24(4):358–67.

[30] Rey G, et al. Genome-wide and phase-specific DNA-binding rhythms of BMAL1 control circadian output functions in mouse liver. PLoS Biol 2011;9(2):e1000595.

[31] Koike N, et al. Transcriptional architecture and chromatin landscape of the core circadian clock in mammals. Science 2012;338(6105):349–54.

[32] Vollmers C, et al. Circadian oscillations of protein-coding and regulatory RNAs in a highly dynamic mammalian liver epigenome. Cell Metab 2012;16(6):833–45.

[33] Crosio C, Cermakian N, Allis CD, Sassone-Corsi P. Light induces chromatin modification in cells of the mammalian circadian clock. Nat Neurosci 2000;3(12):1241–7.

[34] Brownell JE, et al. Tetrahymena histone acetyltransferase A: a homolog to yeast Gcn5p linking histone acetylation to gene activation. Cell 1996;84(6):843–51.

[35] Curtis AM, et al. Histone acetyltransferase-dependent chromatin remodeling and the vascular clock. J Biol Chem 2004;279(8):7091–7.

[36] Doi M, Hirayama J, Sassone-Corsi P. Circadian regulator CLOCK is a histone acetyltransferase. Cell 2006;125(3):497–508.

[37] Naruse Y, et al. Circadian and light-induced transcription of clock gene Per1 depends on histone acetylation and deacetylation. Mol Cell Biol 2004;24(14):6278–87.

[38] Duong HA, Robles MS, Knutti D, Weitz CJ. A molecular mechanism for circadian clock negative feedback. Science 2011;332(6036):1436–9.

[39] Nakahata Y, et al. The NAD + -dependent deacetylase SIRT1 modulates CLOCK-mediated chromatin remodeling and circadian control. Cell 2008;134(2):329–40.

[40] Asher G, et al. SIRT1 regulates circadian clock gene expression through PER2 deacetylation. Cell 2008;134(2):317–28.

[41] Aguilar-Arnal L, Katada S, Orozco-Solis R, Sassone-Corsi P. NAD(+)-SIRT1 control of H3K4 trimethylation through circadian deacetylation of MLL1. Nat Struct Mol Biol 2015;22(4):312–8.

[42] Masri S, et al. Partitioning circadian transcription by SIRT6 leads to segregated control of cellular metabolism. Cell 2014;158(3):659–72.

[43] Nakahata Y, Sahar S, Astarita G, Kaluzova M, Sassone-Corsi P. Circadian control of the NAD+ salvage pathway by CLOCK-SIRT1. Science 2009;324(5927):654–7.

[44] Grimaldi B, et al. The *Neurospora crassa* White Collar-1 dependent blue light response requires acetylation of histone H3 lysine 14 by NGF-1. Mol Biol Cell 2006;17(10):4576–83.

[45] Etchegaray JP, et al. The polycomb group protein EZH2 is required for mammalian circadian clock function. J Biol Chem 2006;281(30):21209–15.

[46] Ripperger JA, Schibler U. Rhythmic CLOCK-BMAL1 binding to multiple E-box motifs drives circadian *Dbp* transcription and chromatin transitions. Nat Genet 2006;38(3):369–74.

[47] Ruesch CE, et al. The histone h3 lysine 9 methyltransferase dim-5 modifies chromatin at frequency and represses light-activated gene expression. G3 2014;5(1):93–101.

[48] Taylor P, Hardin PE. Rhythmic E-box binding by CLK-CYC controls daily cycles in per and tim transcription and chromatin modifications. Mol Cell Biol 2008;28(14):4642–52.

[49] Li N, Joska TM, Ruesch CE, Coster SJ, Belden WJ. The frequency natural antisense transcript first promotes, then represses, frequency gene expression via facultative heterochromatin. Proc Natl Acad Sci USA 2015;112(14):4357–62.

[50] Katada S, Sassone-Corsi P. The histone methyltransferase MLL1 permits the oscillation of circadian gene expression. Nat Struct Mol Biol 2010;17(12):1414–21.

[51] King DP, et al. Positional cloning of the mouse circadian clock gene. Cell 1997;89(4):641–53.

[52] Raduwan H, Isola AL, Belden WJ. Methylation of histone H3 on lysine 4 by the lysine methyltransferase SET1 protein is needed for normal clock gene expression. J Biol Chem 2013;288(12):8380–90.

[53] Miller T, et al. COMPASS: a complex of proteins associated with a trithorax related SET domain protein. Proc Natl Acad Sci USA 2001;98(23):12902–7.

[54] DiTacchio L, et al. Histone lysine demethylase JARID1a activates CLOCK-BMAL1 and influences the circadian clock. Science 2011;333(6051):1881–5.

[55] Nam HJ, et al. Phosphorylation of LSD1 by PKCalpha is crucial for circadian rhythmicity and phase resetting. Mol Cell 2014;53(5):791–805.

[56] Shi Y, et al. Histone demethylation mediated by the nuclear amine oxidase homolog LSD1. Cell 2004;119(7):941–53.

[57] Metzger E, et al. LSD1 demethylates repressive histone marks to promote androgen-receptor-dependent transcription. Nature 2005;437(7057):436–9.

[58] Strahl BD, et al. Set2 is a nucleosomal histone H3-selective methyltransferase that mediates transcriptional repression. Mol Cell Biol 2002;22(5):1298–306.

[59] Kizer KO, et al. A novel domain in Set2 mediates RNA polymerase II interaction and couples histone H3 K36 methylation with transcript elongation. Mol Cell Biol 2005;25(8):3305–16.

[60] Li J, Moazed D, Gygi SP. Association of the histone methyltransferase Set2 with RNA polymerase II plays a role in transcription elongation. J Biol Chem 2002;277(51):49383–8.

[61] Schaft D, et al. The histone 3 lysine 36 methyltransferase, SET2, is involved in transcriptional elongation. Nucleic Acids Res 2003;31(10):2475–82.

[62] Li B, et al. Histone H3 lysine 36 dimethylation (H3K36me2) is sufficient to recruit the Rpd3s histone deacetylase complex and to repress spurious transcription. J Biol Chem 2009;284(12):7970–6.

[63] Carrozza MJ, et al. Histone H3 methylation by Set2 directs deacetylation of coding regions by Rpd3S to suppress spurious intragenic transcription. Cell 2005;123(4):581–92.

[64] Joshi AA, Struhl K. Eaf3 chromodomain interaction with methylated H3-K36 links histone deacetylation to Pol II elongation. Mol Cell 2005;20(6):971–8.

[65] Zhou Z, et al. Suppression of WC-independent frequency transcription by RCO-1 is essential for *Neurospora* circadian clock. Proc Natl Acad Sci USA 2013;110(50):E4867–74.

[66] Jones MA, et al. Jumonji domain protein JMJD5 functions in both the plant and human circadian systems. Proc Natl Acad Sci USA 2010;107(50):21623–8.

[67] Belden WJ, Lewis ZA, Selker EU, Loros JJ, Dunlap JC. CHD1 remodels chromatin and influences transient DNA methylation at the clock gene frequency. PLoS Genet 2011;7(7):e1002166.

[68] Dubruille R, Murad A, Rosbash M, Emery P. A constant light-genetic screen identifies KISMET as a regulator of circadian photoresponses. PLoS Genet 2009;5(12):e1000787.

[69] Cha J, Zhou M, Liu Y. CATP is a critical component of the *Neurospora* circadian clock by regulating the nucleosome occupancy rhythm at the frequency locus. EMBO Rep 2013;14(10):923–30.

[70] Belden WJ, Loros JJ, Dunlap JC. Execution of the circadian negative feedback loop in *Neurospora* requires the ATP-dependent chromatin-remodeling enzyme CLOCKSWITCH. Mol Cell 2007;25(4):587–600.

[71] Awad S, Ryan D, Prochasson P, Owen-Hughes T, Hassan AH. The Snf2 homolog Fun30 acts as a homodimeric ATP-dependent chromatin-remodeling enzyme. J Biol Chem 2010;285(13):9477–84.

[72] Neves-Costa A, Will WR, Vetter AT, Miller JR, Varga-Weisz P. The SNF2-family member Fun30 promotes gene silencing in heterochromatic loci. PLoS One 2009;4(12):e8111.

[73] Rowbotham SP, et al. Maintenance of Silent chromatin through replication requires SWI/SNF-like chromatin remodeler SMARCAD1. Mol Cell 2011;42(3):285–96.

[74] Menet JS, Pescatore S, Rosbash M. CLOCK:BMAL1 is a pioneer-like transcription factor. Genes Dev 2014;28(1):8–13.

[75] Sancar C, et al. Combinatorial control of light induced chromatin remodeling and gene activation in *Neurospora*. PLoS Genet 2015;11(3):e1005105.

[76] Boeger H, Griesenbeck J, Strattan JS, Kornberg RD. Nucleosomes unfold completely at a transcriptionally active promoter. Mol Cell 2003;11(6):1587–98.

[77] Reinke H, Horz W. Histones are first hyperacetylated and then lose contact with the activated PHO5 promoter. Mol Cell 2003;11(6):1599–607.

[78] Almer A, Rudolph H, Hinnen A, Horz W. Removal of positioned nucleosomes from the yeast PHO5 promoter upon PHO5 induction releases additional upstream activating DNA elements. EMBO J 1986;5(10):2689–96.

[79] Aguilar-Arnal L, et al. Cycles in spatial and temporal chromosomal organization driven by the circadian clock. Nat Struct Mol Biol 2013;20(10):1206–13.

[80] Lieberman-Aiden E, et al. Comprehensive mapping of long-range interactions reveals folding principles of the human genome. Science 2009;326(5950):289–93.

[81] Zhao H, et al. PARP1- and CTCF-mediated interactions between active and repressed chromatin at the lamina promote oscillating transcription. Mol Cell 2015;59(6):984–97.

[82] Chen ST, et al. Deregulated expression of the PER1, PER2, and PER3 genes in breast cancers. Carcinogenesis 2005;26(7):1241–6.

[83] Hsu MC, Huang CC, Choo KB, Huang CJ. Uncoupling of promoter methylation and expression of Period1 in cervical cancer cells. Biochem Biophys Res Commun 2007;360(1):257–62.

[84] Schernhammer ES, et al. Rotating night shifts and risk of breast cancer in women participating in the nurses' health study. J Natl Cancer Inst 2001;93(20):1563–8.

[85] Davis S, Mirick DK, Stevens RG. Night shift work, light at night, and risk of breast cancer. J Natl Cancer Inst 2001;93(20):1557–62.

[86] Schernhammer ES, et al. Night-shift work and risk of colorectal cancer in the nurses' health study. J Natl Cancer Inst 2003;95(11):825–8.

[87] Fu L, Pelicano H, Liu J, Huang P, Lee C. The circadian gene Period2 plays an important role in tumor suppression and DNA damage response in vivo. Cell 2002;111(1):41–50.

[88] Dang Y, Li L, Guo W, Xue Z, Liu Y. Convergent transcription induces dynamic DNA methylation at disiRNA loci. PLoS Genet 2013;9(9):e1003761.

[89] Azzi A, et al. Circadian behavior is light-reprogrammed by plastic DNA methylation. Nat Neurosci 2014;17(3):377–82.

[90] Lim AS, et al. 24-hour rhythms of DNA methylation and their relation with rhythms of RNA expression in the human dorsolateral prefrontal cortex. PLoS Genet 2014;10(11):e1004792.

[91] De Jager PL, et al. Alzheimer's disease: early alterations in brain DNA methylation at ANK1, BIN1, RHBDF2 and other loci. Nat Neurosci 2014;17(9):1156–63.

[92] Kondratov RV, Kondratova AA, Gorbacheva VY, Vykhovanets OV, Antoch MP. Early aging and age-related pathologies in mice deficient in BMAL1, the core componentof the circadian clock. Genes Dev 2006;20(14):1868–73.

CHAPTER 17

3D Nuclear Architecture and Epigenetic Memories: Regulators of Phenotypic Plasticity in Development, Aging and Cancer

B.A. Scholz*, L. Millán-Ariño**, A. Göndör*

*Department of Microbiology, Tumor and Cell Biology, Karolinska Institutet, Stockholm, Sweden
**Department of Medicine, Karolinska University Hospital, Stockholm, Sweden

17.1 INTRODUCTION

Cell type and differentiation stage–specific changes in gene expression patterns are governed by the collaboration between cell fate–determining transcription factors, signal-induced transcription factors, and specific chromatin states resulting from the combinatorial presence of DNA and histone modifications, chromatin architectural proteins, as well as noncoding RNAs [1]. Previous chapters of this book have illustrated in great detail that changes in reversible chromatin modifications enable the cell to change its transcriptome in response to environmental cues (Chapters 1–11). At the same time, elaborate mechanisms have evolved to enable the stable inheritance of epigenetic marks through mitotic and sometimes even meiotic cell divisions [2]. These mechanisms ensure the maintenance of developmental decisions and the propagation of the effects of environmental signals long after the initial stimulus has ceased to exist. Heritable and reversible chromatin modifications thus fine-tune two fundamental principles of development: phenotypic plasticity and its counterpoint, developmental robustness [3]. Here, "phenotypic plasticity" refers to the ability of cells or the organism to change its phenotype when challenged by changes in the environment. "Developmental stability," on the other hand, entails robustness against extrinsic or intrinsic perturbations to ensure a stable and reproducible developmental outcome [3–5].

Phenotypic plasticity and robustness thus have important consequences on cellular and organismal functions, and can be advantageous, adaptive in some situations and harmful in others. Altered epigenetic flexibility has emerged as a key mediator of maladaptive responses in multifactorial diseases, such as cancer [3]. For example, increased plasticity of the epigenome is considered to be at the core of cancer development, because it leads to the continuous regeneration

CONTENTS

417

Chromatin Regulation and Dynamics. http://dx.doi.org/10.1016/B978-0-12-803395-1.00017-4

of phenotypic heterogeneity among cancer cells [3,6]—upon which selection mechanisms can act to fuel cancer growth. The molecular mechanism(s) of epigenetic flexibility and its deregulation by gene–environment interactions in complex diseases remains, however, poorly understood. Recently, a model [3,7] has been proposed, in which the epigenome and its organisation within the 3D space of the nucleus influence phenotypic flexibility by regulating the probability of transitions between gene activity and inactivity, and by affecting the level of transcriptional noise or stochastic fluctuations in transcript abundance.

In this chapter we summarize the evidence suggesting that the 3D architecture of the nucleus influences cellular phenotypes and adaptive responses by regulating the formation and stability of epigenetic memories. We will begin by describing the basic principles that regulate 3D genome folding and our current understanding of transcriptional regulation in 3D. We will discuss recent observations documenting that transient physical interactions between distant regulatory elements, such as enhancers, promoters, and insulators, influence the probability of transcription and diversify the transcriptome [8–10]. We will then examine how these interactions modulate phenotypic variation during differentiation and in diseases [3,9,11,12].

The probability of such chromatin crosstalk is also greatly influenced by the overall organization of the genome in the nucleus. This organization is exemplified by the formation of chromosome territories that display cell type–specific preference for chromosomal neighborhoods promoting or antagonizing possibilities for interactions [13]. Moreover, chromosome territories are radially oriented based on gene density and chromatin states, partitioning the gene-poor and developmentally repressed portion of the genome close to the lamina or to the nucleolus [10]. Gene dense and transcriptionally active regions, on the other hand, tend to occupy more internal positions within the nuclear space. This is further translated into a functional compartmentalization within the nucleus, where highly active regions tend to cluster together and form dynamic transcription factories in the nuclear interior, whereas inactive genes colocalize in transcriptionally repressive areas, for example, at the nuclear periphery [10,14]. Such organization leads to the partial spatial separation between transcriptionally permissive and repressive subnuclear environments, which constrains 3D chromatin crosstalk between active and inactive chromatin domains [7]. We will continue by discussing how such compartmentalization of nuclear functions might promote the maintenance of cell type–specific chromatin states, expression patterns, and restrict cellular phenotypic plasticity [3,7,10]. Conversely, chromatin movements between active and inactive compartments have been linked to dynamic as well as stable transitions between active and repressed chromatin states, and to changes in gene activity [7]. Finally, we discuss the hypothesis that stochastic chromatin movements resulting in functional interactions increase nongenetic cell-to-cell variation in the transcriptome to modulate the degree of cellular phenotypic plasticity during development and are the target of pathologic reprogramming in ageing and cancer [7].

17.2   3D GENOME ORGANIZATION AND CELL IDENTITY

High-throughput assays designed to map transient chromatin fiber interactions have uncovered that the genome establishes dynamic physical contacts between regulatory elements that are located far apart on the linear chromatin fiber [15]. The most frequent interactions involve chromatin regions that are located within the same chromosome, and result in the formation of local chromatin loops in *cis*. Some regions, however, display large-scale movements and can form recurrent, transient contacts between different chromosomes. These *trans* interactions tend to involve genomic loci that are expressed in a coordinated, alternative or monoallelic fashion [9,11,12]. By fine-tuning and coordinating genomic functions, such 3D chromosome crosstalk in *cis* and in *trans* modulates phenotypic variation and adaptive responses [9,11,12]. Regulatory elements that are often involved in chromatin fiber interactions thus include enhancers and their target promoters, which can be located even a million base pairs distal to each other, or even on different chromosomes [16]. Furthermore, unscheduled gene activation is prevented by so-called insulator elements that can prevent enhancer–promoter communication if placed in between the enhancer and the target gene [10]. Given that enhancers regulate cell type–specific gene expression patterns, one of the major questions in chromatin biology concerns the mechanism by which enhancer–promoter interactions are specified and reorganised during differentiation and in diseases.

17.2.1   Enhancer Usage in 3D and the Regulation of Cell Type–Specific Gene Expression

Enhancer elements coordinate inputs from developmental and oncogenic pathways, as well as signals from the local chromatin structure to regulate the probability and variability of transcriptional bursts at their target genes [17–19]. The molecular mechanism of enhancer–promoter interactions and the determinants of the cell type- and differentiation stage–specific features of this process are just being elucidated. Global analyses of chromatin fiber contacts have thus mapped long-range, cell type–specific enhancer–promoter interactions [16]—suggesting that stochastic chromatin fiber movements and transient stabilization of chromatin fiber interactions form the basis of transcriptional regulation. The frequency of enhancer–promoter interactions might therefore regulate the frequency and/or duration of productive transcription, and thereby modulate the gene expression variability in a cell population.

Rare, but functional interactions between regions located on different chromosomes have, for example, been shown to result in variegated gene expression, providing a selectable feature within a cell population [20]. In contrast, stable, cell type- and differentiation stage–specific transcription is regulated by the convergence of multiple enhancer elements on specific, shared target genes in *cis*. This pattern ensures robust promoter activation by increasing the local concentration of factors promoting the assembly of preinitiation complex and the release of Pol II from pausing. An intriguing form of enhancer specificity is exemplified by the so called fail-safe or split enhancers that are spatially separated enhancer units, which need to be simultaneously active for the induction of gene expression, conferring tight and specific regulation of genes that might become oncogenic if deregulated [21]. Finally, clustering of multiple enhancer elements covering tens or hundreds of kilobases have evolved in the vicinity of cell fate–determining genes to enable their stable expression and the maintenance of cellular phenotypes [22]. The subunits of these so-called superenhancers collaborate and dynamically interact with each other and with other enhancers to integrate signals from multiple cell fate–determining pathways regulating the different individual enhancer subunits, resulting in a combinatorial signal output that increases the probability of transcription at their target genes [17].

Formation and decommissioning of superenhancers are hypersensitive to fluctuations in the levels of transcriptional regulators and chromatin factors, such as bromodomain containing 4 (BRD4) and the Mediator complex—an essential cofactor regulating enhancer–promoter contact, chromatin remodelling, and release of Pol II from pausing [23]. For example, activation of the proinflammatory cytokine tumor necrosis factor (TNF) leads to the redistribution of Mediator complex from cell fate–specifying superenhancers to enhancers that activate inflammatory genes, leading to the downregulation of genes that define cell identity [23]. Furthermore, the location and the activity of superenhancers are also affected by the cellular microenvironment of the stem cell niche [24]. Hence, the genomic positions of superenhancers are reversibly reorganized in follicular stem cells when they localize outside of their stem cell niche during wound healing, or under in vitro culture conditions.

The probability of functional contacts between enhancers and promoters is regulated by factors that act as the molecular ties of enhancer–promoter interactions [25–28], and by constrains of chromatin mobility, which is influenced by the 3D organization of the neighboring chromatin context [29]. Consistent with the role of enhancers in defining cellular states, downregulation of or mutations in the molecular ties of enhancer–promoter interactions and chromatin architectural proteins impair differentiation [30], counteract the reprogramming of differentiated cells into an induced pluripotent state [31] and affect tumor cell–specific gene expression programs and survival [32]. For example, downregulation of a member of the cohesin complex, a protein complex stabilizing enhancer–promoter contacts, interferes with the establishment of intrachromosomal loops in the *OCT1* locus and with the generation of induced pluripotent stem (iPS) cells [31]. Furthermore, tethering the self-association domain of a looping factor, lim domain binding 1 (LDB1), to the β-globin gene promoter is sufficient to establish an interaction with the globin locus control region, and to induce transcription initiation in the absence of the lineage-specific transcription factor GATA1 [33]. Other domains of the LDB1 protein are then necessary for the subsequent recruitment of transcription factors and chromatin remodelling activities [34] for a transition to productive elongation. Moreover, liver-specific knockout of LDB1 promotes the development of liver cancer in mice exposed to carcinogens [32]. Finally, enhancer RNAs (eRNAs) that mark active enhancers have also been shown to regulate chromatin loop formation between enhancers and promoters by promoting the recruitment and enhancing the kinase activity of the Mediator complex that regulates Pol II activity and binds to chromatin remodelling enzymes and transcription factors necessary for efficient transcription [35]. Mutations in the mediator subunit 12 (MED12) cause developmental defects, and also abolish eRNA–MED12 interactions, further supporting the importance of eRNAs in shaping cell type-specific gene expression patterns.

Apart from the molecular ties mediating chromatin contacts, the organization of the genome into topologically associated domains (TADs) also affects the probability of enhancer–promoter interactions by providing constrains for chromatin movements (see also Chapter 1) [36]. Importantly, the boundary strength of TADs is influenced by the presence of chromatin architectural proteins or genome organizers, such as the CCCTC-binding factor (CTCF) [36]. Certain naturally occurring sequence variations in humans and engineered mutations in mice that prevent CTCF binding to a TAD boundary are causally linked with limb developmental defects due to the emergence of ectopic interactions between distal enhancers and genes regulating limb development [29]. As CTCF often binds to DNA in a methylation-sensitive manner, TAD boundaries might become reprogrammed also in tumors due to the frequent deregulation of DNA methylation patterns. Altered TAD boundaries, in turn, might deregulate the mobility of oncogenic superenhancers and interfere with the stability of cell type–specific gene expression patterns [7]. Given the tight link between 3D genome organization and transcription, it will be thus important to determine how 3D chromatin cross talk is reprogrammed during transitions between various cellular states in development and neoplasia.

17.2.2 Insulators

Insulators are named after their ability to prevent functional contacts between enhancers and promoters if located in between them [8,9,37]. Similarly to enhancers, insulators regulate gene expression by influencing the probability of interactions between distant regulatory elements (Fig. 17.1) [8]. So far, the only well-characterized insulator of the mouse genome is the *H19* imprinting control region (ICR), which regulates the parent of origin-specific monoallelic

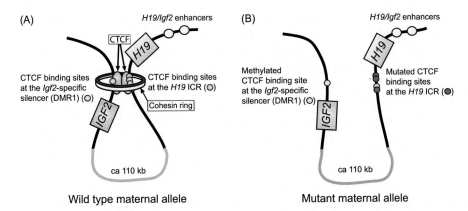

FIGURE 17.1 Insulators regulate gene expression by organizing 3D chromosome cross talk between regulatory elements.

(A) *The H19 imprinting control region (ICR) regulates the parent of origin-specific monoallelic expression of H19 and Igf2.* On the maternal allele, CCCTC-binding factor *(CTCF)*-binding sites at the *H19* ICR (indicated by four blue circles) mediate an interaction in *cis* with a CTCF-binding site in the DMR1 silencer (pink circle) to position *Igf2* in a repressive loop, separated from the neighboring enhancers (yellow circles). The involvement of the cohesin ring is hypothetical [40] (B) *CTCF-binding sites within the maternal H19 ICR regulate the chromatin states of both the H19 ICR and the DMR1 silencer.* Mutations in three of the four CTCF-binding sites (red circles) and the consequent loss of CTCF binding to the maternal *H19* ICR leads to biallelic *Igf2* expression, and abolishes the interaction between the *H19* ICR and DMR1. This is paralleled by de novo DNA methylation gain both at the mutant CTCF binding sites within the *H19* ICR and at the CTCF-binding site in DMR1, and by loss of CTCF binding to the DMR1 (brown circle), likely inhibiting its silencer activity (see text).

expression of a pair of imprinted genes, *Ifg2* and *H19* (see also Chapter 1) [8,38]. The monoallelic expression of *Igf2* from the paternal allele and *H19* from the maternal allele is thus regulated by the maternal allele–specific binding of the insulator protein, CTCF, to its binding site cluster within the *H19* ICR. On the maternal allele, the CTCF-bound insulator prevents communication between the *Igf2* promoter and its upstream enhancers. On the paternal allele, the *H19* ICR is methylated, which not only prevents CTCF binding but also spreads to the promoter of *H19*, leading to its consequent silencing [8,38]. Moreover, the functional insulator might indirectly contribute to the inactive state of the maternal *Igf2* allele by forming an interaction with a silencer, also called differentially methylated region one (DMR1), that is active only on the maternal allele and when it is in an unmethylated state. This interaction between the *H19* ICR and the silencer was suggested to be necessary for the maintenance of silencer activity (Fig. 17.1), because deletion of the CTCF-binding sites within the maternal *H19* ICR led not only to the loss of this interaction, but also to the de novo methylation and consequent inactivation of the silencer, and to biallelic *Igf2* expression [8]. The essential role of the silencer in repressing maternal *Igf2* expression is further illustrated by the findings that its deletion leads to bialelleic *Igf2* expression despite the presence of an intact insulator [39].

Although the role of insulators in regulating the probability of gene expression and cell-to-cell variation in gene activity genome wide is largely unexplored, the insulator strength has recently been linked to the number of intact CTCF-binding sites within the *H19* ICR [41]. Hence, out of the 4 CTCF-binding sites within the mouse *H19* ICR, CTCF binding to site number 2 is sufficient to provide insulator function. However, the reduced CTCF occupancy leads to variable relaxation of the monoallelic *Igf2* expression and to the partial, variable activation of the maternal *Igf2* allele [41]. The role of CTCF in 3D genome organization has gained substantial attention with the discovery that convergent orientation of CTCF-binding sites predicts the formation of chromatin loops within the genome [41]. Clustered and single CTCF-binding sites might thus have context-dependent functions, and potentially regulate the levels of expression variability genome wide both within and outside the insulator concept.

17.2.3 Chromatin Crosstalk, Transcriptional Noise and Cell Fate Decisions

A picture is thus emerging where the interplay between chromatin structure, chromatin architectural proteins, and lineage determining, as well as signal-dependent transcription factor networks regulate enhancer activity and cellular responsiveness to environmental stimuli that regulate cell fate decisions [36]. The role of chromatin in this process is highlighted by the discovery that enhancers regulating cell differentiation in early development are often premarked in human embryonic stem cells [42] with a specific chromatin signature. Hence, poised enhancers that regulate gastrulation, mesoderm formation and neurulation carry chromatin regulators, such as p300 and Brahma-related gene-1 (BRG1), histone H3 lysine 4 mono-methylation (H3K4me1) marks and display low nucleosome density already in embryonic stem cells. Active enhancers, on the other hand, that regulate genes expressed in stem cells and in the epiblast carry—in addition to the aforementioned chromatin features— also H3K27ac marks [42], which is linked to the production of eRNAs and gene activity [43]. Decommissioning of stem cell–specific active enhancer chromatin is equally important for proper differentiation, and is mediated *via* the recruitment of the lysine-specific demethylase 1 (LSD1) and the nucleosome remodeling and deacetylase (NuRD) complex to enhancers [44]. Epigenetic priming of enhancers has also been observed during the diversification of endodermal lineages, where poised chromatin signature predicted responsiveness to inductive cues and developmental competence. Chromatin states at regions overlapping with future active enhancers thus guided the recruitment of the pioneer transcription factors forkhead box A1 and 2 (FOXA1,2), followed by the recruitment of lineage-specifying transcription factors, such as pancreatic and duodenal homeobox 1 (PDX1), to specify cell fate [45]. Similarly, cell type–specificity of signal-dependent transcription factors is guided, at least in part, by the existence of poised enhancers [45]. Importantly, such poised state often involves the existence of chromatin fiber interactions between poised enhancers and their silent target genes before overt activation [46,47]. Preexisting enhancer–promoter interactions thus mark the genes that efficiently respond to TNF alpha–induced gene expression [46]. An intriguing class of poised enhancers is represented by those elements that carry the permissive H3K4me1 marks and also interact with polycomb-repressed genes [48]. The functional consequences of transcription factor binding to such chromatin organization, which is found for example at the repressed myogenic differentiation 1 (*MYOD1*) gene promoter in nonexpressing cells, depends on the identity of transcription factors involved and their binding partners. Hence, whereas exogenous expression of OCT4 led to the formation of bivalent chromatin marks at the *MYOD1* promoter, binding of MYOD1 to its own enhancer induced the deposition of activating chromatin marks and gene activation [48]. In summary, although lineage-determining pioneer transcription factors and cell type–specific combinations of signal-dependent DNA-binding factors play important roles in the establishment of chromatin states at enhancers, the mechanism of this process and the formation of poised enhancer states before overt activation remain still poorly understood [43,49].

Importantly, the initial phase of lineage commitment in primary stem and progenitor cell compartments of the bone marrow occurs in the absence of lineage-determining transcription factors, and is governed by uncoordinated and stochastic activation of lineage-affiliated genes [50]. Establishment of coherent, cell type–specific gene expression thus constitutes a second phase of lineage commitment, which follows an initially divergent expression pattern. An intriguing possibility is that stochastic chromatin fiber interactions might contribute to the establishment of poised chromatin during development and tumorigenesis. Although 3D chromatin organization is surprisingly stable at the mega base scale [47], even TAD boundaries can be extensively rewired by cellular stress, such as heat shock [51]. Experiments in *Drosophila* have thus demonstrated that heat-induced loss of TAD boundaries were likely caused by the redistribution of chromatin architectural proteins, and led to the emergence of novel long-range enhancer–promoter interactions. Although the heat shock–induced enhancer–promoter interactions were nonfunctional and silenced by Polycomb proteins during temperature stress, they documented that TAD boundaries can be reprogrammed by environmental cues to alter global enhancer-promoter interaction patterns. Long-range interactions have also been proposed to mediate the deposition of chromatin remodelling activities on interacting chromatin regions [52], suggesting that chromatin fibre interactions might influence chromatin states pleiotropically in *cis* and in *trans*. For example, although targeting of the ISW2 chromatin remodelling factor to DNA requires transcription factors, its genomic-binding sites often lack transcription factor–binding sites and instead reflect the 3D folding of chromatin [52]. It will be important therefore to establish how cell-to-cell variations in gene expression patterns in the initial phase of lineage commitment relate to stochastic 3D organization of chromatin structure.

This endeavor requires the invention of sensitive and quantitative methods that enable the examination of chromatin fiber interactions in small, transient cell populations. Understanding how the 3D organization of the genome is reprogrammed and stabilized during cell state transitions will likely also shed new light on the mechanisms that act as barriers for dedifferentiation [7]. Interestingly, fibroblasts of patients with the premature ageing syndrome, Hutchinson–Gilford progeria syndrome (HGPS), show remarkable resistance against oncogene-induced transformation and loss of fibroblast-specific gene expression patterns. This feature has been linked with the resistance of these cells to oncogene-induced rewiring of BRD4 binding to fibroblast-specific enhancers [53]. Importantly, BRD4 is also protective against oncogenic transformation of fibroblasts from healthy individuals, which might reflect the heritable nature of BRD4 binding to chromatin and its ability to maintain cell type–specific enhancer–promoter cross talk [53,54]. In line with these observations, BRD4 expression in solid tumors contributes to good prognosis and protection against dedifferentiation and metastasis formation. In liquid tumors, on the other hand, BRD4 expression correlates with poor prognosis, suggesting that it has context-dependent functions, which is likely linked to its different binding partners and cellular chromatin organization [53].

17.3 COMPARTMENTALIZATION OF NUCLEAR FUNCTIONS IN 3D

The probability of enhancer-promoter crosstalk and gene activity has also been linked to subnuclear localization and the overall compartmentalization of the genome within the 3D nucleus (Fig. 17.2). In differentiated cells, transcriptionally active or poised chromatin thus tends to occupy the nuclear interior, where transcription and RNA processing is coordinated in transcription factories and nuclear speckles [55]. Repressed chromatin states, on the other hand, are particularly enriched at the nuclear periphery and around the nucleolus, and form the so-called lamina-associated or nucleolus-associated domains (LADs or NADs), respectively [56,57]. LADs contain not only the constitutively silenced, AT-rich regions of the genome, but also include developmentally silenced genes, thereby contributing to the robust maintenance of cell type–specific phenotypes [56]. Identifying the factors that target certain genomic regions and enzymatic activities to different subnuclear positions is therefore pivotal for our understanding of differentiation and the regulation of cell type–specific gene expression. In the following paragraphs we will first examine how heterochromatin compartments form at the lamina and at the nucleolus, and discuss how genomic regions are targeted to these repressive compartments. Finally, we will explore the role of chromatin mobility between transcriptionally repressive and permissive environments in the regulation of expression variability and phenotypic plasticity.

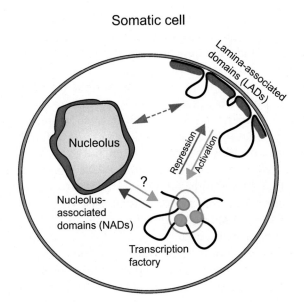

FIGURE 17.2 Maintenance of epigenetic memories in the 3D nuclear architecture.
Repressed chromatin domains dynamically localize to the nuclear periphery and to the nucleolus forming lamina-associated domains (*LADs*, red clouds) and nucleolus-associated domains (*NADs*, red clouds), respectively (see text). Transcriptionally permissive chromatin regions localize to the nuclear interior (transcription factory indicated by green circles). Repression of certain developmentally regulated genes involves the recruitment of these genomic loci to the nuclear periphery (red arrow), where repressive chromatin modifiers coordinate the establishment of multiple layers of epigenetic modification to ensure stable repression. Activation of such repressed genes is preceded by their release to the nuclear interior (green arrows). Moreover, variegated gene expression has also been linked to the stochastic recruitment of genes to the lamina in a cell population (red and green arrows). Recruitment of developmentally regulated genes to and from the nucleolus [58] is hypothetical (red and green arrows with question mark). LADs localize to the nuclear periphery in a dynamic manner (see text). Maintenance of repressed states at LADs involves the stochastic recruitment of LADs to the lamina after each cell division, a time window when LADs can also be recruited to the nucleolus and NADs can localize to the lamina (red dashed arrows). Tendency for spatial separation between active and inactive chromatin domains and the respective chromatin modifiers has been suggested to decrease transcriptional noise and promote the maintenance of epigenetic memories [7] (see text).

17.3.1   Formation of Heterochromatin Compartment at the Nuclear Periphery During Differentiation

Whereas the chromatin of pluripotent embryonic stem cells [59] is hyperdynamic and fills the available nuclear space rather homogenously, differentiated cells possess a highly compartmentalized nuclear architecture. This is exemplified by the formation of the transcriptionally repressive environment at the lamina and around the nucleolus, which parallels the acquisition of heterochromatin at gene-poor genomic regions and at developmentally repressed genes during differentiation [60,61]. The hyperdynamic nature of chromatin in pluripotent cells has been argued to support increased developmental potential [59], whereas the spatial separation of repressed and active chromatin has been suggested to stabilize developmental decisions [7], and to provide an additional layer of transcriptional regulation by chromatin mobility between transcriptionally permissive and repressive environments [7].

17.3.1.1 The Nuclear Periphery is Enriched in Repressed Chromatin Domains

The localisation of genomic regions to the nuclear periphery is generally linked with transcriptional repression (Fig. 17.2). Molecular maps of chromatin–lamina interactions [62,63] have thus uncovered that LADs overlap with repressed, gene-poor, AT-rich regions that tend to localize to the nuclear envelope already in embryonic stem cells. These regions form the so-called constitutive LADs (cLADs) that are present at the lamina in many different cell types. Apart from cLADs, LADs contain also many developmentally repressed genes, which are recruited to the periphery in a cell type–specific manner, and form the so-called facultative LADs (fLADs) [64]. In mammalian cells, around 40% of the genome is thus covered by LADs with a size ranging from ~10 Kbp to ~10 Mbp. In differentiated cells, a significant fraction of LADs overlaps with large domains that are enriched in repressive H3K9me2 histone modifications, also called large organized chromatin K9 modifications (LOCKs) [65]. The size and frequency of such H3K9me2 LOCKs increases during the differentiation of embryonic stem cells to ensure the stable repression of gene-poor regions and to coordinate cell type–specific repression [65]. Interestingly, downregulation of OCT4 and loss of pluripotency during mouse embryonic development is accompanied by the formation of compact chromatin structures at the nuclear periphery, which likely correspond to the emerging H3K9me2 LOCKs [60]. Such repressive, compact chromatin environment at the lamina has recently been also visualized in colon cancer cells by a novel technique called *chromatin in situ proximity* assay (ChRISP) [66,67] that translates chromatin fiber density and colocalization between histone modifications and specific genomic regions into a fluorescent signal.

Interestingly, reducing the levels of H3K9me2/3 marks by the inhibition of histone H3K9-methyltrtansferases G9a/GLP and SETDB1 greatly enhanced the efficiency of iPS cell generation, suggesting that repressed LOCKs might act as barriers against dedifferentiation [68]. Indeed, H3K9me3 marks can initiate the formation of multiple layers of self-reinforcing repressive modifications by providing a platform for heterochromatin protein 1 (HP1) binding, which in turn recruits histone- and DNA methyl-transferases enabling not only the spreading of heterochromatin states in *cis*, but also the generation of repressive chromatin states that are heritable during cell division [6]. How the spreading of heterochromatin into neighboring active chromatin domains is prevented is still poorly understood. Candidate factors involve features that are enriched at LAD borders, such as CTCF-binding sites, promoters that are oriented away from the LADs and cytosine-phosphate-guanine (CpG) islands that tend to be protected from gaining DNA methylation [62]. Furthermore, LADs, especially the outer third of LADs adjacent to the LAD borders, are also enriched in histone H3 lysine 27 tri-methylatlion (H3K27me3) modification [62], a repressive chromatin mark that has been shown to functionally collaborate with H3K9me2/3 modifications to maintain the stable repressed state of the inactive X chromosome [69]. Key questions address how deposition of such modifications is orchestrated at the nuclear periphery, and which factors target genomic regions to the lamina.

17.3.1.2   Recruitment of Repressive Chromatin Modifiers and LADs to the Lamina

The acquisition of repressive chromatin modifications and peripheral localization appear to be functionally linked [70]. On one hand, the nuclear envelope can attract repressive chromatin modifiers via protein–protein interactions to coordinate and maintain silenced states at the nuclear periphery. On the other hand, repressive chromatin states and chromatin modifiers seem to collaborate in targeting LADs to the lamina. The components of the lamina play thus essential roles both in 3D genome organization and in the regulation of gene expression patterns.

17.3.1.2.1   Recruitment of Chromatin Modifiers to the Lamina

The composition of the nuclear lamina changes dynamically during differentiation and within cell types [71–74]. It is composed of a meshwork of proteins formed by intermediate filaments and membrane associated proteins. The main structural components are type V intermediate filament proteins, A- and B-type lamins. The nuclear lamina contains also other proteins that span the inner nuclear membrane, such as lamin B receptor (LBR), emerin, LEM domain containing 2 (LEMD2), and lamina-associated polypeptide 2 beta (LAP2β). These components attract and, in some instances, also activate repressive chromatin modifiers, such as the lysine-specific histone demethylase LSD1 [75], the H3K9me2 histone methyltransferase G9a [76], histone deacetylase 3 (HDAC3) [77,78], and members of the nuclear receptor corepressor (N-CoR) complex [79], which together maintain a repressive environment at the nuclear envelope [70,80]. These chromatin-modifying enzymes play important roles in regulating the balance between self-renewal and differentiation [81,82], the efficiency of reprogramming into iPS states [83], age-related chromatin changes [84,85] and cancer development [85,86]—indicating that their function is tightly linked to the spatiotemporal compartmentalization of the genome in the 3D space of the nucleus [7].

17.3.1.2.2   Recruitment of LADs to the Lamina by Sequence-Specific DNA-Binding Factors

Recruitment of genomic regions to the repressive environment of the nuclear envelope is promoted by collaboration between repressive histone marks, sequence-specific transcription factors [87], chromatin components [88], and proteins of the nuclear envelope [70]. Attraction of LADs to the nuclear periphery thus can be partially explained by their DNA sequence composition. A study in mice has revealed that repeated GAGA motives in LADs direct lamina association by binding to the transcriptional repressor c-Krox, that forms a complex with HDAC3 and Lap2β [89]. Furthermore, the subtelomeric 80 bp D4Z4 repeated sequence also positions an adjacent telomere to the nuclear periphery in a lamin A- and CTCF-dependent manner [90]. However, other DNA sequences may also be important for perinuclear localization. As indicated earlier, mammalian constitutive LADs are enriched in A/T-rich isochores [91], which might also regulate peripheral targeting. Sequence-specific DNA-binding proteins recognizing A/T-rich sequences may thus act as molecular bridges between the DNA and the nuclear lamina. Finally, the transcription factor Ying Yang 1 (YY1), which is associated with both transcriptional activation and repression, has also been found to promote association of LADs to the nuclear periphery in concert with lamin A/C, H3K27me3, and H3K9me2/3 marks [87].

17.3.1.2.3 Recruitment of LADs to the Lamina by Lamin Proteins

The first high-throughput experiments scoring for factors involved in genome positioning [92,93] confirmed that localization of genomic regions to the lamina is indeed an active process, which depends on multiple components of the nuclear envelope, histone modifiers, histone modifications, and RNA-binding proteins. Moreover, these studies also showed that functions in chromatin positioning could be uncoupled from transcriptional repressor activity.

The importance of lamins and the integrity of the nuclear envelope in the peripheral positioning of genomic loci have thus been convincingly demonstrated. Mutations or downregulation of the level of lamin proteins cause the relocalization of specific loci into the nuclear interior in several different species [94–96]. Chromatin tethering to the lamina may thus occur through the binding of nuclear lamina components to particular chromatin states or to other chromatin proteins. One of the main candidates for chromatin tethering to the nuclear periphery is the LBR, which binds to lamin B1 (LMNB1) and also interacts with the chromatin components HP1 and MeCP2 [97,98], as well as with histone H4 lysine 20 di-methylation (H4K20me2) marks [99]. Moreover, LBR expression levels affect nuclear organization. The collaboration between lamin A/C and LBR in tethering heterochromatin to the lamina has been illustrated by experiments showing that engineered absence of both lamin A/C and LBR leads to the formation of inverted nuclear architecture with heterochromatin localizing to the interior of the nucleus [100]. Such inverted nuclear architecture and the absence of lamin A/C and LBR are characteristic features of the rod cells located in the retina of nocturnal mammals, and serves the purpose to minimize the scattering of light by heterochromatin foci and improve night vision [101]. Ectopic expression of LBR in these cells causes peripheral relocalization of heterochromatin [100,101], further highlighting the role of LBR in 3D genome organization. Interestingly, the cell type-specific presence or absence of LBR appears to regulate the targeting of certain heterochromatic regions to different repressive subnuclear environments. For example, each mouse olfactory sensory neuron (OSN) expresses only one allele of the ~2800 olfactory receptor (OR) alleles [102]. Repressed OR alleles located on different chromosomes are known to form heterochromatic aggregates in the interior of the nucleus of OSNs, which normally lack LBR. Ectopic expression of LBR, however, relocates the inactive OR genes from the nuclear interior to the nuclear periphery, and interferes with the monoallelic expression of ORs [102]. Taken together, these studies highlight the importance of lamins and LBR in 3D genome organization and peripheral tethering of chromatin.

17.3.1.2.4 Recruitment of LADs to the Lamina by Chromatin Modifications and Chromatin Components

Localization of genomic regions to the nuclear lamina also depends on the existence of specific histone posttranslational modifications and the activity of certain chromatin-modifying enzymes, which confer affinity to the chromatin fiber to favor interaction with the nuclear envelope. Repressive chromatin marks are thus tightly linked with peripheral positioning and influence the integrity of the nuclear envelope. For example, in human and mouse cells deposition of H3K9me1 in the cytoplasm by Prdm3 and Prdm16 histone monomethyl-transferases is required not only for subsequent heterochromatin formation in the nucleus, but also for the integrity of the nuclear lamina [103]. In mouse fibroblasts and in HeLa cells, both H3K9me2 modifications (deposited by G9a/GLP heterodimers) and H3K9me3 marks (deposited by suppressor of variegation 3-9 homolog 1/2 (SUV39H1,2)) have been shown to promote peripheral targeting [104]. Furthermore, in a human fibrosarcoma cell line G9a activity has been reported to have a crucial role in positioning LADs to the nuclear periphery [105,106]. H3K9me2/3 and the responsible histone methyl transferases are linked with peripheral chromatin positioning also in *Caenorhabditis elegans* [107]. Finally, histone deacetylation may also play a role in this process, as class I HDACs have been implicated not only in gene repression, but also in the positioning of heterochromatin to the nuclear lamina in *Drosophila* [77,80,108,109].

Apart from chromatin marks, specific histone variants also appear to play important roles in 3D genome organization. They confer different structural and functional properties to the nucleosome, and provide a distinct landscape for interactions with other chromatin proteins [110]. Thus, it is not surprising that specific histone variants are enriched in LADs and are associated with the nuclear lamina. For example, macroH2A1 is enriched at the border of LADs in mouse liver cells, and is required for the maintenance of chromatin compaction at the nuclear periphery and for the peripheral localization of LADs [111]. Moreover, specific histone H1 variants are preferentially enriched at LADs and NADs in different cell types, implicating these variants in the tethering genomic regions to the periphery [112–114]. Nonetheless, the extent to which the presence of these variants affects genomic positioning still needs to be resolved.

17.3.1.3 Stochastic Chromatin Movements to the Lamina and Gene Expression Variability

Subnuclear localization of certain genomic loci can dynamically change during differentiation and in response to environmental signals with consequences on transcriptional activity (Fig. 17.2). Localization to the nuclear periphery has thus been proposed to enable stable gene repression by decreasing the probability of transcriptional activation and the level of transcriptional noise, thereby stabilizing cellular phenotypes and cell fate decisions [7]. Transcriptionally active genes destined to undergo developmentally stable repression often relocate from the interior of the nucleus to the lamina (Fig. 17.2). Seminal examples are represented by the relocalization of repressed pluripotency genes from the nuclear interior to the nuclear periphery during differentiation, and cell type–specific recruitment developmentally silenced genes to the lamina [63,115,116]. Developmentally regulated gene activation, on the other hand, can be accompanied or preceded by the regulated release of the facultative LAD covering the activated gene from the lamina into the nuclear interior (Fig. 17.2) [117]. The immunglobulin haevy chain and kappa loci, for example, are repressed and frequently positioned to the lamina in non-B cells and in hematopoietic progenitors. In pro-B cells, however, they are positioned away from the nuclear periphery and aqcuire a compact conformation in preparation for V-D-J recombination and gene activation [117].

Stochastic and dynamic recruitment of genes to the nuclear periphery and other repressive nuclear environments also contributes to cell-to-cell variation in gene expression, that is, variegated silencing (Fig. 17.2). This phenomenon includes, for example, the allelic exclusion of T cell receptors (TCR) [118], which limits the number of antigen receptors to a single allele per cell. In developing thymocytes, the germline T cell receptor b (*Tcrb*) genes thus localize to the lamina and to the pericentric heterochromatin in a stochastic and monoallelic manner. Peripheral localization in turn has been shown to suppress recombination [118,119].

How positioning at the nuclear lamina affects gene expression and other enzymatic activities is still an open question. This issue was addressed by experiments where a locus was artificially tethered to or released from the nuclear periphery [80,120,121]. Although, there are exceptions to the rule, these observations document that there is a tight correlation between recruitment to the periphery and repression of the attachment site, as well as the repression of nearby genomic regions [120,122,123]. The effects of local chromatin context and subnuclear localization on gene expression has been examined also with reporter genes randomly integrated into the genome [91]. Twenty-seven thousand reporters integrated thus randomly into mouse embryonic stem cells (mESC) revealed that although position per se does not always reflect a transcriptional state, integration into LADs tends to be associated with gene silencing. Importantly, however, position alone is not sufficient to explain gene expression or silencing, and transcription per se does not position a locus away from the periphery.

17.3.2   Heterochromatin Compartment at the Perinucleolar Space

LADs extensively overlap with NADs that localize to the repressive shell surrounding the nucleoli [124]. Moreover, a given LAD associated with the nuclear periphery in one cell cycle might be found in the vicinity of the nucleolus in another cell cycle (Fig. 17.2) [105]. LADs and NADs thus represent a similar chromatin environment with comparable properties [124,125], and contribute to the organization of silent chromatin states within the nucleus. The precise mechanism of the establishment and maintenance of repressed states in these compartments, however, needs to be further elucidated.

The repressive compartment surrounding the nucleolus is established by the pericentric heterochromatin located in the vicinity of rRNA genes on the acrocentric chromosomes, and by the repressed portion of the rRNA genes looping out from the active compartment [57]. Such a chromatin arrangement—which isolates the transcriptionally active nucleolus from the rest of the genome—might be disrupted via increased rRNA expressionor or by the erosion of pericentric heterochromatin of acrocentric chromosomes. Importantly, downregulation of rRNA expression and acquisition of heterochromatin at the rRNA genes appear to be essential for the differentiation of embryonic stem cells [126]. The hyperdynamic chromatin of pluripotent cells is generally permissive for transcription, and is mostly devoid of heterochromatin, which features contribute to the developmental potential of embryonic stem cells. The emergence of repressed chromatin at the rDNA loci during embryonic stem cell differentiation has been shown to represent an important first step in the formation of repressive compact chromatin structures genome wide. This step is thus necessary for the expansion of LOCKs throughout the genome and for the formation of repressed chromatin states at the nuclear periphery during differentiation [126]. In turn, inhibition of heterochromatin formation at rDNA interferes with the expansion of LOCKs and also with proper differentiation. How heterochromatin formation at rRNA genes promotes the formation of repressive chromatin environment at the lamina and elsewhere in the genome is not completely understood. Relocation of repressed NADs after cell division either to the nucleolus or to the nuclear periphery might, however, underlie such a process (Fig. 17.2). The heterochromatin compartments at the lamina and at the nucleolus might thus play interchangeable and interconnected roles in the maintenance of cell type–specific gene expression in differentiated cells.

Although heterochromatin shows a tendency to associate with the nuclear periphery, and tethering of genes to the periphery can promote silencing, peripheral localization is neither necessary nor sufficient for silencing [70]. Moreover, it is important to note, that chromatin at the nuclear periphery is not always in a repressed transcriptional state. In yeast, chromatin surrounding nuclear pores can be highly active, and enriched in genes that are actively exporting mRNA to the cytoplasm in order to be translated [127]. A closer look at the microstructure of LOCKs in mammalian cells also reported more than 2500 small regions with very low H3K9me2 levels, termed euchromatic islands, embedded within LOCKs. These islands were enriched in DNase I hypersensitive sites, CTCF binding, and displayed tissue-specific differential methylation [128]. Although LOCKs highly correlate with LADs, it remains to be seen if euchromatic islands also associate with specific microenvironments, such as nuclear pores, at the nuclear periphery.

17.3.3 Nuclear Pores in the Regulation of Developmental Programs and Adaptation

17.3.3.1 *Nuclear Pores: Platforms for Gene Activation and Gene Repression*

Studies on the nuclear pore complex (NPC) and its components have during the last decades uncovered novel functions of this complex in the organization of chromatin structure and in the regulation of gene expression (as reviewed in [129–131]. Nucleoporins (Nups) thus possess also transport-independent functions and have a dual role in controlling both transcriptional activation and repression. Moreover, NPCs and their components associated to chromatin exhibit an essential role in the regulation of 3D genome organization, formation of transcriptional memory and maintenance of genome stability (as reviewed in References. [127,129,130]). In line with the above functions, Nups have been implicated in the regulation of pluripotency, differentiation, as well as the plasticity of phenotypes during adaption to environmental changes [132,133].

17.3.3.1.1 Gene Activation

Association of Nups with genes can occur both at the nuclear pore and in the interior of the nucleus with differential consequences on transcription. Genome-wide interaction maps and localization studies of several Nups (Nup98, Nup50, and Nup62) in *Drosophila* have revealed that they interact with genes both at the NPC and in the nuclear interior. Nups in the nucleoplasm predominantly associate with transcriptionally active genes, particularly those involved in developmental regulation and cell cycle progression. Moreover, depletion of Nup98 has led to the downregulation of its target genes [134], documenting its role in transcriptional activation. Nups were documented to dynamically bind also to the human genome, uncovering an unexpected role for NPC components in regulating gene expression programs during human embryonic stem cell differentiation [135]. In embryonic stem cells, chromatin-bound NUP 98 marked a subset of active and silent genes, while NUP 98 binding in neuronal progenitors was enriched only at active genes. Interestingly, genes destined to become active at later stages of neuronal development were occupied by NUP 98 already at an early differentiation phase, and preferentially localized to the NPC in a poised state for activation. This is in contrast to NUP98-bound genes active already at the early stages of differentiation, which occupied internal positions within the nucleus [135]. As overexpression of a truncated, dominant negative version NUP 98 interfered with the upregulation of a set of genes important for neuronal differentiation, NUP 98 binding appears to be not sufficient but necessary for transcriptional activation and proper differentiation [135].

17.3.3.1.2   Formation of Transcriptional Memories

Advancing our understanding of epigenetic flexibility and adaptation, Light et al have uncovered that NUP 98—gene interactions can also serve as transcriptional memory of previous gene activation [136]. Such a memory has been documented to exist at the interferon gamma (IFN-γ)-inducible genes, which respond to stimuli stronger and quicker in cells that were previously exposed to IFN-γ. The mechanism of memory formation involves NUP98-mediated retention of H3K4me2 modifications and poised Pol II binding at the transcriptionally inactive promoters exposed to prior stimuli. Although such transcriptional memory is evolutionarily conserved and present already in yeast, the mechanism and the mediators of this process are slightly different in different organisms [136]. For example, whereas transcriptionally inactive but poised IFN-γ targets tend to localize to the nuclear interior in human cells, they are recruited to the nuclear pores *via* NUP100 in yeast.

Other mechanisms of transcriptional memory involve the formation of gene loops between the 5′ and 3′ ends of certain genes [127]. Experiments in yeast thus uncovered that tethering the 5′ or the 3′ end, or both ends of a subset of genes to the NPC through the interactions with different Nups stabilizes the formation of gene loop structures between gene promoters and terminators [137]. Chromatin loops at the NPC were suggested to serve as a platform for the assembly of transcriptional and mRNA processing machineries or to demarcate active chromatin domains, thereby promoting the initial steps and later events of transcription and the formation of transcriptional memory [138–141].

17.3.3.1.3   Silencing

Apart from memory formation and transcriptional activation, there are several examples implicating Nup–chromatin associations in gene silencing. Genome-wide studies of NUP-chromatin binding in mammalian cells documented that a significant fraction of NUP-binding sites overlap with regions carrying repressive histone modifications and associate with the nuclear periphery [62,142,143]. In yeast, Nup170 interacts with silencers and chromatin remodeling enzymes to repress ribosomal protein genes and subtelomeric genes, and promotes the association of telomeres to the nuclear envelope [144]. In mESC, Nup153 maintains stem cell pluripotency by promoting Polycomb-mediated epigenetic silencing of genes that would induce differentiation. Depletion of Nup153 thus leads to the derepression of developmentally regulated genes and to the induction of early differentiation despite the presence of pluripotency factors [133]. Finally, in budding yeast, galactose-induced genes are recruited to the nuclear periphery via pore components upon transcriptional activation. Interestingly, localization to the nuclear periphery attenuates the induction of the GAL gene locus, and leads to rapid repression after initial gene activation, uncovering a function for the nuclear periphery in the repression of endogenous GAL gene expression. These results suggest a model in which recruitment to the repressive environment at the nuclear periphery contributes to the formation of a negative feedback loop that enables the GAL genes to quickly respond to changes in the environment [145].

17.3.3.2 NPC and Nups and Their Contribution to the "Gene Gating" Mechanism

Nups have been shown to control gene expression also at the posttranscriptional level. The *"gene gating* hypothesis,*"* formulated by Guenter Blobel already in 1985, posits that yeast NPCs and their components can serve as gene-gating organelles by coordinating transcription, mRNA processing, and nuclear export [139]. In this paragraph we will first summarize the evidence that Nups and/or the physical tethering of active gene loci to the NPCs increase the efficiency of transcriptional activation, mRNA processing and nuclear export. Second, we will examine the factors that interact with Nups to coordinate these processes at the nuclear pore.

17.3.3.2.1 The Gene Gating Principle

The *gene gating* hypothesis is supported by a series of studies documenting that certain inducible genes are recruited to the NPCs upon transcriptional activation, demonstrating that NPCs contribute to the 3D organization of the yeast genome [146,147]. Furthermore, genome-wide approaches have documented that Nups preferentially associate with transcriptionally active genes. Moreover, the transcriptional activation of many genes is paralleled by their frequent relocation to the NPCs, indicating that contact with pores might contribute to gene activation at these sites [141,143,148,149]. In line with this reasoning, artificial targeting of a reporter gene to NPCs has been shown to increase its transcriptional activity [140]. Evidence is thus emerging that certain Nups can increase the transcription of their target genes, and a subset of these target genes are frequently recruited to the envelope, while maintaining or increasing their activity at the nuclear periphery.

Several findings exist, however, which seem to contradict the *gene gating* hypothesis and suggest that it does not represent a general mechanism of gene regulation. For example, inhibiting the recruitment of the yeast *Gal1* and *HSP104* genes to the NPC does not affect their expression level, suggesting that gating of these genes to the pore is not essential for their activation. These findings thus highlight that nuclear localization to NPCs does not always contribute to increased gene expression. The relocalization of *Gal1* and *HSP104* to NPCs might thus be a consequence rather than a cause of their transcriptional activity [149] [150]. Moreover, other studies documented that chromatin movements within the 3D architecture of the nucleus can regulate transcription in a multistep process. For example, certain developmentally regulated genes first move from the repressive environment of the nuclear periphery to transcription factories in the nucleoplasm upon activation, followed by their gating back to the NPC for efficient mRNA export [127,151,152]. Finally, several experiments have uncovered yet other mechanisms of NPC action, and suggest that the recruitment of inducible genes to NPCs might fine-tune the expression kinetics by altering the rates of initial transcriptional induction and transcriptional reactivation [153,154].

All these observations illustrate that the mechanism of *gene gating* might not be universal. Gene–NPC associations might thus be particularly important for inducible genes, which need to rapidly reach high expression levels in the cytoplasm in response to environmental signals, which process could be facilitated by the close positioning of these genes to the NPC. While gene gating has been described in yeast, it is yet to be determined whether or not similar principles regulate a subset of mammalian genes as well. Interestingly, factors interacting with Nups in yeast to coordinate efficient transcription, mRNA, processing and nuclear export (see below) have evolutionary conserved functions and protein interaction partners, suggesting that similar coordination between these processes might operate also in the highly complex mammalian genomes.

17.3.3.2.2 Nup-Binding Factors Involved in the Coordination of Transcription, mRNA Processing and Nuclear Transport

The efficient nuclear export of mature mRNAs requires the accumulation and coordinated action of a number of specific factors on nascent transcripts. A central component in this process is the TREX-2 (transcription elongation and RNA export-2) complex, which functions in mRNA export and interacts with other complexes including RNA-polymerase II and the NPCs [155,156]. Moreover, the yeast TREX-2 subunit Sac3 associates with the Med31 subunit of the Mediator complex, which interaction not only regulates the assembly of the complex, but also promotes the consequent activity of RNA–Pol-II during the early stages of transcription [155–157]. These experiments thus pinpoint TREX-2, an evolutionary conserved complex that associates with the NPCs also in humans [158], as the coordinator of nuclear localization, transcription and mRNA export in yeast.

Several examples document a connection between the transcriptional export machinery and the association of Nups to highly transcribed gene loci. In yeast, Nup1 has been shown to bind to chromatin via the TREX-2 complex, which is necessary for the localization of inducible Gal genes to NPC [157,159,160]. As noted earlier, TREX-2 interacts with and stabilizes the composition of the Mediator complex [155]. Interestingly, loss of TREX-2–Mediator association prevents localization of Gal genes to NPCs, demonstrating that the TREX-2–Mediator association is functionally linked with Nup–chromatin binding

[157,159,160]. These examples illustrate that interactions between TREX 2, the Mediator complex, and components of the NPC provide a platform for the coordination of transcription, nuclear export, and 3D genome organization – thereby ensuring that a set of inducible genes undergoing active transcription are in close proximity to the NPCs to enable efficient mRNA export. Another example of the coordination between export and transcription at the NPC is represented by the regulation of tRNA expression. Cell cycle phase–specific increase in RNA–pol III-mediated tRNA transcription induces the repositioning of tDNA loci to NPCs—a process that requires a selected set of Nups, the nuclear exportin Los1 and cohesin. The tDNA–NPC association thus coordinates the peak of transcription with efficient nuclear export of pre-tRNAs [161].

The impact of NPC–chromatin and Nup–chromatin interactions at specific genomic loci are thus not restricted to transcriptional regulation, but also include transcriptional memory formation and global chromatin organization to modulate gene expression dynamics during development and in response to environmental cues [162]. The overall mechanism mediating and/or controlling Nup–chromatin interactions and the impact of chromatin-bound Nups on the transcriptional process are not fully characterized. Several observations have proposed that intranuclear Nups bound to chromatin might function as transport factors promoting the rapid and efficient shuttling of genes within different sub-compartments of the nucleus [162]. It is, however, still poorly understood what regulates the mobility of Nups to and from the NPCs. In mouse myoblast cells, for example, Nup50 is required for proper differentiation, and can be found both in the nucleoplasm and incorporated into NPCs. Interestingly this dynamic shuttling of Nup50 between the NPC and the nuclear interior requires active transcription by RNA–polymerase-II [163]. The complexity of Nup–chromatin interactions was further illustrated in *Drosophila*, where a regulatory feedback mechanism between the NPC subunits Nup62, Nup93, and Nup155 controls the dynamic attachment of chromatin to the nuclear envelope [164]. In this model, Nup155 not only promotes the tethering of chromatin to the periphery, but also binds Nup93 to form a platform for the recruitment of Nup62 that antagonizes chromatin tethering. This negative feedback loop will likely modulate the dynamics of gene expression response during adaptation to environmental cues in diverse cellular and developmental contexts [164]. Nups can thus form evolutionary conserved protein interaction hubs connecting the transcriptional machinery with the transcriptional export machinery at highly transcribed gene loci, and regulate gene expression from yeast to mammals during development and adaptation to the environment. It is not surprising therefore that mutations or deregulated expression of Nups have been linked with a diverse array of tissue specific as well as systemic human diseases and aging [165–167].

17.4   CHROMATIN MOBILITY BETWEEN NUCLEAR COMPARTMENTS AND PHENOTYPIC PLASTICITY

Although the role of chromatin fiber interactions and 3D genome organization in developmental decisions remains a largely uncharted territory [168,169], the importance of transcriptional noise in pluripotency, early steps of lineage specification, and phenotypic plasticity is increasingly being recognized (Fig. 17.3A and B) [3,170]. In this section, we will illustrate via two paradigms of phenotypic plasticity how changes in 3D genome organization

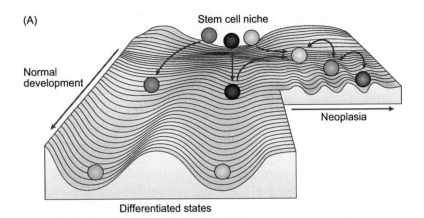

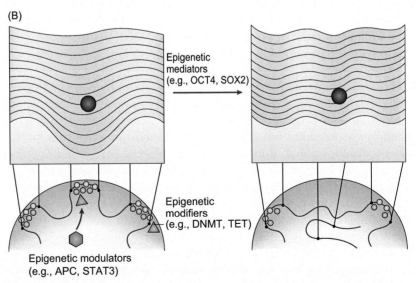

FIGURE 17.3 Epigenetic mediators collaborate with epigenetic modulators and epigenetic modifiers to regulate the phenotypic plasticity of tumor cells.

might contribute to phenotypic change. The first example is represented by epithelial-to-mesenchymal transition (EMT), which is facilitated by transcriptional noise and is paralleled by the loss of 3D chromatin compaction at the lamina [3]. In the second example we discuss the circadian plasticity of the 3D nuclear architecture and its contribution to the spatiotemporal coordination of transcription, a process that requires an interplay between 3D genome organizers and the circadian clock [106].

17.4.1 Reprogramming of 3D Genome Organization During EMT

As described earlier, the nuclear envelope plays a complex role in the maintenance of cell type–specific gene expression patterns. Recent studies demonstrated that cLADs are highly conserved between cell types, while fLADs dynamically associate with the nuclear periphery during the course of differentiation to coordinate cell type–specific repression at the laminay [64,116]. LOCKs/LADs located at the nuclear periphery were thus suggested to "lock in" developmental decisions by ensuring the stable maintenance of silent chromatin states and gene expression patterns [3].

(A) The Waddington landscape of development is adapted to compare phenotypic plasticity and developmental potential during normal differentiation (left side of the image) and in tumor development (right side of the image). The increased developmental potential of somatic stem cells (grey balls) and tumor stem cells (orange balls) is indicated by a position on the top of the hill. Phenotypic plasticity of a cell population as a whole is mediated by cellular heterogeneity, which is caused, in part, by cell-to-cell variations in epigenetic states (different shades of grey or orange). During differentiation, cells are canalized toward specific mature cell fates (brown and green balls) with lower developmental potential and phenotypic plasticity (deepening of the valleys or canalization). Epigenetic instability in tumor stem cells (CSCs) (orange balls), manifested as the erosion of LOCKs and the emergence of hypomethylated blocks, leads to phenotypic heterogeneity within the tumor (different shades of red). Mechanisms underlying increased heterogeneity include increased transcriptional noise (shallow valleys) and the emergence of cells states with different degrees of maturation and stochastic switches between diverse cell-states (arrows between valleys). The deregulated epigenome together with constantly changing selective pressures (e.g., inflammation, carcinogen exposures, aging) enables the selection of the fittest clones and tumor evolution. (B) During this process, epigenetic modifiers, modulators collaborate with mediators that shape the Waddington landscape of tumor development as described earlier. Epigenetic modulators (open circles) fine-tune and coordinate the activity of epigenetic modifiers (open triangles) early on during tumor development, leading to the ectopic expression of epigenetic mediators. Epigenetic mediators recruit epigenetic modifiers and modulators to reprogram chromatin modifications (blue circles), lamin proteins (yellow circles), chromosome cross talk and genome organization (new loop on right) to destabilize cellular states (shallow valleys, right hand image), and to increase phenotypic plasticity in neoplastic or preneoplastic cells.
Source: Reused with permission from Feinberg AP, Koldobskiy MA, Gondor A. Epigenetic modulators, modifiers and mediators in cancer aetiology and progression. Nat Rev Genet 2016;17(5):284–99.

The reprogramming of cellular phenotypes and the accompanying large-scale changes in gene expression patterns likely require flexible higher-order chromatin structures that enable adaptations to changes in the microenvironment. In line with this reasoning, signals that induce EMT are also able to alter 3D chromatin compaction at the nuclear periphery and affect the levels of repressive and activating histone modifications genome wide [3]. TGFβ-induced EMT is thus preceded by global loss of H3K9me2 modifications, gain of H3K4me3 and H3K36 modifications specifically at LOCKs, and the loss of chromatin compaction at the lamina [171]. Changes in chromatin structure during EMT are paralleled by increased transcription at genes that regulate RAS signaling, cell migration, and enzymatic functions relevant for the mesenchymal phenotype. Importantly, chromatin states during EMT resemble that of embryonic stem cells, where lack of large H3K9me2 LOCKs enables the maintenance of pluripotency and increased developmental potential [7,65,171]. Such similarity in chromatin states might provide mechanistic explanation for earlier observations suggesting that EMT phenotypes induce the acquisition of stem cell characteristics [7,65,171]. While stemness has been linked to increased transcriptional noise, decrease in the level of repressive H3K9me2 modification at LADs/LOCKs [172] has been hypothesized to increase transcriptional noise during TGFß-induced EMT, and thereby contribute to the destabilisation of epithelial cell identity [3].

Further studies will be required to dissect the mechanism by which TGFß reprograms peripheral heterochromatin compaction and cellular phenotypes, and whether this process also involves the relocalisation of LADs from the lamina to the nuclear interior. Initial observations established a role in this process for the dual-specific histone demethylase, LSD1 that can demethylate either H3K9me2 or H3K4me2, depending on its interacting partners [171]. As 3D chromatin compaction itself contributes to transcriptional repression [173], it will be important to decipher the molecular mechanism of the establishment and reprogramming of chromatin compaction at the lamina. One candidate factor regulating chromatin compaction is the tumor suppressor SMCHD1 that indirectly bridges the H3K27me3- and H3K9me3-enriched regions of the inactive X chromosome [174]. It remains to be seen, however, whether its functions extend to the regulation of chromatin structure at the nuclear periphery. Indeed, very little is known about the factors that regulate the 3D folding of LADs, and fine-tune the functional and structural cross talk between different repressive chromatin modifications within LADs to influence the robustness of cellular gene expression patterns [69].

17.4.2   Circadian Plasticity of 3D Genome Organization and Phenotypic Change

Circadian rhythm of physiology and behavior represents a paradigm of adaptation to changes in the environment, and includes, for example, rhythmic changes in hormonal levels, metabolism, and many other functions (reviewed in Chapter 16). It is not surprising therefore that perturbation of circadian regulation predisposes to a host of diseases, such as cardio-metabolic diseases and cancer. Despite efforts to understand the basic principles of circadian regulation, we still remain largely ignorant of its inner workings.

Circadian changes in physiology can be interpreted as seminal examples of adaptive plasticity. Cell-autonomous clocks organized by autoregulatory intertwined feedback loops, time delays and posttranslational modifications thus generate daily oscillation in the transcriptome and proteome [175–178]. External time cues, such as light and feeding, are necessary to synchronize the phasing of cell-autonomous clocks with the geophysical time within an organism or a cell population [179]. At the same time, periodic oscillations in the levels of gene products that regulate cell signalling and metabolism [180] generate timing-dependent combinatorial patterns in the proteome of single cells that may synergize or antagonize the effects of the environmental cues of the cell [181].

Circadian transcription significantly contributes to daily oscillations in protein levels, and is regulated by circadian chromatin transitions [182]. Interestingly, synchronization of circadian chromatin transitions by external time cues has recently been linked to the coordinated and transient recruitment of circadian loci to the repressive environment of the lamina in human cells [106]. Recruitment of circadian genes to the lamina thus preceded the acquisition of repressive chromatin modifications at circadian genes and promoted their gradual transcriptional attenuation. Circadian genes were subsequently released from the lamina in a repressed state to pave the way fro a new round of activation in the interior of the nucleus. Regulators of chromatin mobility to and from the lamina included CTCF, a master regulator of 3D genome organization, and its protein interaction partner PARP1. These factors control not only circadian changes in 3D chromatin crosstalk between active and inactive chromatin environments, but also regulate circadian transcriptional oscillations. Circadian chromatin transitions and their entrainment by external time cues thus appear to require interplay between 3D genome organizers and the central oscillator.

Importantly, transcriptional clocks are absent from embryonic stem cells and robustly emerge only in differentiated cells [183], potentially linked with the establishment of large LOCKs [65] at the lamina. Tissue-specific stem cells, on the other hand, display an intermediate phenotype with heterogeneous circadian output, where cells within the stem cell niche might reside in opposite phases of the circadian cycle [181,184]. Lack of circadian synchrony between individual cellular clocks in the follicular stem cell niche has been shown to contribute to cell-to-cell variations in the responsiveness of follicular stem cells toward differentiation cues, thereby regulating the balance between dormancy and differentiation [181]. Hence, downregulation of the negative circadian regulator, PER1, enhances responsiveness to WNT signalling, and decreases the proportion of dormant stem cells within the hair follicle bulge. On the other hand, lack of the positive regulator of circadian transcription, BMAL1, promotes responses to TGF-β, and leads to a premature aging phenotype with inefficient stem cell renewal in the follicular stem cell niche. At the same time, absence of BMAL1 counteracts formation of squamous tumors in mice with an EGFR mutant-heterozygous background. Of note, there was a significant decrease in the amount of tumor-initiating cells in animals lacking BMAL1, compared to controls [181]. These observations raise the question whether the observed erosion of repressive chromatin at the nuclear periphery in tumor cells might indirectly compromise circadian transcriptional attenuation and the responsiveness of tumor stem/progenitor cells to environmental cues that promote dormancy.

17.5   DEREGULATED 3D GENOME ORGANIZATION AND CANCER EVOLUTION

Malignant tumors represent an extreme form of phenotypic plasticity where individual tumor cells display heterogeneity, for example, in size, shape, behavior, metastatic potential, and therapy resistance [7]. Such variability in phenotype has been proposed to drive tumor development, as it provides a survival advantage for the tumor under constantly changing selective pressure [7,185]. Phenotypic heterogeneity stems in part from epigenetic instability that continuously regenerates the observed heterogeneity within the evolving tumor tissue (Fig. 17.3A) [6,7,185]. Epigenetic disruption has thus been argued to be at the core of tumor evolution, which contributes to major tumor characteristics, such as the emergence of cancer stem cell states with self-renewal capacity, metastatic potential, therapy resistance, and the observed metastable phenotypic heterogeneity representing different degrees of cellular differentiation and transcriptional noise [7]. In this model, epigenetic disruption and the consequently arising cell-to-cell variability in gene expression emerges early and remains present throughout the different stages of tumor development, and enables selection mechanisms to drive the survival of the fittest clone once mutations occur in gate keeper genes (Fig. 17.3A) [7].

Factors driving epigenetic instability have recently been functionally classified as epigenetic modulators, epigenetic modifiers, or epigenetic mediators (Fig. 17.3B) [7]. Epigenetic modifiers include writers, readers, and erasers of the epigenome, which are often mutated at later stages of tumor development [7]. Epigenetic mediators, on the other hand, drive the emergence of cancer stem cells and increase phenotypic plasticity [7]. They include, for example, ectopically expressed pluripotency gene products, such as OCT4 and NANOG, the expression of which has already been linked to bad prognosis and increased metastatic potential in different tumor types [7]. Epigenetic mediator genes are rarely mutated themselves, but are often targets of epigenetic modifiers leading to their ectopic expression. As they are the key mediators of malignant state, epigenetic mediators represent important therapeutic targets. Finally, epigenetic modulators include signaling or metabolic pathway members that impinge on epigenetic modifiers and mediators to propagate the effects of environmental carcinogens, injury, and inflammation to the epigenome [7]. They are often mutated in cancer and include, for example, oncogenes and tumor suppressor genes, such as H-RAS and P53. Importantly, epigenetic changes observed in tumors often increase phenotypic plasticity, to confer a survival advantage on tumor cells in a changing microenvironment [7]. In the following section, we will examine how disruption of 3D genome organization and chromosome crosstalk contributes to epigenetic instability that drives phenotypic plasticity in tumors (Fig. 17.3B).

17.5.1 Deregulated Enhancer Usage

Tumor-specific chromatin organization might undermine differentiation in part by affecting enhancer activity and the specificity of enhancer–promoter interactions [7]. Since enhancers regulate cell type–specific gene expression patterns, a key question is how enhancer usage is altered in tumor cells to inhibit differentiation and to increase phenotypic plasticity. Recent years have uncovered that oncogenic signalling factors acting as epigenetic modulators can direct chromatin modifiers to specific genomic loci to induce the de novo formation of oncogenic super-enhancers that drive oncogene expression [186]. Importantly, the sensitivity of oncogenic super-enhancers for the availability of certain chromatin proteins, such as BRD4 and Mediator components, provided novel pharmaceutical targets to inhibit cancer growth [186]. However, unspecific inhibition of BRD4 is expected to interfere with both oncogenic and cell fate-determining super-enhancer functions, preventing proper differentiation. It is tempting to speculate that the acquisition of oncogenic superenhancers might also sequester BRD4 away from cell fate–determining super-enhancers to indirectly destabilise the expression of genes that maintain differentiated cells states. Specific epigenetic decommissioning of oncogenic superenhancers might therefore enable the reactivation of differentiation pathways in tumors to inhibit uncontrolled growth and to decrease the phenotypic plasticity of tumors [7].

Proper cell type–specific enhancer–promoter targeting is also regulated by the compartmentalization of the genome into TADs [7]. As mentioned earlier, CTCF, the chromatin architectural protein implicated in the formation of TAD boundaries, frequently binds to DNA in a methylation-sensitive manner [7]. DNA methylation variation in tumors might reprogram enhancer usage and increase expression variability by altering TAD structures [7]. Finally, epigenetic mediators, such as OCT4, NANOG, and SOX2 are well known to form self-reinforcing positive feedback loops in pluripotent cells by activating each other's enhancers [7]. Reactivation of any of these factors could potentially lead to the activation of oncogenic enhancer circuitries to promote stemness and the emergence of cancer stem cell states [7]. The precise documentation of the genome-wide impact of oncogenic super-enhancers on 3D chromatin crosstalk and TAD structure will, however, require the invention of novel methodologies that enable the detection of interactions between different combinations of enhancer subunits covering large genomic regions and the rest of the genome in a quantitative manner.

17.5.2   Perturbed Nuclear Compartmentalization

Altered enhancer usage, gene expression patterns and increased gene expression variability in tumor cells are also linked with extensive perturbations of the 3D nuclear architecture and the global chromatin context. Epigenetic variation and gene expression variation in tumor precursor lesions and in tumors samples are thus particularly high within large chromatin domains that tend to be DNA-hypomethylated in multiple tumors [7,187,188]. These large domains emerge at early stages of cancer development [189–191], and overlap with domains of increased expression variability containing genes that regulate functions important for cancer growth and metastasis [187]. Moreover, the degree of methylation variation in premalignant lesions correlates with the risk for developing cancer later in life [189,190], suggesting that epigenetic variability is causally linked with cancer development [7]. Interestingly, hypomethylated blocks overlap with LADs and LOCKs present in differentiated cells [7,191]. Based on these observations, it has recently been suggested that the molecular mechanism of increased phenotypic plasticity and stochastic epigenetic variation in cancer involves the deregulation of nuclear compartmentalization, impaired separation of active and inactive chromatin environments, and/or altered chromatin mobility between these environments [7].

How deregulation of the 3D nuclear architecture in tumor precursor lesions and in tumors contributes to increased gene expression variation is not completely understood (Fig. 17.3). Since nuclear compartmentalization has been proposed to decrease the level of transcriptional noise [3,7], it is tempting to speculate that alterations of 3D genome organization in tumors would be selected upon because they promote phenotypic plasticity. In support of this hypothesis [7], even heritable epigenetic marks are reversible modifications, the levels of which can be rapidly altered by the concerted action of chromatin erasers and writers. As mentioned earlier, the high concentration of repressive chromatin modifiers at the lamina and the existence of self-reinforcing crosstalk between repressive histone modifications, chromatin-binding proteins and DNA-methylation likely reduce the probability of unscheduled gene activation within LADs at the periphery. On the other hand, relocation of LADs from the nuclear periphery to the interior of the nucleus not only accompanies developmental stage–specific gene activation, but also affects the stringency of repression at the relocated LADs in general [105,192]. Single cell studies of chromatin–lamina association have thus discovered that only 30% of the LADs are positioned at the periphery in each cell, at any given time. Moreover, LAD positioning is not inherited after mitosis, but instead LADs are stochastically reshuffled between the nuclear interior, nuclear periphery and the heterochromatin compartment around the nucleolus [105]. Maintenance of the repressive modifications at LADs was thus suggested to rely on stochastic reestablishment of chromatin–lamina interactions in the G1 phase of the cell cycle in cycling cells [105,192]. Combination of live cell imaging of dynamic LAD localization with measurements of H3K9me2/me3 enrichment has indeed shown that LADs transiently localized to the nuclear interior carry lower levels of repressive histone modifications, than LADs located at the nuclear periphery. Impaired stochastic recruitment of inactive chromatin domains to repressive compartments might thus lead to the stochastic erosion LOCKs, leading to variable reactivation of genes located within these domains [7].

The disruption of the various repressive nuclear environments is a hallmark of diverse tumor types and involves the variable loss of heterochromatin not only from the nuclear periphery [193], but also from the space around the nucleolus [194], centromeres and the inactive X chromosome of female cells [195]. The impairment of multiple compartments might be functionally linked with the dynamic mobility of chromatin between different subnuclear environments and with the increased plasticity of chromatin in tumor cells [7]. The emergence of transcriptionally permissive perinucleolar compartments in tumor cells, which has been linked to poor prognosis and metastasis formation [194], might thus indirectly deregulate also other heterochromatic compartments in the nucleus. It would be therefore important to examine whether reestablishment of the repressive environment at the nucleolus by the pharmacological inhibition of rRNA expression would counteract expression variability of genes within LOCKs to reduce cancer cell heterogeneity.

Finally, blocks of hypomethylated DNA in tumors also overlap with regions that loose H3K9me2 modifications in various cancer cell lines [3]. These changes are similar to the loss of LOCKs during EMT in epithelial cells on the path to a mesenchymal phenotype, suggesting that cancer cells might acquire phenotypic plasticity by initiating EMT-related changes in chromatin states at the lamina [3]. It has recently been hypothesized that failure to coordinate spatiotemporal cross talk between repressive chromatin factors at the lamina might underlie the emergence of cell states with metastable gene expression patterns and impaired differentiation (Fig. 17.3B) [7]. Cells that are not able to establish or maintain self-reinforcing crosstalk between repressive epigenetic modifications that block dedifferentiation might ectopically reexpress epigenetic mediators that increase phenotypic plasticity and increase the chance for cancer stem cell states to arise [7]. Understanding the impact of deregulated nuclear compartmentalization on stochastic gene expression variation requires, however, novel techniques that can compare cell-to-cell variations of 3D chromatin organization to transcriptional variability in small cell populations or single cells [7].

17.6 DEREGULATION OF HETEROCHROMATIN COMPARTMENTS DURING SENESCENCE AND AGING

Although the aging process itself represents one of the strongest risk factors for cancer, the complex mechanism by which age-related chromatin changes might contribute to tumor development is not well understood. [7]. Decline of functions during aging and in senescent cells in vitro has been associated with altered epigenetic plasticity and with the emergence of megabase-scale hypomethylated blocks of DNA [196–201]that are also preferentially hypomethylated in various cancers. Since the hypomethylated blocks overlap with cLADs, these epigenetic perturbations are likely linked to alterations of the 3D nuclear architecture and lamina composition. A shared hallmark of aging and senescence is, for example, LMNB1 downregulation [84,202,203]. Decrease in the level of LMNB1 expression contributes to the loss of H3K9me2 modifications at the nuclear periphery and to the reactivation of inflammatory genes located in LOCKs/LADs [204,205]. Physiological relevance of lamin loss-induced epigenetic alterations has been examined in *Drosophila* fat body, a major immune organ in flies [204]. Age-related decrease of lamin B expression in the fat body thus induces systemic inflammation that affects local immune responses in the gut and leads to gut hyperplasia— a feature of aging flies. The role of the lamina in aging-related phenotypes is also underscored by the existence of disease-causing *LMNA* mutations that underlie, for example, the laminopathy-associated premature aging syndrome, Hutchinson-Gilford progeria syndrome (HGPS) [206]. The mutation in *LMNA* interferes with proper lamin A processing and leads to the accumulation of its permanently farnesylated form at the lamina, which results in chromatin alterations. The mechanism by which chromatin is affected involves the altered localization of LMNB1 and LAPs, leading to the loss of repressive chromatin modifiers and repressed regions from the lamina [53,207]. Moreover, HPGS cells display global reduction in the levels of LMNB1, HP1α, H3K9me3 and H3K27me3 [53,207–209]. Global chromatin changes have been also linked to the downregulation of the subunits of the NuRD chromatin remodelling complex and to the significantly reduced activity of HDAC1, which is also characteristic hallmark of aged cells [210]. Importantly, despite the high frequency of mutations and impaired DNA repair, the premature aging phenotype of HPGS fibroblasts is reversible, demonstrating the dominant role of the mutant lamin A protein and the importance of reversible chromatin modifications in these processes [53,207].

Alterations of 3D genome organization in conditions with perturbed heterochromatin structure and lamina composition has recently been documented using Hi-C (see Chapter 1). There are thus significant differences in 3D chromatin cross talk both within the active and the repressed chromatin compartments of the genome in senescent and HGPS cells [211]. As mentioned earlier, TADs constitute domains of coordinated gene expression with extensive enhancer–promoter communication within but not between domains [211]. Repressed chromatin is similarly organized into distinct domains of self-interacting, compact structures [211]. The most prominent feature of senescent cells and cells of HGPS patients is altered lamina composition and loss of compact heterochromatin structures from the nuclear periphery. In line with this observation, both phenotypes displayed reduced interaction frequency in *cis* within TADs, in particular at regions corresponding to LADs, and an increase in cross-boundary interactions [212]. Furthermore, inter-LAD regions that gained lamin B1 binding in senescent cells showed opposite behavior with increased TAD boundary strength. In addition, senescent cells harbor extensive long-range contacts between repressed states located on different chromosomes that reflect the formation of the characteristic senescence associated heterochromatin foci [212]. Finally, primary cultures of HGPS fibroblast display lower levels and redistribution of the repressive H3K27me3 mark and loss of spatial separation between active and inactive chromatin compartments as measured by Hi-C [208]. It remains to be determined how such alterations of 3D chromatin compartmentalization and crosstalk affect age-related decline of functions at the level of the organism, and modulate epigenetic flexibility and phenotypic plasticity at the cellular level during normal and premature aging, as well as in senescence. Moreover, it will be important to explore how chromatin alterations might pave the way for the increased tumor risk observed with advanced age, and to reconcile these observations with the in vitro transformation resistance of the cells of HGPS patients.

17.7 CONCLUSIONS

Chromatin-based processes modulate the level of cellular phenotypic plasticity, which is deregulated during aging and in diseases, such as cancer (Fig. 17.3) [7]. The 3D nuclear architecture and genome organization have been proposed to act as epigenetic modulators to affect the stability of chromatin states by providing platforms for the spatiotemporal coordination of epigenetic modifier activities [7]. Epigenetic mediators thus are likely to disrupt such organization in order to reprogram epigenetic memories during the transition toward cellular states with increased developmental potential and features of stemness in tumor development [7]. Important questions address therefore (1) the mechanism of epigenetic plasticity in stem cells, (2) the identity of epigenetic barriers that prevent dedifferentiation and maintain cell type–specific stable cell states during development, and (3) their deregulation during aging and in diseases [7].

Establishment of chromatin domains with stable repression during differentiation relies on self-reinforcing cross talk between multiple layers of heritable, repressive epigenetic modifications spreading over large distances [6,7]. Such repressed domains are separated from transcriptionally permissive regions by boundary elements, leading to the domain organization of chromatin states. This process is likely facilitated by the functional and physical separation between transcriptionally permissive and repressive environments within the 3D space of the nucleus. Identifying the factors that target genomic regions and epigenetic modifiers to repressive subnuclear compartments will thus likely deepen our understanding of the mechanisms underlying lineage choice and differentiation. Although the lamina is mostly recognized as a repressive compartment, it will be important to dissect further the various chromatin environments that might be present at the nuclear periphery. For example, very little is know about the formation and function of the chromatin states associated to nuclear pores in mammalian cells. Finally, an important topic not discussed in this review concerns how polycomb-mediated silencing compartments communicate with other repressive compartments and affect phenotypic plasticity genome wide. Indeed, loss of polycomb function has already been linked to gene expression variability and phenotypic heterogeneity in cancer [213].

Stable cell type–specific gene expression is regulated also by cell type–specific enhancer–promoter cross talk within active chromatin domains [16]. It is still an enigma how enhancer usage is regulated and fine-tuned by the physical constrains of 3D chromatin structures. As multiple enhancers might impinge on a specific gene, and a set of enhancers can regulate the expression of multiple genes [16], 3D maps of enhancer–enhancer and enhancer–promoter interactomes will likely be required for the understanding of transitions between enhancer-associated chromatin states preceding gene expression change during differentiation. Understanding how cell-to-cell variation in 3D genome organization and enhancer usage is linked to gene expression variability and phenotypic plasticity will require the development of novel techniques that can quantitatively measure chromatin fiber interaction frequencies in small cell populations and single cells.

Abbreviations

3D	Three-dimensional
LDB1	Lim domain binding 1
GATA1	GATA-binding protein 1
eRNAs	Enhancer RNAs
MED12	Mediator subunit 12
TADs	Topologically associated domains
CTCF	CCCTC-binding factor
p300	E1A-binding protein P300
H3K4me1	Histone H3 lysine 4 mono-methylation
H3K27me3	Histone H3 lysine 27 tri-methylatlion
H3K27ac	Histone H3 lysine 27 acetylation
NuRD	Nucleosome remodeling deacetylase
MYOD1	Myogenic differentiation 1
BRD4	Bromodomain containing 4
LOCKs	Large organized chromatin K9 modifications

LAP2β	Lamina-associated polypeptide 2 beta
cKrox	ZBTB7B, zinc finger and BTB domain containing 7
HDAC3	Histone deacetylase 3
LBR	Lamin B receptor
OR	Olfactory receptor
Tcrb	T cell receptor B
NPC	Nuclear pore complex
Nup	Nucleoporin
IFN-γ	Interferon gamma
HP1	Heterochromatin protein 1

Acknowledgment

This work was supported by grants to AG from Karolinska Institutet.

References

[1] Burton A, Torres-Padilla ME. Chromatin dynamics in the regulation of cell fate allocation during early embryogenesis. Nat Rev Mol Cell Biol 2014;15(11):723–34.

[2] Almouzni G, Cedar H. Maintenance of epigenetic information. Cold Spring Harb Perspect Biol 2016;8(5).

[3] Pujadas E, Feinberg AP. Regulated noise in the epigenetic landscape of development and disease. Cell 2012;148(6):1123–31.

[4] Raj A, van Oudenaarden A. Nature, nurture, or chance: stochastic gene expression and its consequences. Cell 2008;135(2):216–26.

[5] Debat V, David JR. Analysing phenotypic variation: when old-fashioned means up-to-date. J Biosci 2002;27(3):191–3.

[6] Ohlsson R, Kanduri C, Whitehead J, Pfeifer S, Lobanenkov V, Feinberg AP. Epigenetic variability and the evolution of human cancer. Adv Cancer Res 2003;88:145–68.

[7] Feinberg AP, Koldobskiy MA, Gondor A. Epigenetic modulators, modifiers and mediators in cancer aetiology and progression. Nat Rev Genet 2016;17(5):284–99.

[8] Kurukuti S, Tiwari VK, Tavoosidana G, Pugacheva E, Murrell A, Zhao Z, et al. CTCF binding at the H19 imprinting control region mediates maternally inherited higher-order chromatin conformation to restrict enhancer access to Igf2. Proc Natl Acad Sci U S A 2006;103(28):10684–9.

[9] Sandhu KS, Shi C, Sjolinder M, Zhao Z, Gondor A, Liu L, et al. Nonallelic transvection of multiple imprinted loci is organized by the H19 imprinting control region during germline development. Genes Dev 2009;23(22):2598–603.

[10] Gondor A, Ohlsson R. Chromosome crosstalk in three dimensions. Nature 2009;461(7261):212–7.

[11] Spilianakis CG, Lalioti MD, Town T, Lee GR, Flavell RA. Interchromosomal associations between alternatively expressed loci. Nature 2005;435(7042):637–45.

[12] Thanos D, Maniatis T. Virus induction of human IFN beta gene expression requires the assembly of an enhanceosome. Cell 1995;83(7):1091–100.

[13] Parada LA, McQueen PG, Misteli T. Tissue-specific spatial organization of genomes. Genome Biol 2004;5(7):R44.

[14] Bickmore WA, van Steensel B. Genome architecture: domain organization of interphase chromosomes. Cell 2013;152(6):1270–84.

[15] Rao SS, Huntley MH, Durand NC, Stamenova EK, Bochkov ID, Robinson JT, et al. A 3D map of the human genome at kilobase resolution reveals principles of chromatin looping. Cell 2014;159(7):1665–80.

[16] Li G, Ruan X, Auerbach RK, Sandhu KS, Zheng M, Wang P, et al. Extensive promoter-centered chromatin interactions provide a topological basis for transcription regulation. Cell 2012;148(1-2):84–98.

[17] Hnisz D, Schuijers J, Lin CY, Weintraub AS, Abraham BJ, Lee TI, et al. Convergence of developmental and oncogenic signaling pathways at transcriptional super-enhancers. Mol Cell 2015;58(2):362–70.

[18] Bahar Halpern K, Tanami S, Landen S, Chapal M, Szlak L, Hutzler A, et al. Bursty gene expression in the intact mammalian liver. Mol Cell 2015;58(1):147–56.

[19] Lam A, Deans TL. A noisy tug of war: the battle between transcript production and degradation in the liver. Dev Cell 2015;33(1):3–4.

[20] Noordermeer D, de Wit E, Klous P, van de Werken H, Simonis M, Lopez-Jones M, et al. Variegated gene expression caused by cell-specific long-range DNA interactions. Nat Cell Biol 2011;13(8):944–51.

[21] Smith E, Shilatifard A. Enhancer biology and enhanceropathies. Nat Struct Mol Biol 2014;21(3):210–9.

[22] Whyte WA, Orlando DA, Hnisz D, Abraham BJ, Lin CY, Kagey MH, et al. Master transcription factors and mediator establish super-enhancers at key cell identity genes. Cell 2013;153(2):307–19.

[23] Schmidt SF, Larsen BD, Loft A, Nielsen R, Madsen JG, Mandrup S. Acute TNF-induced repression of cell identity genes is mediated by NFκB-directed redistribution of cofactors from super-enhancers. Genome Res 2015;25(9):1281–94.

[24] Adam RC, Yang H, Rockowitz S, Larsen SB, Nikolova M, Oristian DS, et al. Pioneer factors govern super-enhancer dynamics in stem cell plasticity and lineage choice. Nature 2015;521(7552):366–70.

[25] Soutourina J, Wydau S, Ambroise Y, Boschiero C, Werner M. Direct interaction of RNA polymerase II and mediator required for transcription in vivo. Science 2011;331(6023):1451–4.

[26] Clark AD, Oldenbroek M, Boyer TG. Mediator kinase module and human tumorigenesis. Crit Rev Biochem Mol Biol 2015;50(5):393–426.

[27] Ong CT, Corces VG, Enhancers:. Emerging roles in cell fate specification. EMBO Rep 2012;13(5):423–30.

[28] Allen BL, Taatjes DJ. The Mediator complex: a central integrator of transcription. Nat Rev Mol Cell Biol 2015;16(3):155–66.

[29] Lupianez DG, Kraft K, Heinrich V, Krawitz P, Brancati F, Klopocki E, et al. Disruptions of topological chromatin domains cause pathogenic rewiring of gene-enhancer interactions. Cell 2015;161(5):1012–25.

[30] Utami KH, Winata CL, Hillmer AM, Aksoy I, Long HT, Liany H, et al. Impaired development of neural-crest cell-derived organs and intellectual disability caused by MED13L haploinsufficiency. Hum Mutat 2014;35(11):1311–20.

[31] Zhang H, Jiao W, Sun L, Fan J, Chen M, Wang H, et al. Intrachromosomal looping is required for activation of endogenous pluripotency genes during reprogramming. Cell Stem Cell 2013;13(1):30–5.

[32] Teufel A, Maass T, Strand S, Kanzler S, Galante T, Becker K, et al. Liver-specific Ldb1 deletion results in enhanced liver cancer development. J Hepatol 2010;53(6):1078–84.

[33] Deng W, Lee J, Wang H, Miller J, Reik A, Gregory PD, et al. Controlling long-range genomic interactions at a native locus by targeted tethering of a looping factor. Cell 2012;149(6):1233–44.

[34] Krivega I, Dale RK, Dean A. Role of LDB1 in the transition from chromatin looping to transcription activation. Genes Dev 2014;28(12):1278–90.

[35] Lai F, Orom UA, Cesaroni M, Beringer M, Taatjes DJ, Blobel GA, et al. Activating RNAs associate with Mediator to enhance chromatin architecture and transcription. Nature 2013;494(7438):497–501.

[36] Gomez-Diaz E, Corces VG. Architectural proteins: regulators of 3D genome organization in cell fate. Trends Cell Biol 2014;24(11):703–11.

[37] Gondor A. Dynamic chromatin loops bridge health and disease in the nuclear landscape. Semin Cancer Biol 2013;23(2):90–8.

[38] Klenova E, Ohlsson R. Poly(ADP-ribosyl)ation epigenetics. Is CTCF PARt of the plot? Cell Cycle 2005;4(1):96–101.

[39] Constancia M, Dean W, Lopes S, Moore T, Kelsey G, Reik W. Deletion of a silencer element in Igf2 results in loss of imprinting independent of H19. Nat Genet 2000;26(2):203–6.

[40] Parelho V, Hadjur S, Spivakov M, Leleu M, Sauer S, Gregson HC, et al. Cohesins functionally associate with CTCF on mammalian chromosome arms. Cell 2008;132(3):422–33.

[41] Guibert S, Zhao Z, Sjolinder M, Gondor A, Fernandez A, Pant V, et al. CTCF-binding sites within the H19 ICR differentially regulate local chromatin structures and *cis*-acting functions. Epigenetics 2012;7(4):361–9.

[42] Rada-Iglesias A, Bajpai R, Swigut T, Brugmann SA, Flynn RA, Wysocka J. A unique chromatin signature uncovers early developmental enhancers in humans. Nature 2011;470(7333):279–83.

[43] Heinz S, Romanoski CE, Benner C, Glass CK. The selection and function of cell type-specific enhancers. Nat Rev Mol Cell Biol 2015;16(3):144–54.

[44] Whyte WA, Bilodeau S, Orlando DA, Hoke HA, Frampton GM, Foster CT, et al. Enhancer decommissioning by LSD1 during embryonic stem cell differentiation. Nature 2012;482(7384):221–5.

[45] Wang A, Yue F, Li Y, Xie R, Harper T, Patel NA, et al. Epigenetic priming of enhancers predicts developmental competence of hESC-derived endodermal lineage intermediates. Cell Stem Cell 2015;16(4):386–99.

[46] Jin F, Li Y, Dixon JR, Selvaraj S, Ye Z, Lee AY, et al. A high-resolution map of the three-dimensional chromatin interactome in human cells. Nature 2013;503(7475):290–4.

[47] Phillips-Cremins JE, Sauria ME, Sanyal A, Gerasimova TI, Lajoie BR, Bell JS, et al. Architectural protein subclasses shape 3D organization of genomes during lineage commitment. Cell 2013;153(6):1281–95.

[48] Taberlay PC, Kelly TK, Liu CC, You JS, De Carvalho DD, Miranda TB, et al. Polycomb-repressed genes have permissive enhancers that initiate reprogramming. Cell 2011;147(6):1283–94.

[49] Van Bortle K, Corces VG. Lost in transition: dynamic enhancer organization across naive and primed stem cell states. Cell Stem Cell 2014;14(6):693–4.

[50] Pina C, Fugazza C, Tipping AJ, Brown J, Soneji S, Teles J, et al. Inferring rules of lineage commitment in haematopoiesis. Nat Cell Biol 2012;14(3):287–94.

[51] Li L, Lyu X, Hou C, Takenaka N, Nguyen HQ, Ong CT, et al. Widespread rearrangement of 3D chromatin organization underlies polycomb-mediated stress-induced silencing. Mol Cell 2015;58(2):216–31.

[52] Yadon AN, Singh BN, Hampsey M, Tsukiyama T. DNA looping facilitates targeting of a chromatin remodeling enzyme. Mol Cell 2013;50(1):93–103.

[53] Fernandez P, Scaffidi P, Markert E, Lee JH, Rane S, Misteli T. Transformation resistance in a premature aging disorder identifies a tumor-protective function of BRD4. Cell Rep 2014;9(1):248–60.

[54] Dey A, Chitsaz F, Abbasi A, Misteli T, Ozato K. The double bromodomain protein Brd4 binds to acetylated chromatin during interphase and mitosis. Proc Natl Acad Sci U S A 2003;100(15):8758–63.

[55] Deng B, Melnik S, Cook PR. Transcription factories, chromatin loops, and the dysregulation of gene expression in malignancy. Semin Cancer Biol 2013;23(2):65–71.

[56] Reddy KL, Feinberg AP. Higher order chromatin organization in cancer. Semin Cancer Biol 2012;.

[57] Guetg C, Santoro R. Formation of nuclear heterochromatin: the nucleolar point of view. Epigenetics 2012;7(8):811–4.

[58] Fedoriw AM, Starmer J, Yee D, Magnuson T. Nucleolar association and transcriptional inhibition through 5S rDNA in mammals. PLoS Genet 2012;8(1):e1002468.

[59] Meshorer E, Yellajoshula D, George E, Scambler PJ, Brown DT, Misteli T. Hyperdynamic plasticity of chromatin proteins in pluripotent embryonic stem cells. Dev Cell 2006;10(1):105–16.

[60] Ahmed K, Dehghani H, Rugg-Gunn P, Fussner E, Rossant J, Bazett-Jones DP. Global chromatin architecture reflects pluripotency and lineage commitment in the early mouse embryo. PLoS One 2010;5(5):e10531.

[61] Fussner E, Ahmed K, Dehghani H, Strauss M, Bazett-Jones DP. Changes in chromatin fiber density as a marker for pluripotency. Cold Spring Harb Symp Quant Biol 2010;75:245–9.

[62] Guelen L, Pagie L, Brasset E, Meuleman W, Faza MB, Talhout W, et al. Domain organization of human chromosomes revealed by mapping of nuclear lamina interactions. Nature 2008;453(7197):948–51.

[63] Pickersgill H, Kalverda B, de Wit E, Talhout W, Fornerod M, van Steensel B. Characterization of the *Drosophila melanogaster* genome at the nuclear lamina. Nat Genet 2006;38(9):1005–14.

[64] Meuleman W, Peric-Hupkes D, Kind J, Beaudry JB, Pagie L, Kellis M, et al. Constitutive nuclear lamina-genome interactions are highly conserved and associated with A/T-rich sequence. Genome Res 2013;23(2):270–80.

[65] Wen B, Wu H, Shinkai Y, Irizarry RA, Feinberg AP. Large histone H3 lysine 9 dimethylated chromatin blocks distinguish differentiated from embryonic stem cells. Nat Genet 2009;41(2):246–50.

[66] Chen X, Yammine S, Shi C, Tark-Dame M, Gondor A, Ohlsson R. The visualization of large organized chromatin domains enriched in the H3K9me2 mark within a single chromosome in a single cell. Epigenetics 2014;9(11):1439–45.

[67] Chen X, Shi C, Yammine S, Gondor A, Ronnlund D, Fernandez-Woodbridge A, et al. Chromatin in situ proximity (ChrISP): single-cell analysis of chromatin proximities at a high resolution. Biotechniques 2014;56(3):117–8. 120–124.

[68] Sridharan R, Gonzales-Cope M, Chronis C, Bonora G, McKee R, Huang C, et al. Proteomic and genomic approaches reveal critical functions of H3K9 methylation and heterochromatin protein-1gamma in reprogramming to pluripotency. Nat Cell Biol 2013;15(7):872–82.

[69] Boros J, Arnoult N, Stroobant V, Collet JF, Decottignies A. Polycomb repressive complex 2 and H3K27me3 cooperate with H3K9 methylation to maintain heterochromatin protein 1alpha at chromatin. Mol Cell Biol 2014;34(19):3662–74.

[70] Lemaitre C, Bickmore WA. Chromatin at the nuclear periphery and the regulation of genome functions. Histochem Cell Biol 2015;144(2):111–22.

[71] Gruenbaum Y, Foisner R. Lamins: nuclear intermediate filament proteins with fundamental functions in nuclear mechanics and genome regulation. Annu Rev Biochem 2015;84:131–64.

[72] Korfali N, Wilkie GS, Swanson SK, Srsen V, de Las Heras J, Batrakou DG, et al. The nuclear envelope proteome differs notably between tissues. Nucleus 2012;3(6):552–64.

[73] Schirmer EC, Foisner R. Proteins that associate with lamins: many faces, many functions. Exp Cell Res 2007;313(10):2167–79.

[74] Worman HJ. Nuclear lamins and laminopathies. J Pathol 2012;226(2):316–25.

[75] Schooley A, Moreno-Andres D, De Magistris P, Vollmer B, Antonin W. The lysine demethylase LSD1 is required for nuclear envelope formation at the end of mitosis. J Cell Sci 2015;128(18):3466–77.

[76] Montes de Oca R, Shoemaker CJ, Gucek M, Cole RN, Wilson KL. Barrier-to-autointegration factor proteome reveals chromatin-regulatory partners. PLoS One 2009;4(9):e7050.

[77] Demmerle J, Koch AJ, Holaska JM. The nuclear envelope protein emerin binds directly to histone deacetylase 3 (HDAC3) and activates HDAC3 activity. J Biol Chem 2012;287(26):22080–8.

[78] Somech R, Shaklee S, Geller O, Amariglio N, Simon AJ, Rechavi G, et al. The nuclear-envelope protein and transcriptional repressor LAP2beta interacts with HDAC3 at the nuclear periphery, and induces histone H4 deacetylation. J Cell Sci 2005;118(Pt 17):4017–25.

[79] Holaska JM, Wilson KL. An emerin "proteome": purification of distinct emerin-containing complexes from HeLa cells suggests molecular basis for diverse roles including gene regulation, mRNA splicing, signaling, mechanosensing, and nuclear architecture. Biochemistry 2007;46(30):8897–908.

[80] Finlan LE, Sproul D, Thomson I, Boyle S, Kerr E, Perry P, et al. Recruitment to the nuclear periphery can alter expression of genes in human cells. PLoS Genet 2008;4(3):e1000039.

[81] Melcer S, Hezroni H, Rand E, Nissim-Rafinia M, Skoultchi A, Stewart CL, et al. Histone modifications and lamin A regulate chromatin protein dynamics in early embryonic stem cell differentiation. Nat Commun 2012;3:910.

[82] Adamo A, Sese B, Boue S, Castano J, Paramonov I, Barrero MJ, et al. LSD1 regulates the balance between self-renewal and differentiation in human embryonic stem cells. Nat Cell Biol 2011;13(6):652–9.

[83] Ma DK, Chiang CH, Ponnusamy K, Ming GL, Song H. G9a and Jhdm2a regulate embryonic stem cell fusion-induced reprogramming of adult neural stem cells. Stem Cells 2008;26(8):2131–41.

[84] Sadaie M, Salama R, Carroll T, Tomimatsu K, Chandra T, Young AR, et al. Redistribution of the Lamin B1 genomic binding profile affects rearrangement of heterochromatic domains and SAHF formation during senescence. Genes Dev 2013;27(16):1800–8.

[85] Zane L, Sharma V, Misteli T. Common features of chromatin in aging and cancer: cause or coincidence? Trends Cell Biol 2014;24(11):686–94.

[86] Shankar SR, Bahirvani AG, Rao VK, Bharathy N, Ow JR, Taneja R. G9a, a multipotent regulator of gene expression. Epigenetics 2013;8(1):16–22.

[87] Harr JC, Luperchio TR, Wong X, Cohen E, Wheelan SJ, Reddy KL. Directed targeting of chromatin to the nuclear lamina is mediated by chromatin state and A-type lamins. J Cell Biol 2015;208(1):33–52.

[88] Guarda A, Bolognese F, Bonapace IM, Badaracco G. Interaction between the inner nuclear membrane lamin B receptor and the heterochromatic methyl binding protein, MeCP2. Exp Cell Res 2009;315(11):1895–903.

[89] Zullo JM, Demarco IA, Piqué-Regi R, Gaffney DJ, Epstein CB, Spooner CJ, et al. DNA sequence-dependent compartmentalization and silencing of chromatin at the nuclear lamina. Cell 2012;149(7):1474–87.

[90] Ottaviani A, Schluth-Bolard C, Rival-Gervier S, Boussouar A, Rondier D, Foerster AM, et al. Identification of a perinuclear positioning element in human subtelomeres that requires A-type lamins and CTCF. EMBO J 2009;28(16):2428–36.

[91] Akhtar W, de Jong J, Pindyurin AV, Pagie L, Meuleman W, de Ridder J, et al. Chromatin position effects assayed by thousands of reporters integrated in parallel. Cell 2013;154(4):914–27.

[92] Gonzalez-Sandoval A, Towbin BD, Kalck V, Cabianca DS, Gaidatzis D, Hauer MH, et al. Perinuclear anchoring of H3K9-methylated chromatin stabilizes induced cell fate in *C. elegans* embryos. Cell 2015;163(6):1333–47.

[93] Shachar S, Voss TC, Pegoraro G, Sciascia N, Misteli T. Identification of gene positioning factors using high-throughput imaging mapping. Cell 2015;162(4):911–23.

[94] Kohwi M, Lupton JR, Lai SL, Miller MR, Doe CQ. Developmentally regulated subnuclear genome reorganization restricts neural progenitor competence in *Drosophila*. Cell 2013; 152(1-2):97–108.

[95] Shevelyov YY, Lavrov SA, Mikhaylova LM, Nurminsky ID, Kulathinal RJ, Egorova KS, et al. The B-type lamin is required for somatic repression of testis-specific gene clusters. Proc Natl Acad Sci U S A 2009;106(9):3282–7.

[96] Towbin BD, Meister P, Pike BL, Gasser SM. Repetitive transgenes in *C. elegans* accumulate heterochromatic marks and are sequestered at the nuclear envelope in a copy-number- and lamin-dependent manner. Cold Spring Harb Symp Quant Biol 2010;75:555–65.

[97] Babbio F, Castiglioni I, Cassina C, Gariboldi MB, Pistore C, Magnani E, et al. Knock-down of methyl CpG-binding protein 2 (MeCP2) causes alterations in cell proliferation and nuclear lamins expression in mammalian cells. BMC Cell Biol 2012;13:19.

[98] Olins AL, Rhodes G, Welch DB, Zwerger M, Olins DE, Lamin B. Receptor: multi-tasking at the nuclear envelope. Nucleus 2010;1(1):53–70.

[99] Hirano Y, Hizume K, Kimura H, Takeyasu K, Haraguchi T, Hiraoka Y. Lamin B receptor recognizes specific modifications of histone H4 in heterochromatin formation. J Biol Chem 2012;287(51):42654–63.

[100] Solovei I, Wang AS, Thanisch K, Schmidt CS, Krebs S, Zwerger M, et al. LBR and lamin A/C sequentially tether peripheral heterochromatin and inversely regulate differentiation. Cell 2013;152(3):584–98.

[101] Solovei I, Kreysing M, Lanctot C, Kosem S, Peichl L, Cremer T, et al. Nuclear architecture of rod photoreceptor cells adapts to vision in mammalian evolution. Cell 2009;137(2):356–68.

[102] Clowney EJ, LeGros MA, Mosley CP, Clowney FG, Markenskoff-Papadimitriou EC, Myllys M, et al. Nuclear aggregation of olfactory receptor genes governs their monogenic expression. Cell 2012;151(4):724–37.

[103] Pinheiro I, Margueron R, Shukeir N, Eisold M, Fritzsch C, Richter FM, et al. Prdm3 and Prdm16 are H3K9me1 methyltransferases required for mammalian heterochromatin integrity. Cell 2012;150(5):948–60.

[104] Bian Q, Khanna N, Alvikas J, Belmont AS. Beta-Globin *cis*-elements determine differential nuclear targeting through epigenetic modifications. J Cell Biol 2013;203(5):767–83.

[105] Kind J, Pagie L, Ortabozkoyun H, Boyle S, de Vries SS, Janssen H, et al. Single-cell dynamics of genome-nuclear lamina interactions. Cell 2013;153(1):178–92.

[106] Zhao H, Sifakis EG, Sumida N, Millan-Arino L, Scholz BA, Svensson JP, et al. PARP1- and CTCF-mediated interactions between active and repressed chromatin at the lamina promote oscillating transcription. Mol Cell 2015;59(6):984–97.

[107] Towbin BD, Gonzalez-Aguilera C, Sack R, Gaidatzis D, Kalck V, Meister P, et al. Step-wise methylation of histone H3K9 positions heterochromatin at the nuclear periphery. Cell 2012;150(5):934–47.

[108] Demmerle J, Koch AJ, Holaska JM. Emerin and histone deacetylase 3 (HDAC3) cooperatively regulate expression and nuclear positions of MyoD, Myf5, and Pax7 genes during myogenesis. Chromosome Res 2013;21(8):765–79.

[109] Mllon BC, Cheng H, Tselebrovsky MV, Laviov SA, Nenasheva VV, Mikhaleva EA, et al. Role of histone deacetylases in gene regulation at nuclear lamina. PLoS One 2012;7(11):e49692.

[110] Turinetto V, Giachino C. Histone variants as emerging regulators of embryonic stem cell identity. Epigenetics 2015;10(7):563–73.

[111] Fu Y, Lv P, Yan G, Fan H, Cheng L, Zhang F, et al. MacroH2A1 associates with nuclear lamina and maintains chromatin architecture in mouse liver cells. Sci Rep 2015;5:17186.

[112] Izzo A, Kamieniarz-Gdula K, Ramirez F, Noureen N, Kind J, Manke T, et al. The genomic landscape of the somatic linker histone subtypes H1.1 to H1. 5 in human cells. Cell Rep 2013;3(6):2142–54.

[113] Mayor R, Izquierdo-Bouldstridge A, Millan-Arino L, Bustillos A, Sampaio C, Luque N, et al. Genome distribution of replication-independent histone H1 variants shows H1.0 associated with nucleolar domains and H1X associated with RNA polymerase II-enriched regions. J Biol Chem 2015;290(12):7474–91.

[114] Millan-Arino L, Islam AB, Izquierdo-Bouldstridge A, Mayor R, Terme JM, Luque N, et al. Mapping of six somatic linker histone H1 variants in human breast cancer cells uncovers specific features of H1.2. Nucleic Acids Res 2014;42(7):4474–93.

[115] Ikegami K, Egelhofer TA, Strome S, Lieb JD. *Caenorhabditis elegans* chromosome arms are anchored to the nuclear membrane via discontinuous association with LEM-2. Genome Biol 2010;11(12):R120.

[116] Peric-Hupkes D, Meuleman W, Pagie L, Bruggeman SW, Solovei I, Brugman W, et al. Molecular maps of the reorganization of genome-nuclear lamina interactions during differentiation. Mol Cell 2010;38(4):603–13.

[117] Kosak ST, Skok JA, Medina KL, Riblet R, Le Beau MM, Fisher AG, et al. Subnuclear compartmentalization of immunoglobulin loci during lymphocyte development. Science 2002;296(5565):158–62.

[118] Schlimgen RJ, Reddy KL, Singh H, Krangel MS. Initiation of allelic exclusion by stochastic interaction of Tcrb alleles with repressive nuclear compartments. Nat Immunol 2008;9(7):802–9.

[119] Chan EA, Teng G, Corbett E, Choudhury KR, Bassing CH, Schatz DG, et al. Peripheral subnuclear positioning suppresses Tcrb recombination and segregates Tcrb alleles from RAG2. Proc Natl Acad Sci U S A 2013;110(48):E4628–37.

[120] Reddy KL, Zullo JM, Bertolino E, Singh H. Transcriptional repression mediated by repositioning of genes to the nuclear lamina. Nature 2008;452(7184):243–7.

[121] Therizols P, Illingworth RS, Courilleau C, Boyle S, Wood AJ, Bickmore WA. Chromatin decondensation is sufficient to alter nuclear organization in embryonic stem cells. Science 2014;346(6214):1238–42.

[122] Kumaran RI, Spector DL. A genetic locus targeted to the nuclear periphery in living cells maintains its transcriptional competence. J Cell Biol 2008;180(1):51–65.

[123] Finlan LE, Bickmore WA. Porin new light onto chromatin and nuclear organization. Genome Biol 2008;9(5):222.

[124] van Koningsbruggen S, Gierlinski M, Schofield P, Martin D, Barton GJ, Ariyurek Y, et al. High-resolution whole-genome sequencing reveals that specific chromatin domains from most human chromosomes associate with nucleoli. Mol Biol Cell 2010;21(21):3735–48.

[125] Nemeth A, Conesa A, Santoyo-Lopez J, Medina I, Montaner D, Peterfia B, et al. Initial genomics of the human nucleolus. PLoS Genet 2010;6(3):e1000889.

[126] Savic N, Bar D, Leone S, Frommel SC, Weber FA, Vollenweider E, et al. lncRNA maturation to initiate heterochromatin formation in the nucleolus is required for exit from pluripotency in ESCs. Cell Stem Cell 2014;15(6):720–34.

[127] Arib G, Akhtar A. Multiple facets of nuclear periphery in gene expression control. Curr Opin Cell Biol 2011;23(3):346–53.

[128] Wen B, Wu H, Loh YH, Briem E, Daley GQ, Feinberg AP. Euchromatin islands in large heterochromatin domains are enriched for CTCF binding and differentially DNA-methylated regions. BMC Genomics 2012;13:566.

[129] Ptak C, Wozniak RW. Nucleoporins and chromatin metabolism. Curr Opin Cell Biol 2016;40:153–60.

[130] Ptak C, Aitchison JD, Wozniak RW. The multifunctional nuclear pore complex: a platform for controlling gene expression. Curr Opin Cell Biol 2014;28:46–53.

[131] Akhtar A, Gasser SM. The nuclear envelope and transcriptional control. Nat Rev Genet 2007;8(7):507–17.

[132] D'Angelo MA, Gomez-Cavazos JS, Mei A, Lackner DH, Hetzer MW. A change in nuclear pore complex composition regulates cell differentiation. Dev Cell 2012;22(2):446–58.

[133] Jacinto FV, Benner C, Hetzer MW. The nucleoporin Nup153 regulates embryonic stem cell pluripotency through gene silencing. Genes Dev 2015;29(12):1224–38.

[134] Kalverda B, Pickersgill H, Shloma VV, Fornerod M. Nucleoporins directly stimulate expression of developmental and cell-cycle genes inside the nucleoplasm. Cell 2010;140(3):360–71.

[135] Liang Y, Franks TM, Marchetto MC, Gage FH, Hetzer MW. Dynamic association of NUP98 with the human genome. PLoS Genet 2013;9(2):e1003308.

[136] Light WH, Freaney J, Sood V, Thompson A, D'Urso A, Horvath CM, et al. A conserved role for human Nup98 in altering chromatin structure and promoting epigenetic transcriptional memory. PLoS Biol 2013;11(3):e1001524.

[137] Tan-Wong SM, Wijayatilake HD, Proudfoot NJ. Gene loops function to maintain transcriptional memory through interaction with the nuclear pore complex. Genes Dev 2009;23(22):2610–24.

[138] Casolari JM, Brown CR, Drubin DA, Rando OJ, Silver PA. Developmentally induced changes in transcriptional program alter spatial organization across chromosomes. Genes Dev 2005;19(10):1188–98.

[139] Blobel G. Gene gating: a hypothesis. Proc Natl Acad Sci U S A 1985;82(24):8527–9.

[140] Menon BB, Sarma NJ, Pasula S, Deminoff SJ, Willis KA, Barbara KE, et al. Reverse recruitment: the Nup84 nuclear pore subcomplex mediates Rap1/Gcr1/Gcr2 transcriptional activation. Proc Natl Acad Sci U S A 2005;102(16):5749–54.

[141] Brickner JH, Walter P. Gene recruitment of the activated INO1 locus to the nuclear membrane. PLoS Biol 2004;2(11):e342.

[142] Brown CR, Kennedy CJ, Delmar VA, Forbes DJ, Silver PA. Global histone acetylation induces functional genomic reorganization at mammalian nuclear pore complexes. Genes Dev 2008;22(5):627–39.

[143] Casolari JM, Brown CR, Komili S, West J, Hieronymus H, Silver PA. Genome-wide localization of the nuclear transport machinery couples transcriptional status and nuclear organization. Cell 2004;117(4):427–39.

[144] Van de Vosse DW, Wan Y, Lapetina DL, Chen WM, Chiang JH, Aitchison JD, et al. A role for the nucleoporin Nup170p in chromatin structure and gene silencing. Cell 2013;152(5): 969–83.

[145] Green EM, Jiang Y, Joyner R, Weis K. A negative feedback loop at the nuclear periphery regulates GAL gene expression. Mol Biol Cell 2012;23(7):1367–75.

[146] Taddei A, Van Houwe G, Hediger F, Kalck V, Cubizolles F, Schober H, et al. Nuclear pore association confers optimal expression levels for an inducible yeast gene. Nature 2006;441(7094):774–8.

[147] Taddei A, Gasser SM. Structure and function in the budding yeast nucleus. Genetics 2012;192(1):107–29.

[148] Schmid M, Arib G, Laemmli C, Nishikawa J, Durussel T, Laemmli UK. Nup-PI: the nucleo-pore-promoter interaction of genes in yeast. Mol Cell 2006;21(3):379–91.

[149] Cabal GG, Genovesio A, Rodriguez-Navarro S, Zimmer C, Gadal O, Lesne A, et al. SAGA in-teracting factors confine sub-diffusion of transcribed genes to the nuclear envelope. Nature 2006;441(7094):770–3.

[150] Dieppois G, Iglesias N, Stutz F. Cotranscriptional recruitment to the mRNA export re-ceptor Mex67p contributes to nuclear pore anchoring of activated genes. Mol Cell Biol 2006;26(21):7858–70.

[151] Egecioglu D, Brickner JH. Gene positioning and expression. Curr Opin Cell Biol 2011;23(3):338–45.

[152] Oeffinger M, Zenklusen D. To the pore and through the pore: a story of mRNA export kinet-ics. Biochim Biophys Acta 2012;1819(6):494–506.

[153] Hampsey M, Singh BN, Ansari A, Laine JP, Krishnamurthy S. Control of eukaryotic gene expression: gene loops and transcriptional memory. Adv Enzyme Regul 2011;51(1):118–25.

[154] Texari L, Dieppois G, Vinciguerra P, Contreras MP, Groner A, Letourneau A, et al. The nuclear pore regulates GAL1 gene transcription by controlling the localization of the SUMO prote-ase Ulp1. Mol Cell 2013;51(6):807–18.

[155] Rubin JD, Taatjes DJ. Molecular biology: mediating transcription and RNA export. Nature 2015;526(7572):199–200.

[156] Kohler A, Schneider M, Cabal GG, Nehrbass U, Hurt E. Yeast Ataxin-7 links histone deubiq-uitination with gene gating and mRNA export. Nat Cell Biol 2008;10(6):707–15.

[157] Schneider M, Hellerschmied D, Schubert T, Amlacher S, Vinayachandran V, Reja R, et al. The nuclear pore-associated TREX-2 complex employs mediator to regulate gene expression. Cell 2015;162(5):1016–28.

[158] Umlauf D, Bonnet J, Waharte F, Fournier M, Stierle M, Fischer B, et al. The human TREX-2 complex is stably associated with the nuclear pore basket. J Cell Sci 2013;126(Pt 12):2656–67.

[159] Jani D, Valkov E, Stewart M. Structural basis for binding the TREX2 complex to nuclear pores, GAL1 localisation and mRNA export. Nucleic Acids Res 2014;42(10):6686–97.

[160] Fischer T, Strasser K, Racz A, Rodriguez-Navarro S, Oppizzi M, Ihrig P, et al. The mRNA export machinery requires the novel Sac3p-Thp1p complex to dock at the nucleoplasmic entrance of the nuclear pores. EMBO J 2002;21(21):5843–52.

[161] Chen M, Gartenberg MR. Coordination of tRNA transcription with export at nuclear pore complexes in budding yeast. Genes Dev 2014;28(9):959–70.

[162] Ibarra A, Hetzer MW. Nuclear pore proteins and the control of genome functions. Genes Dev 2015;29(4):337–49.

[163] Buchwalter AL, Liang Y, Hetzer MW. Nup50 is required for cell differentiation and exhibits transcription-dependent dynamics. Mol Biol Cell 2014;25(16):2472–84.

[164] Breuer M, Ohkura H. A negative loop within the nuclear pore complex controls global chro-matin organization. Genes Dev 2015;29(17):1789–94.

[165] Capelson M, Hetzer MW. The role of nuclear pores in gene regulation, development and disease. EMBO Rep 2009;10(7):697–705.

[166] Cronshaw JM, Matunis MJ. The nuclear pore complex: disease associations and functional correlations. Trends Endocrinol Metab 2004;15(1):34–9.

[167] Raices M, D'Angelo MA. Nuclear pore complex composition: a new regulator of tissue-specific and developmental functions. Nat Rev Mol Cell Biol 2012;13(11):687–99.

[168] Dowen JM, Fan ZP, Hnisz D, Ren G, Abraham BJ, Zhang LN, et al. Control of cell identity genes occurs in insulated neighborhoods in mammalian chromosomes. Cell 2014;159(2):374–87.

[169] Li Y, Rivera CM, Ishii H, Jin F, Selvaraj S, Lee AY, et al. CRISPR reveals a distal super-enhancer required for Sox2 expression in mouse embryonic stem cells. PLoS One 2014;9(12):e114485.

[170] Chang HH, Hemberg M, Barahona M, Ingber DE, Huang S. Transcriptome-wide noise controls lineage choice in mammalian progenitor cells. Nature 2008;453(7194):544–7.

[171] McDonald OG, Wu H, Timp W, Doi A, Feinberg AP. Genome-scale epigenetic reprogramming during epithelial-to-mesenchymal transition. Nat Struct Mol Biol 2011;18(8):867–74.

[172] Chen X, Yammine S, Shi C, Tark-Dame ML, Göndör A, Ohlsson R. The visualization of large organized chromatin domains enriched in the H3K9me2 mark within a single chromosome in a single cell. Epigenetics 2014;.

[173] Vallot C, Herault A, Boyle S, Bickmore WA, Radvanyi F. PRC2-independent chromatin compaction and transcriptional repression in cancer. Oncogene 2015;34(6):741–51.

[174] Nozawa RS, Nagao K, Igami KT, Shibata S, Shirai N, Nozaki N, et al. Human inactive X chromosome is compacted through a PRC2-independent SMCHD1-HBiX1 pathway. Nat Struct Mol Biol 2013;20(5):566–73.

[175] Asher G, Schibler U. Crosstalk between components of circadian and metabolic cycles in mammals. Cell Metab 2011;13(2):125–37.

[176] Bass J. Circadian topology of metabolism. Nature 2012;491(7424):348–56.

[177] Sahar S, Sassone-Corsi P. Regulation of metabolism: the circadian clock dictates the time. Trends Endocrinol Metab 2012;23(1):1–8.

[178] Takahashi JS, Hong HK, Ko CH, McDearmon EL. The genetics of mammalian circadian order and disorder: implications for physiology and disease. Nat Rev Genet 2008;9(10):764–75.

[179] Balsalobre A, Damiola F, Schibler U. A serum shock induces circadian gene expression in mammalian tissue culture cells. Cell 1998;93(6):929–37.

[180] Balsalobre A, Marcacci L, Schibler U. Multiple signaling pathways elicit circadian gene expression in cultured Rat-1 fibroblasts. Curr Biol 2000;10(20):1291–4.

[181] Janich P, Pascual G, Merlos-Suarez A, Batlle E, Ripperger J, Albrecht U, et al. The circadian molecular clock creates epidermal stem cell heterogeneity. Nature 2011;480(7376):209–14.

[182] Benegiamo G, Brown SA, Panda S. RNA dynamics in the control of circadian rhythm. Adv Exp Med Biol 2016;907:107–22.

[183] Yagita K, Horie K, Koinuma S, Nakamura W, Yamanaka I, Urasaki A, et al. Development of the circadian oscillator during differentiation of mouse embryonic stem cells in vitro. Proc Natl Acad Sci U S A 2010;107(8):3846–51.

[184] Brown SA. Circadian clock-mediated control of stem cell division and differentiation: beyond night and day. Development 2014;141(16):3105–11.

[185] Feinberg AP, Ohlsson R, Henikoff S. The epigenetic progenitor origin of human cancer. Nat Rev Genet 2006;7(1):21–33.

[186] Loven J, Hoke HA, Lin CY, Lau A, Orlando DA, Vakoc CR, et al. Selective inhibition of tumor oncogenes by disruption of super-enhancers. Cell 2013;153(2):320–34.

[187] Hansen KD, Timp W, Bravo HC, Sabunciyan S, Langmead B, McDonald OG, et al. Increased methylation variation in epigenetic domains across cancer types. Nat Genet 2011;43(8):768–75.

[188] Berman BP, Weisenberger DJ, Aman JF, Hinoue T, Ramjan Z, Liu Y, et al. Regions of focal DNA hypermethylation and long-range hypomethylation in colorectal cancer coincide with nuclear lamina-associated domains. Nat Genet 2012;44(1):40–6.

[189] Teschendorff AE, Jones A, Fiegl H, Sargent A, Zhuang JJ, Kitchener HC, et al. Epigenetic variability in cells of normal cytology is associated with the risk of future morphological transformation. Genome Med 2012;4(3):24.

[190] Teschendorff AE, Widschwendter M. Differential variability improves the identification of cancer risk markers in DNA methylation studies profiling precursor cancer lesions. Bioinformatics 2012;28(11):1487–94.

[191] Timp W, Bravo HC, McDonald OG, Goggins M, Umbricht C, Zeiger M, et al. Large hypomethylated blocks as a universal defining epigenetic alteration in human solid tumors. Genome Med 2014;6(8):61.

[192] Kind J, Pagie L, de Vries SS, Nahidiazar L, Dey SS, Bienko M, et al. Genome-wide maps of nuclear lamina interactions in single human cells. Cell 2015;163(1):134–47.

[193] Reddy KL, Feinberg AP. Higher order chromatin organization in cancer. Semin Cancer Biol 2013;23(2):109–15.

[194] Pollock C, Huang S. The perinucleolar compartment. Cold Spring Harb Symp Quant Biol 2010;2(2):a000679.

[195] Chaligne R, Popova T, Mendoza-Parra MA, Saleem MA, Gentien D, Ban K, et al. The inactive X chromosome is epigenetically unstable and transcriptionally labile in breast cancer. Genome Res 2015;25(4):488–503.

[196] Cruickshanks HA, McBryan T, Nelson DM, Vanderkraats ND, Shah PP, van Tuyn J, et al. Senescent cells harbour features of the cancer epigenome. Nat Cell Biol 2013;15(12): 1495–506.

[197] Hannum G, Guinney J, Zhao L, Zhang L, Hughes G, Sadda S, et al. Genome-wide methylation profiles reveal quantitative views of human aging rates. Mol Cell 2013;49(2):359–67.

[198] Heyn H, Li N, Ferreira HJ, Moran S, Pisano DG, Gomez A, Distinct DNA, et al. Methylomes of newborns and centenarians. Proc Natl Acad Sci U S A 2012;109(26):10522–7.

[199] Horvath S. DNA methylation age of human tissues and cell types. Genome Biol 2013;14(10):R115.

[200] Vandiver AR, Irizarry RA, Hansen KD, Garza LA, Runarsson A, Li X, et al. Age and sun exposure-related widespread genomic blocks of hypomethylation in nonmalignant skin. Genome Biol 2015;16:80.

[201] Yuan T, Jiao Y, de Jong S, Ophoff RA, Beck S, Teschendorff AE. An integrative multi-scale analysis of the dynamic DNA methylation landscape in aging. PLoS Genet 2015;11(2):e1004996.

[202] Chen H, Zheng X, Zheng Y. Lamin-B in systemic inflammation, tissue homeostasis, and aging. Nucleus 2015;6(3):183–6.

[203] Shah PP, Donahue G, Otte GL, Capell BC, Nelson DM, Cao K, et al. Lamin B1 depletion in senescent cells triggers large-scale changes in gene expression and the chromatin landscape. Genes Dev 2013;27(16):1787–99.

[204] Chen H, Zheng X, Zheng Y. Age-associated loss of lamin-B leads to systemic inflammation and gut hyperplasia. Cell 2014;159(4):829–43.

[205] Tran JR, Chen H, Zheng X, Zheng Y. Lamin in inflammation and aging. Curr Opin Cell Biol 2016;40:124–30.

[206] Arancio W, Pizzolanti G, Genovese SI, Pitrone M, Giordano C. Epigenetic involvement in Hutchinson–Gilford progeria syndrome: a mini-review. Gerontology 2014;60(3):197–203.

[207] Liu GH, Barkho BZ, Ruiz S, Diep D, Qu J, Yang SL, et al. Recapitulation of premature ageing with iPSCs from Hutchinson–Gilford progeria syndrome. Nature 2011;472(7342):221–5.

[208] McCord RP, Nazario-Toole A, Zhang H, Chines PS, Zhan Y, Erdos MR, et al. Correlated alterations in genome organization, histone methylation, and DNA–lamin A/C interactions in Hutchinson–Gilford progeria syndrome. Genome Res 2013;23(2):260–9.

[209] Loi M, Cenni V, Duchi S, Squarzoni S, Lopez-Otin C, Foisner R, et al. Barrier-to-autointegration factor (BAF) involvement in prelamin A-related chromatin organization changes. Oncotarget 2016;7(13):15662–77.

[210] Pegoraro G, Kubben N, Wickert U, Gohler H, Hoffmann K, Misteli T. Ageing-related chromatin defects through loss of the NURD complex. Nat Cell Biol 2009;11(10):1261–7.

[211] Nora EP, Dekker J, Heard E. Segmental folding of chromosomes: a basis for structural and regulatory chromosomal neighborhoods? BioEssays 2013;35(9):818–28.

[212] Chandra T, Ewels PA, Schoenfelder S, Furlan-Magaril M, Wingett SW, Kirschner K, et al. Global reorganization of the nuclear landscape in senescent cells. Cell Rep 2015;10(4):471–83.

[213] Wassef M, Rodilla V, Teissandier A, Zeitouni B, Gruel N, Sadacca B, et al. Impaired PRC2 activity promotes transcriptional instability and favors breast tumorigenesis. Genes Dev 2015;29(24):2547–62.

Index

Printed in the United States
By Bookmasters